THE UNIVERSE

THE UNIVERSE

IAIN NICOLSON
PATRICK MOORE

MACMILLAN PUBLISHING COMPANY
A Division of Macmillan, Inc.
New York

Senior Editor Lawrence Clarke
Art Editor John Ridgeway
Editor Peter MacDonald

Advisors
Professor Jack Meadows, University of Leicester
Professor Martin Rees, University of Cambridge

Planned and produced by:
Equinox (Oxford) Ltd
Littlegate House
St Ebbe's Street
Oxford OX1 1SQ

Published by:
Macmillan Publishing Company
A Division of Macmillan, Inc.
886 Third Avenue,
New York, N.Y. 10022

Library of Congress Catalog Card Number: 85–13887

Origination by Scantrans, Singapore

Text films set by Peter MacDonald

Printed in Spain

printing number
1 2 3 4 5 6 7 8 9 10

Library of Congress Cataloging in Publication Data

Nicolson, Iain.
The Universe

Bibliography: p.
Includes index.
1. Astronomy—Popular works.
II. Moore, Patrick.
II. Title.
QB44.2.N55 1985
523 85-13887
ISBN 0–02–922110–2

Introductory pictures (pages 1-8)
1 A typical spiral galaxy, NGC 2997 (⧫ page 200)
2-3 A collection of nebulae, Rho Ophiuchi (⧫ page 183)
Right A dark nebula, the Horsehead (⧫ page 186)
7 Nebulosity (in false-color) around a star, Eta Carinae (⧫ page 167)
8 Emissions (in false-color) from the Sun's corona (⧫ page 110)

Contents

Introduction

Astronomy has a strong claim to be considered the oldest of the sciences. Important astronomical observations were already being made thousands of years ago, including some which are still of value for modern astronomy. What has changed down the years is the reason for studying astronomy. Until the last century, most of the motivation came from a need to determine time (for agriculture, religious services, etc) and position (for navigation and surveying). All these purposes were highly practical. Add to this that, up to the 17th century, the domain of astronomy overlapped considerably with astrology, and you can see that past generations of astronomers could offer a whole range of essential services.

In the 20th century, most of these applications have disappeared. Time, for example, is almost always found from clocks, which can now give greater accuracy than astronomical determination. Yet, despite this, astronomy has grown rapidly during the present century: there are more astronomers alive now, both professional and amateur, than ever before. What is the reason?

The fascination of astronomy

From its very beginning, alongside all these practical applications, astronomy has been studied for what it can tell us about the Universe in which we live. The sheer excitement and interest of astronomy was apparent even in early times. One of the greatest astronomers of antiquity, Ptolemy, put it this way:

"I know that I am mortal and ephemeral, but when I scan the crowded circling spirals of the stars I no longer touch the Earth with my feet, but side by side with Zeus I take my fill of ambrosia, the food of the gods."

Modern astronomers would phrase it differently, but they recognize the feeling. There can be very few people who do not feel this fascination of astronomy. Astronomical discoveries are news, even though they have no direct impact on human life. This has long been true. In the latter part of the 18th century, William Herschel discovered a new planet (later called Uranus). It made him immediately famous. George III became his patron, so allowing him to throw over his job as a musician and establish himself as a full-time astronomer. Herschel was supported to study astronomy for its own sake. The one requirement was that he kept his patron informed of anything of interest happening in the heavens.

Astronomers nowadays follow in the Herschel tradition. They are employed to find out more about the Universe, rather than to give practical advice. What has changed is the nature of patronage. Most astronomers are now supported by state funding – that is, by the taxpayers – rather than by individuals. Modern astronomers feel that it is just as important for taxpayers today to hear about the exciting work they are supporting as it was for previous generations of individual patrons to be told.

It is, of course, true that the more people who are interested in astronomy, the better it will be for the future of the subject. However, taxpayers do not usually have a direct say in how their money is spent. In the present financial climate, this makes astronomy rather vulnerable. On the one hand, only the state has enough money to finance the costly instrumentation required for modern astronomy. On the other, its lack of practical application makes astronomy an obvious target for financial cutbacks. It is some commment on the intrinsic fascination of astronomy that it has survived so well up to the present.

Lessons from the past

From the historical viewpoint, astronomy is not only one of the oldest of disciplines, it was also the very first to become scientific in the modern sense of that word. When you talk about science now, most people have a picture of an activity where observation and experiment mix with theory in a fairly systematic way to produce new knowledge. In particular, there is a feeling that science is a process which can both explain the world around us and predict some of the future happenings in it. Astronomy reached this stage of explanation and prediction in Western Europe during the 17th century. It was followed in succeeding centuries by the other sciences.

Both scientists and historians have long wondered what was so special about that place and time. There have been many attempted answers, some of which are very relevant to modern astronomy. For example, 17th-century Europe allowed relatively easy communication between scientists in different countries: something which is still regarded as essential for astronomy today.

It follows that, in its early days, astronomy blazed a trail for the other sciences. Subsequently, as these developed, astronomy found it hard to start learning from them. In the 19th century, attempts to understand the Universe became closely entwined with new developments in physics and chemistry, and so it continues today. Astronomers therefore sometimes distinguish between "astronomy", which concerns the positions and movements of celestial bodies, and "astrophysics", which concerns their physical and chemical properties. As this book will show, the two branches are tightly linked; but it will also show that astronomers nowadays are overwhelmingly concerned with astrophysical investigations. The "big" contemporary questions of astronomy are of the type: How did the Universe begin? How did the Solar System originate? and so on. To answer such questions – and we are still in the early stages of providing solutions – requires information from a vast range of studies.

The principles behind this book

Despite the complexity of these questions, the Universe does present itself for study in a logical order. There are two basic principles to guide us. The first is that our ignorance of the Universe usually increases with distance. The objects which are closer to the Earth are also brighter, and so we can derive more information from them. The second principle is that the Universe is built up in a systematic way. Near at hand we have a solar system built up of planets; on a larger scale, our Sun joins with other stars to form the Galaxy; then our Galaxy joins with others to form the Local Group of galaxies; and so on.

The first principle implies that the best way of tackling astronomy, when you are learning about it, is to start near home – in the Solar System – and work outwards. The second suggests that we should do this thematically – by intercomparing similar types of object. The way this book is laid out reflects these two principles. If you scan the contents list you will see that we start with the Solar System and move steadily outwards to the limits of the Universe. At the same time, we bring together the different bodies at each stage to see what they can tell us about their part in the Universe.

The picture you are given here is, naturally, as up-to-date as can be managed. But clearly this is not the end of the story. During its history, observational astronomy has suffered from a number of limitations: the most important has been the necessity of examining the Universe from below a blanket of obscuring atmosphere. Modern developments are allowing us to break through many of these restrictions. In conseqence, astronomical knowledge is now growing at an unprecedented rate.

At many places in this book it would be possible to insert a marker saying "Watch this space". You can be certain that something new will be discovered there before you are much older. Astronomy may be considered the oldest of the sciences, but it is also the most rapidly changing science in the 20th-century.

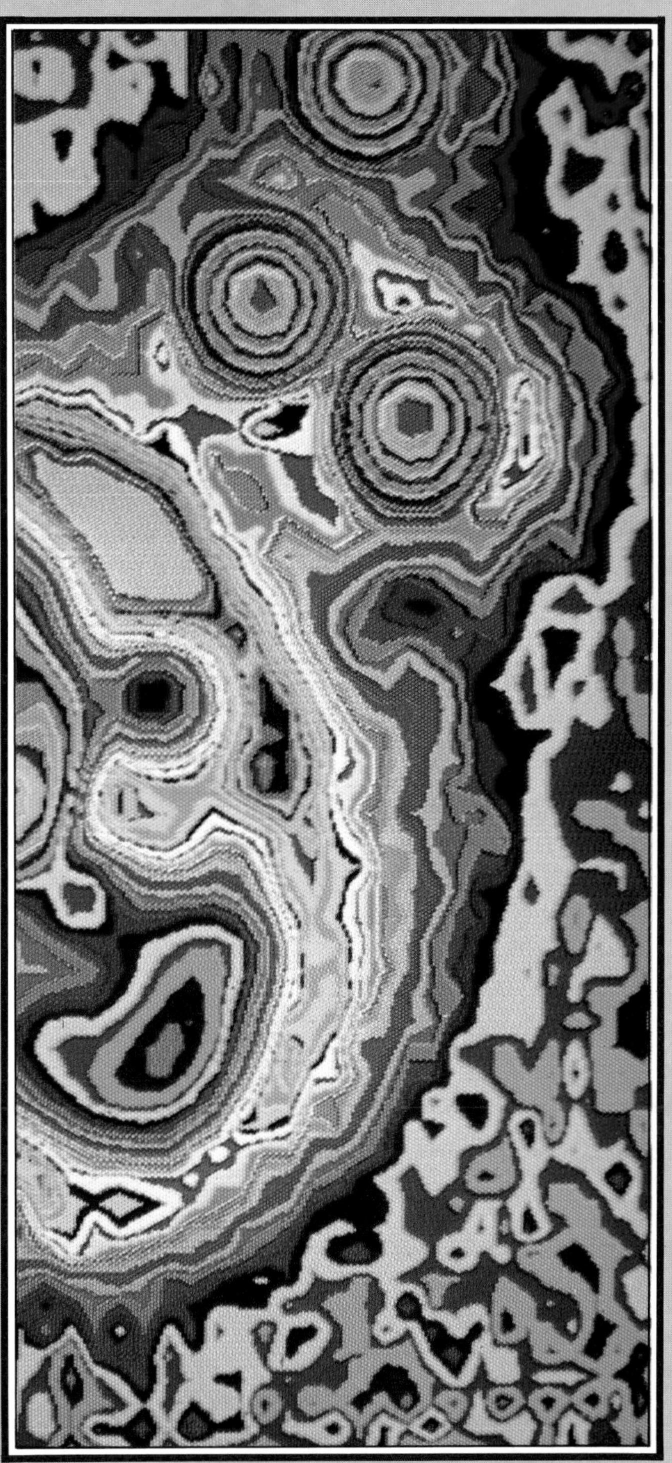

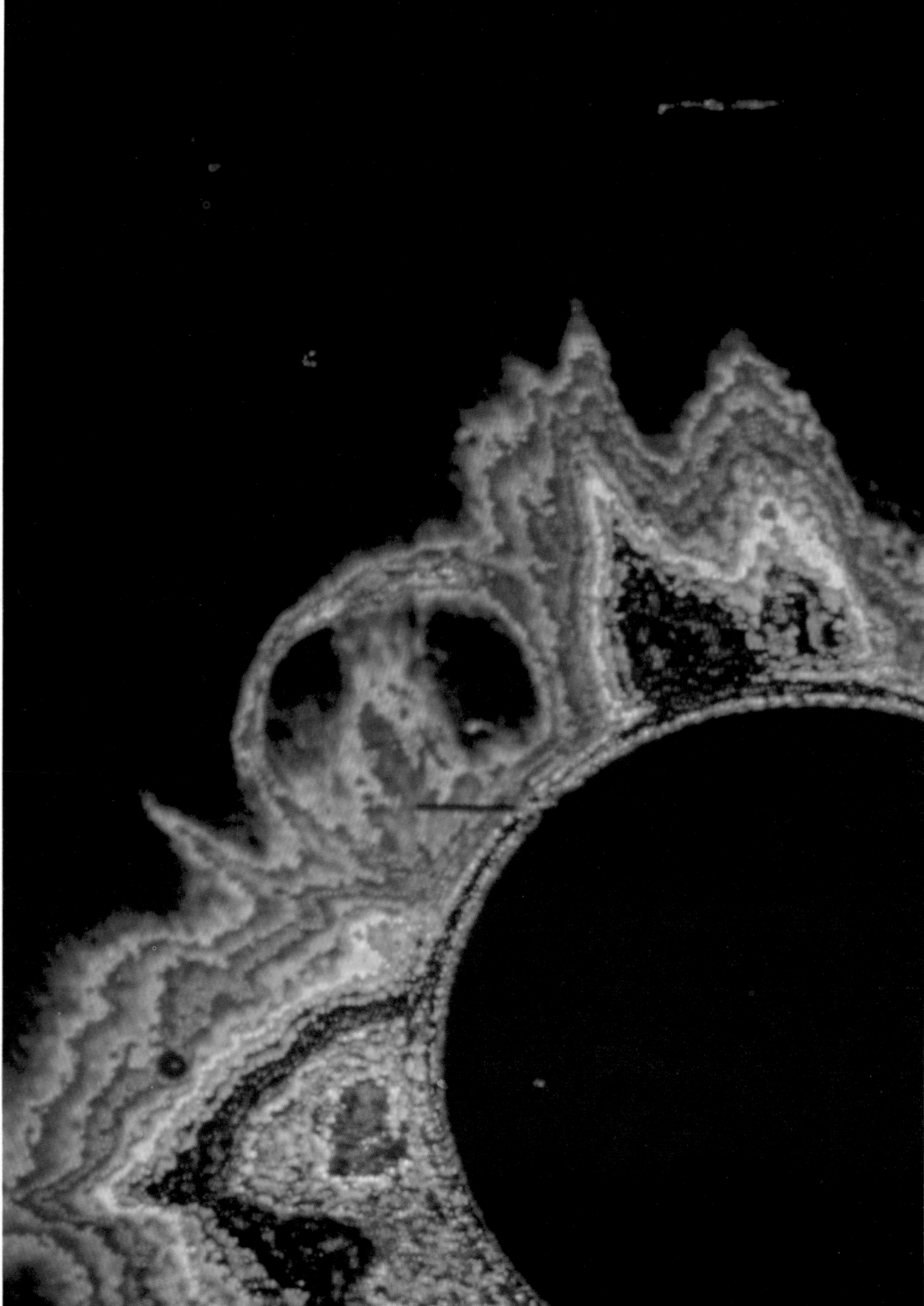

Planets and Orbits

1

Earth, Sun and Solar System introduced...Describing the motions of the planets...The exceptional orbits of Mercury and Pluto...The terrestrial planets...The Jovian planets...Comparative planetary data...Their moons, or satellites...Eclipses of the Sun and Moon... PERSPECTIVE...Early speculation on Earth's shape and position...Copernicus, Galileo and other founders of modern astronomy...Later discoveries of planets

The Earth is a small planet traveling in its annual path, or orbit, round a very ordinary middle-aged star, the Sun. Nine planets orbit the Sun, and of these seven are themselves orbited by satellites known as moons. These planets and moons, together with a host of minor bodies and some gas and dust, make up the Solar System.

The Sun, a self-luminous globe of gas, is by far the biggest member of the system. Its diameter of 1,392,000km is 109 times that of the Earth and nearly ten times that of the largest planet, Jupiter. It is 330,000 times as massive as the Earth, and about 740 times as massive as all the planets put together. Because the Sun is so massive it exerts a powerful gravitational attraction, and this is the force that holds the system together, controlling the motions of all the planets and minor bodies within it.

The names of the planets (in order of increasing distance from the Sun) are Mercury, Venus, Earth, Mars, Jupiter, Saturn, Uranus, Neptune and Pluto. Some astronomers believe there may be a tenth planet far beyond Pluto's orbit. The largest minor bodies, the asteroids, lie mainly between the orbits of Mars and Jupiter. Each planet's orbital path is elliptical rather than perfectly circular, and its distance from the Sun therefore varies during its yearly orbit. The point of closest approach to the Sun is called perihelion and the farthest point aphelion. The average, or mean, distance between the Earth and the Sun is 149,597,870km, and this distance, known as the astronomical unit (AU), is a useful yardstick for comparing planetary orbits. The mean distances of the other planets range from 0·39AU for Mercury to 39·44AU for Pluto. The distance of Pluto varies between 29·6 and 49·2AU, and at perihelion it is closer than Neptune to the Sun.

◄ **The Sun is shown orbiting the Earth between Venus and Mars in this stylized illustration of the Ptolemaic system, published in Nürnberg in 1493, 50 years before the idea of a centrally placed Earth was seriously challenged.**

From flat Earth to round

Looking around him, a casual observer knowing nothing about science would assume the Earth was flat, allowing for local irregularities such as hills and valleys. It was quite natural for early civilizations to regard the world as not only flat, but also stationary, with the whole sky moving around it and completing one revolution every 24 hours.

Even the early Greek philosophers believed in a central, non-rotating flat Earth. Thales of Miletus (c. 624-537 BC), who is generally regarded as the first of the great Greek scientists, taught that the Earth was a flat disk, floating on water like a cork.

The eventual realization by the Greeks that the Earth must be a globe followed logically from observation. For instance, the bright star Canopus can be seen from Alexandria, but not from Athens, where it never rises above the horizon. And when the shadow of the Earth falls upon the Moon, producing a lunar eclipse, it is seen to be curved – so that the Earth's surface must also be curved. Acceptance of this was a major step forward, but the Greeks still could not dethrone the Earth from its position in the center of the universe.

The Earth-centered universe

A few of the philosophers did in fact challenge the concept of the Earth-centered or "geocentric" universe. Aristarchus of Samos (c. 310-230 BC) was bold enough to suggest that the Earth moved round the Sun in a period of one year. He could provide no proof, however, and the Sun-centered or "heliocentric" theory was disregarded for many centuries. Instead the Greeks developed a theory involving what were known as "epicycles". They thought that all celestial bodies must move in perfect circles, since the circle was the "perfect" form and nothing short of perfection could be allowed in the heavens. However, it was quite clear that the "wandering stars", or planets, which had been recognized from very early times, did not move smoothly and regularly against the starry background. The Greeks therefore assumed that although a planet moved round the Earth, it also described a small circle or epicycle, the center of which moved round the Earth in a perfect circle. As more and more irregularities came to light, further epicycles had to be added, until the whole system became a complicated cosmic maze.

The Ptolemaic system (named after Ptolemy, last of the great astronomers of the classical period, who lived around AD 150) survived for centuries in spite of its complexity, because it did not conflict with observed evidence, and there was no reason to reject it.

The Arab tradition

It was only with the rise of Arab civilization in the 7th century that systematic observation began once more, and then it was mainly for astrological purposes (until the 17th century, astrology was still regarded as a true science). The Arabs built improved measuring instruments and calculated the apparent motions of the known planets with impressive accuracy. Specialized observatories were built, the last and most elaborate by Ulugh Beigh at Samarkand in 1433.

Except for Mercury and Pluto the orbits of the planets are in much the same plane

The outer Solar System

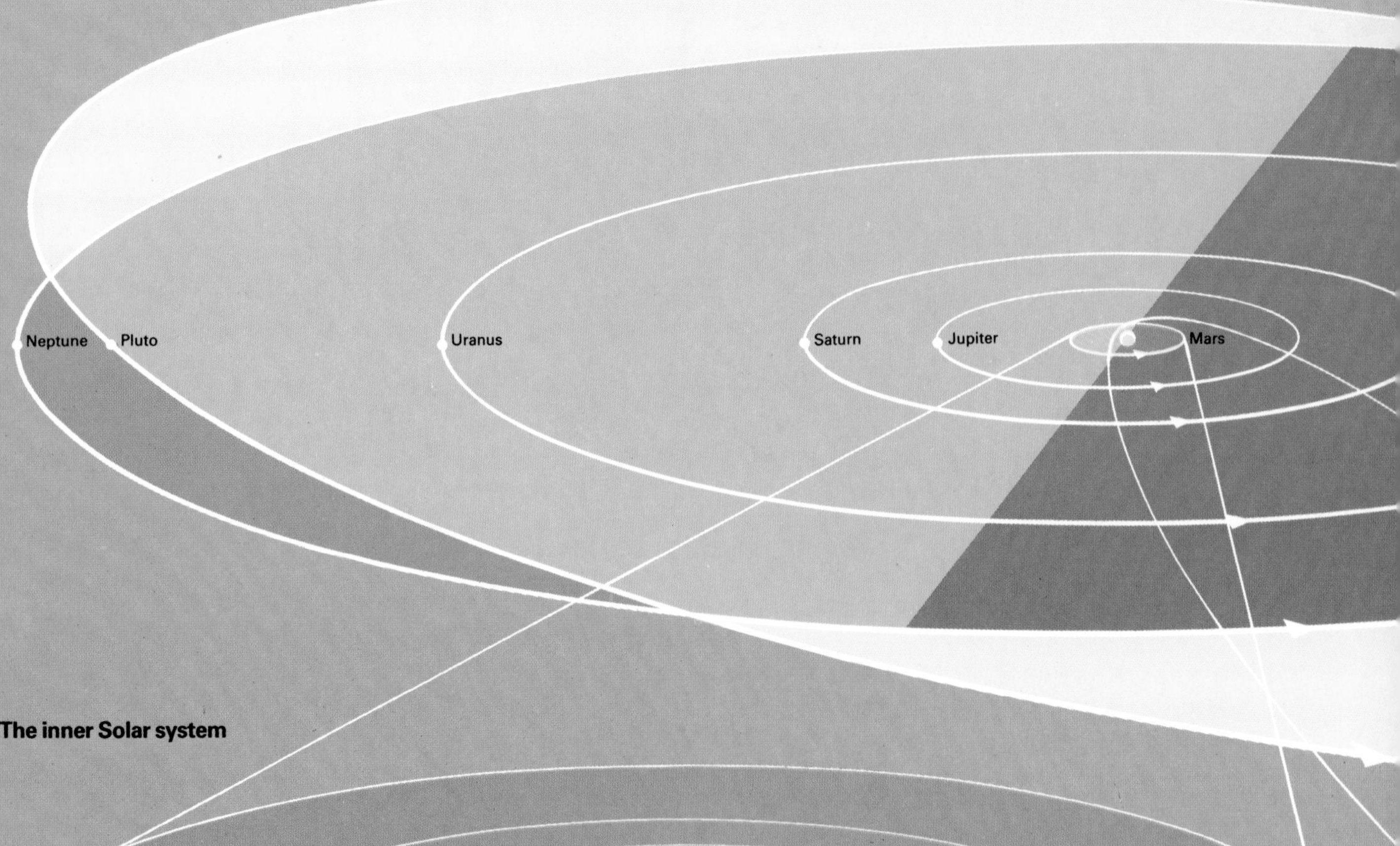

Neptune Pluto Uranus Saturn Jupiter Mars

The inner Solar system

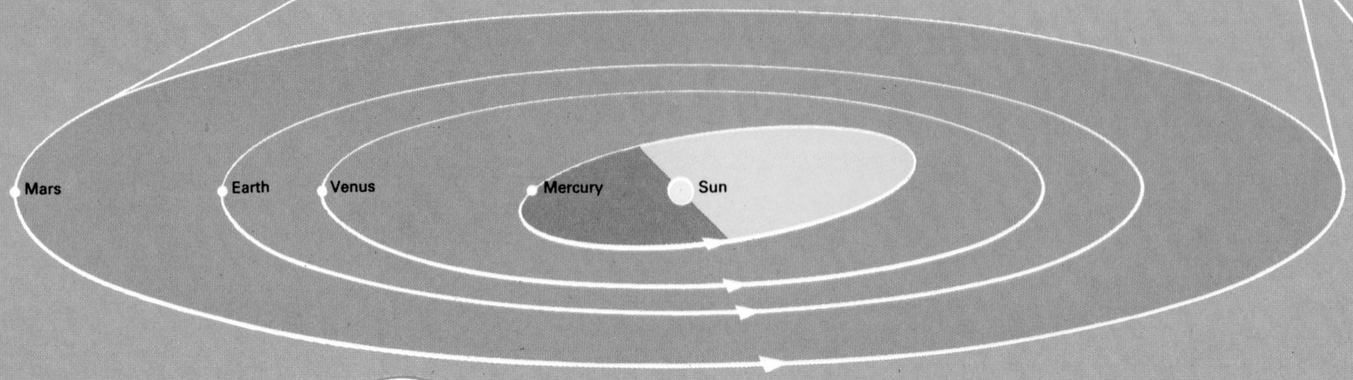

Mars Earth Venus Mercury Sun

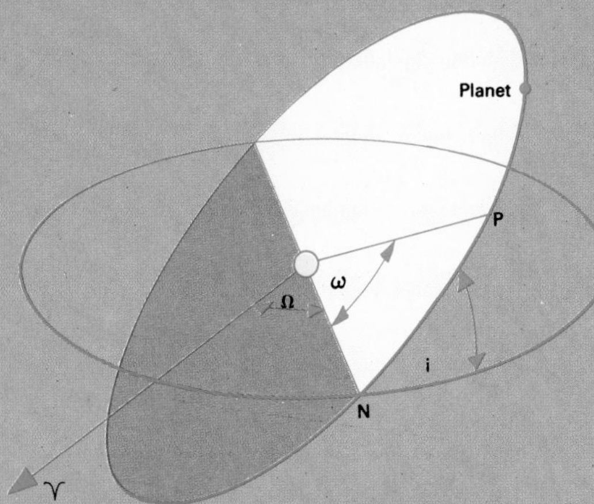

▲ ***The orientation of a planetary orbit is described by three angles: i (inclination) – the angle between the plane of the Earth's orbit (the ecliptic) and that of the planet; Ω – the angle between a fixed direction in space (♈) and the point N where the orbit crosses from south to north of the ecliptic; and ω – the angle between N and perihelion P (the point at which the planet is closest to the Sun).***

Although the planets all move in elliptical orbits, most of these are nearly circular. The extent to which an ellipse is elongated – that is, its eccentricity – is given by a figure between 0 (for a circle) and 1 (for a parabola): the greater the eccentricity the flatter or more elongated the ellipse. Among the planets, Mercury and Pluto have distinctly elliptical orbits with eccentricities of 0·206 and 0·250 respectively. Most elliptical of all are the orbits of the comets, nearly all of which extend far beyond the outermost planets (most astronomers believe that there is a vast reservoir of comets with orbits extending to a distance of some 40,000AU from the Sun).

The planets all travel round the Sun in the same direction, and this motion is called direct. Most of their orbits lie within a few degrees of the ecliptic (the plane of the Earth's orbit). Again, Mercury and Pluto are exceptions, their orbits being tilted, or inclined to the ecliptic, by angles of 7° and 17° respectively. Some asteroids have more steeply tilted orbits, and the orbital inclinations of long-period comets range from 0° to 180°. Those with inclinations greater than 90° move in retrograde orbits (in the opposite direction to the planets).

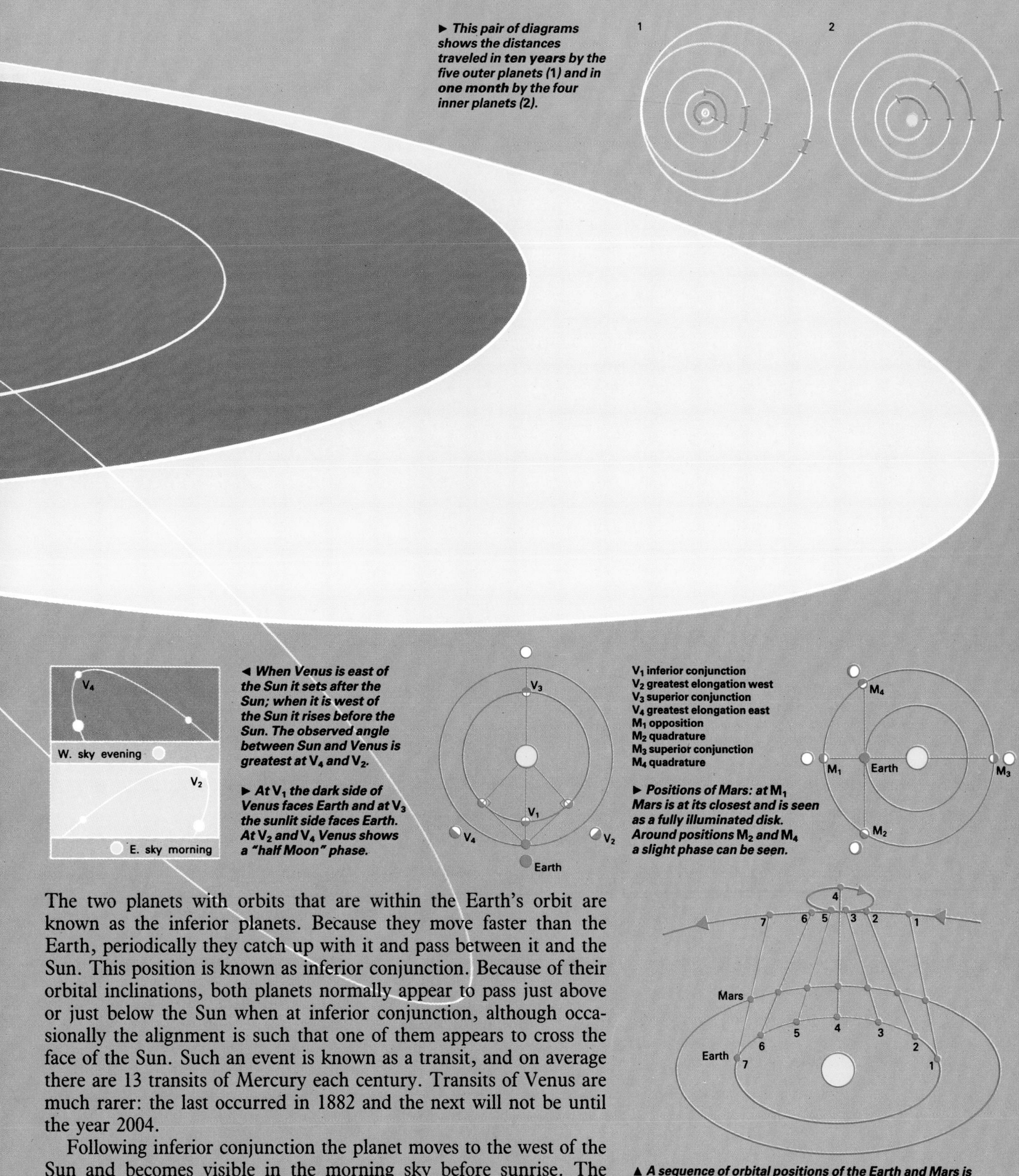

► This pair of diagrams shows the distances traveled in ten years by the five outer planets (1) and in one month by the four inner planets (2).

◄ When Venus is east of the Sun it sets after the Sun; when it is west of the Sun it rises before the Sun. The observed angle between Sun and Venus is greatest at V_4 and V_2.

► At V_1 the dark side of Venus faces Earth and at V_3 the sunlit side faces Earth. At V_2 and V_4 Venus shows a "half Moon" phase.

V_1 inferior conjunction
V_2 greatest elongation west
V_3 superior conjunction
V_4 greatest elongation east
M_1 opposition
M_2 quadrature
M_3 superior conjunction
M_4 quadrature

► Positions of Mars: at M_1 Mars is at its closest and is seen as a fully illuminated disk. Around positions M_2 and M_4 a slight phase can be seen.

▲ A sequence of orbital positions of the Earth and Mars is shown in this diagram, together with the apparent motion of Mars in the sky. Most of the time Mars moves directly – right to left – relative to the background stars, but near opposition, when the Earth is overtaking it, Mars appears to move backwards (in retrograde motion) and traces out a loop in the sky.

The two planets with orbits that are within the Earth's orbit are known as the inferior planets. Because they move faster than the Earth, periodically they catch up with it and pass between it and the Sun. This position is known as inferior conjunction. Because of their orbital inclinations, both planets normally appear to pass just above or just below the Sun when at inferior conjunction, although occasionally the alignment is such that one of them appears to cross the face of the Sun. Such an event is known as a transit, and on average there are 13 transits of Mercury each century. Transits of Venus are much rarer: the last occurred in 1882 and the next will not be until the year 2004.

Following inferior conjunction the planet moves to the west of the Sun and becomes visible in the morning sky before sunrise. The angle between the Sun and the planet is known as the elongation, and increases to a maximum of 28° for Mercury and 47° for Venus. Thereafter the angle narrows until the planet passes behind the Sun at superior conjunction. It then reemerges on the east side.

Every planet except Venus rotates on its axis in the same direction

The planets which have orbits outside that of the Earth are known as the superior planets, and their behavior is somewhat different. Traveling more slowly than Earth, a superior planet is periodically overtaken by it. On such an occasion the Sun, Earth and planet lie in a straight line with the Earth in the middle. The planet is said to be at opposition and is then best placed for observation, because it is at its closest to Earth as well as being visible all night. After opposition the planet falls behind the Earth and eventually passes behind the Sun (superior conjunction) before emerging again to the west of the Sun and becoming visible from Earth in the morning sky.

The time interval between two similar alignments of Sun, Earth and planet (for example, between two conjunctions or two oppositions) is called the synodic period of the planet.

The terrestrial planets

The four innermost planets are collectively known as the terrestrial planets because they have some features in common with the Earth. All are small bodies of relatively high density, composed mainly of rocky minerals and metals. Mercury has just under 40 percent of the Earth's diameter. Its axial rotation period (the time it takes to rotate once on its axis) of 58·7 days is two-thirds of its orbital period of 87·97 days, and this leads to some interesting seasonal effects. Venus is almost the same size as the Earth, but is a hot and hostile world, permanently shrouded in cloud. It rotates on its axis in a retrograde sense once every 243 days. The third planet, Earth, has a surface largely covered in water and is the only planet on which life is known to exist. It has a large moon which is about one-quarter of the diameter of the planet. Mars has about half the diameter of the Earth and about one-tenth its mass. It is a chilly world with a thin atmosphere, but is nevertheless the least hostile of the planets apart from Earth itself. Even so, it now seems highly unlikely that Mars supports even the most basic forms of life.

The Jovian planets

The next four are giant planets, often called the Jovian planets because in some respects they all resemble Jupiter. They are much less dense than the terrestrial planets, and are composed mainly of hydrogen and helium. Jupiter has 11 times the diameter of the Earth and is 318 times as massive; indeed it is 2·5 times as massive as all the other planets put together. It has the shortest rotation period of any planet (9 hours and 50 minutes at the equator) and consequently spins so fast that it bulges at the equator and is flattened at the poles, the equatorial diameter exceeding the polar diameter by some 8,500km. Saturn is nearly twice as far as Jupiter from the Sun. It has 9 times the Earth's diameter and 95 times its mass. Less dense on average than water, it spins rapidly and bulges markedly at the equator. Its most remarkable feature, however, is its extensive and complex system of rings.

Uranus, the next planet, is 19 times the Earth's distance from the Sun, about four times the Earth's diameter, and pursues its lonely orbit in a period of some 84 years. Neptune is similar in size to Uranus, but slightly more massive and significantly denser. It is 30AU from the Sun and takes 164 years to complete each orbit.

Icy Pluto is the last of the known planets. Although smaller than the Moon, Pluto is accompanied by its own relatively large satellite, Charon. It takes 248 years to complete each circuit of the Sun.

Scientific Revolutionaries

▲ *Copernicus: a 16th-century woodcut.*

▼ ***The Sun-centered Copernican system as published in 1543, shortly before Copernicus himself died. Martin Luther declared that "this fool seeks to overturn the whole art of astronomy". His words typified the reaction of the Church.***

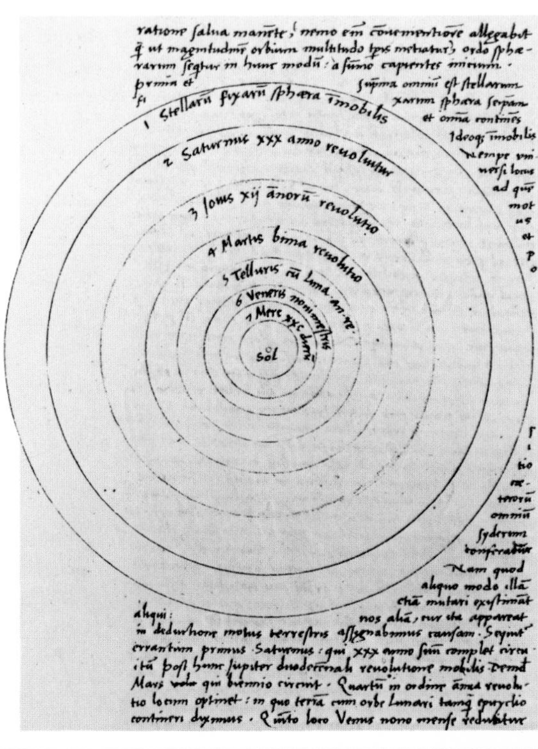

Tycho Brahe

Copernicus

It was not until some fourteen centuries after the death of Ptolemy that his geocentric theory was seriously questioned – by a Polish cleric, Mikolaj Kopernik (1473-1543), always known by his Latinized name of Copernicus. Early in his career he had begun to doubt the Ptolemaic theory, and he realized that many of the objections could be overcome simply by removing the Earth from its position in the center of the Solar System and replacing it with the Sun. His great book "De Revolutonibus Orbium Celestum" was probably more or less complete by 1530, but he was reluctant to publish it, because he knew that the Church would be bitterly hostile to any idea that the Earth was not the most important body in the universe. The book finally appeared in 1543, and even then the publisher, Osiander, added a preface – without Copernicus' authorization – to the effect that the theory was not intended to be taken literally, and was merely offered as an aid to mathematical computations of the movements of the planets. There was indeed strong opposition, and it is also true that Copernicus made many mistakes. In particular, he retained the concept of perfectly circular orbits, making his theory almost as cumbersome as the one it replaced.

Tycho Brahe

The next important figure was the Danish astronomer Tycho Brahe (1546-1601), one of the most colorful characters in the history of science. In his student days he fought a duel and lost part of his nose, which he replaced with "gold, silver and wax". He was first drawn to astronomy in 1572, when he observed a supernova (an exploding star) in the constellation of Cassiopeia. Tycho established an observatory at Hven, an island in the Baltic, with support from the Danish court. He had no telescopes, but his measuring instruments were much the best of their time, and between 1576 and 1596 he drew up an accurate star catalog as well as measuring the movements of the planets with remarkable precision. He paid particular attention to Mars, which was fortunate because Mars has an orbit which is rather less circular than those of the Earth or Venus.

Tycho realized that the old Ptolemaic system simply did not work. Yet he did not believe that the Earth was in motion, and he worked out a compromise system according to which the planets moved round the Sun, while the Sun itself moved round the Earth. He died suddenly in 1601, leaving his observations to his last assistant, a German named Johannes Kepler.

▲ Tycho's observatory at Hven, where most of his important work was carried out. After Tycho left Denmark in 1596 the observatory was never used again.

▼ One of Tycho's instruments was this zodiacal armillary sphere, used for sighting and following the motions of the stars, as well as the planets.

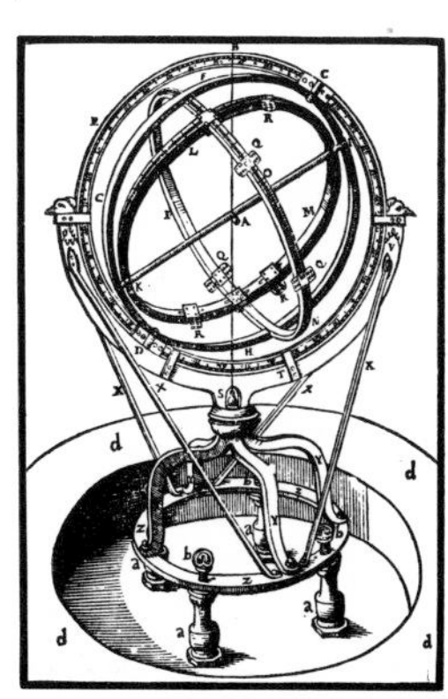

Observation versus Revelation

▲ **Galileo – a contemporary portrait.**

Galileo Galilei

The telescope had been discovered in Holland in 1608 (possibly before), and when news of it reached Padua, Galileo Galilei (1564-1642), Professor of Mathematics at the university there, made one for himself. His first observations with it, made in the winter of 1609-10, resulted in a series of spectacular discoveries. Galileo saw the mountains of the Moon and the four bright satellites of the planet Jupiter. Everything he saw confirmed his already strong belief that the Sun was the center of the Solar System. The satellites of Jupiter were particularly significant: there could be no doubt that they moved round Jupiter, so that at the very least there had to be more than one center of movement in the planetary system.

There was also the question of the phases of Venus. According to the old Ptolemaic theory Venus could not possibly show a full sequence of phases from new to full – yet it did. This too confirmed that even if Galileo was not right, Ptolemy was still unquestionably wrong.

Galileo published his findings in a book, "Sidereus Nuncius" (The Starry Messenger), which was well received although some churchmen were skeptical. When Galileo went to Rome in 1611 he was made welcome – not only by Cardinal Barberini, an old and influential friend, but also by the Pope. Meanwhile he had moved from Venice to Florence to become Chief Mathematician at the University of Pisa and Philosopher to the Grand Duke. This proved to be a mistake; the atmosphere at Florence was much less liberal than it had been in Venice.

▼ **Galileo's wash drawings on the left show the phases of the Moon; the notes and sketches on the right are his entry for 28 January 1613, relating to Neptune, which he thought was a fixed star.**

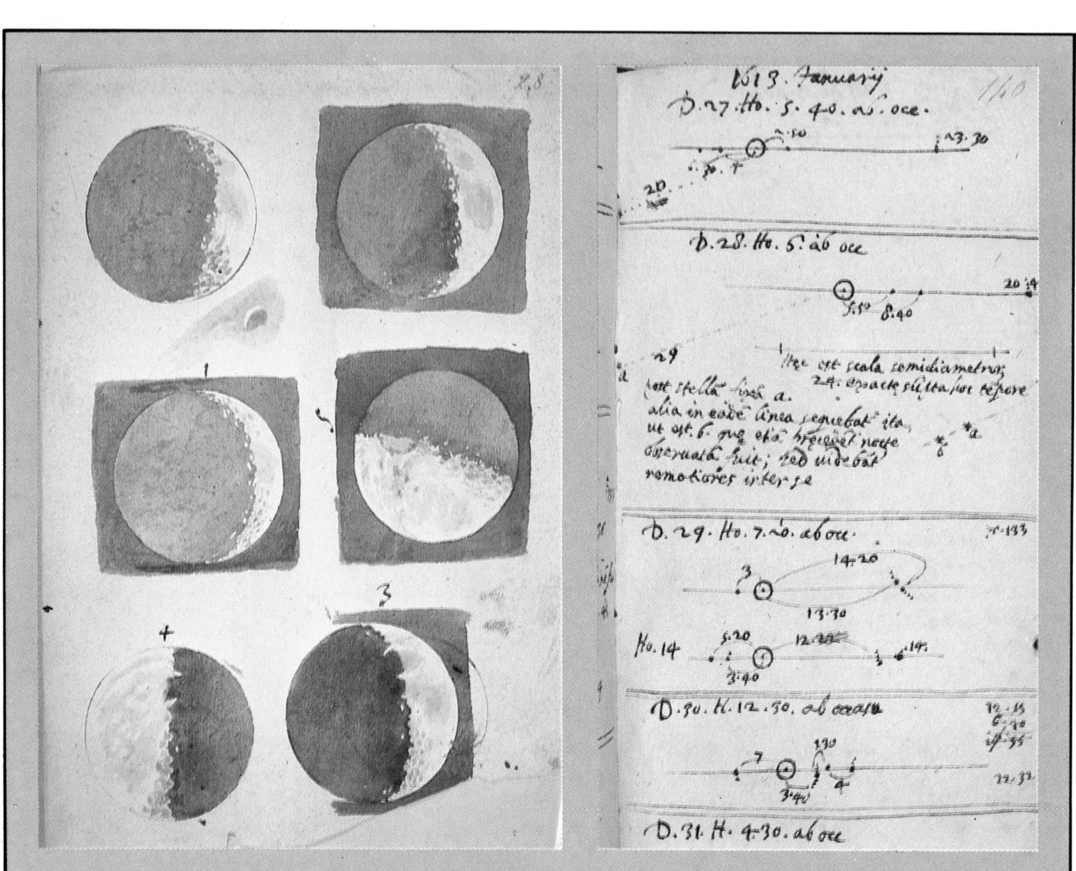

▲ **These two telescopes made by Galileo have a field of view of 17 arc minutes. That on the left has a magnification of 14, the one on the right 20.**

One of Galileo's enemies, Tommaso Caccini, a Dominican, sent the Holy Office what he saw as an incriminating letter from Galileo to his friend Castelli. In Rome one Cardinal, Bellarmine, wrote that the theory of a moving Earth "...injures our holy faith and makes the sacred Scriptures false." In 1616 Pope Paul himself intervened and, through Bellarmine, ordered Galileo to cease teaching the heresy of a moving Earth.

The situation changed when Galileo's old friend Cardinal Barberini became Pope Urban VIII. Encouraged, in 1630 Galileo produced a major book in the form of a dialog between three characters, Salviati, who favors the Copernican theory, Simplicio, who opposes it and Sagredo, who is more or less neutral. Simplicio is made to look decidedly naive and even ridiculous, so that in effect the "Dialog" is a work of open propaganda for the theory of a central Sun and a moving Earth.

Galileo sent the manuscript to Rome, where it was read by the Holy Office and passed for publication. It finally came out in February 1632.

Then the storm broke. Pope Urban VIII, a very different person from the old Cardinal Barberini, concluded that he was being ridiculed – in fact, that Galileo's Simplicio represented him. The aging scientist was called back to Rome, arrested on a clearly fabricated charge of heresy, tried and condemned. In June 1633 Galileo was forced to recant – to "curse, abjure and detest" the false theory that the Earth moved round the Sun – after which he was kept under close supervision at his villa for the rest of his life.

► Galileo invented a device called a Jovilabe to calculate the position of Jupiter's satellites. The four largest are represented by the numbered circles.

▼ Galileo's drawings of Jupiter and its four largest satellites.

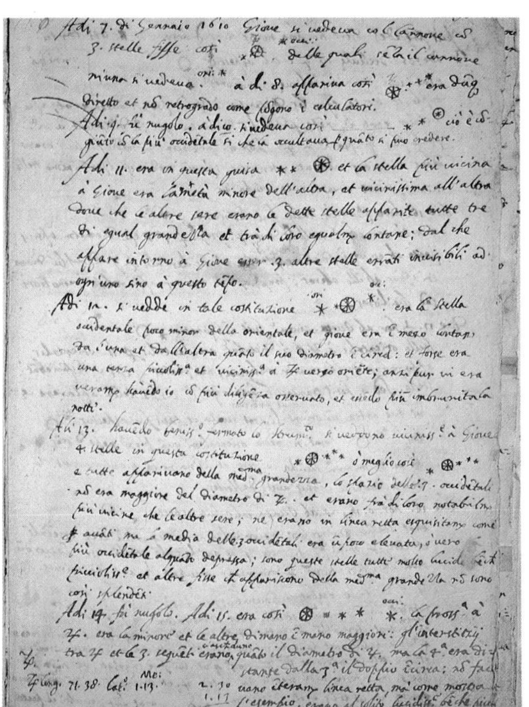

Mathematics and Universal Law

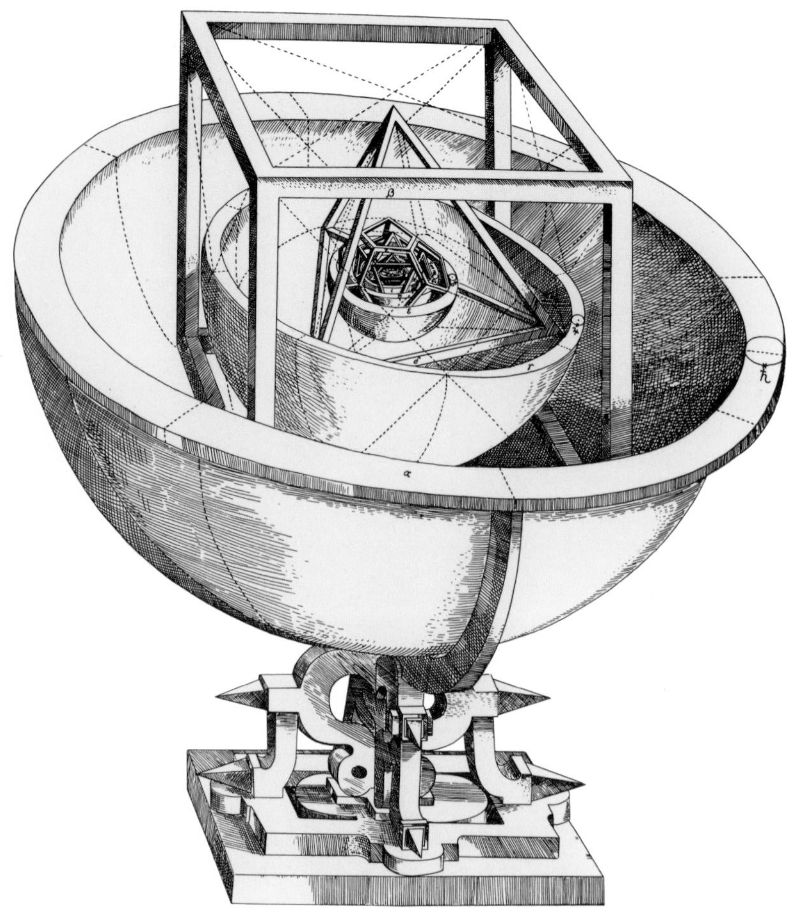

Johannes Kepler

Belonging both to the past and to the near-present, Kepler (1571-1630) is a curious figure. A brilliant mathematician, he was also very much a mystic, and some of his theories sound strange today. In particular he placed great faith in the "five regular solids" which, he believed, could be accommodated between the orbits of the planets. He certainly practiced astrology, although whether he had much real faith in it is not clear. His background was unfortunate: his father was an idle, shiftless adventurer who eventually deserted his family and disappeared, while his mother, apparently a most sinister-looking woman, was once arrested on a charge of witchcraft.

Kepler had implicit faith in the accuracy of Tycho Brahe's observations, but despite all his efforts he could not reconcile them with any accepted theory of planetary motions. He concentrated chiefly upon Mars, and finally made the all-important discovery that the orbits of the planets were elliptical rather than perfectly circular. Once this had been established, the great Danish observer's measurements fell perfectly into place.

Although he was not persecuted in the way that Galileo had been, Kepler had an unhappy life. He moved from place to place and was always short of money. Indeed, he died in 1630 while on a journey to collect some salary owing to him.

◄ A contemporary illustration depicts Kepler's belief that the planetary orbits reflected an arrangement of concentric geometrical solids. Kepler was convinced by this theoretical model and always regarded it as the greatest of his discoveries. In 1596 he announced his theory in a book, copies of which he sent to both Tycho and Galileo. Tycho's reply was encouraging, recognizing a mathematician of great skill; Galileo was more polite, saying that he was looking forward to reading it.

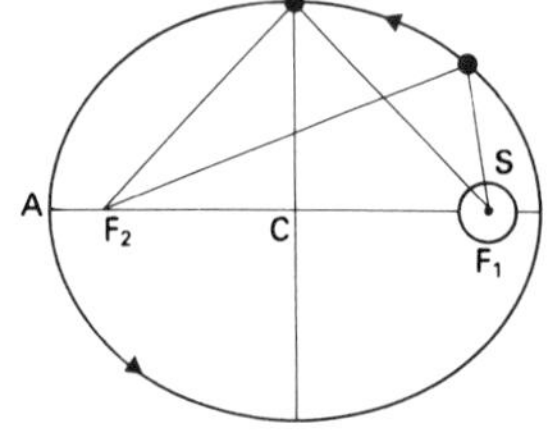

▲ Kepler's first law shows that each planet moves round the Sun in an elliptical orbit with the Sun (S) at one focus (F_1) of the ellipse. The sum of the distances from two fixed points – foci (F_1,F_2) is constant.** P **is perihelion,** A **aphelion,** C **center,** AP **the major axis, and** CA=CP **the semimajor axis.

◄ Johannes Kepler formulated the three laws of planetary motion. The first two were published in 1609 and the third in 1618; together they formed the basis of all subsequent work. They apply to any small body in the gravitational field of a more massive one.

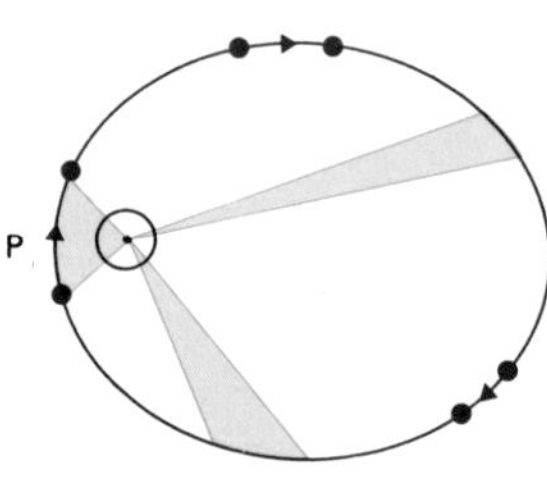

▲ According to Kepler's second law, the radius vector (Sun-planet line) sweeps out equal areas in equal times. A planet's movement during three equal time intervals is shown: the shaded areas are all equal since the planet moves fastest when nearest the Sun.

► Kepler's third law shows that the farther a planet is from the Sun, the longer it takes to complete one orbit, and the relationship between the two values is fixed. The square of the orbital period is directly proportional to the cube of the mean distance.

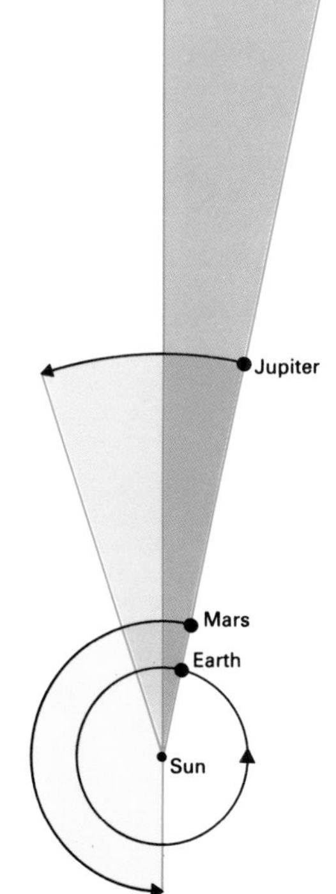

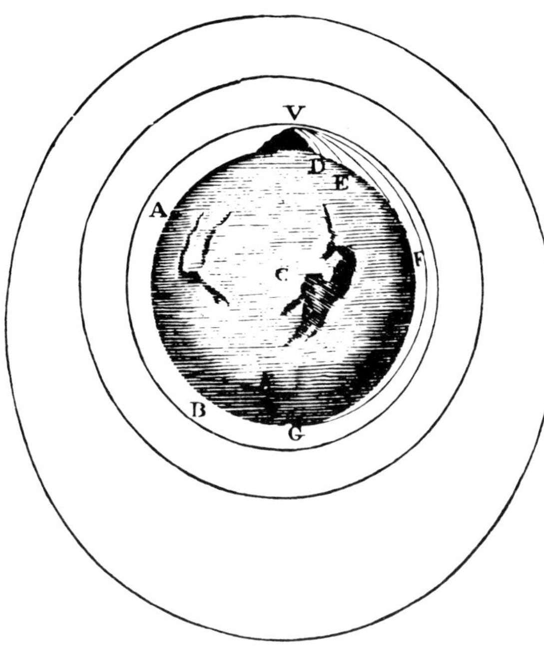

► *Isaac Newton at forty-six, shortly after the publication of his "Principia". His greatness was recognized both by his fellow scientists, who elected him president of the Royal Society in 1703, and by his country, when he was knighted in 1705. Shortly before he died he summed up his life: "...I seem to have been merely a child playing on the seashore, diverting myself in now and then finding a pebble more smooth or a shell more beautiful than others, whilst before me the great ocean of Truth lay all undiscovered."*

◄ *Newton's diagram shows what happens to an object thrown from a great height and at different speeds. The laws governing the speed of objects falling to the ground also determine the motion of the planets.*

Sir Isaac Newton

Isaac Newton (1642-1727) was born in the year of Galileo's death, in the Lincolnshire village of Woolsthorpe. He showed an early aptitude for mathematics and became a student at Cambridge University, although during the Plague he returned to Woolsthorpe and there laid the foundations of much of his future work. (The story of the falling apple is quite possibly true.) It was also at Woolsthorpe that he passed light through a prism, splitting it up into the colors of the rainbow.

After the Plague Newton returned to Cambridge, and in 1671 he presented the recently founded Royal Society with the first reflecting telescope ever made.

In 1687 he produced the great work "Principia". In this volume, among other things, he set out his law of universal gravitation – that all bodies in the universe attract each other with a force which depends on their masses and which diminishes with the square of their separations. For two bodies of m_1 and m_2 the force (F) of attraction between them is represented by:

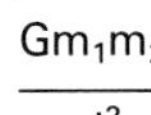

$$\frac{Gm_1m_2}{d^2}$$

where G is the gravitational constant and d is the distance between the two masses. Newton showed that the same force that governed the fall of apples, also governed the orbit of the Moon and the motions of all the planets as they orbit the Sun.

Newton's creative work did not end in 1687. He undertook much more, including another major book "Opticks" which was published in 1704. However, it is for the "Principia" that he will always be remembered.

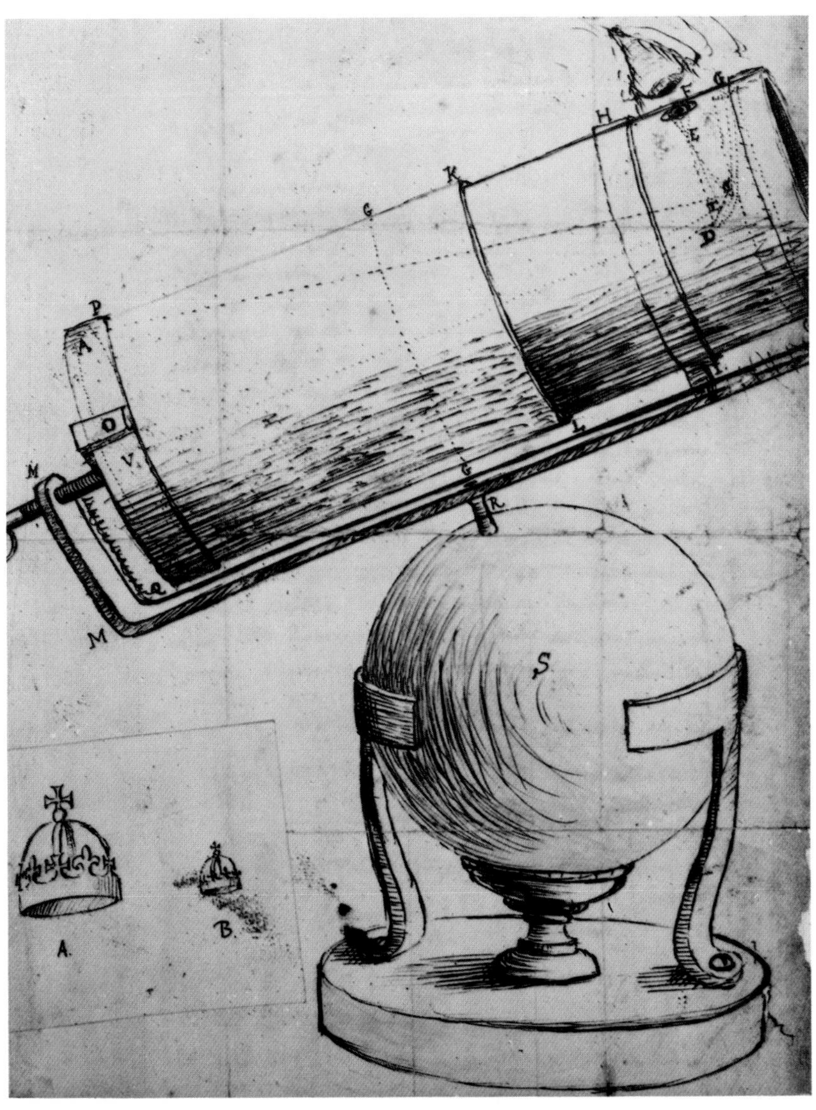

◄ ***This sketch of Newton's reflecting telescope belongs to the Royal Society, and depicts the telescope itself with inset details of a weathercock ornament seen through it (A) and through a refracting telescope (B) about 65cm in length.***

All but two of the Solar System's nine planets are accompanied by moons

	Pluto	Neptune	Uranus	Saturn	Jupiter	Mars	Earth	Venus	Mercury	Moon	Sun
Mean distance (AU)	39.44	30.0580	19.1818	9.5388	5.2028	1.5237	1.0000	0.7233	0.3871	—	—
Mean distance (millions of km)	5,900.0	4,496.7	2,869.6	1,427.01	778.34	227.94	149.60	108.21	59.91	—	—
Eccentricity	0.250	0.0086	0.0473	0.0556	0.0485	0.0934	0.0167	0.0068	0.2056	0.0549	—
Inclination to ecliptic	17°12′	1°46′	0°46′	2°29′	1°18′	1°50′	—	3°23′	7°00′	5°09′	—
Sidereal period (days)	90,465.0	60,190.5	30,684.8	10,759.20	4332.59	686.980	365.256	224.701	87.969	27.322	—
Equatorial diameter (km)	2,500	49,500	51,800	120,000	142,800	6,794	12,756	12,104	4,878	3,476	1,392,530
Polar diameter (km)	2,500	47,400	49,000	108,000	134,200	6,759	12,714	12,104	4,878	3,476	1,392,530
Equatorial rotation	6.3d	18.2hr	16.3hr	10.2hr	9.8hr	24.62hr	23.93hr	243d	58.65d	27.32d	24.6d
Mass (kg)	1.6×10^{22}	1.028×10^{26}	8.6978×10^{25}	5.684×10^{26}	1.899×10^{27}	6.4191×10^{23}	5.9742×10^{24}	4.8689×10^{24}	3.3022×10^{23}	7.3483×10^{22}	1.9891×10^{30}
Density (water = 1)	1-2	1.77	1.27	0.70	1.32	3.94	5.52	5.24	5.43	3.34	1.41

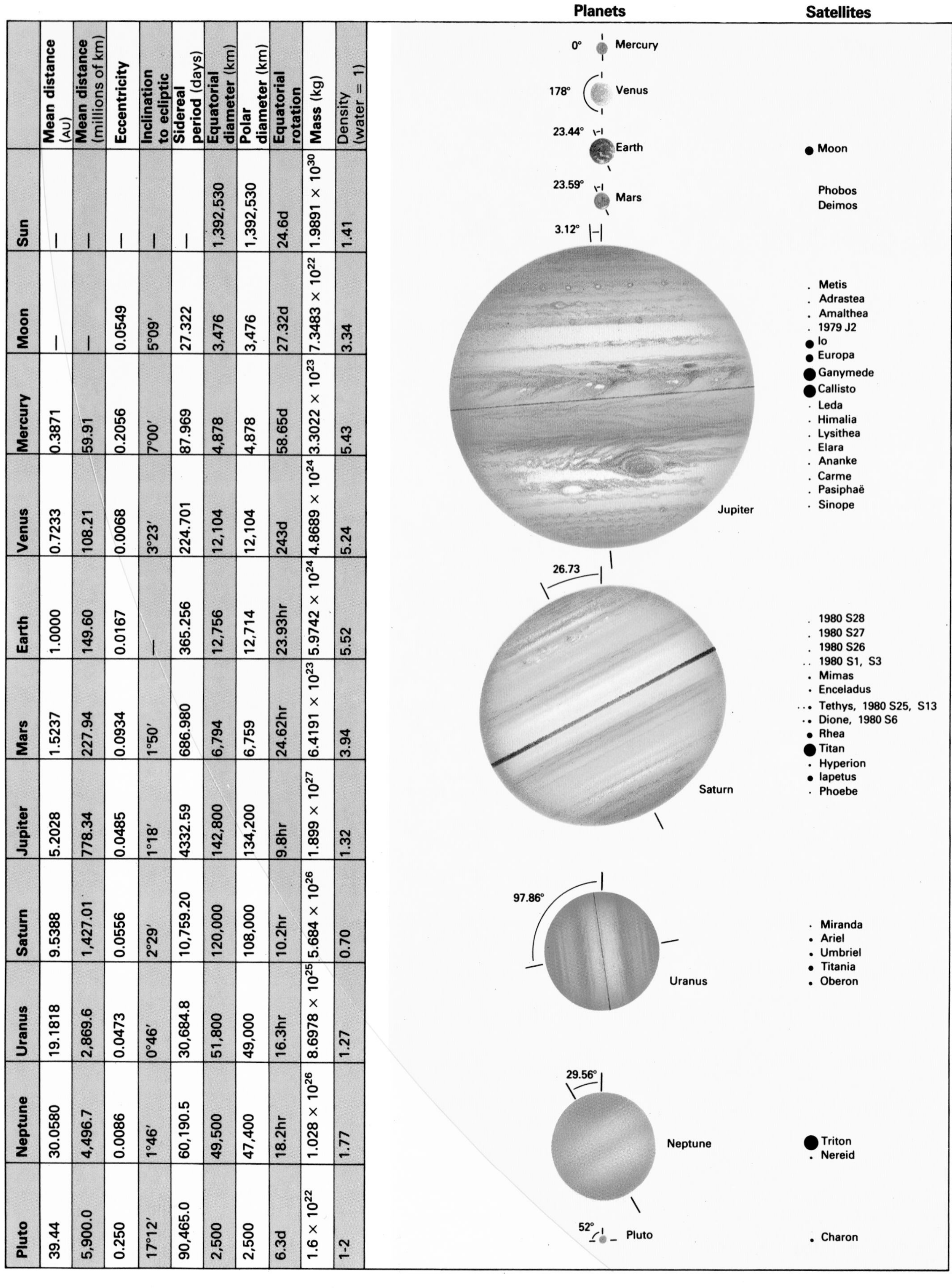

The Earth's satellite

The Moon, a barren, airless, cratered world, lies at a mean distance of 384,392km from Earth and has an orbital period of 27·3 days. Its orbit is elliptical, the distance from Earth varying between 356,410km at perigee (closest point) and 406,680km at apogee (farthest point). This is a variation of over 10 percent, and there is a corresponding variation in the Moon's apparent size. It rotates on its axis in the same period of time as it takes to travel round the Earth, so that the same face is always turned towards the Earth. This phenomenon is known as "captured" or "synchronous" rotation (➧ page 73).

Because the Moon shines by reflecting sunlight, at any instant one hemisphere is lit while the other is dark. When it is "new" the Moon lies in line with the Sun, and the hemisphere that faces Earth is dark. After New Moon it moves to the east of the Sun and as the angle between Sun and Moon increases, more of the illuminated hemisphere becomes visible and the appearance of the Moon changes from a thin crescent to a fully illuminated disk (Full Moon) when it is opposite the Sun. Thereafter the Moon again begins to approach the Sun in the sky, returning eventually to New Moon in 29·5 days.

Eclipses of the Sun and Moon

Since the Moon's orbit is inclined to the ecliptic by about 5°, the New Moon usually passes above or below the Sun. However, if the new phase occurs when the Moon is close to one of the nodes of its orbit (the points where its orbit meets the ecliptic) it crosses the face of the Sun, giving rise to a solar eclipse. If the alignment is exact, the Sun is completely covered and there is a total eclipse. (By coincidence the Sun and Moon appear virtually the same size in the sky. Although the Sun is 400 times the diameter of the Moon, it is also 400 times farther away.) Outside the narrow region from which a total eclipse can be observed, partial eclipse occurs. If the Moon is near apogee the solar disk is not completely covered and a ring of sunlight remains round the dark lunar disk. This event is called an annular eclipse.

A lunar eclipse occurs when the Full Moon passes into the shadow cast by the Earth instead of passing above or below the shadow.

Eclipses

Sun
Sun
Moon
Earth
Earth
Moon
1 2 3

▲ If the Moon covers the whole disk of the Sun, a total eclipse (1) is seen from the umbra (area of deepest shadow) and a partial eclipse (2) from adjoining areas. When the Moon is at apogee an annular eclipse (3) occurs.

▲ When the Moon enters the Earth's penumbra its brightness is reduced. If it brushes the umbra it will be partially eclipsed and the Earth's shadow can be seen. At total eclipse the Moon lies completely within the Earth's umbra.

▼ The upper diagram shows how at New Moon and Full Moon the shadow of the Moon usually passes above or below the Earth, and vice versa. If New or Full Moon occurs when the Moon is near one of the nodes of its orbit, then its shadow falls on the Earth, causing an eclipse of the Sun; alternatively it passes into the Earth's shadow, when an eclipse of the Moon is seen (lower diagram).

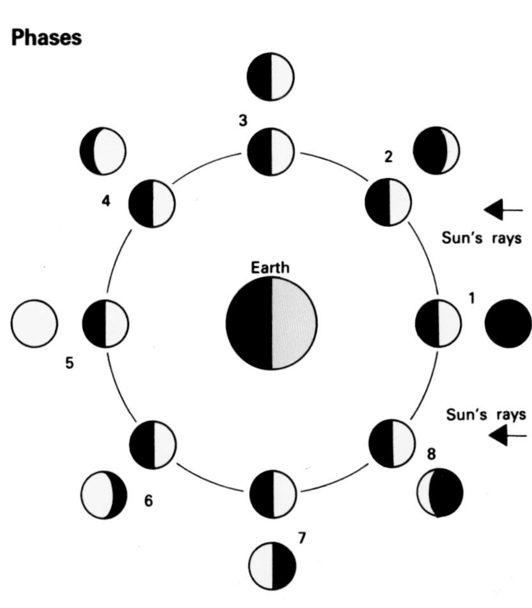

► The phase of the Moon – the part of its disk which is illuminated as seen from the Earth – depends on the angle between Sun, Earth and Moon. The Earth-facing hemisphere is dark at New Moon (1). Thereafter the phase goes through the following sequence: (2) crescent, (3) first quarter, (4) gibbous, (5) Full Moon, (6) gibbous, (7) last quarter and (8) crescent before returning to New Moon.

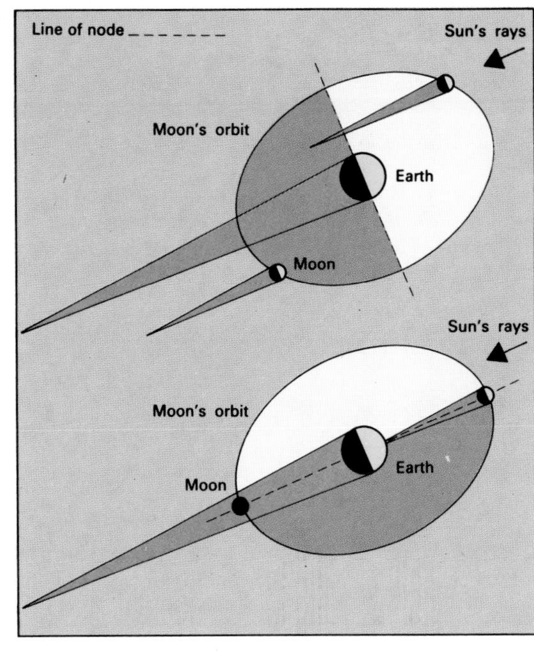

See also
Inside the Planets 33-40
Planetary Landscapes 41-56
Planetary Atmospheres 57-72
Moons and Ring Systems 73-88
Overview of the universe 129-32

The growth of the Solar System: Uranus...

It came as a distinct surprise to scientists when, in 1781, William Herschel discovered a new planet moving well beyond the orbit of Saturn, and taking 84 years to complete one circuit of the Sun. Indeed, Herschel himself did not immediately recognize it as a planet – he believed it to be a comet.

Records showed that the new planet had been observed earlier, although without being identified as a planet. The first Astronomer Royal, John Flamsteed, had seen it in 1690 and had taken it for a star, designating it No. 34 in Taurus (The Bull) – which is why 34 Tauri is no longer to be found on star maps. A Frenchman, Pierre le Monnier, had seen it eight times in December 1768 and January 1769, but overlooked its gradual movement, and thus missed his chance of immortality.

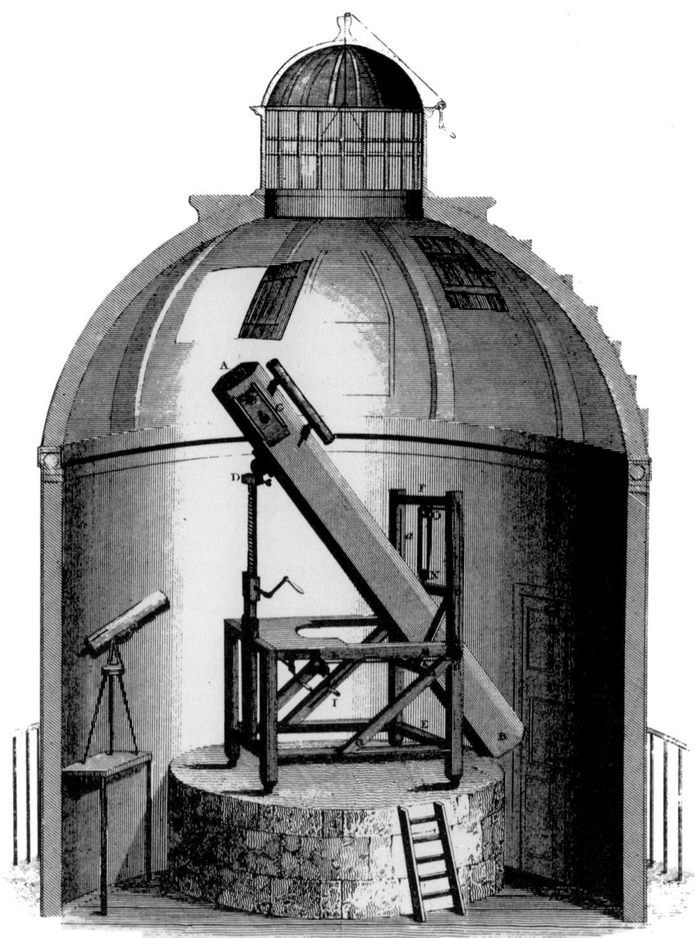

...Neptune...

However, Uranus did not move quite as expected. It deviated from its computed path, and as early as 1834 an amateur English astronomer, the Rev. T. J. Hussey, suggested that the cause might be perturbations by another, more distant planet.

He even wrote to Sir George Airy, who was later to become Astronomer Royal. But Airy did not believe that, even if Hussey were right, it would ever be possible to track down an unknown planet.

In 1841 the problem was taken up by John Couch Adams, a Cambridge undergraduate. By 1845 Adams was fairly sure that he had calculated the position of the planet which was perturbing Uranus, and he sought an interview with Airy, by then Astronomer Royal. This proved so difficult to arrange that Adams became discouraged, and nothing was done.

Meanwhile, Urbain Le Verrier, a French astronomer and mathematician, had turned his attention to the same problem, and come to much the same conclusion. One of his memoirs reached Airy in December 1845, and finally Airy asked James Challis, Professor of Astronomy at Cambridge, to search in the position indicated. Challis, however, had no good star chart of the region, and he began his task by laboriously plotting all the stars. In fact, he recorded the planet twice almost as soon as he began observing, but did not check his results. In Berlin, Johann Gottfried Galle and Heinrich Louis d'Arrest, using Le Verrier's results, identified Neptune on their first observing night.

Neither Adams nor Le Verrier took much part in the acrimonious controversy that ensued over who should claim the credit for the discovery, and they are today recognized as co-discoverers of the planet.

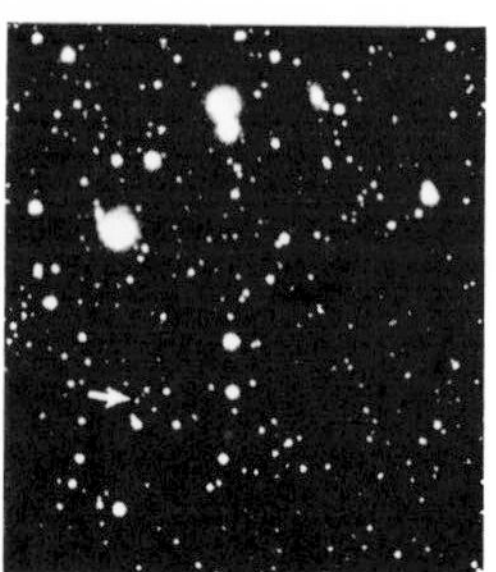

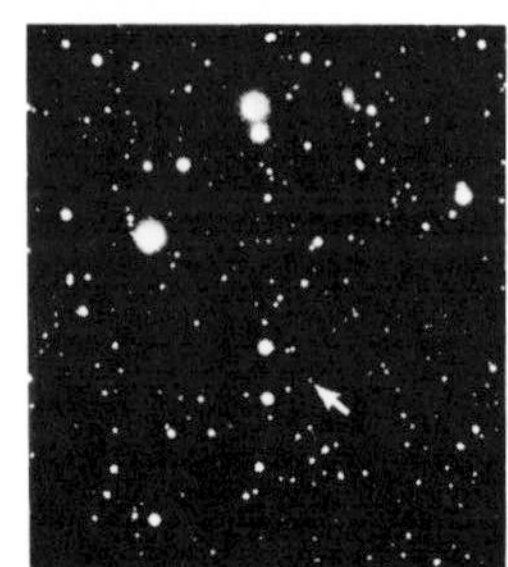

▲ William Herschel discovered Uranus in 1781 with this telescope.

◄ Pluto (arrowed) was identified from these photographic plates.

▼ Le Verrier discovers Neptune and Adams observes Le Verrier, according to one French cartoonist.

...and Pluto

The discovery of Neptune still did not completely satisfy astronomers, however, and there were suggestions that there might be yet another planet awaiting discovery. Percival Lowell, founder of the Lowell Observatory at Flagstaff, Arizona, worked out a possible position for "Planet X" but failed to find it, despite a careful search. It was finally located by Clyde Tombaugh on photographic plates taken at the Lowell Observatory in 1930, fourteen years after Lowell's death.

Observing the Planets

Various ways of establishing the size, mass and distance of planets...Looking at the planets... Telescopes...Rocket-launched spacecraft...The planetary probe...Pictures from space...Planning an interplanetary voyage...PERSPECTIVE...An ancient attempt to measure the size of the Earth...Historical estimates of the distance from the Earth to the Sun... Famous observatories...Pioneers of the space age

Man's understanding of the Solar System, and of the universe at large, rests on his ability to measure accurately the most basic properties of astronomical bodies, their size and mass – and also their distance. The size of an object is related in a simple way to its distance, and the angle it subtends at the eye (the angle between two imaginary lines drawn from the observer's eye to the edges of the object as it appears). It is therefore quite easy to calculate the actual size of a planet if its distance and its apparent size are known.

There are also other methods of determining size. Occasionally a planet, satellite or asteroid will pass in front of a star and briefly obscure it from view. Timing such an event, which is called an occultation, yields the diameter of the occulting body. It is also possible to work out the size of a distant object from its infrared radiation, using an established formula which relates the quantity of energy emitted to the surface area and temperature of the body emitting it.

Newton's law of gravitation provides the key to measuring the mass of a planet. The speed at which a satellite or moon moves in its orbit is governed by the force of attraction exerted by the planet, and this depends on its mass. If the distance between the two is known, therefore, a moon's speed reveals the mass of the planet. Recently, accurate values for the masses of planets and satellites have been obtained from their gravitational influence on passing spacecraft.

How large is the Earth?

Modern astronomers are accustomed to thinking in terms not of millions of kilometers, but of millions of millions of kilometers. The human mind cannot begin to visualize such enormous distances: it simply has to accept them. The early astronomers naturally had no reason to know this, and Heraclitus of Ephesus, one of the Greek philosophers (c. 540 BC), believed that the diameter of the Sun was about 30cm and that its actual size was the same as its apparent size.

Over 200 years later, Eratosthenes of Cyrene was more successful in measuring the size of the Earth. From one of the books in the great scientific library at Alexandria, Eratosthenes learned that at the time of the summer solstice – the longest day in northern latitudes – the Sun was directly overhead, or at zenith, in the town of Syene (the modern Aswân, near where the Nile crosses the Tropic of Cancer), so that at noon its rays would shine directly into a well without casting a shadow. At the same moment in Alexandria, however, the Sun was not overhead: it was 7° away from the zenith.

A full circle contains 360° and 7 is about one-fiftieth of 360. Eratosthenes therefore reasoned that if the Earth was a globe, its circumference must be 50 times the distance from Alexandria to Syene. He measured this and then worked out that the distance round the Earth must be about 39,984km.

There is some doubt about this figure because Eratosthenes gave his result in stadia, and the precise length of one stadion is not known. However, he was certainly correct to within a few hundred kilometers.

Modern measurements have shown that the Earth is not a true sphere. Its rotation causes a small bulge at the equator, making the equatorial diameter approximately 41km greater than the polar diameter.

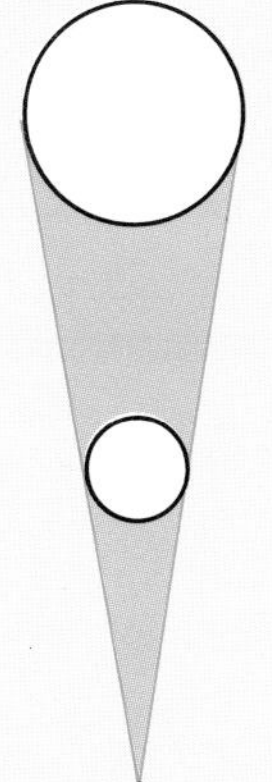

◀ The apparent diameter of a body depends on its true diameter and its distance. To an observer at A, the two objects shown here appear the same size, but one is twice as distant and hence twice as large.

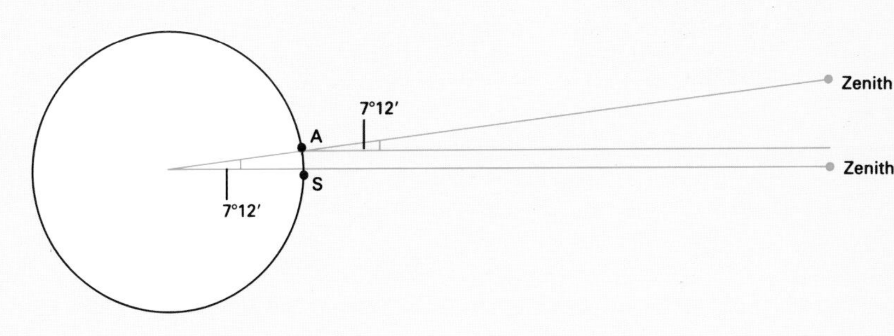

▶ At noon with the Sun overhead at Syene (S) it was 7°12′ away from the zenith at Alexandria (A), so the angle between A and S was also 7°12′. The circumference of the Earth can be deduced from the known distance AS.

Over the centuries, the Earth-Sun distance has expanded immensely in the minds of men. Of the methods of measurement described on page 22, 1-4 used the half Moon method; 5, 7 and 8 used parallax and 6 the transits of Venus. Radar, 9, has settled the matter.

Estimate	Earth-Sun distance
1 Aristarchus (*c.* 270 BC)	4,800,000km
2 Ptolemy (*c.* AD 150)	8,000,000km
3 Copernicus (1543)	3,200,000km
4 Kepler (1618)	22,500,000km
5 Cassini (1672)	138,730,000km
6 Encke (1823)	153,303,000km
7 Galle (1875)	148,290,000km
8 Spencer Jones (1931)	149,645,000km
9 radar (1976)	149,597,870km

How far away? The measurement of distance has always been the astronomer's first concern

How far is the Sun?

The first scientific attempts at measuring the distance of the Earth from the Sun seem to have been made by Aristarchus of Samos, about 270 BC. Although the method he used was sound in theory, his estimated distance was 4,800,000km. Ptolemy gave the distance as 8,000,000km; Copernicus reduced this to only 3,200,000km. Kepler's estimate was 22,500,000km. In 1672 the Italian astronomer Giovanni Domenico Cassini gave a result of the right order: using the parallax of the planet Mars, he arrived at an estimate of 138,730,000km.

Measuring the parallax of Mars was not easy because the planet appears as a definite disk, whereas the minor planets or asteroids appear as points of light. In 1875 the German astronomer Johann Gottfried Galle used one of the asteroids, Flora, to derive a value of 148,290,000km. In 1931 Sir Harold Spencer Jones used another asteroid, Eros, to derive a new figure: 149,645,000km, just slightly too great.

These methods have now all been superseded by measurements based on the radar ranging of Venus, giving the accepted value for the astronomical unit as 149,597,870km.

The traditional method of distance measurement makes use of an effect called parallax: against a distant background (the stars, say) a relatively close object (such as a planet or asteroid) will show a slight shift in position when viewed from different points on the Earth's surface. The distance between the observers is known, and from this the distance of the object can be calculated by simple trigonometry. The method becomes progressively less reliable, however, the farther away the object.

Within the Solar System, the much more precise technique of radar ranging is now available to complement the parallax method. A pulse of microwave radiation of up to 100,000 watts is sent from a radio antenna towards the target body. The target reflects a weak "echo" to the antenna. The signal travels at 300,000km per second (the speed of light), so the distance to the object and back is equal to this figure multiplied by the time interval between transmission of the signal and reception of its echo. The range of the target is half this distance.

This method has been used for the Sun and the planets out to Saturn, and has given the exact value of the astronomical unit (the mean distance between the Sun and the Earth – the standard unit of distance in the Solar System) to within one kilometer.

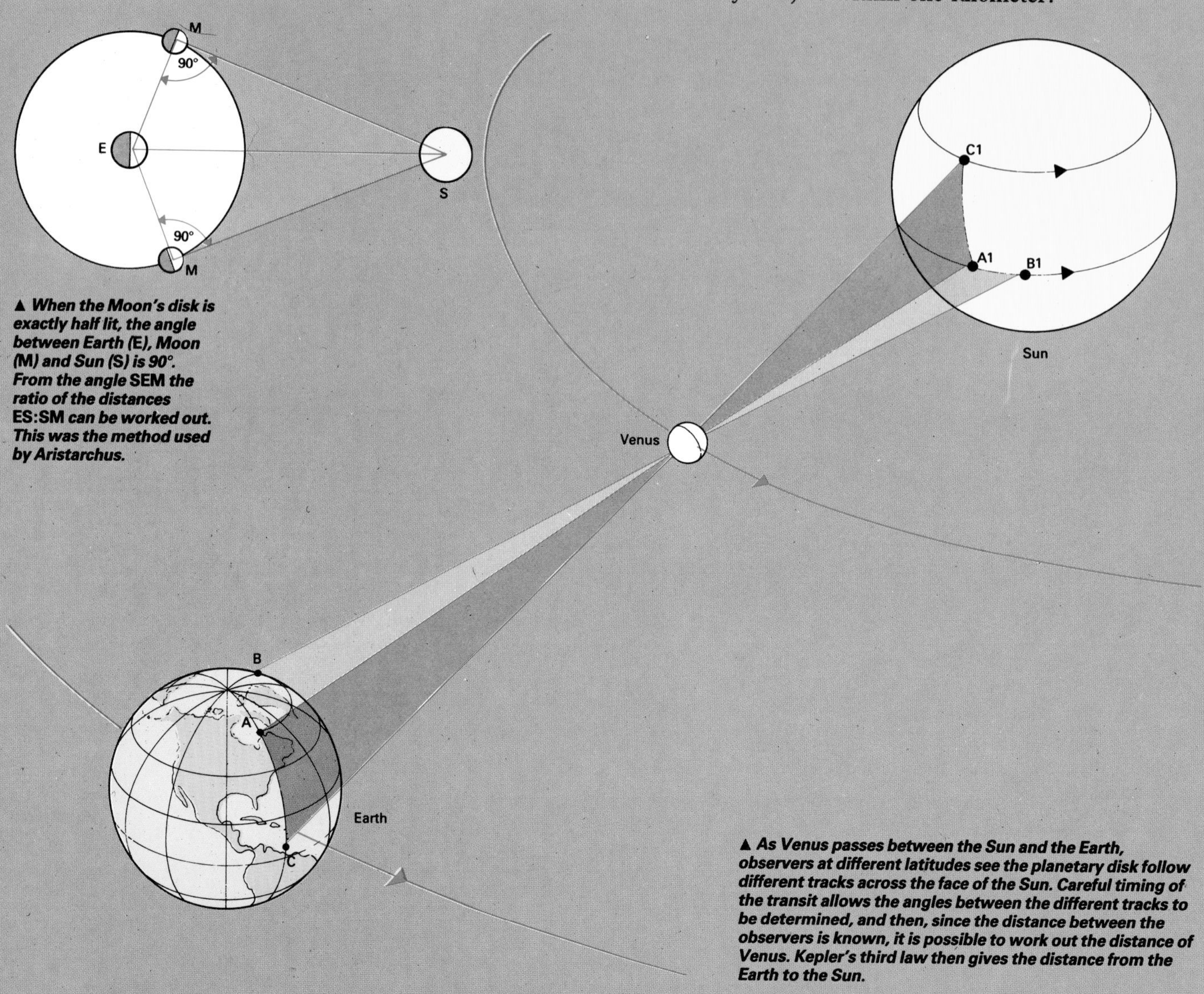

▲ When the Moon's disk is exactly half lit, the angle between Earth (E), Moon (M) and Sun (S) is 90°. From the angle SEM the ratio of the distances ES:SM can be worked out. This was the method used by Aristarchus.

▲ As Venus passes between the Sun and the Earth, observers at different latitudes see the planetary disk follow different tracks across the face of the Sun. Careful timing of the transit allows the angles between the different tracks to be determined, and then, since the distance between the observers is known, it is possible to work out the distance of Venus. Kepler's third law then gives the distance from the Earth to the Sun.

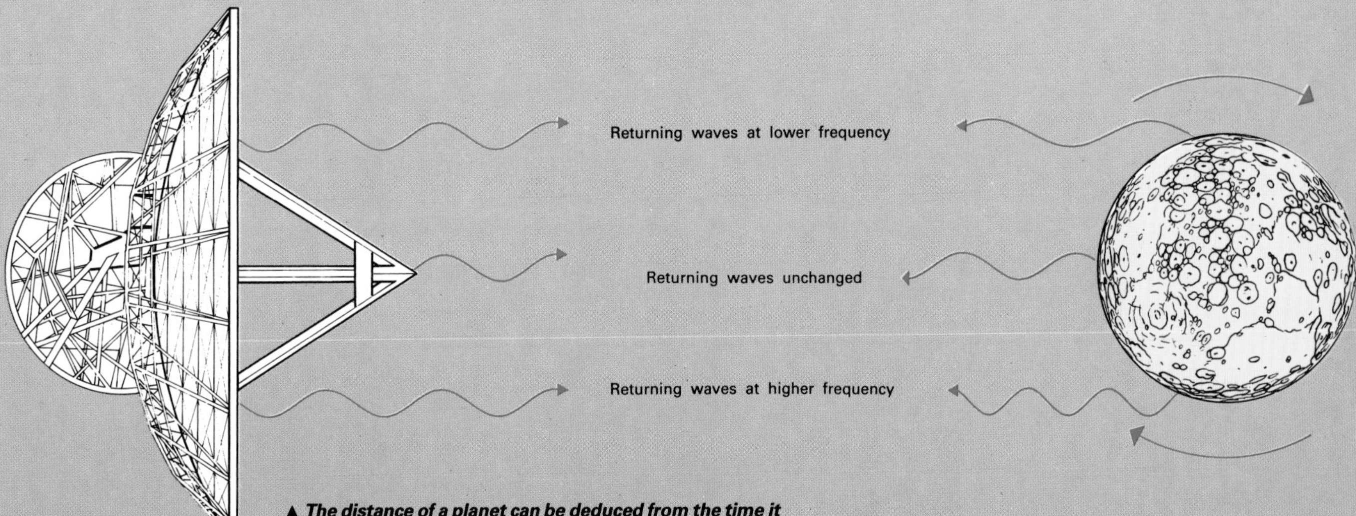

▲ *The distance of a planet can be deduced from the time it takes the "echo" of a microwave pulse sent from Earth to come back again. Minute differences in timing enable researchers to draw up surface maps of the planet. Furthermore, a rotating planet will send back a signal of shorter wavelength than the original from its approaching side, and longer from its receding side. An analysis of the difference gives the rotation speed.*

▼ *Precise measurements of the distance to the Moon are made from laser beams directed at reflectors left on the surface by Apollo astronauts. The distance is in fact increasing at a rate of about 4·5 cm per year.*

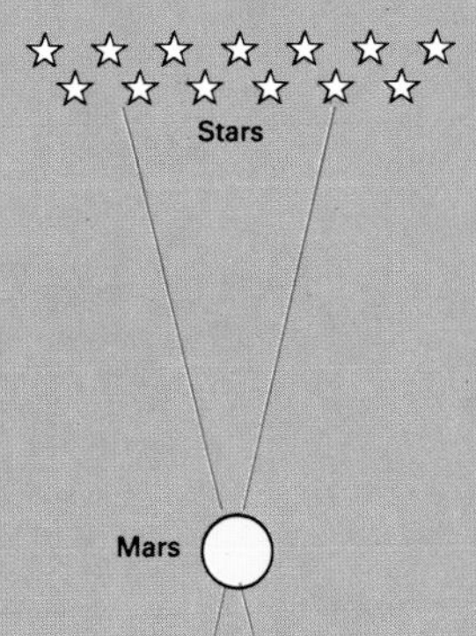

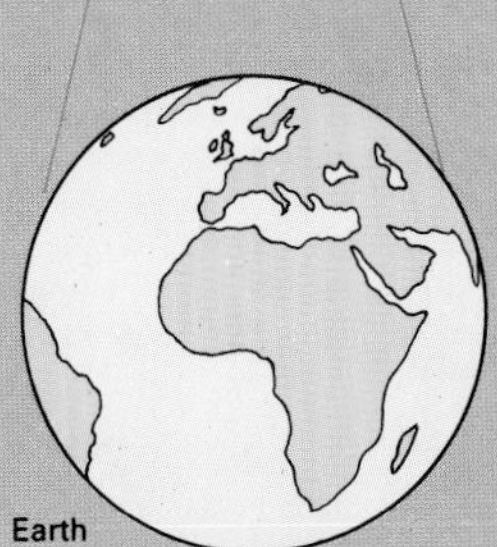

▲ *The distance to a nearby planet can be obtained from the small shift in its position (parallax) against the background of stars, as seen by two observers a great distance apart.*

Distances within the Solar System

Kepler's third law established a relationship between the distance of a planet from the Sun, and the time taken for it to complete one revolution. Using this, it is possible to make a scale map of the Solar System, giving the distances of the planets in astronomical units. Then, if the actual distance of one planet is known, it will be possible to calculate the rest. In 1672 Giovanni Domenico Cassini realized that he could find the distance of Mars from the Sun, and then, because he knew that Mars took 687 days to complete its orbit, he could use Kepler's third law to calculate the other planetary distances.

Mars comes to within 59,000,000km of Earth – closer than any other planet except Venus. Its apparent position against the starry background will not then be the same to two observers at widely separated positions on the Earth. The difference in angle (parallax) enables the real distance of the planet to be calculated. Modern measurements give the mean distance between Mars and the Sun as 1·524 astronomical units.

Transits of Venus

The next method of working out planetary distances used transits of Venus. Since Venus is closer to the Sun than the Earth is, it occasionally passes directly between the two, appearing in transit as a black disk. Transits do not happen often – in fact, they occur in pairs separated by eight years, after which there are none for over a century. Exact timings of a transit can reveal the distance of Venus from the Earth. Observations were made of transits in 1761 and 1769, but the results were affected by a phenomenon known as the black drop – as Venus moves on the the Sun's disk it seems to draw a dark strip after it, and when this vanishes the transit has already begun. Even so, the estimated value of the astronomical unit was revised upwards to 153,303,000km.

Reaching for the Planets

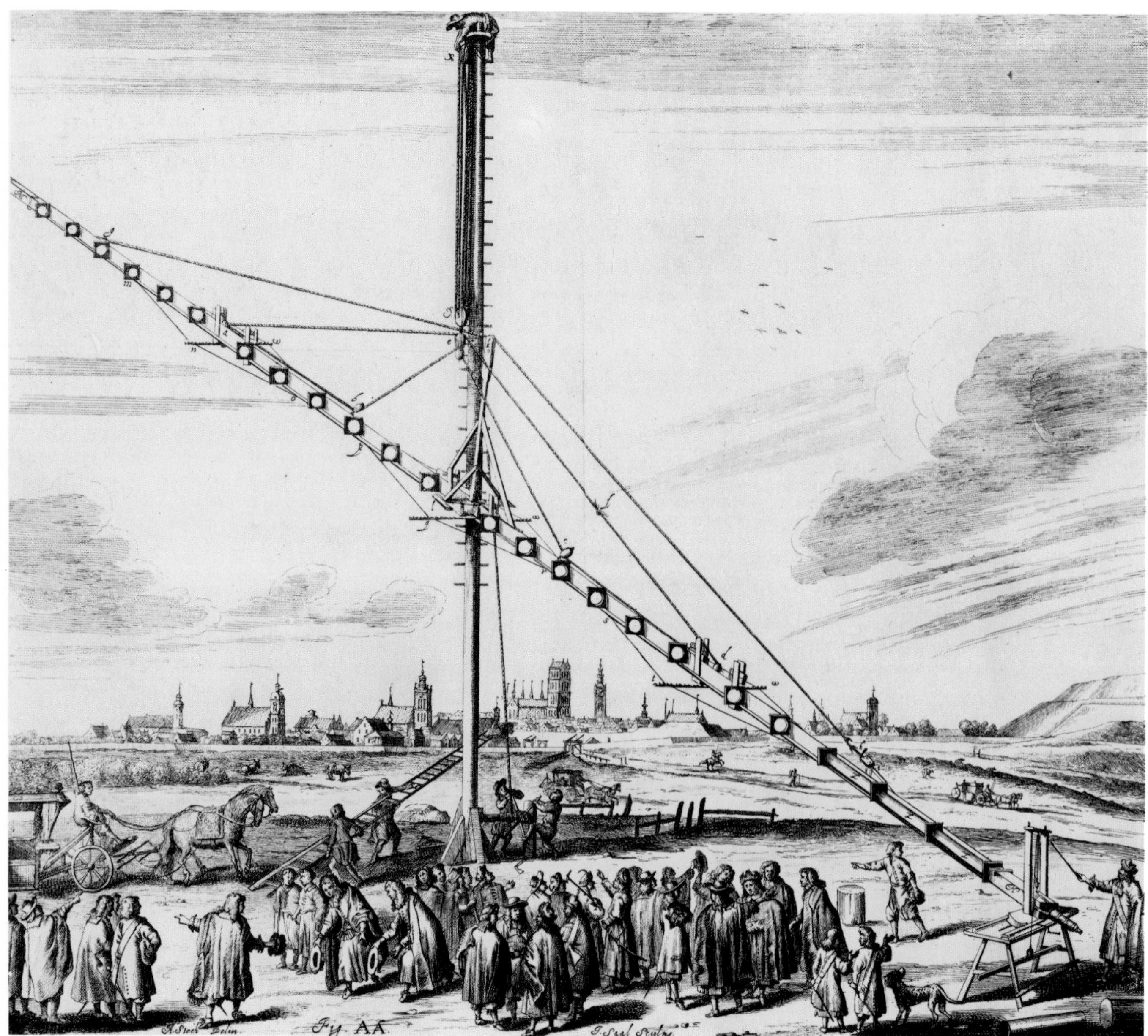

▲ **Johann Hevelius, who constructed this long "aerial telescope" at his private observatory near Danzig, was primarily interested in the Moon, and he drew up good lunar maps with the aid of such instruments.**

Early telescopes

Before the invention of the telescope, astronomers had detected the existence of planetary bodies but had no means of learning anything about the planets themselves. Galileo is generally regarded as the first telescopic observer, and although in fact an Englishman named Thomas Harriot, one-time tutor to Sir Walter Raleigh, had drawn a telescopic map of the Moon some months before Galileo began his work, it remains true that Galileo was the first to use telescopes in systematic studies of the sky.

Naturally, there were limitations. Early refracting telescopes suffered from chromatic aberration – the presence of false color in the image. The only known remedy for this was to use very thin lenses, and these are necessarily of very long focal length. Refractors of this kind were therefore long and unwieldy, some of them having focal lengths of over 60m. Even so they enabled observers such as Johannes Hevelius (1611-1687) of Danzig (now Gdansk in Poland) and the Italian Jesuit Giovanni Battista Riccioli (1598-1671) to map the Moon.

Since his discovery in 1666 that light of different colors is refracted in different amounts, Isaac Newton had dismissed the refractor in telescope design. In 1671 he presented to the Royal Society the first reflecting telescope, which collects its light with a mirror instead of a lens. The light rays are only reflected, not refracted, and so chromatic aberration is not a problem.

In 1758 John Dollond built the first telescope using the principle that combining different kinds of glass and lens shapes could cancel out the aberration. His telescopes were immediately successful, even though they were limited in size to 10cm diameter because he did not know how to cast lenses any larger. As techniques improved, however, larger refracting telescopes were built, culminating in 1879 with the 101·6 cm refractor at Yerkes, near Chicago, still the largest of its kind.

Early observatories

The first national observatories were founded largely to aid in navigation. The Royal Greenwich Observatory, established in 1675 by order of King Charles II, had the express purpose of drawing up a new star catalog for the use of British seamen. Sir Christopher Wren, a former professor of astronomy, designed the buildings, and the first Astronomer Royal, John Flamsteed, eventually produced a catalog of stars that became the standard work for many years.

The first great modern observatory founded for the express purpose of planetary observation was the Lowell Observatory at Flagstaff, Arizona. It was entirely due to Percival Lowell (1855-1916), formerly an Oriental scholar and diplomat, who turned his attention to astronomy, with a particular interest in Mars. In 1877 Giovanni Virginio Schiaparelli (1835-1910) had first drawn the complicated network of canals, which Lowell believed to be artificial and worth building a large telescope to study.

After extensive site-testing he decided upon Flagstaff. From 1894 until his death in 1916 Lowell was actively concerned with Mars, but his belief that the planet was inhabited affected the reputation of the Observatory. It was not until the discovery of Pluto in 1930, by Clyde Tombaugh, using a 33cm refractor obtained specifically for the purpose, that the Lowell was recognized as one of the world's major astronomical institutions.

▲ The observatory of Pic du Midi is sited at an altitude of 2860m, in the French Pyrenees; it is equipped with a 56cm refractor. It has a diverse program but is famous for photographs and drawings of the Solar System, which, prior to the space age, were the best of their kind.

▼ The 101.6 cm refractor at Yerkes, near Chicago, was completed in 1879 and is still the largest refracting telescope in the world. It gathers 40,000 times as much light as the naked eye. This picture shows Albert Einstein with the staff of the observatory in 1921.

Light can be regarded as waves which travel through space at a speed of 300,000km per second, and in some respects they have properties similar to those of water waves. The distance between the crest of one wave and that of the next is the wavelength, and this is measured in units of a billionth of a meter, called nanometers or nm. The number of wave crests per second passing an observer is the frequency.

Visible light spans a range of wavelengths from just under 400nm to around 700nm, and the eye distinguishes the different wavelengths within this range as colors, violet and blue being the shortest wavelengths and orange and red the longest. However, the visible waves represent only a narrow band within the vast range of wavelengths making up the elecromagnetic spectrum. Light of shorter wavelengths than the shortest visible is called ultraviolet, still shorter waves are called X-rays and the shortest of all are termed gamma rays. Going in the other direction, waves longer than the visible ones are infrared, still longer are the microwaves and longest of all are the radio waves. Infrared "light" is emitted by warm objects, including those at room temperature, which "shine" brightly at a wavelength of around 10,000nm. Study of infrared radiation from space provides information about matters such as the temperatures of planets.

The Earth's atmosphere prevents most wavelengths from reaching ground level, either by absorbing them or by reflecting them back into space. But most visible light does get through (unless it is cloudy), and so do micro/radio waves from about one centimeter to a few tens of meters. Apart from these and a small amount of infrared, practically nothing else penetrates to the Earth's surface. Observers of the planets and of deep space alike are seriously hampered by the atmosphere. However, instruments on board orbiting satellites, being outside the atmosphere, are capable of studying the whole range of wavelengths, thus overcoming the limitations placed on Earth-based observers.

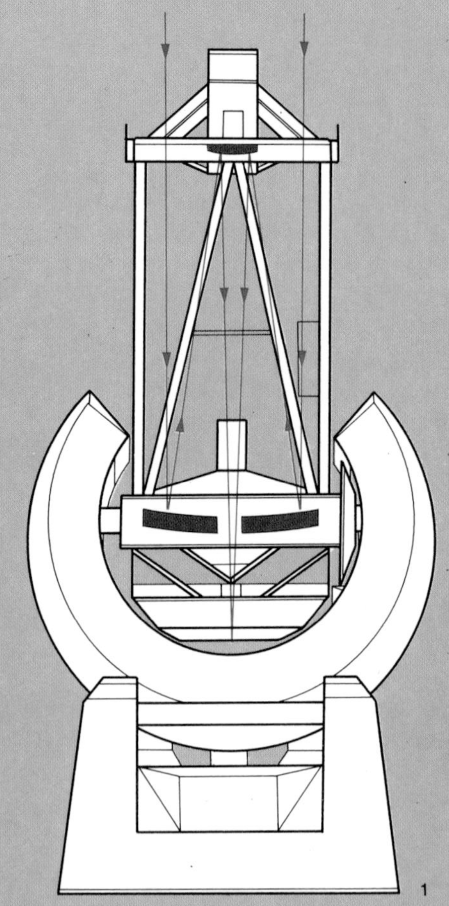

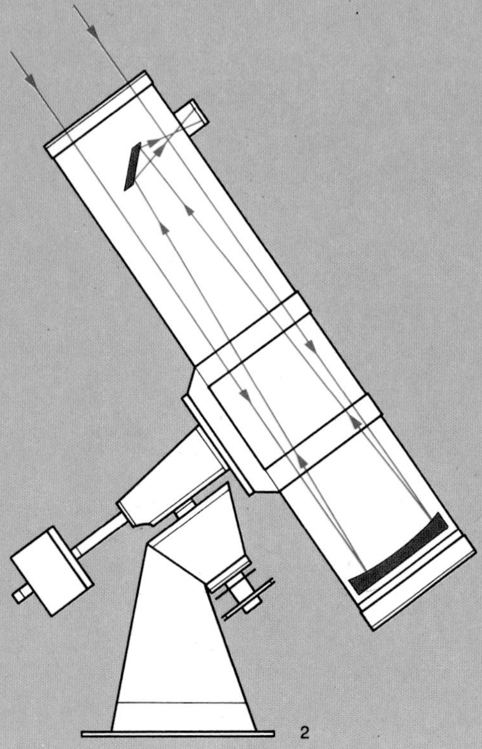

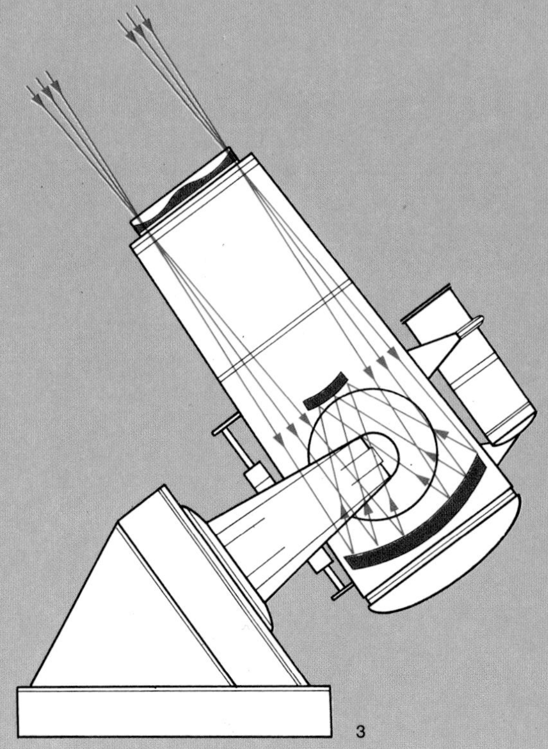

***1** The primary mirror of a Cassegrain reflector is of parabolic cross-section. The converging cone of light is reflected from a convex secondary mirror, through a hole in the primary, to the eyepiece or instrument platform.*

***2** In the Newtonian design light reflected from the concave primary mirror is then reflected from a flat secondary to the side of the tube, where the eyepiece is located.*

***3** A Schmidt telescope has a specially shaped lens at the front of the tube to ensure that light rays reflected from the concave (spherical) primary mirror are brought to a focus in the same plane, where a photographic plate is placed. The instrument is used to photograph wide areas of sky.*

Resolving detail

The basic instrument for studying the surfaces and atmospheres of bodies in the Solar System is the optical telescope.

The telescope has two principal functions – to collect light and to resolve details. The power of a telescope to gather light, and hence its ability to reveal objects too faint to be seen by the unaided eye, depends upon the surface area of the objective or primary mirror. This, in turn, is proportional to the square of the telescope's aperture (the clear diameter of the lens or mirror). In other words, a telescope of 2 meters aperture collects four times as much light as one of 1 meter aperture.

The resolving power (R) is a factor of crucial importance to the planetary astronomer. Technically it is defined as the minimum angular separation of two equal point sources of light (or two points on a planetary surface) which can just be distinguished as separate points. The smaller the value of R, the more detail the telescope can reveal. R can be calculated using Dawes' rule: R=0·11/D, where D is the diameter of the aperture in meters and R is the resolving power in seconds of arc, or arcsecs.

The examples on the right show a planet and two close stars (top). Viewed by a telescope of 50mm aperture (centre) each little box or "pixel" represents the smallest resolvable feature. With a 200mm aperture (bottom) the pixels are smaller and more detail is seen.

In practice the turbulence of the atmosphere interferes with incoming light rays, and this prevents telescopes from ever attaining their theoretical resolving power.

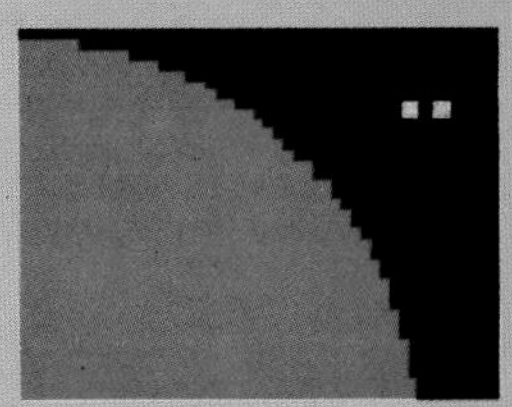

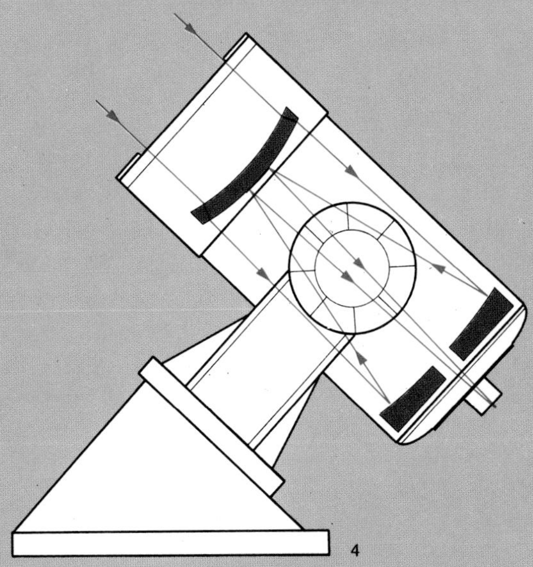

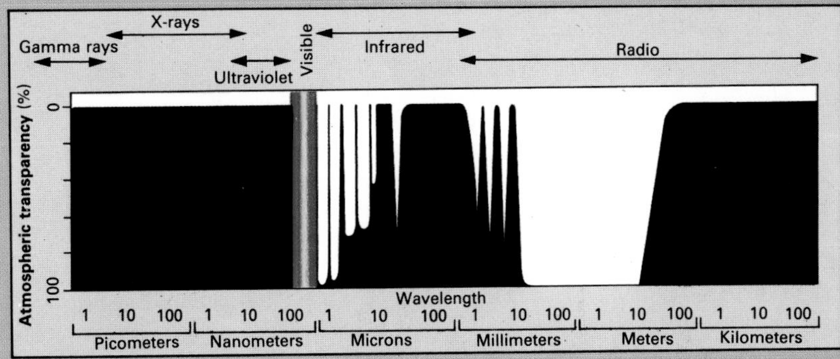

▲ *On this chart the full range of electromagnetic wavelengths is plotted against atmospheric transparency, showing by how much each type of radiation penetrates to ground level.*

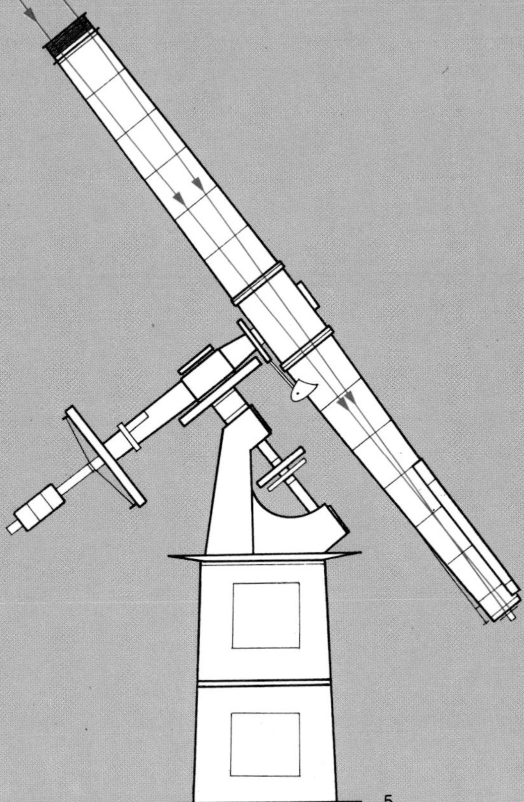

4 *The Maksutov, using a Cassegrain system, has a curved corrector plate which allows light rays to be focused more accurately, making detailed astronomical photography possible.*

5 *A refracting telescope has a large objective (a lens facing the object being watched) of long focal length, which brings light to a focus where the image may be magnified by an eyepiece.*

6 *A radio telescope collects long-wave radiation which the human eye does not detect. In one type a dish antenna reflects radio waves to a receiver placed at the focus.*

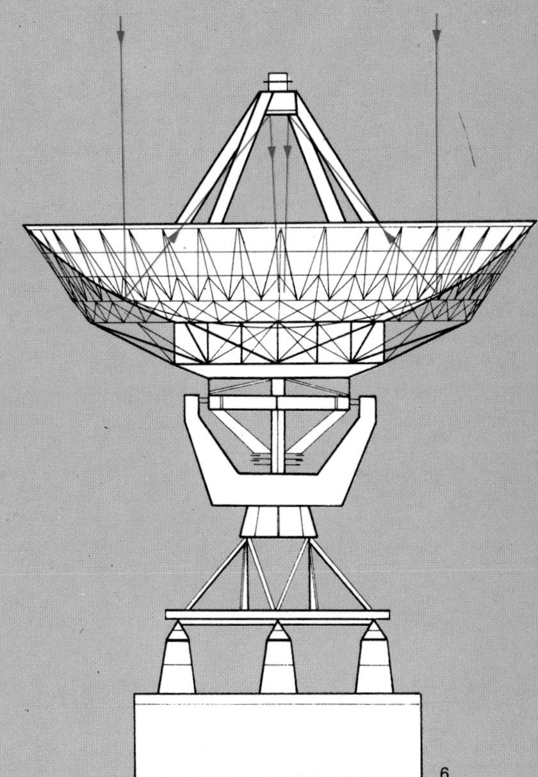

▲ The Russian schoolmaster Konstantin Tsiolkovskii pointed out in 1903 that the only way to put a vehicle in space was by means of rocket propulsion. Most people thought either that it would never be achieved at all, or that the vehicle would be "shot" into space from a huge cannon like that envisaged by Jules Verne.

Rockets into orbit

Most of the detailed knowledge that scientists now have about the planets has been obtained by space-research methods. The idea of space travel itself is old – as long ago as AD 180 the Greek satirist Lucian of Samosata wrote his "True History", in which a party of sailors, caught up in a violent waterspout, land on the Moon – but it was not until modern times that the concept was treated at all seriously. In 1865 Jules Verne wrote a famous novel in which his travelers were fired moonward from the mouth of a huge cannon, but this idea was clearly unworkable, both because the projectile would be destroyed by friction against the atmosphere and because the shock of departure would certainly prove fatal to the adventurers.

The first genuinely scientific ideas about space travel were put forward by a Russian school teacher, Konstantin Tsiolkovskii (1857-1935), who published the fundamental work on the subject in 1903.

In Germany, an amateur research team of which Wernher von Braun (b. 1912) was a member began to experiment with liquid-propellant rockets. In the 1930s the Nazi government realized that rockets could be used as weapons. Von Braun and his colleagues were taken to Peenemünde, an island in the Baltic, where they developed the highly sophisticated V-2 weapons used to bombard England during the final stages of the war. Subsequently von Braun and others went to America, where they turned their attention to the problems of interplanetary flight.

Their plan was to launch an artificial satellite into orbit by rocket power. However, the Russians took the lead when, on 4 October 1957, they launched Sputnik 1. This was football-sized and carried little apart from a radio transmitter, but it marked the true beginning of the Space Age. A year later von Braun's team in the USA launched their first successful satellite, Explorer 1.

► On 16 March 1926 Dr. Robert Hutchings Goddard, a New England professor, launched the first liquid-propellant rocket into the atmosphere. The real value of Goddard's work was not fully appreciated until later, when scientists found that much of it was protected by patents.

The ability to send spacecraft to fly by, enter orbit round, or land upon the planets, and to sample interplanetary space directly, has had the greatest impact on our understanding of the Solar System.

The key to the exploration of space is the rocket – the only form of propulsion which can operate in the near-perfect vacuum of space. The rocket operates on the principle of reaction, whereby if hot gases are expelled from a nozzle in one direction, the rocket itself is pushed in the opposite direction. The rocket continues to accelerate until it runs out of fuel, its final speed depending on the speed of the expelled gases (exhaust velocity) and the mass ratio (the combined weight of the rocket plus fuel divided by the weight of the rocket alone).

In order to escape from the Earth's gravity and move out into space, a rocket must reach a minimum speed known as escape velocity. If its final speed is less than this, it will fall back to Earth. Escape velocity at the Earth's surface is 11·1km per second, or around 40,000km per hour. The exhaust velocity of a present-day rocket is considerably slower than this, so that it can only overcome the gravitational pull of the Earth if it has a high mass ratio. Spacecraft designers have met this need by stacking several rocket stages on top of one another (or grouping a number of them around a central core), each successive stage being jettisoned as its fuel is spent.

Landers, probes and spacelabs

Since the early days of rocketry the technology of space flight has advanced at remarkably rapid pace. To date, probes have been sent to Mercury, Venus, the Moon, Mars, Jupiter and Saturn; spacecraft have been sent into orbit round Venus, the Moon and Mars, and landers have touched down on these three. Men have been to the Moon and brought back samples, and remote controlled vehicles have roamed the lunar surface. If all goes well, before the end of the 1980s probes will have encountered Uranus and Neptune and at least two comets, and Jupiter will have been visited by an orbiter and atmospheric probe (the Galileo mission). The remainder of the century should see more missions to a wider range of celestial objects, perhaps a resumption of manned flights to the Moon, followed, possibly at the turn of the century, by a manned mission to Mars. Telescopes in orbit round the Earth will extend both planetary and deep space monitoring activities and robot vehicles may return samples of Martian material to an Earth-orbiting laboratory. With the reusable Space Shuttle it is feasible to put sophisticated instruments such as the Space Telescope into orbit round the Earth.

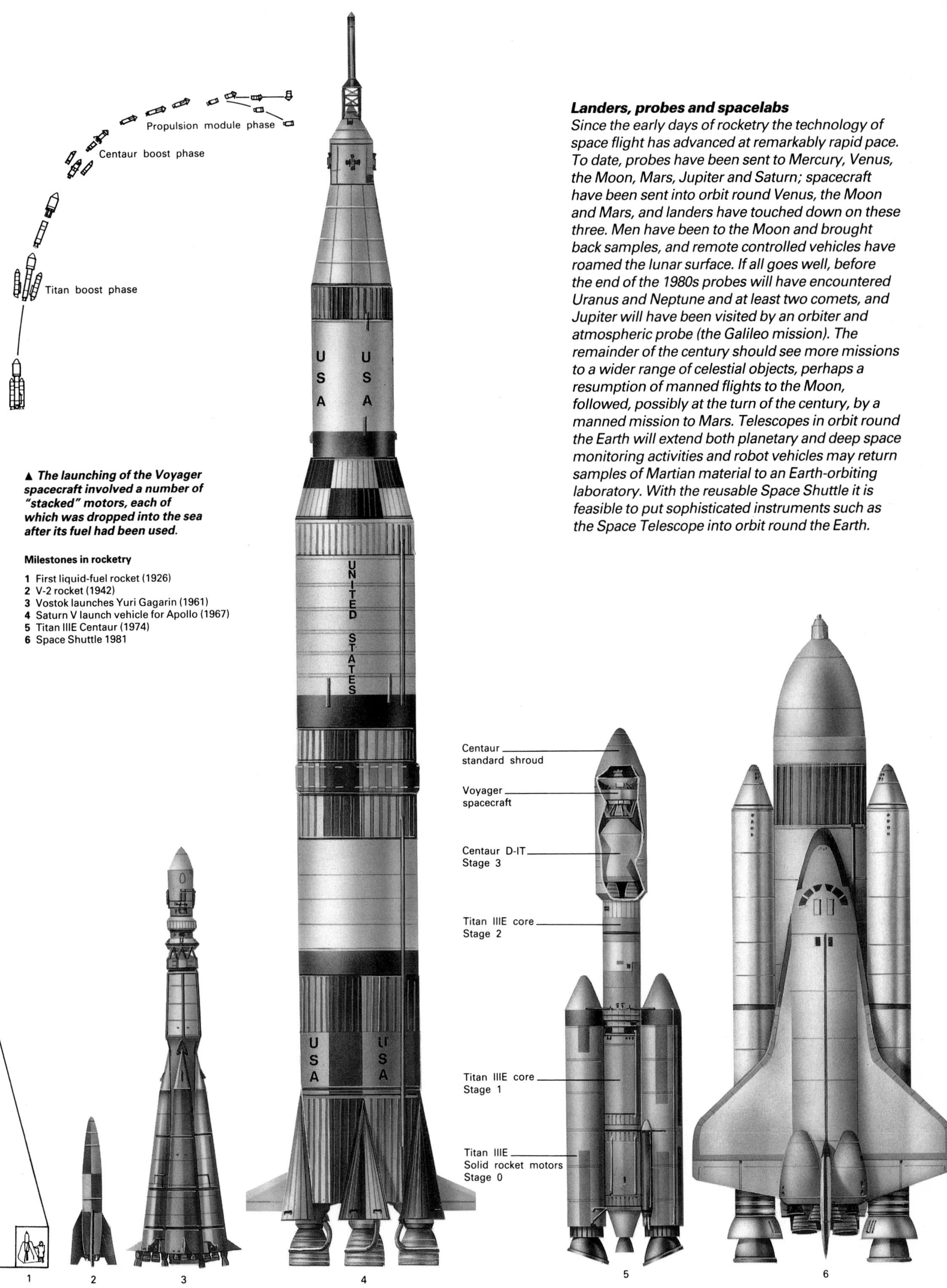

▲ ***The launching of the Voyager spacecraft involved a number of "stacked" motors, each of which was dropped into the sea after its fuel had been used.***

Milestones in rocketry

1 First liquid-fuel rocket (1926)
2 V-2 rocket (1942)
3 Vostok launches Yuri Gagarin (1961)
4 Saturn V launch vehicle for Apollo (1967)
5 Titan IIIE Centaur (1974)
6 Space Shuttle 1981

► ***The ultimate "Grand Tour" to the giants. The Voyager 2 spacecraft was dragged along by the powerful gravitational pull of Jupiter and then hurled on ahead towards its next target – Saturn. The chance alignment of the outer planets means that the probe can also visit Uranus and Neptune.***

The planetary probes carry a wide range of instruments. In addition to television cameras these include magnetometers which measure magnetic fields around planets and in interplanetary space; charged particle counters which measure particles of the solar wind, the radiation belts of planets, and particles from beyond the Solar System; microwave and infrared detectors which measure the temperatures of planetary surfaces and atmospheres; and spectrometers which, by measuring the spectrum of light reflected from planets and analyzing the wavelengths absorbed on the surface or in the atmosphere, provide information about their chemical composition. The fading and deflection of a radio beam transmitted by spacecraft passing behind a planet has yielded valuable information about planetary atmospheres (this is known as a "radio occultation experiment"). Finally, landing vehicles can directly analyze surface materials and measure phenomena such as winds.

There are two means of providing electrical power for spacecraft systems. For those operating in the inner Solar System, out as far as Mars, sunlight is plentiful and can be converted by solar panels into electrical energy. Deep space probes carry a miniature nuclear power plant (a nuclear thermoelectric generator) which converts the heat released by the decay of a radioactive material such as plutonium into electricity.

Spacecraft have remarkably weak transmitters – 28 watts sufficed for Voyager and only 8 watts for Pioneer 10. Nevertheless the radio antennae of the Deep Space Network, which includes three 64m dishes in the USA, Australia and Spain, are capable of detecting Pioneer's feeble signals at a range of more than 4·5 billion kilometers. The same dishes are capable of transmitting commands to the spacecraft at a power of up to 400,000 watts – thus the mission controllers have no difficulty in making themselves "heard".

Pictures from space

Onboard television cameras shoot the same scene through a range of color filters. Each picture is divided into a large number of tiny picture elements, or "pixels", and the degree of darkness of each pixel converted into a binary number. (The binary system uses a base of two, rather than 10 as in the decimal system, so that all numbers are made up from a combination of the digits 0 and 1.) This data is transmitted to Earth as "bits" of information, each bit representing one binary digit. Each picture received consists of 5 million bits of information and these are sent at a rate of 115,200 bits per second. The color-filtered pictures are reconstructed, then combined to give a full-color result.

Voyager 2

1 High-gain antenna for communications
2 Television cameras
3 Cosmic ray detector
4 Ionized gas detector
5 Infrared radiation detector
6 Micrometeorite shield
7 Electricity generators
8 Electronics compartments
9 Fuel tank

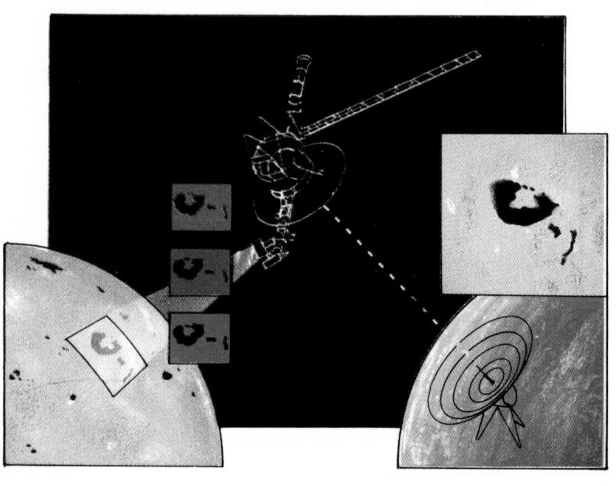

◄ ***The Voyager spacecraft transmitted black-and-white images of Jupiter's surface back to receiving stations on Earth. Each black-and-white image was taken through three color filters – red, blue and green. When combined at the receiving station an apparently "natural" full-color image was produced.***

See also
Planets and Orbits 9-20
The Sun's Influence 113-20
The Origin of the Solar System 121-8
Observing the Universe 137-44
Communication and Travel 245-8

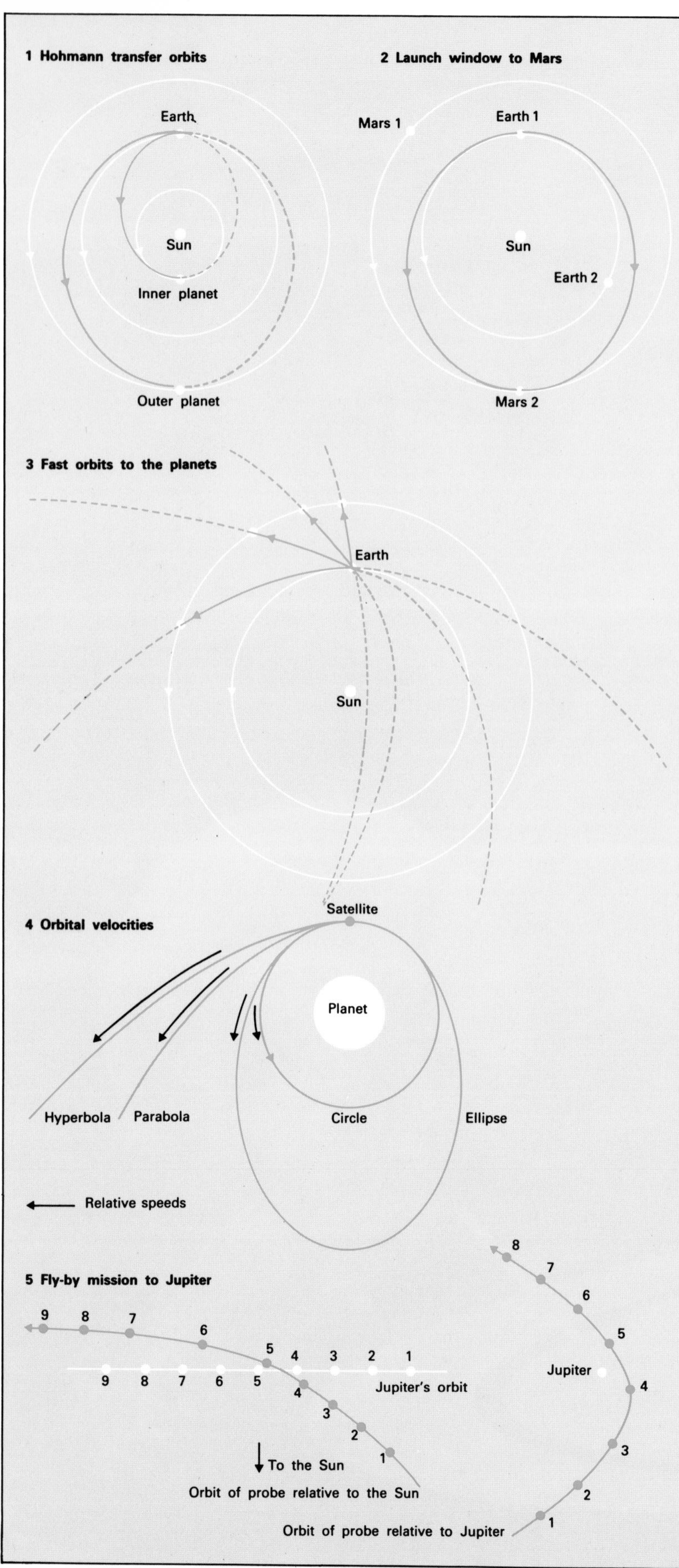

Planning an interplanetary voyage

The most economical way for a spacecraft to reach another planet is to follow an elliptical path called a Hohmann transfer orbit (1), after Walter Hohmann, who published the concept in 1925. To reach one of the superior planets a spacecraft is launched in the same direction as the Earth is moving, so that – relative to the Sun – it travels at a speed equal to the orbital velocity of the Earth plus its own speed relative to the Earth. Because it is moving faster the spacecraft's orbit will diverge outwards from the Earth's, following an ellipse which, if correctly judged, will just reach the orbit of Mars when the spacecraft is at aphelion. Provided the launch is correctly timed – the launch window – both spacecraft and planet will reach this point at the same moment, and a rendezvous will occur (2). By consuming more fuel the spacecraft can follow a flatter curve which reaches the target sooner (3), but no rocket yet constructed can travel fast enough to follow a straight-line path directly out from the Earth to Mars or any other planet.

To reach an inferior planet, the spacecraft is launched in the opposite direction to the Earth's orbital motion. Its speed relative to the Sun is equal to that of the Earth minus its own speed relative to the Earth. The spacecraft is then moving more slowly than the Earth and follows an ellipse which takes it closer in towards the Sun.

If the spacecraft is to enter a closed orbit round its target its motor must be fired to reduce its speed to below the escape velocity of the planet. Usually the motor is fired at the point of closest approach, and this brakes the spacecraft sufficiently for it to be captured by the planet's gravity. At a given distance from the planet there is a particular speed known as circular velocity which will allow the spacecraft to follow a circular orbit. In principle the spacecraft could be placed into a circular orbit, but in practice if its final speed differs even minutely from circular velocity, its orbit will be an ellipse. For a flyby mission no more fuel need be burnt; the spacecraft will pass its target at a speed greater than the planet's escape velocity following an open curve called a hyperbola (4).

In making a close encounter with (or following a hyperbolic orbit round) a planet such as Jupiter, a spacecraft can use that planet's gravitational field to gain or lose energy, and so change course, increasing or decreasing its velocity relative to the Sun (5). By this process (the "gravitational slingshot" or "interplanetary billiards" technique) a spacecraft can be sent farther out towards the fringes of the Solar System or closer in towards the Sun without using any more fuel. A spacecraft that leaves the Earth with just enough speed to reach the orbit of Jupiter will, by the time it gets there, be traveling much more slowly than the giant planet. As Jupiter overtakes the craft, its powerful gravitational pull will drag it along and hurl it on ahead at a much higher speed than it had before. Such an encounter can accelerate a spacecraft beyond the escape velocity of the Solar System. The first time this happened was in the case of Pioneer 10, which made a close encounter with Jupiter in 1973 and passed beyond the planet Pluto in 1983, en route for interstellar space.

Inside the Planets

3

Overview...Interior of the Earth...How such information is obtained...Interiors of the terrestrial planets – and the Moon...Planetary magnetic fields and their possible cause...An unexplained anomaly... Bulging equators, flattened poles...Liquid interiors of the gaseous giants...The iceberg planet... PERSPECTIVE...Indirect evidence...Early notions about planetary interiors

Each planet has a distinct chemical composition and physical structure. The four low-density giants contain large amounts of hydrogen and helium, differing fractions of "ices" and small fractions of rocks and metals. The four terrestrial planets consist almost entirely of rocky materials and metals – such as iron and nickel – in differing proportions. Each planet contains several distinct concentric layers.

The terrestrial planets

The interior of the Earth consists of three main layers. On the outside is a crust, only 10-40km thick and composed of relatively light rocks, which sits on top of a denser rocky "mantle", 2,900km thick. Below the mantle is a dense iron-nickel core, 3,400km in radius. Temperature increases with depth from about 290K at the surface to 700K at the base of the crust, and to more than 4,000K in the core. The principal source of the Earth's internal heat is energy released by the radioactive decay of elements such as uranium and thorium. The heating and melting of the interior, early in the Earth's history, allowed the heavy metals to sink to the center and the light rocks to rise to the surface, hence the present structure. The mean density of continental rocks is 2,670 kilograms per cubic meter, compared with about 13,600 kilograms per cubic meter at the center of the core.

The core consists of a solid center and an outer liquid region. The mantle contains three regions. The outermost layer, or lithosphere, is rigid and less than 100km thick. Below this lies the more plastic asthenosphere. The deepest mantle layer, the mesosphere, is more rigid than the asthenosphere.

Each of the other terrestrials has a similar structure but there are important differences. Mercury has the largest core in proportion to its radius, Venus has a slightly smaller core than the Earth, and, apart from the Moon, Mars seems to have the smallest, least dense core.

What shock waves reveal

More detailed analysis of planetary interiors is possible if any seismic activity can be studied. This includes shock waves that travel along the surface and through the interior of a planet as a result of internal events such as earthquakes, and external events such as meteoritic impacts or man-made explosions. Two types of waves, known as P- and S-waves (primary and secondary waves), travel through the body of the planet and yield information about its deep structure from the way in which they are refracted or reflected as they reach layers of different densities. The fact that P-waves travel through both solids and liquids but S-waves can only pass through solids reveals whether or not all or part of a planet's interior is liquid in nature. The study of these waves by a global network of seismic stations has yielded a detailed model of the Earth's interior. A similar, although less detailed, analysis of the Moon was made by seismometers left on its surface by the Apollo astronauts. The only other planet to receive such instruments is Mars. However, only one Viking seismometer worked and the results were inconclusive, except to show that Mars is rather quiet seismically.

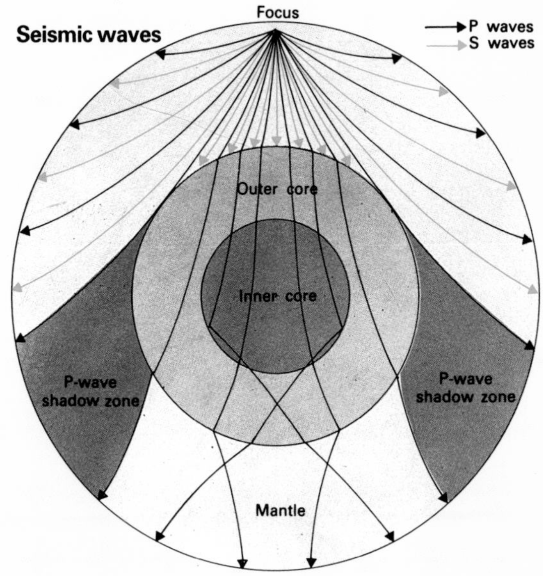

Clues to composition

There is a simple technique by which astronomers are able to deduce the chemical composition of a planet. The first clue is its mean density, a figure obtained by dividing the mass by the volume. Mean density depends on chemical composition and on the extent to which material is compressed by gravity. The greater the mass of a planet, the greater the compression in its interior.

If the mean density is similar to that of the Earth (5,500kg/m^3 – that is, 5·5 times the density of water), the composition is probably similar too, which means primarily metals (iron and nickel) and silicates (rock-forming minerals). On the other hand, if a massive body has a mean density similar to that of water (1,000kg/m^3), then, taking compression into account, it must be composed mainly of the lightest chemical elements. The mean density of the Sun is 1,400kg/m^3 and its composition by mass is probably about 73 percent hydrogen, 25 percent helium and 2 percent heavier elements. Jupiter and Saturn (with densities of 1,300kg/m^3 and 700kg/m^3 respectively) are also composed mainly of hydrogen and helium but the less massive giants, Uranus and Neptune, contain proportionally more "icy" materials such as water, methane and ammonia.

Mercury, Venus and Earth have similar mean densities, but Mars, with a mean density of 3,940kg/m^3, resembles the Moon (3,340kg/m^3) in this respect, and must differ structurally from the other terrestrial planets.

▲ P- and S-waves pass through the Earth's interior from the site, or "focus", of an earthquake or man-made explosion. Their speeds vary with depth, and they follow curved paths. S-waves travel only in solids: they cannot pass through the liquid core so there is a "shadow zone" at the surface where they are not received. P-waves can travel through liquids but they are refracted as they cross the boundary of the core, and this creates a smaller shadow zone within which these waves are not detected by seismometers.

The dense core of Mercury is probably larger than the whole globe of the Moon

The Earth

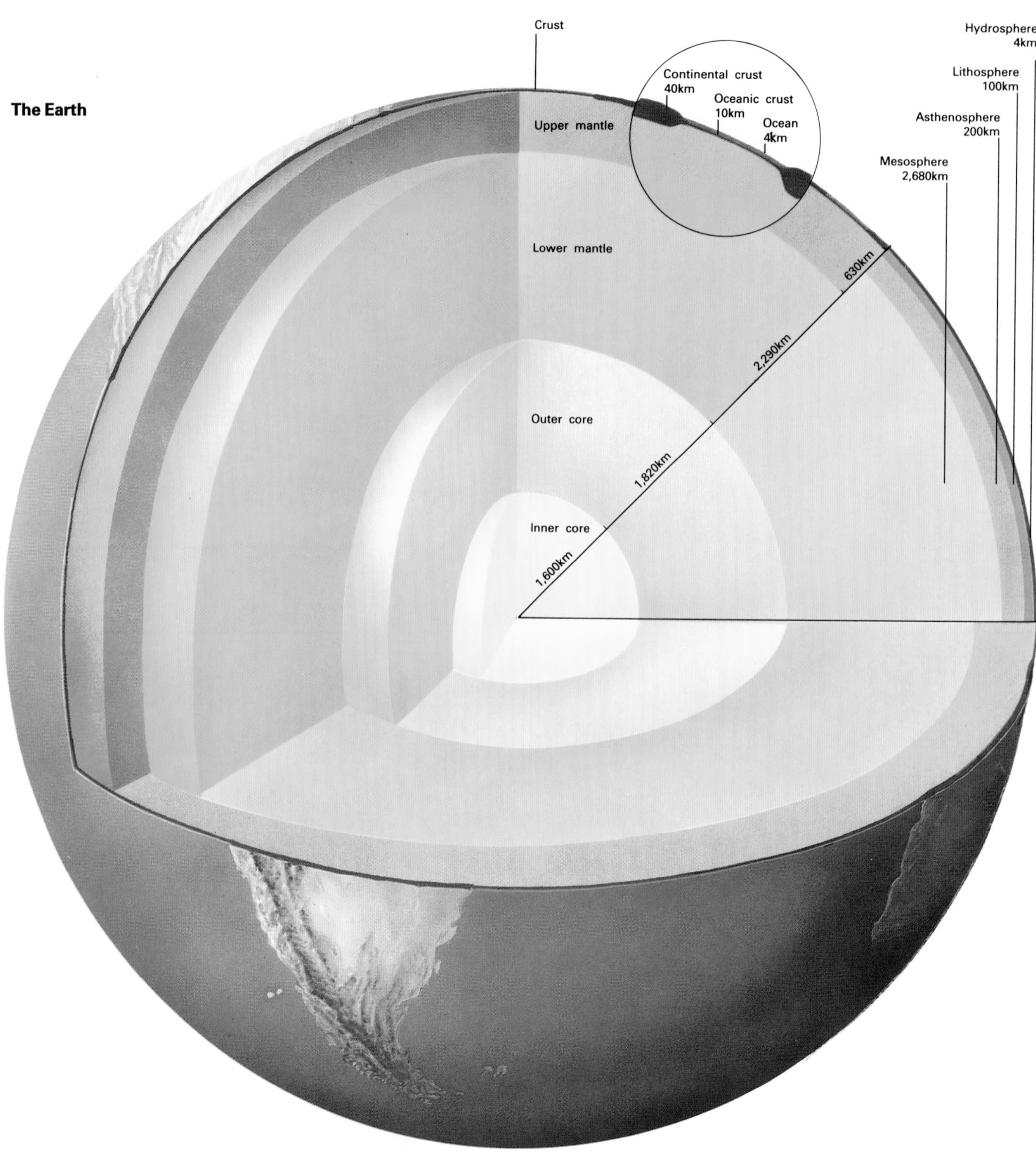

The basic structure of the Earth reveals a core, mantle and crust, on top of which lies the hydrosphere (surface water and oceans). The region outside the core may be subdivided according to the rigidity of its material into mesosphere, asthenosphere (which is weaker and more plastic) and lithosphere, or according to composition and density into lower mantle, upper mantle and (least dense) crust.

Moon

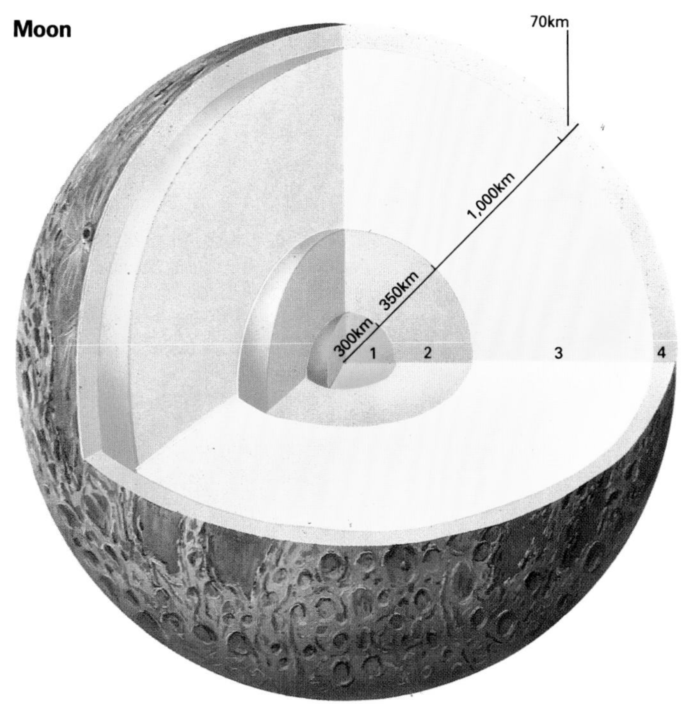

▲ *The Moon may contain a small iron-rich core (1) surrounded by a partially molten zone (2). The rigid mantle (3) is about 1,000km thick, while the crust (4) varies in thickness from about 60km to some 75km. D[illegible]-seated moonquakes occur in the lower part of the mantle. The lava which filled the surface basins came from a depth of several hundred kilometers.*

Mercury

▲ *Despite its small size, Mercury has nearly the same density as the Earth, and this suggests that it must be about twice as rich in iron as the Earth. The iron-nickel core (1) probably extends to a radius of about 1,800km – about 75 percent of the radius of the planet – and contains nearly 80 percent of its mass. Above this is a rocky mantle (2) and a lighter crust (3).*

Venus

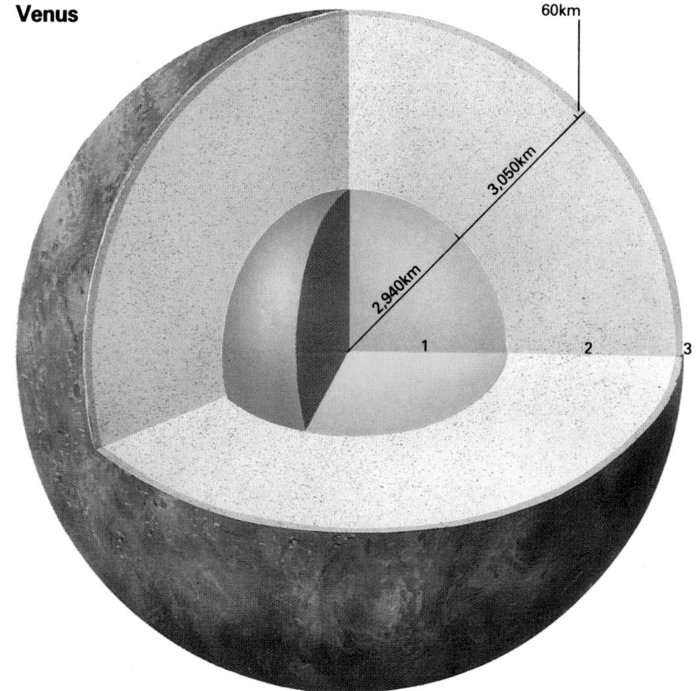

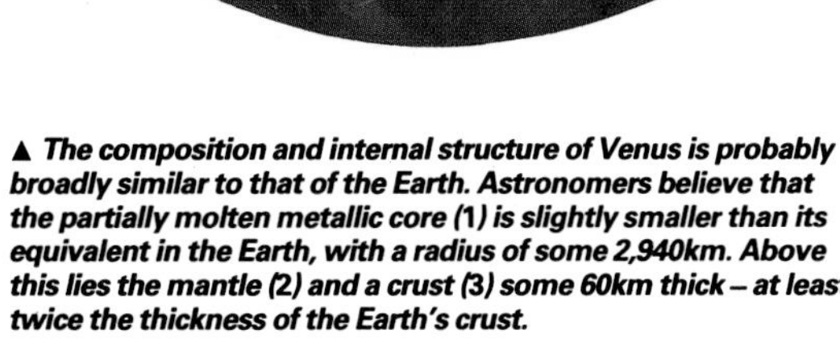

▲ *The composition and internal structure of Venus is probably broadly similar to that of the Earth. Astronomers believe that the partially molten metallic core (1) is slightly smaller than its equivalent in the Earth, with a radius of some 2,940km. Above this lies the mantle (2) and a crust (3) some 60km thick – at least twice the thickness of the Earth's crust.*

Mars

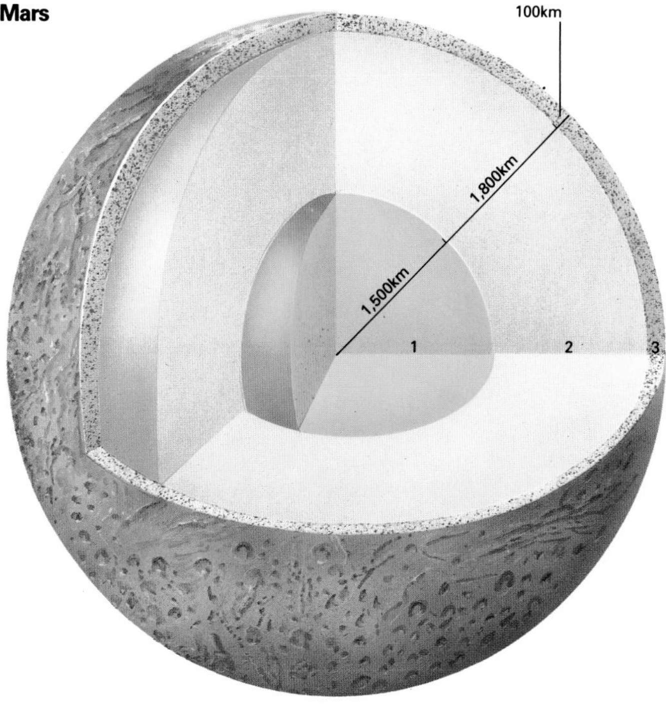

▲ *The mean density of Mars is about 30 percent lower than that of the Earth, and therefore the planet cannot have a large metallic core. Many alternatives have been suggested. The one illustrated here has a core (1) of iron and iron compounds some 1,500km in radius, a silicate mantle (2) about 1,800km thick, and a crust with a thickness of about 100km (3).*

Though regarded as the Earth's twin, Venus – surprisingly – has no detectable magnetic field

The structure of the Moon

The lunar interior has been investigated in some detail by means of heat flow experiments and seismometers placed on the surface during the Apollo missions. The heat flow experiments showed that the interior is hot, reaching as high as 1,500K. The seismometers measured natural moonquakes and the tremors produced by meteorites and man-made impacts. The crust consists typically of a 20km layer of basalt on top of a 20-40km layer of rock; it is about 15km thicker on the side which faces away from the Earth. The rigid lithosphere extends down to about 1,000km. Below this lies a partially molten asthenosphere and a core which may or may not be molten and probably comprises a mixture of iron and iron sulfide.

Planetary magnetic fields

The Earth has a magnetic field with a strength at its surface of between 3×10^{-5} and 6×10^{-5} tesla. It is a dipole field – one which behaves like a simple bar magnet – and compass needles line up along the lines of force connecting the north and south magnetic poles. The direction of the magnetic field changes with time, and at present the north magnetic pole lies at a latitude of 78·5°N, in northwest Greenland.

Of the terrestrial planets, Venus and Mars have exceedingly weak fields, but Mercury has a field strength at its surface of nearly one percent of the Earth's field. Although the presence of a large metallic core is conducive to the presence of a magnetic field, the planet's slow rotation seems to rule out a dynamo mechanism, and the origin of Mercury's field remains a mystery. Jupiter and Saturn have much more powerful fields than the Earth, and astronomers expect that Uranus and Neptune will also prove to have strong magnetic fields.

The significance of a magnetic field

The conditions necessary for a planet to have a magnetic field include the existence in the interior of a fluid, electrical conductor kept in a state of circulation by the effects of planetary rotation and convection currents. Theory suggests that if planetary material initially had a weak magnetic field (for example, a galactic magnetic field that permeated the material from which all the planets formed), the motion of the conducting fluid through the magnetic field would generate an electrical current which would sustain and amplify the original weak field. For as long as the circulation continued, the magnetic field would remain, being sustained by a dynamo-like process operating in, for example, a liquid metallic core. The presence or absence of a magnetic field thus gives a further clue about the nature of planetary interiors.

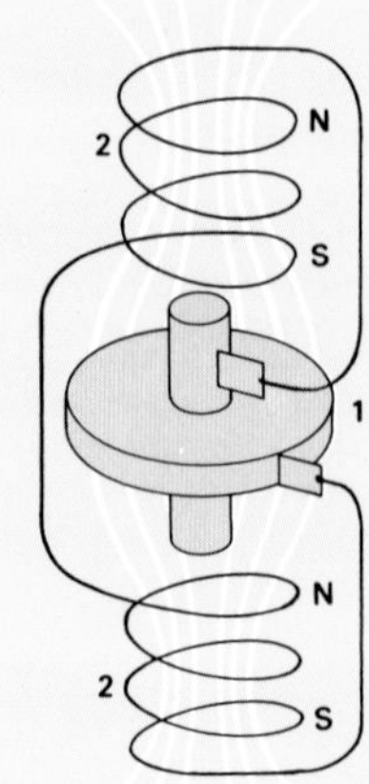

◄ If a metal disk rotates in a magnetic field, the resulting force pushes electrons towards the center of the disk or along a wire attached at the center. Unlike an ordinary dynamo with two magnets, a simple self-exciting dynamo uses a coil (2) with the metal disk (1). Once the disk starts spinning, the flow of electrons sets up a current through the coil. The resulting system generates a magnetic field as long as the disk remains spinning.

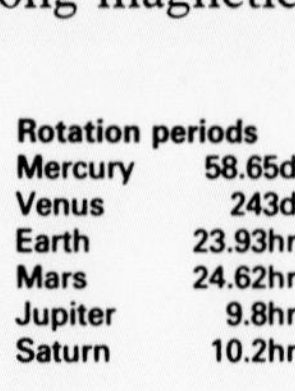

Rotation periods	
Mercury	58.65d
Venus	243d
Earth	23.93hr
Mars	24.62hr
Jupiter	9.8hr
Saturn	10.2hr

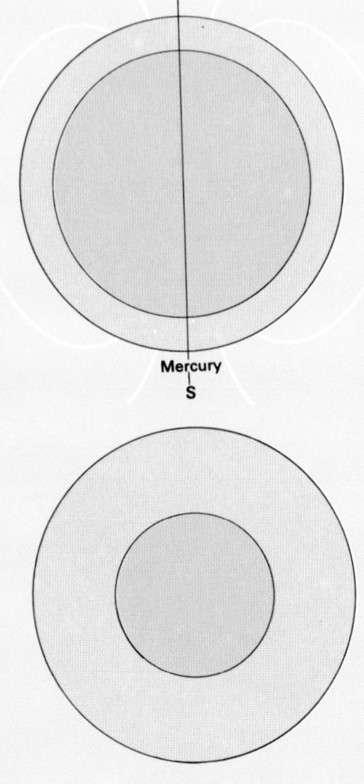

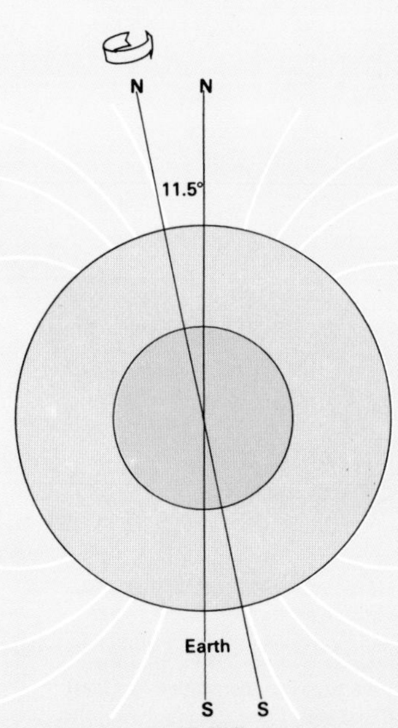

The Earth's magnetic axis (the line passing through north and south magnetic poles) is inclined to the axis of rotation by about 11·5°. Mercury has a weak field closely aligned with its rotational axis. Jupiter and Saturn have powerful fields tilted at the angles shown; both are directed in the opposite sense to that of the Earth, each having its south magnetic pole in the northern hemisphere. The field strength at the cloud tops on Saturn is similar to that at the Earth's surface, while the strength at the Jovian cloud tops is 20 times greater. Since both planets are much larger than the Earth, their overall fields are much more powerful. The electrical currents needed to sustain a magnetic field are carried by the circulation of a conducting material such as iron in the Earth and "metallic hydrogen" in Jupiter and Saturn. Mars does not have a suitable core, while Venus rotates too slowly to carry an electrical current at the core. Although Mercury has a large metallic core, it rotates slowly; its measurable magnetic field remains a mystery.

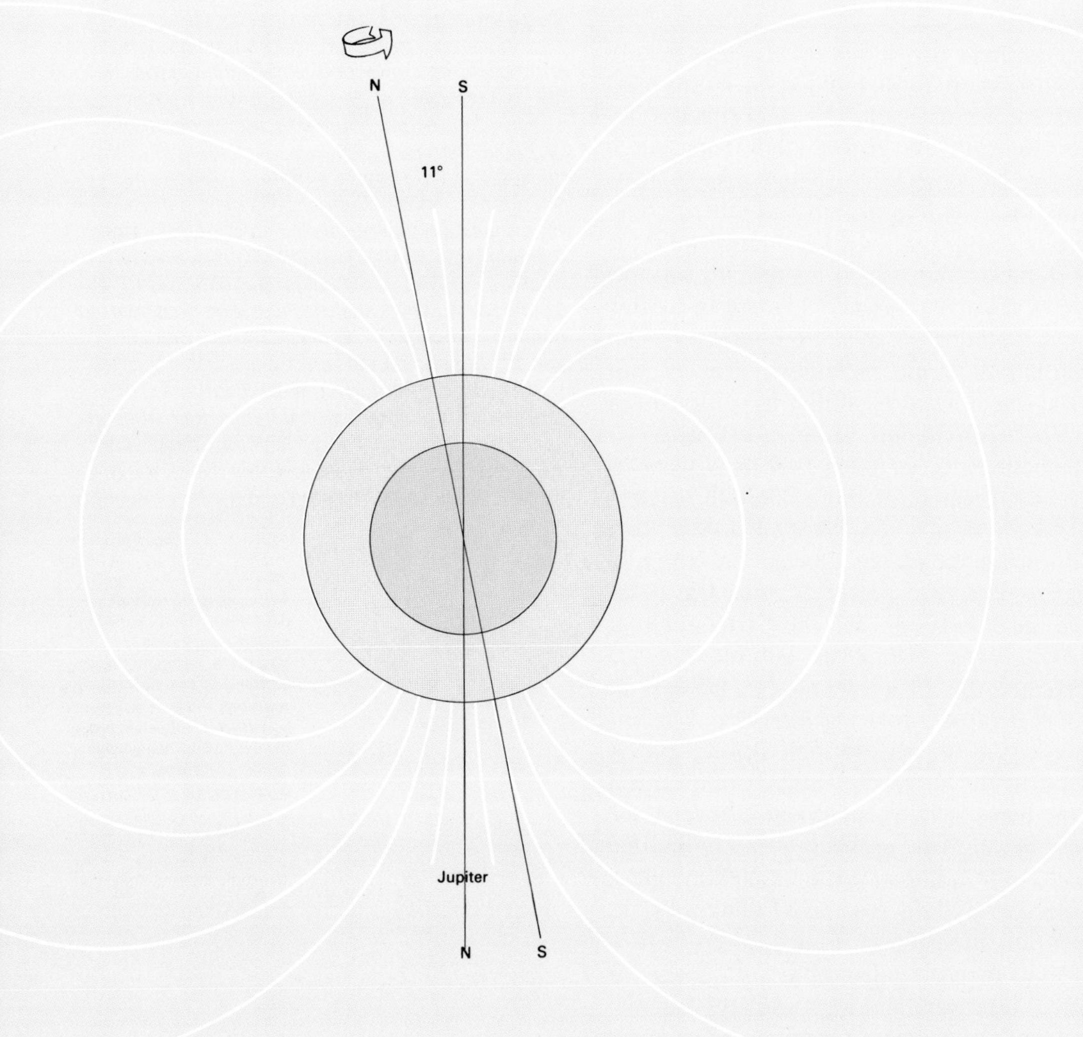

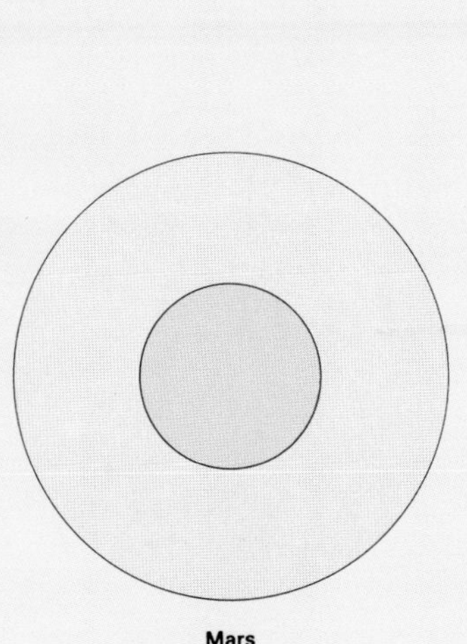

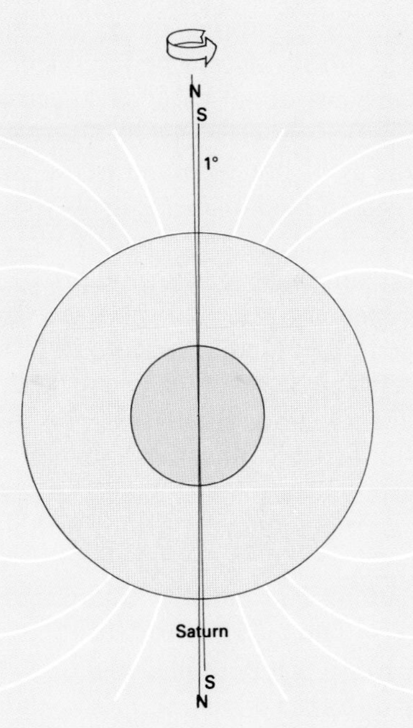

Of the four giant planets, only Uranus seems to have no internal heat source

The giant planets – Jupiter and Saturn

Jupiter, the largest and most massive of the planets, is completely different in structure and composition from the Earth. Composed primarily of hydrogen and helium, it does not have a solid surface like that of the terrestrial planets. Beneath the visible cloud tops lies a hydrogen-rich atmosphere about 1,000km thick, below which there is a deep ocean of liquid molecular hydrogen. At a depth of some 25,000km, where the pressure is about 3 million Earth-atmospheres, hydrogen molecules are broken down, their electrons moving around and conducting heat and electrical currents easily. Hydrogen in this state behaves like a metal, and is known as metallic hydrogen. The bulk of Jupiter's interior consists of liquid metallic hydrogen, but there is probably a central rocky-metallic core, similar in composition to the Earth's, with a mass of between 10 and 30 Earth-masses.

Jupiter radiates into space about twice as much heat as it receives from the Sun and has a central temperature of about 30,000K. Astronomers do not fully understand the source of Jupiter's internal heat, but it may be heat generated during the planet's formation which is still leaking into space. The presence of a powerful magnetic field, more than 10 times stronger at the cloud-tops than the Earth's field at ground level and extending its influence over a vast volume of space (♦ pages 36-7), is readily accounted for by a dynamo-like process in the metallic hydrogen zone of this rapidly rotating planet.

Saturn has a significantly lower mean density than Jupiter and is flattened to a greater extent, but it has a broadly similar composition and structure. Estimates of the mass of its rocky-metallic core range from 3 to 20 Earth-masses. This core is probably surrounded by liquid metallic hydrogen which, at a radius of some 28,000km, gives way to an ocean of liquid molecular hydrogen some 32,000km thick, on top of which is a hydrogen-rich atmosphere. Like Jupiter, Saturn emits about twice as much heat as it receives from the Sun. Saturn is much less massive than Jupiter, however, and it is unlikely that the heat source is left-over heat from the planet's formation. The most plausible theory is that droplets of helium, formed in Saturn's cooler atmosphere, sink towards the core, releasing energy as they do so.

The farthest planets

Uranus and Neptune have yet to be investigated by planetary probes, and there is still some uncertainty about basic data such as diameter, mean density and rotation period for each planet. Uranus seems to be marginally the larger of the two, but Neptune is the more massive and has the higher mean density. Each planet is likely to have a rocky-metallic core of about 3 Earth-masses, surrounded by a deep envelope of water, methane and ammonia "ices", and a deep atmosphere composed mainly of hydrogen, helium and methane. By mass, the relative proportions of core to ices to atmosphere are probably about 20:65:15. A puzzling discrepancy between the two planets is that Neptune, like Jupiter and Saturn, emits about twice as much heat as it receives from the Sun, but as yet Uranus has shown no evidence of a similar internal heat source.

The last of the known planets, enigmatic Pluto, with a diameter of only 2,500km, is significantly smaller than the Moon. Its mass, deduced from the orbital motion of its satellite Charon, is only about one-fifth of the Moon's mass. The mean density of Pluto is similar to that of water, and it seems likely that both planet and satellite are cosmic icebergs made mainly of frozen water, ammonia and methane.

Shape and structure

The oblateness, or flattening, of a planet provides another clue about its composition and structure. This is a measure of the extent to which a planet's rotation causes it to "bulge" at the equator. The degree of flattening depends on the rate of rotation, the fluidity of the planet, and the extent to which mass is concentrated towards its center. The dense, slowly rotating terrestrial planets show little polar flattening, but Jupiter and Saturn, low-density bodies each with a rotation period of about 10 hours, are conspicuously oblate, the equatorial radius exceeding the polar radius by 6 percent and 10 percent respectively.

The shape of a planet and its distribution of mass distorts its gravitational field, which affects the motion of any orbiting satellites. The greater the concentration of mass towards the center of the planet, the less the flattening and the less the effect on satellite orbits.

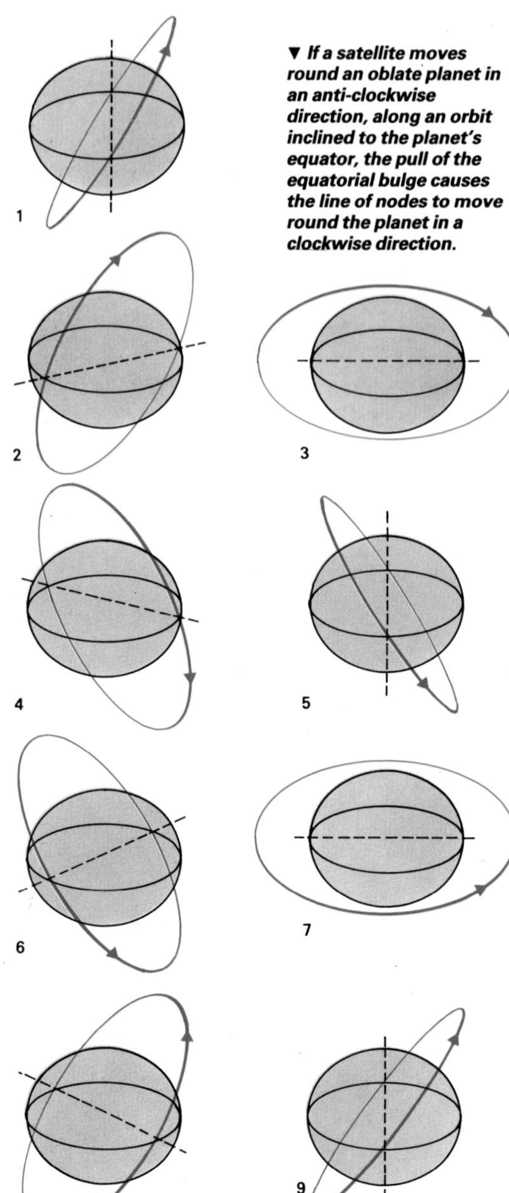

▼ If a satellite moves round an oblate planet in an anti-clockwise direction, along an orbit inclined to the planet's equator, the pull of the equatorial bulge causes the line of nodes to move round the planet in a clockwise direction.

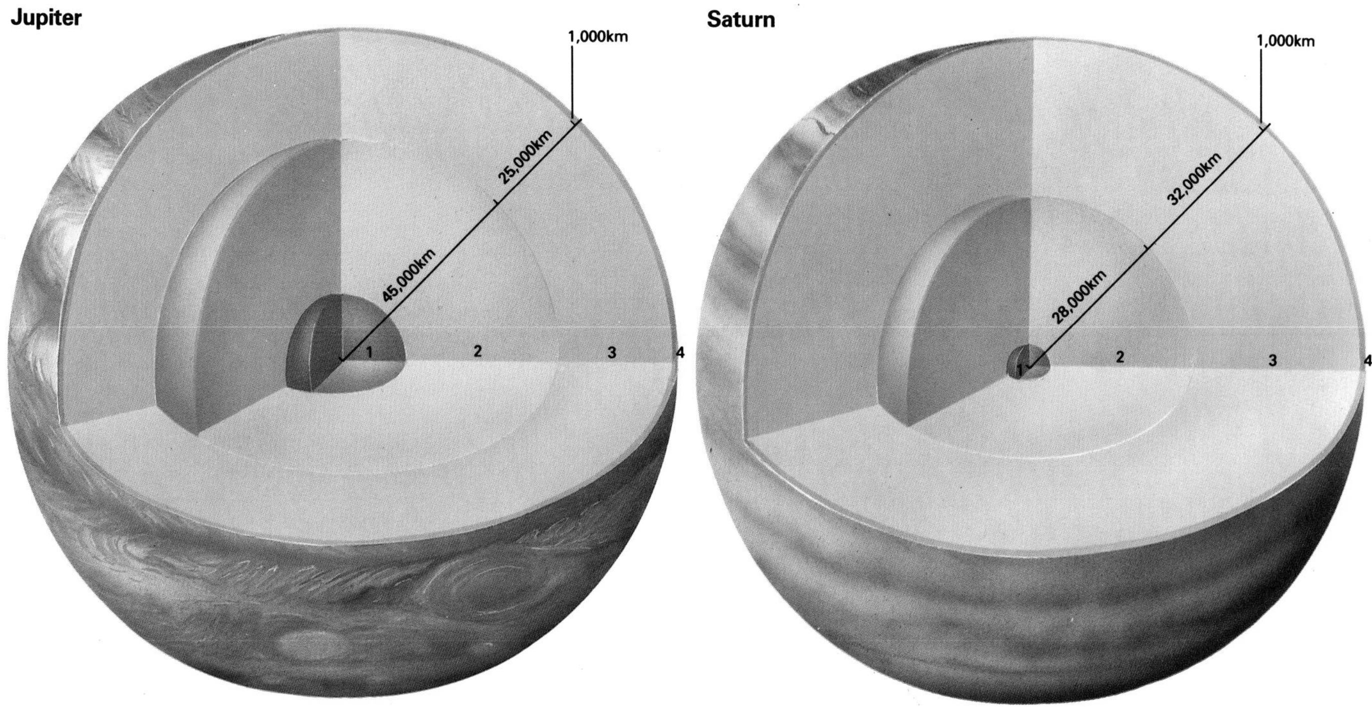

▲ *Jupiter may contain a compact iron-silicate core (1) in a zone of liquid metallic hydrogen (2) which extends to a radius of about 45,000km. Above this there is a layer of liquid molecular hydrogen (3) some 25,000km thick, then the 1,000km deep hydrogen-rich atmosphere (4). Temperature at the center is about 30,000K and pressure is 100 million atmospheres.*

▲ *Like Jupiter, Saturn may contain a compact iron-silicate core (1) embedded in a liquid metallic hydrogen zone (2) extending to a radius of some 28,000km. The liquid molecular hydrogen zone (3), probably 32,000km deep, lies below the hydrogen-rich atmosphere (4). Saturn's core temperature may be 12,000K and pressure 8 million atmospheres.*

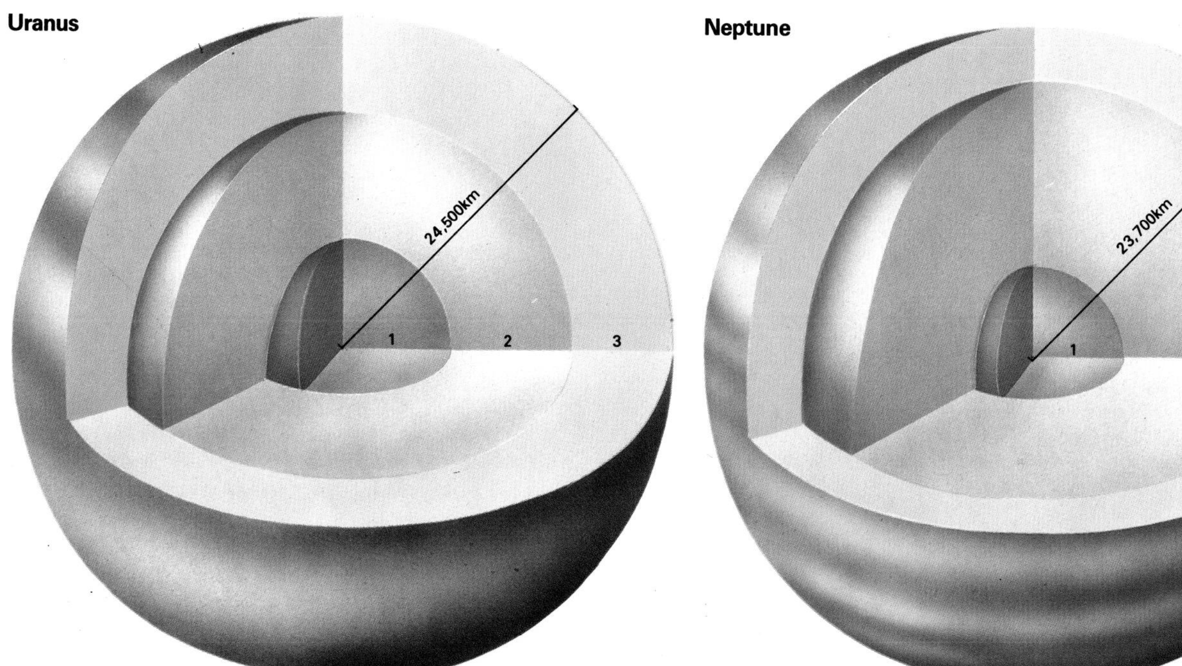

▲ *Planetary geologists think that Uranus has an iron-silicate core (1) of about three Earth-masses which is somewhat larger than the Earth itself. This is probably surrounded by a mantle (2) of water, ammonia and methane ices extending to a radius of some 18,000km, and by a deep atmosphere (3) of hydrogen, helium and methane. The interior may be partly fluid.*

▲ *Neptune is denser than Uranus but probably has a similar general structure, with an iron-silicate core (1), an "icy" mantle (2) and a deep atmosphere (3) rich in hydrogen, helium and methane. Although the core is believed to be solid the planet emits sufficient heat to suggest that the mantle may be fluid, and that motions within it may generate a magnetic field.*

See also
Planetary Landscapes 41-56
Planetary Atmospheres 57-72
Moons and Ring Systems 73-88
Asteroids and Comets 89-96
Inside the Sun 97-104

1

3

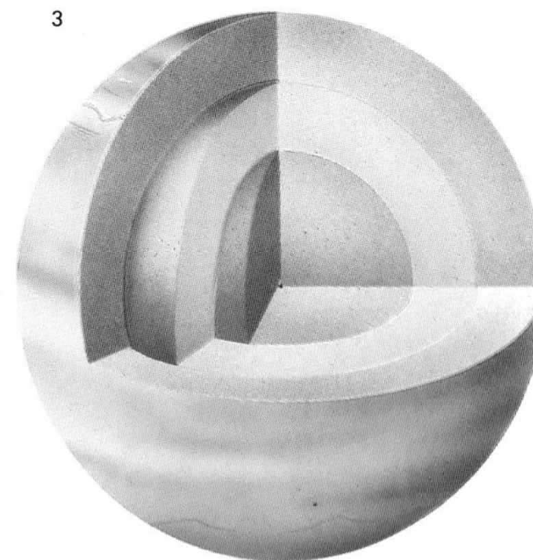

2

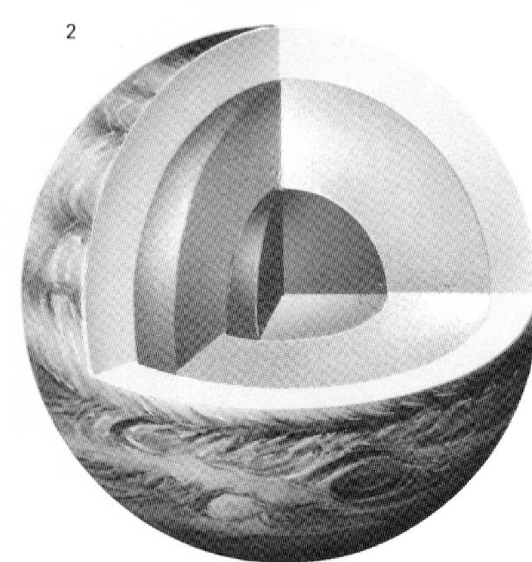

4

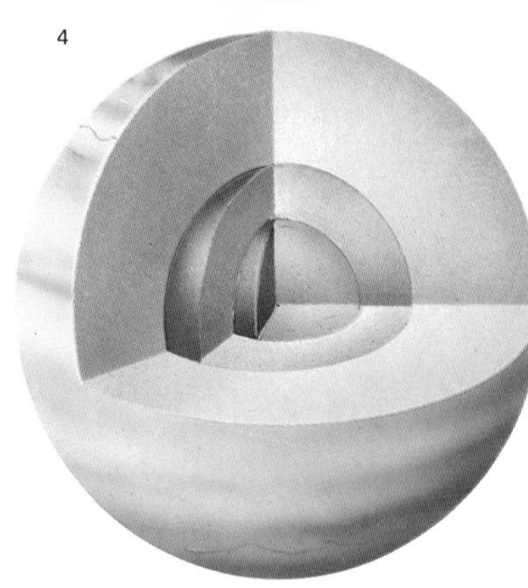

▲ Early theories about the giants. Jeffreys proved that the giant planets were cold, not hot as was widely believed. His 1923 model of Jupiter (1) consisted of a large core and a thin atmosphere, while Wildt in 1934 suggested a smaller, denser core and a deep atmosphere of condensed gases (2). Jeffreys' 1923 model of Saturn (3) was also modified by Wildt in 1934. Wildt gave the planet a rocky and metallic core surrounded by frozen water and carbon dioxide, then by solid hydrogen (4).

► Mr Richard Anthony Proctor, B.A., F.R.A.S., was one of Vanity Fair's "Men of the Day" in the issue of 3 March 1883.

Cold water on a hot theory

Spacecraft can now study the surfaces of the planets directly but, as far as the interiors are concerned, scientists can still only theorize, based on other physical data obtained from Earth, the Moon or spacecraft.

It was always assumed that the terrestrial planets would have compositions not unlike that of the Earth. However, until less than 50 years ago many astronomers believed that the giant planets must be hot worlds, not only on the inside, but also on their gaseous surfaces – rather like primitive suns. Richard Proctor (1837-88), an English astronomer and author, summed up this view in a book published in 1882. With regard to Jupiter and Saturn, he wrote that:

"Over a region hundreds of thousands of square miles in extent, the glowing surface of the planet must be torn by subplanetary forces. Vast masses of intensely hot vapour must be poured forth from beneath, and, rising to enormous heights, must either sweep away the enwrapping mantle of cloud which had concealed the disturbed surface, or must itself form into a mass of cloud".

But were, in fact, the surfaces of the giant planets hot, as Proctor thought? The lengthy investigation carried out in the 1920s by the great British geophysicist Harold Jeffreys (b. 1891) showed conclusively that in the case of the largest planet, Jupiter, the "miniature sun" theory was incorrect.

Early theories about the giants

There is no doubt that the Jovian planets are hot at their cores, but Jeffreys showed that their surfaces were extremely cold. Jeffreys proposed a model in which a giant planet had a rocky core about 46,000km in radius, a mantle of water ice and carbon dioxide 18,000km thick, and an extremely tenuous atmosphere composed largely of hydrogen and extending for about 6,000km.

This theory was modified in 1934 by Rupert Wildt, who was German by birth but worked in America. Since hydrogen is the most abundant element in the universe, it was natural to assume that the giant planets contained a great quantity of it. This was supported in 1932 when T. Dunham, at the Mount Wilson Observatory, used the spectroscope to detect methane and some ammonia in the atmospheres of the giants – both of these are hydrogen compounds. As a result Wildt was able to propose a dense rocky core surrounded by a thick layer of ice, which was in turn overlain by a deep atmosphere of condensed gases.

More recent theories

A new modification was introduced in 1951 by W. Ramsey in England and W. DeMarcus in the US. This time the core had a radius of 61,000km, was made up of hydrogen and was so compressed that it assumed the characteristics of metal. Around this was a layer of liquid hydrogen about 8,900km in depth lying below a relatively shallow atmosphere. The latest view resembles the previous model in some respects, but incorporates a solid central core of rocky materials (silicates).

Planetary Landscapes

Surfaces of the solid rocky planets...How their landscapes were formed...The Moon's dark lava plains and cratered highlands...Formation of the lunar surface...The Earth: plate tectonics and erosion... Mercury's Moon-like landscape...Venus and its searing plains...Mars: ice-caps and massive volcanoes... PERSPECTIVE...Moon-men – a hoax...Mapping the Moon...The Martian "canals"...Life on Mars?

Of the nine major planets only Mercury, Venus, the Earth and Mars have solid rocky surfaces. Pluto probably has an icy surface. The Moon is included here because it has been explored quite thoroughly and, though a satellite, is large enough to be treated as a planet. From Earth the surfaces of Mercury, Mars and the Moon are directly visible, but that of cloud-covered Venus can be studied only by radar.

The terrestrial planets and the Moon display a fascinating variety of surfaces, with the Moon and the Earth contrasting most sharply. The Moon is a dead world whose surface was sculpted by cratering and volcanic activity billions of years ago, and has changed hardly at all since then, while the Earth is an evolving planet continually building up and breaking down its surface features. Mercury is close to the Moon in appearance, Mars is intermediate between the Moon and the Earth, and Venus is yet more Earth-like.

Surface-shaping processes

The surfaces of planets and satellites have been molded by a number of processes, both internal and external. The principal internal process is volcanism – activity that brings molten subsurface material (magma) to the surface (as lava) to build cones, gently-sloping "shield" volcanoes and lava plains. Also important are movements of the mantle due to convection which may fracture the surfaces of planets into distinct plates (➧ page 46).

The major external process is cratering, produced by the impact of bodies such as meteorites, asteroids and the nuclei of comets. The amount of energy released by an impact is enormous. A rocky body about 1km in radius, for example, striking a planet at 20 kilometers per second, would produce a blast comparable to the detonation of more than 500,000 one-megaton nuclear bombs. An impacting body excavates a bowl-shaped crater and throws out a blanket of material, or ejecta, which covers much of the neighboring surface and builds up the crater rim. Larger lumps hurled out by the blast may produce secondary craters. Small craters retain their simple bowl shape but larger ones, depending on the surface gravity of the planet, may slump to produce a series of terraced walls. They may also have mountain peaks in the center. Sometimes magma flows into larger craters and basins to give smooth, lava-floored features.

Astronomers now believe that all the planets and satellites suffered a heavy bombardment soon after their formation (➧ page 124). This cratering record is well preserved on the Moon, Mercury and, to a lesser extent, Mars. Venus has a number of large shallow craters, but the Earth bears the scars of only a few relatively recent impacts, the rest having been obliterated by erosion. Wind, water and ice wears down surface features and the eroded materials are deposited and eventually compacted into sedimentary rocks.

Cratering

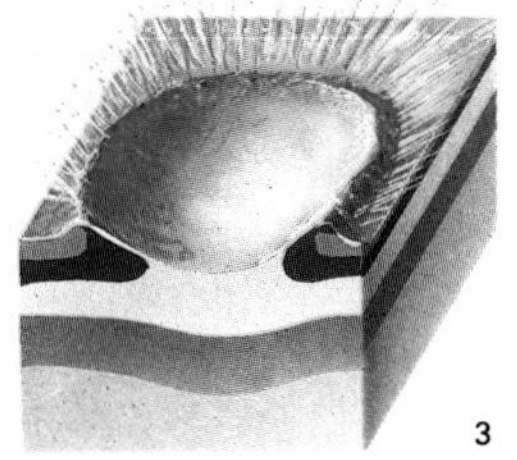

A body striking a planetary surface at high speed (1) blasts a conical stream of matter (ejecta) away from the site of the impact, excavating a bowl-shaped crater (2). Surface matter is ejected before deeper matter. Where ejecta lands to form the crater rim (3) it produces inverted strata.

Volcanism

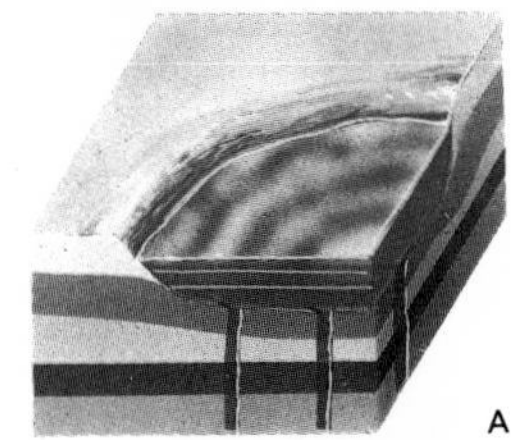

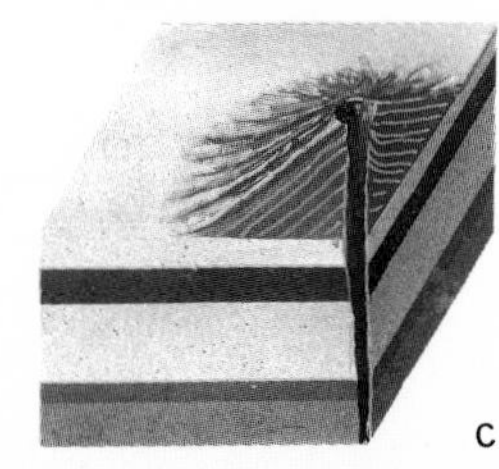

Volcanism brings molten material to the planetary surface with various results. The lunar basins (A) were once filled with lava to form smooth plains – the maria. Highly fluid lava spread (B) to form gently sloping "shield" volcanoes, while less fluid lava (C) built up steep-sided cones.

Clues to the nature of a planet's surface

An important clue to what the planet's surface is like is given by the proportion of sunlight it receives that is reflected back into space. This reflecting power is known as the albedo and is expressed as a number between 1 (for a perfect reflector) and 0 (for a totally matt black surface which absorbs all radiation that falls on it). Atmospheric effects can be considerable. Ice-coated or cloud-covered bodies have high albedos while those with dark, rocky or dusty surfaces have low albedos; thus Venus has an albedo of 0·76 (that is, it reflects 76 percent of the sunlight that falls on its cloud-covered globe), while Mercury and the Moon have albedos of 0·06 and 0·07 respectively. The albedos of Mars and the Earth fall between these two extremes. Measurements of reflectivity are made at a range of wavelengths and angles of illumination.

Other clues as to whether a surface is rocky or dusty, rough or smooth, are given, for example, by infrared measurements of daily temperature changes (dust cools faster than rock) and by radar measurements (rough and smooth surfaces reflect radar pulses differently).

The lunar surface

The surface of the Moon is dominated by two kinds of landscape: the dark lava plains with relatively few craters, known for historical reasons as "seas" or maria (mare in the singular); and the lighter-colored, heavily cratered highlands. Craters range in size from microscopic pits to huge walled depressions up to 250km in diameter. Craters up to about 20km in diameter are bowl-shaped, but larger ones are more complex, many having terraced walls and central peaks. The dark maria cover a large proportion of the Earth-facing hemisphere but are mostly absent from the far side of the Moon, perhaps because the crust is thicker on the far side, which may have inhibited the flow of magma to the surface there.

The lunar surface is overlain by the "regolith", a layer of soil made up of pulverized rocks. Lunar rocks are broadly similar in nature to terrestrial rocks. The maria consist of basalts (dark, fine-grained volcanic rocks) and the highlands of lighter-colored older rocks such as anorthosites, which are richer in aluminium and calcium than the basalts. Many of the highland rocks are breccias – shattered fragments of various types of rock which have become welded together – testifying to the heavy bombardment that the highlands have suffered. The principal difference between terrestrial and lunar rocks is that the latter are richer in refractory elements (those with a high boiling point, such as titanium) and depleted of volatiles (those with a low boiling point). In particular, lunar rocks contain no water or hydrated material. As for their age, rock samples, brought back from the Apollo and Luna missions, date from 3·1 to 3·8 billion years ago in the case of mare basalts and 3·8 to 4·5 billion years ago (older than any terrestrial samples) in the case of highland rocks.

▲ *This lunar crater – Eratosthenes – shows the classic structural features of an impact crater.*

▼ *Harrison H. Schmitt, one of the last men on the Moon, examines a huge split lunar boulder.*

▲ *The Moon, seen from Apollo 11 at a distance of 16,000km. The Mare Crisium is near the center of the picture; to its left is the bright ray-crater Proclus. To the left, near the edge of the disk, is the Mare Serenitatis, and below this the Mare Tranquilitatis, where the Apollo 11 astronauts landed. The regions to the right are on the side of the Moon which is always turned away from the Earth.*

◀ *Samples have now been obtained from different parts of the Moon – from the Apollo landings and unmanned Russian probes. The rocks are essentially volcanic, with little meteoritic material. These are a lunar basalt (left) and an anorthosite (right). Some of the basalts are rich in rare-earth elements and potassium and phosphorus. No unfamiliar elements had been expected and none were found.*

No large craters have been formed on the Moon for at least a thousand million years

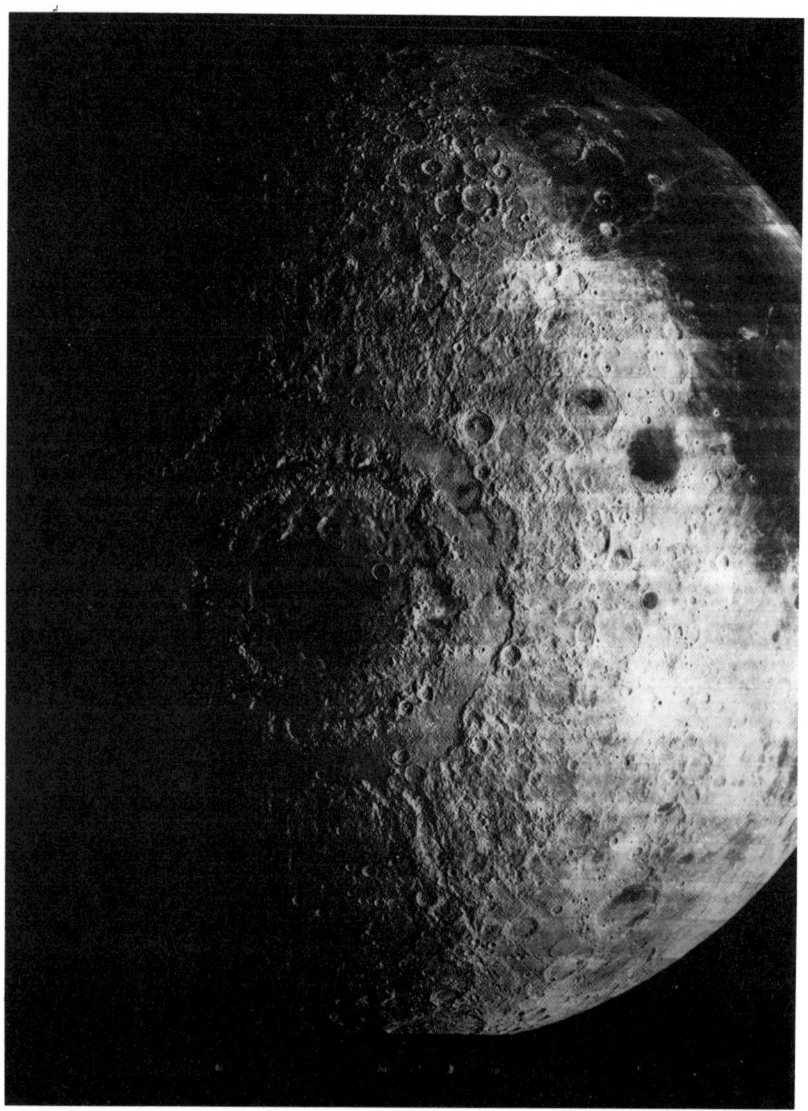

Mapping the Moon

The Moon is the only body in the Solar System whose surface features can be clearly observed from Earth with a small telescope. The first Moon map was drawn in 1609 by an Englishman, Thomas Harriot (1560-1621), one-time tutor to Sir Walter Raleigh. The Italian astronomer Galileo Galilei (1564-1642) attempted to measure and map the heights of lunar mountains in 1610, and his results were quite accurate. In the mid-17th century larger maps of the Moon were produced: in 1651 an Italian Jesuit, Giovanni Riccioli (1598-1671), drew a map based on observations by his pupil Francesco Grimaldi (1618-1663), and introduced the system of naming lunar features after prominent people.

The first really good lunar map was drawn in the 1830s by two Germans, Wilhelm Beer (1797-1850) and Johann von Mädler (1794-1874). Though they used only a small telescope, their map was a masterpiece, and was not surpassed for decades. When it was published, many astronomers felt that, since the Moon was changeless and Beer and Mädler had mapped it so precisely, there was little point in studying it farther. However, the German astronomer Julius Schmidt (1825-1884) continued to observe the Moon, and in 1866 he announced that a small crater, Linné, had disappeared. The effect of this improbable announcement was to revive interest in the Moon, and new maps were produced.

In 1878 Schmidt drew an elaborate map, but this was soon followed by the first photographic atlases. One, in 1904, by the American astronomer William Pickering (1858-1938), showed each region of the Moon under several different illuminations. The last elaborate lunar map before the Space Age was that of Welsh amateur observer Hugh Wilkins (1896-1960). The original, which first appeared in 1946, was 7·6m in diameter, and was accompanied by a detailed description of the entire lunar surface.

A lunar chronology

During the first 200 million years of its existence, the Moon was molten, but the outer part cooled rapidly, becoming solid about 4·4 billion years ago. Over the next few hundred million years, the entire lunar surface suffered heavy bombardment by Solar System debris. The last of the major periods of bombardment, which excavated the mare basins, probably occurred about 3·8 billion years ago and thereafter the cratering rate declined rapidly. Between 3·8 and 3·2 billion years ago the mare basins were flooded by lava. These regions are now made up of rocks that are similar in nature to terrestrial ocean-floor basalts. Parts of the rims of these basins remain in the form of mountain chains such as the Apennines which border the Mare Imbrium.

Since 3·2 billion years ago there have been few substantial changes to the Moon's surface apart from a few impacts which produced conspicuous features such as Tycho and Copernicus. The thickness of the lithosphere, together with the limited amount of mantle convection, has ensured that there is no plate tectonic activity (➧ page 46). Apart from occasional minor gaseous emissions and meteorite impacts, the Moon is a dead and unchanging world.

▲ The imposing Mare Orientale contains little dark lava. It is similar to basins on Mercury, Mars and Jupiter's Callisto.

► This condensed history of the Moon shows how the surface probably appeared before the volcanic phase (1) when magma filled in many of the great basins (2). Later cratering produced the present-day appearance (3).

1

2

3

Moon men

The first observers to use astronomical telescopes believed the Moon to be a world with large oceans on its surface. The lunar "seas" were soon found to be dry – there has, in fact, never been any water on the Moon – but the idea of lunar inhabitants was curiously slow to die. German-born William Herschel (1738-1822), one of the greatest astronomical observers of all time who spent much of his life in England, believed that the ability of the Moon to support life was "an absolute certainty". Later, in 1822, the German observer Franz von Paula Gruithuisen (1774-1852) announced the discovery of a true lunar city, with "dark gigantic ramparts". This turned out to be no more than a completely normal arrangement of low, haphazard ridges.

Some years later, the Danish mathematician Peter Hansen (1795-1874), who was famous for his work on the Moon's movements, put forward a strange idea about the far side which is never visible from Earth. Hansen suggested that the Moon's center of mass might not coincide with its center of figure, so that all the air and water had been drawn round to the hidden side, which might well, therefore, be inhabited. The idea never gained much favor; for one thing, it would certainly have led to an appearance of "twilight" at the cusps of the crescent Moon, and needless to say, nothing of the sort was observed. It was not until 1959 and the round trip of Luna 3 that the far side was seen to be as barren as the Earth-facing side.

The celebrated Moon Hoax

In 1835 John Herschel (1792-1871), son of William, was in South Africa to survey the southern sky. Communications in those days were slow and uncertain, and R. A. Locke, a reporter on the "New York Sun", realized that it would take time to disprove anything outrageous that he published. Accordingly the "Sun" was issued with a series of articles describing how Herschel, using a plausible-sounding though practically impossible method of observing the Moon in great detail, had discovered remarkable forms of life, including bat-men and even "a strange, spherical creature which rolled with great velocity across the pebbly shore". Many people were taken in, though only briefly.

▲ The original caption to this absurd illustration, published in 1835 in the "New York Sun", read as follows: "Lunar animals and other objects, discovered by Sir John Herschel in his Observatory at the Cape of Good Hope and copied from sketches in the Edinburgh Journal of Science. For description, see pamphlet published at the Sun Office." The hoax, perpetrated by a reporter, succeeded – for a while!

The Earth

The main features of the Earth's surface are continental land-masses (about 30 percent) and oceans (about 70 percent). Other striking features are the great mountain chains, and cracks and rifts in the surface such as the San Andreas fault in California and the Great Rift Valley in Africa. Geologists believe that the dominant process shaping the large-scale structures of the Earth's surface is plate tectonics. This is the mechanism whereby convection currents in the molten mantle have broken the lithosphere (upper mantle and crust) into about a dozen rigid plates which drift slowly over the globe.

Where two plates are drifting apart, as in the middle of the Atlantic Ocean, molten mantle material rises at a mid-oceanic ridge and spreads out to form new ocean bed composed of basalt. At present Europe and North America are drifting apart at a rate of about 2cm a year. When two continental plates collide, the crumpling and buckling of crustal material may throw up great ranges of mountains. Where an oceanic plate meets a continental plate, the heavier oceanic plate descends below the lighter continental plate. This process, known as subduction, results in the melting of oceanic plate material, the thrusting up of long mountain chains and the production of deep ocean trenches such as the one running parallel to the Andes off western South America. Volcanic and earthquake activity frequently occurs along plate boundaries.

Plate tectonics and volcanic processes build new landscape features, but there are many erosive processes at work to wear them down and to transport material to new locations. The Earth is thus a dynamic, changing planet.

▼ This computer-generated relief map shows areas lying within the same 500m altitude band. The difference between highest mountain and deepest ocean is about 8km.

▲ The Earth, seen from the weather satellite Meteosat 1.

► The Chilean Andes as seen in winter from the land survey satellite Landsat.

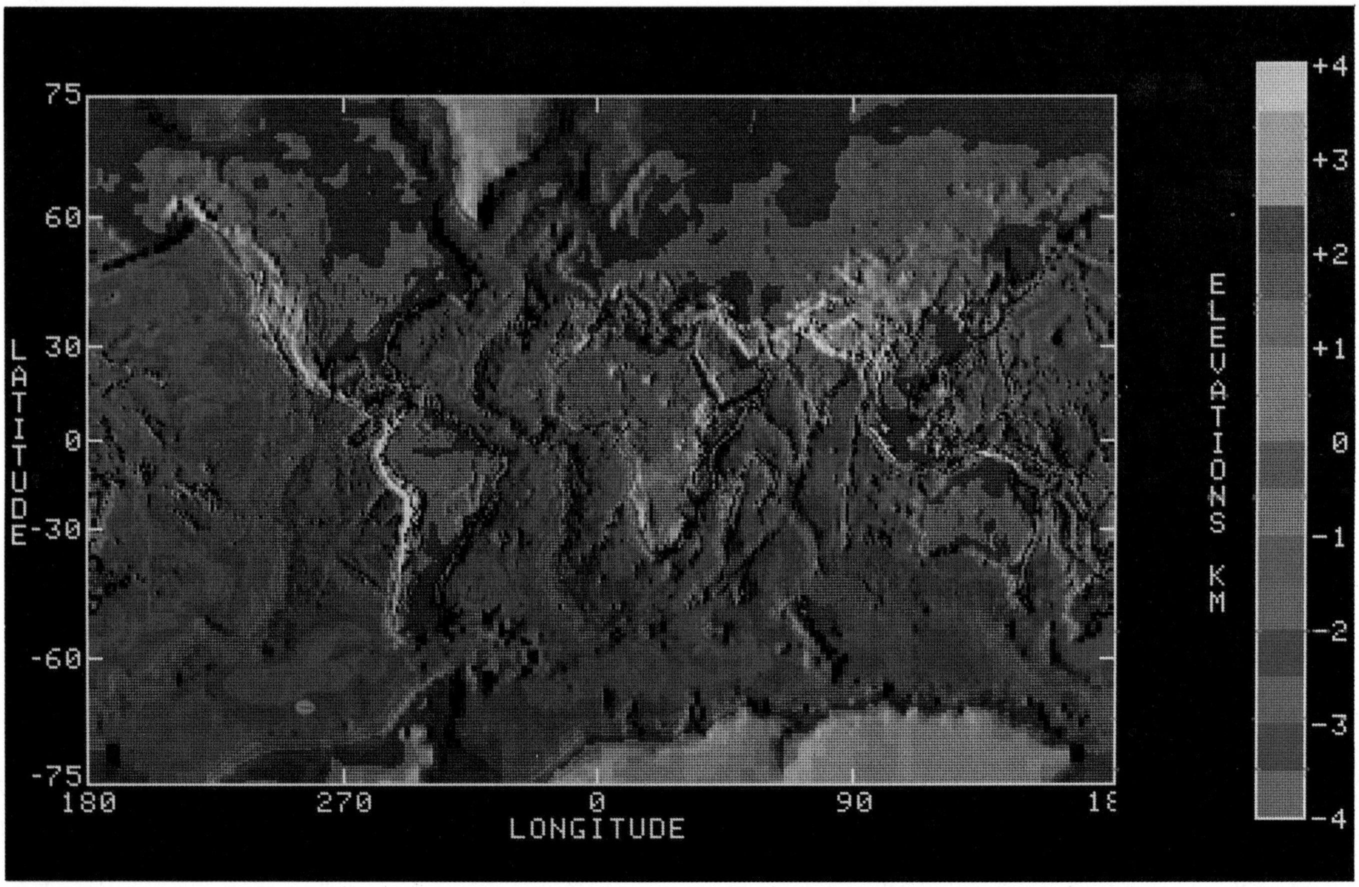

On Mercury temperatures are either extremely hot or cold – there is no "twilight zone"

Mercury

The smallest of the terrestrial planets and the closest to the Sun, Mercury receives about six times as much solar radiation per square meter as the Earth. The daytime temperature is high, 600K (about 330°C) on average, reaching a maximum of about 750K (about 480°C), whereas at night it drops down to 90K (about -180°C). There are two regions of the surface which are most strongly illuminated at alternate perihelia, giving rise to what are termed "hot poles".

The surface of Mercury, like that of the Moon, has cratered highlands and smoother volcanic plains. Intermediate in type between these two regions are the "intercrater" plains, ancient moderately-cratered plains quite unlike the lunar maria. There is one giant concentric ring-shaped basin, the Caloris Basin, some 1,400km across, which is similar to the Mare Orientale on the Moon and was presumably also caused by a giant impact. Lava flows are also visible on Mercury, and the hills and plains are covered in pulverized dust. The planet's history seems to have been similar to that of the Moon, with saturation bombardment and lava flows occurring during the first billion years of its existence. There is no evidence of plate tectonics, but the long lines of cliffs and scarps may be due to wrinkling of the surface as the planet cooled down.

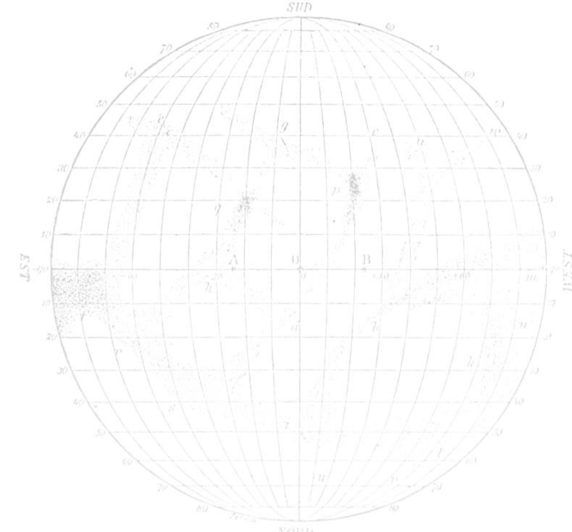

▲ **The first map of Mercury was made by Giovanni Schiaparelli (1835-1910) in Milan. He found the dark patches difficult to distinguish and they later proved illusory**

▼ **This is how Mercury appeared to the onboard television cameras of Mariner 10, (left) six hours before and again (right) six hours after the spacecraft flew past the planet in 1974.**

◄ *This ridge, known as Discovery Scarp, was photographed by Mariner 10. More than 300km long and in places 3km high, it cuts straight through the crater Ramsan. It was probably formed by compressive forces acting on the crust.*

▼ *The Caloris Basin, a huge impact crater on Mercury's surface, has a diameter of 1,400km and shows many similarities with the Mare Orientale on the Moon. Between the ridges and more recently formed craters are large areas of smooth floor – presumably the heat generated by the impact caused the rock to melt.*

Observing the cloudy planet

Venus presented major problems to the early observers, who did not at first realize that all they could see was the top of a cloud layer. In 1726, the Italian astronomer Francesco Bianchini (1662-1729) went so far as to draw a map of the surface showing oceans and continents. In 1789, the German astronomer Johann Schröter (1745-1816) observed what he thought was the top of a mountain, several kilometers high. This led to a dispute with his great contemporary William Herschel, who correctly claimed that Schröter's "enlightened mountain" was non-existent.

The Ashen Light mystery

The Ashen Light is the faint visibility, probably electrical in origin, of the unlit region of the planet when Venus is in the crescent stage. In the pre-space age it gave rise to some imaginative theories. The German astronomer Franz von Paula Gruithuisen (1774-1852) believed that it was due to vast forest fires lit on Venus by the inhabitants to celebrate the accession of a new ruler. It was later suggested that the light could be due to phosphorescent oceans. However, when it became clear that the surface of Venus is always hidden by clouds, theorists came into their own. Particularly notable was the Swedish scientist Svante Arrhenius (1589-1927), who believed that Venus today is in much the same state as the Earth was more than 200 million years ago. He pictured hot swamps with luxuriant vegetation and, quite possibly, amphibious life forms and insects such as giant dragonflies. This led to the attractive conclusion that Venus might evolve in the same way that the Earth had done, so that it would eventually become habitable.

▼ *Illustrations such as this, depicting a weird and exotic flora on the surface of Venus, were common in the 19th century.*

The surface of Venus

Despite the similarity in size and mass of Venus and the Earth, surface conditions on Venus are much harsher. Searingly high temperatures of 750K (about 480°C) are maintained by a dense atmospheric blanket (➧ pages 58-9).

Radar mapping has been carried out by Pioneer and Venera orbiters and by Earth-based radio telescopes. Pioneer results showed that about 70 percent is made up of undulating rolling plains, just under 20 percent is isolated depressed lowlands, and 10 percent is mainly of highland "continents", typically 4 or 5km above mean surface level. There are also many shallow craters up to about 1,000km in diameter, some huge volcanoes, and a number of rift valleys and trenches.

The rolling plains appear to be ancient crust, heavily cratered. The lowlands are smooth and apparently crater-free. They may be basaltic lava flows reminiscent of lunar maria and terrestrial ocean beds. The largest is Atalanta Planitia, a roughly circular depression some 2,500km in diameter and 1·6km deep at its lowest point. Although there is no water on Venus now, there is some evidence to suggest that at a much earlier period in the planet's history there may have been oceans which subsequently evaporated.

The two largest upland areas, or "continents", are Ishtar Terra in the northern hemisphere, and Aphrodite Terra, which lies on and south of the equator. Ishtar, which is about the same size as Australia, rises steeply from the rolling plain and boasts the Maxwell Mountains which, at 11·1km above the mean surface level, are the highest feat-

ures on the planet. Aphrodite Terra is about the size of Africa – about 10,000km long. Another smaller, but interesting upland area is Beta Regio, which has two massive shield volcanoes (Theia Mons and Rhea Mons) and several smaller ones. Volcanic eruptions may be responsible for the dramatic changes which have occurred in the sulfur dioxide content of the atmosphere since Pioneer arrived in 1978.

Mantle motions do not seem to have fractured Venus' thick lithosphere into separate plates. There are, for example, no spreading ridges of material like the mid-oceanic ridges on Earth. However, considerable internal activity is evidenced by giant rift valleys, continents, and the massive volcanic structures. The large volcanic shields are probably located over rising currents of hot mantle material. By sitting over these hot spots for long periods of time they have grown far larger than terrestrial volcanoes (which are carried away from their magma source after a time by plate motion).

The surface rocks investigated by the Venera spacecraft show interesting variety. Granitic rocks were found at the Venera 8 site, and elsewhere other types of basalts, reddish-brown in color, were found. At some locations the rocks were sharp and angular, with little sign of weathering, while at other sites smoother, more eroded rocks and finer soils were found. Also seen were some layered rocks which may be sedimentary. It is not yet clear what erosive forces are at work on Venus, for there is no running water, nor are the temperature changes large, and surface winds appear to be too gentle to produce effective sandblasting.

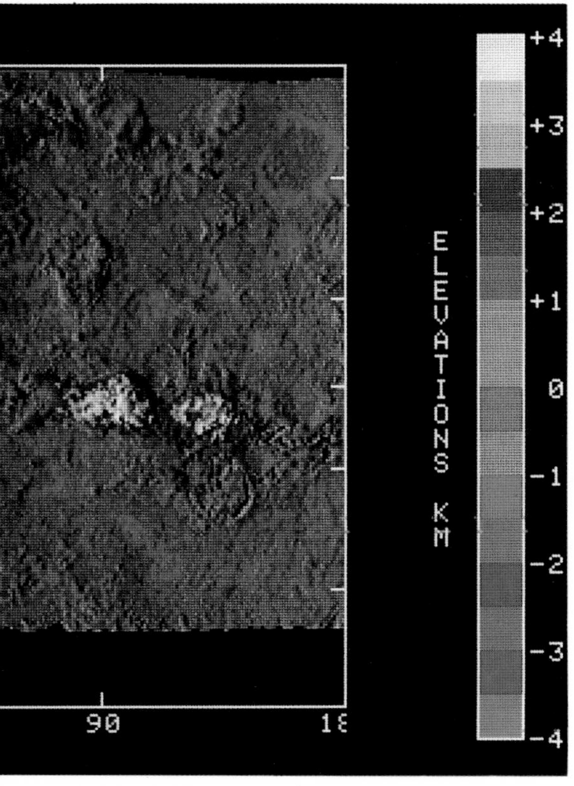

▲ **Radar equipment on the Venus Pioneer Orbiter produced this color-coded relief map of the planet's surface. Areas of each color lie within the same 500m altitude band.**

◀ **The Russian spacecraft Veneras 13 and 14, working in conditions of crushing pressure and intense heat, performed the astonishing feat of transmitting color pictures from Venus.**

▶ **Mariner 10 flew past Venus on 5 February 1974 en route for Mercury. The onboard television cameras recorded this view of the cloudy planet one day and 720,000km later.**

▼ **In this radar image from Arecibo Observatory yellowy areas are rough and blue areas smooth terrain. The distinctive patch to the right is the Maxwell Mountains.**

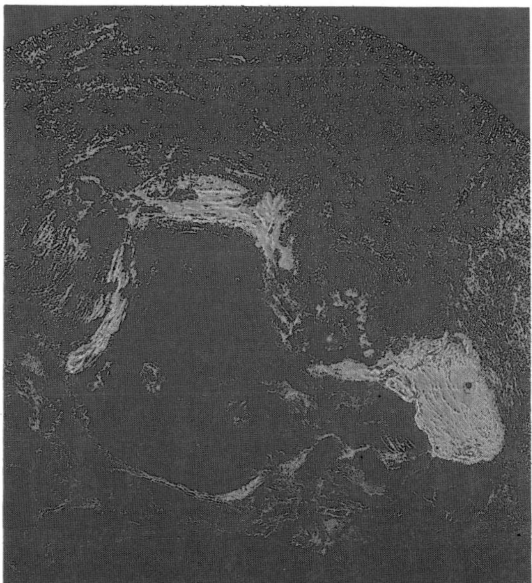

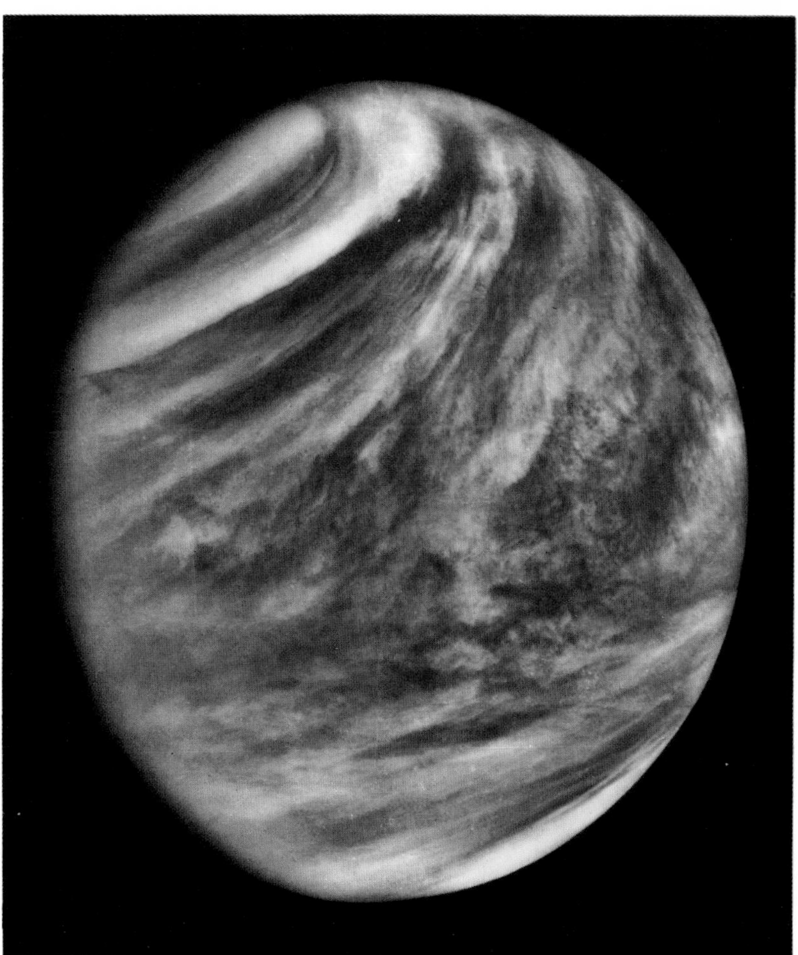

Water rivers must once have flowed on Mars

The surface of Mars

The Martian surface contains both lunar and terrestrial types of terrain. The dominant features are craters, lava plains, giant volcanoes, immense canyons, winding river-like valleys, and polar ice caps which expand and contract with the seasons (➧ pages 62-3). Surface temperatures range from an equatorial maximum of about 300K (27°C) to a minimum of about 135K (−138°C) over the winter pole. The mean value is about 230K (about −40°C).

Heavily cratered older terrain, 1-3km above the mean surface level, is found mainly in the southern hemisphere. It includes a number of large circular basins, the largest of which, Hellas, is some 2,000km in diameter. The region is not as heavily cratered as the Moon, however, and most of the craters are shallower than their lunar counterparts; they have flatter floors, and appear to be partially filled with wind-blown dust and seem to be quite well eroded. The northern hemisphere consists largely of smooth, depressed plains. Some of the smoother plains look like basaltic lava flows and show similar wrinkled "flow fronts" to those found in, for example, the Mare Imbrium on the Moon.

Although no liquid water exists on the Martian surface now, there is ample evidence to suggest that there must have been some, if only for short periods, in the past. There are features that look like dried-up river beds, and teardrop-shaped deposits of material around obstacles are similar to the effects produced by a heavy flow of sediment-laden liquid. Sudden brief periods of flooding may have occurred when the volcanoes were active in the past. Water vapor and other gases poured out by the volcanoes may have increased the density of the atmosphere sufficiently for rain to fall and surface water to flow.

The Martian surface contains about 1 percent by weight of water, and water is also present in the permanent polar ice caps and probably in a subsurface layer of permafrost, like that found in arctic regions on Earth. The thickness of the water-ice caps is uncertain, but a great deal more water probably lies beneath the surface of the planet. The seasonal changes in the extent of the ice caps are due to the melting and freezing of a thin layer of carbon dioxide over the ice.

At the Viking landing sites the surface was dusty and rock-strewn. Analysis of Martian soil by the landers revealed no traces of organic material and, although some of the results are open to question, it does seem that even the most elementary micro-organisms are absent from the dust-strewn surface of the red planet.

▲ *This impressive vista appears when the Martian dust-storms abate. Viking 2 was leaning 8° from perpendicular, which is why the horizon appears to slope.*

◄ *The north polar cap on Mars consists primarily of water-ice. The distant view (far left) shows the cap shrinking as summer approaches. The closer view shows bands of water-ice separated by ice-free slopes at the top, and irregular ice patches below.*

▲ *Argyre Planitia, the large plain in the center, is an ancient impact basin surrounded by heavily cratered terrain. A thin haze, probably of carbon dioxide, highlights the horizon.*

▼ *This "river" is convincing evidence that a fluid once flowed on Mars. The channel is 573km long and 5-6km wide, and resembles terrestrial watercourses. The small tributaries are typical of run-off channels found elsewhere.*

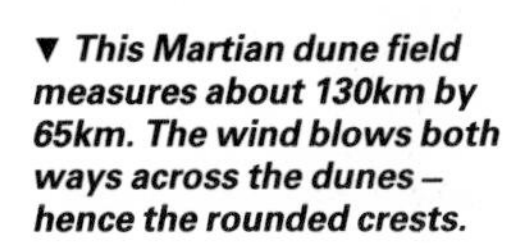

▼ *This Martian dune field measures about 130km by 65km. The wind blows both ways across the dunes – hence the rounded crests.*

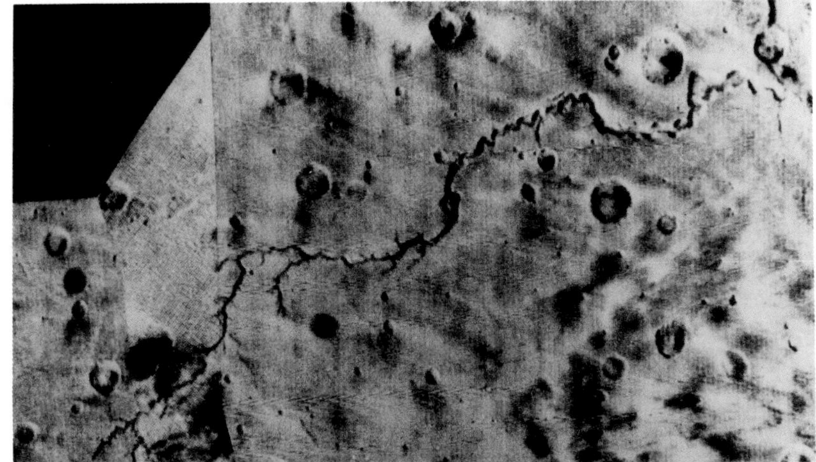

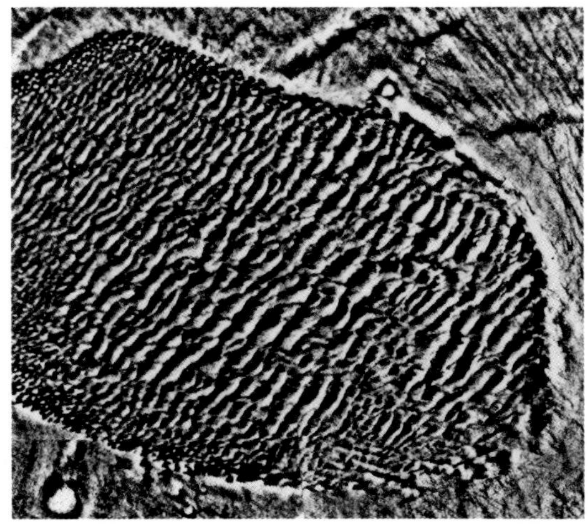

The loftiest volcano on Mars is three times as high as Earth's Mount Everest

Spectacular scenery on Mars

The largest volcano on Mars, named Olympus Mons, is a huge, gently sloping shield volcano, 25km high and some 600km across the base, with a summit caldera (volcanic crater) 80km in diameter. The gentle slopes of this feature and the surrounding lava sheets indicate that the lava from which it was formed must have been highly fluid. The largest comparable feature on Earth, Mauna Kea in Hawaii, stands a mere 9km above the floor of the Pacific Ocean and measures 225km across the base. About 1,000km southeast of Olympus Mons is the Tharsis Ridge, 2,000km long and 9km high; on top of this is a line of three massive shield volcanoes, Arsia Mons, Pavonis Mons and Ascraeus Mons. This ridge marks the edge of a great bulge in the Martian crust which was probably pushed up some 10km above the mean surface level by convection in the mantle.

An enormous canyon system, Valles Marineris, runs along the surface of Mars, just south of the equator for some 4,000km, reaching a maximum width of several hundred kilometers and a depth of 6km. It has numerous tributaries, and massive landslides beneath its cliffs. At the eastern end is a region of collapsed ground, littered with great blocks of rock; this is probably the result of the melting of subsurface ice. Valles Marineris may be a rift valley system which arose in conjunction with the uplifting of the neighboring Tharsis bulge. The mantle motions that produced this were probably too weak to fracture the 200km-thick Martian lithosphere into plates like those on Earth, and the shield volcanoes – like those on Venus – grew to their immense size because they remained stationary over reservoirs of magma.

The striking reddish color, seen in these photographs of the Martian surface, is due to a rust-like iron oxide material known as limonite, which is a hydrated rock.

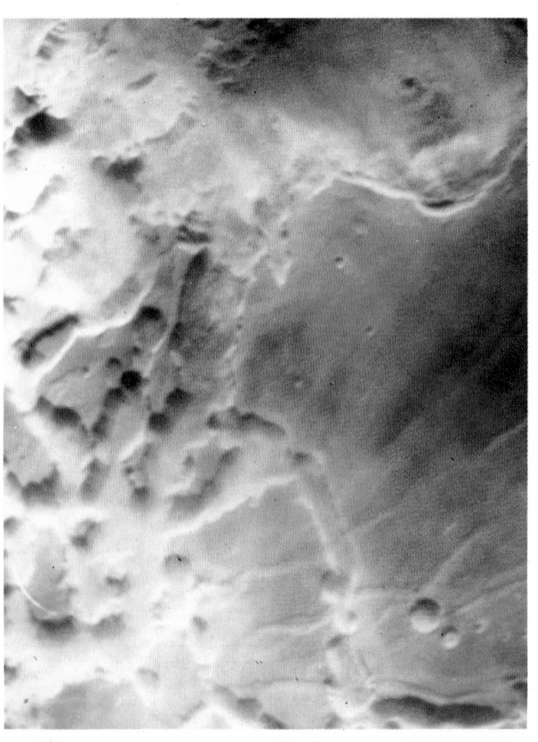

▲ As the Sun rises over the Noctis Labyrinthus area, bright clouds of water-ice become visible in the canyons (lower left). The picture was taken by Viking Orbiter 1 in October 1976.

▼ This contour map of Mars, compiled from information provided by ground-based radar and spacecraft, shows the vertical relief. Each color denotes a 500m altitude band.

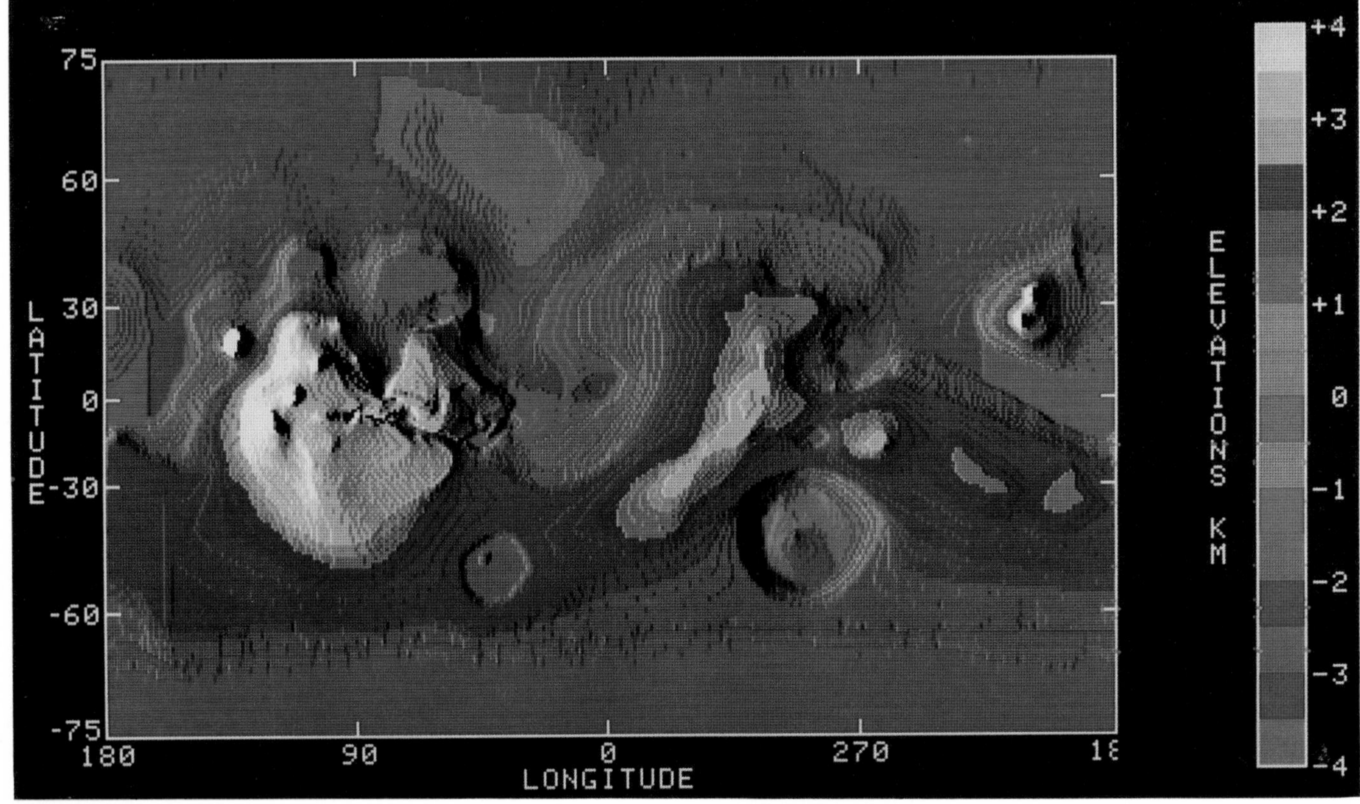

▲ *Three photographs were combined to give this 200° Martian panorama. The surface material is highly reflective, and this has caused flare to degrade the center image.*

▼ *Olympus Mons, the largest mountain on Mars, is far larger than any comparable feature on Earth. It is a giant shield volcano, 25km high, with a base width of 600km.*

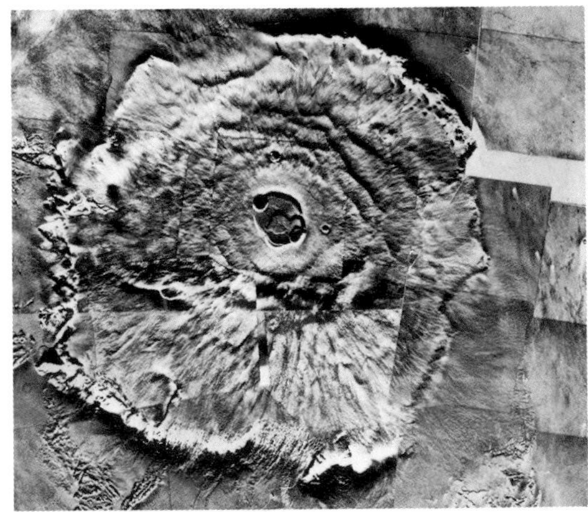

▲ *Slight variations of hue in the lava material suggest that the formation of the Olympus Mons shield took place over a long period of time. The photograph is in false color.*

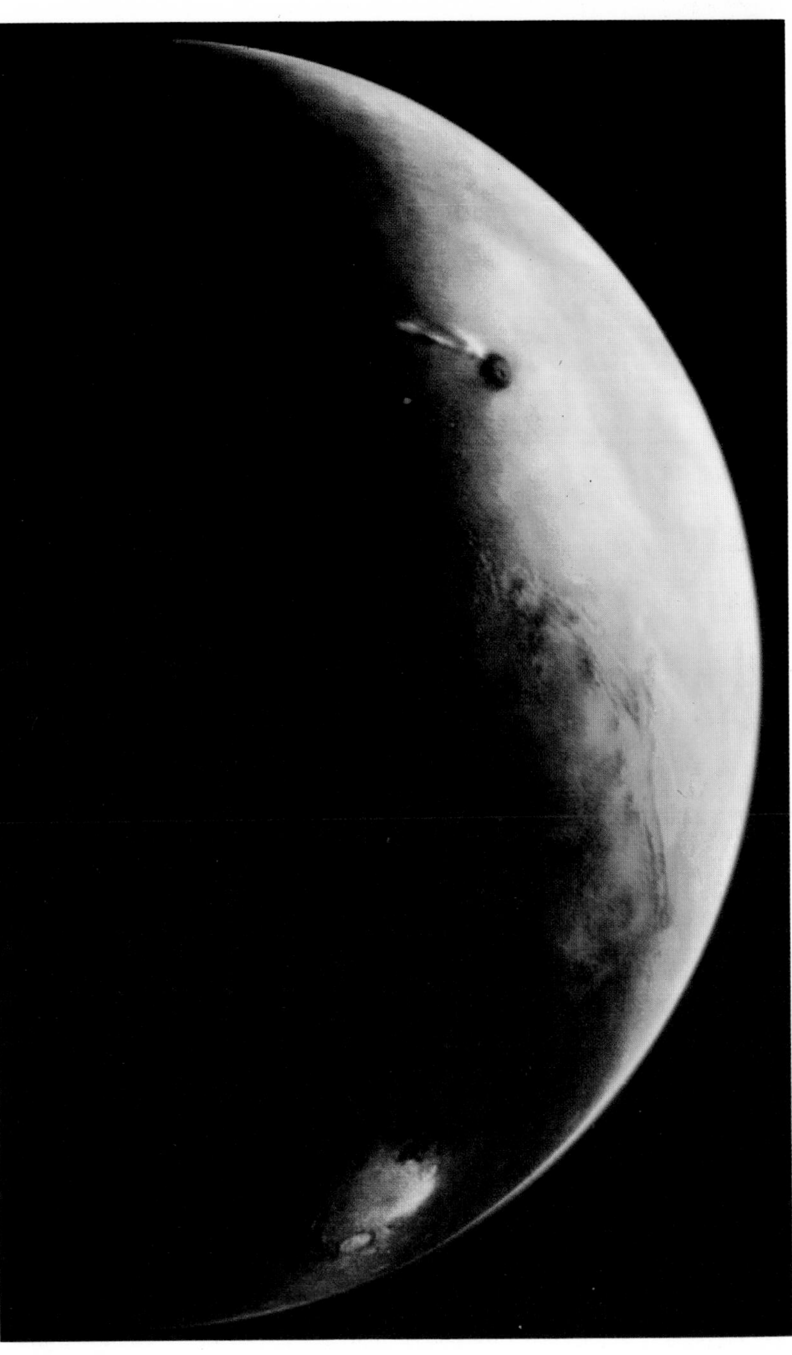

► *The clearest features in this picture of Mars are Ascraeus Mons (near the top), the Valles Marineris canyon system (center) and the crater basin Argyre (near the bottom).*

See also
Observing the Planets 21-32
Inside the Planets 33-40
Planetary Atmospheres 57-72
Moons and Ring Systems 73-88
The Origin of the Solar System 121-8

▶ **"Sunrise over the canals of Mars", a 19-century artist's impression. The idea that a network of artificial canals existed on Mars was completely dispelled by the 1964 Mariner 4 flyby.**

The origins of Martian "life"

Mars, unlike Venus, has a tenuous atmosphere, and usually there is no veiling of the surface, although at times there are planet-wide dust-storms which conceal all surface details, even the polar caps. During the 19th century various maps were produced, showing bright ocher-colored regions and dark areas which were at first thought to be seas. When it became clear that the Martian atmosphere is too thin and too dry to allow the presence of oceans, it was suggested that the dark patches might be old seabeds infilled with primitive organic matter. This theory was not finally disproved until the flight of the first successful Mars probe, Mariner 4, in 1965. Mariner also put an end once and for all to the long-standing controversy about the existence of Martian "canals".

Schiaparelli and the "canals"

The so-called canals were first drawn in detail in 1877 by Giovanni Schiaparelli (1835-1910), a skilled Italian astronomer. Schiaparelli drew long, narrow lines across the Martian "deserts", and called them canali. This is Italian for "channels", but it was inevitably translated as "canals", and before long came the suggestion that they might be artificial in origin. Schiaparelli kept an open mind throughout the great debate, but recorded that he was "very careful not to combat this suggestion, which contains nothing impossible". An American astronomer, Percival Lowell (1855-1916), who built an observatory at Flagstaff in Arizona specially to observe Mars, was much less cautious, and wrote: "That Mars is inhabited by beings of some kind or other is as certain as it is uncertain what those beings may be." His beliefs met with considerable skepticism, however, even during his lifetime.

▼ **G. V. Schiaparelli of Milan drew a new chart of Mars in 1877, and this is a Mercator projection of his drawing. The straight lines which he called "canali" (channels, not canals) are numerous and striking, but Schiaparelli never committed himself to the belief that they were artificial. Percival Lowell, on the other hand, who drew the lower map, was fully convinced that the canals were part of an elaborate irrigation system.**

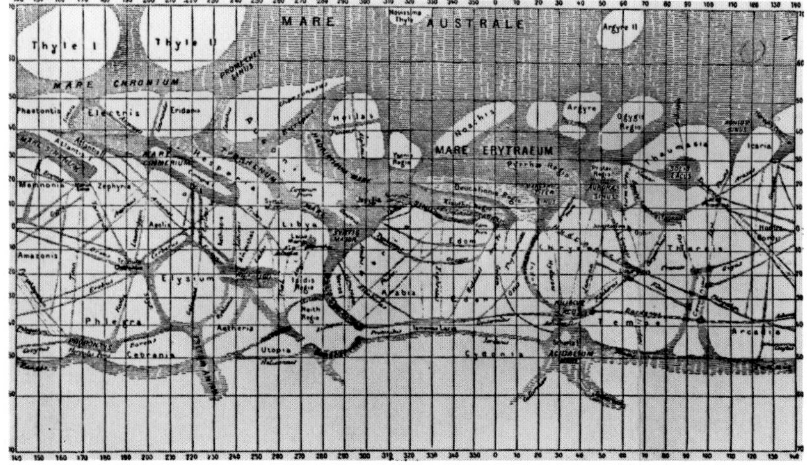

Talking to the Martians

The idea that Mars was inhabited was not new. As long ago as 1802 the great German mathematician Carl Gauss (1777-1855) had suggested signalling to the Martians by drawing vast geometrical patterns in the Siberian tundra. In 1874, the French astronomer Charles Cros (1842-1888) put forward a scheme to focus the Sun's heat onto the Martian deserts by means of a huge burning-glass, swinging the glass around so as to "write" messages in the desert. In 1901, a prize of 100,000 francs was offered in France to the first person to establish communication with extraterrestrial beings. Mars was excluded from the competition because it was thought too easy to contact Martian inhabitants!

What the Vikings found

By the start of the Space Age the canal network had been relegated to the realms of science fiction. Space-probe photographs confirmed that canals do not exist in any form, and that they were due simply to tricks of the eye.

Other established ideas about conditions on Mars were also found to be wrong. Contrary to expectations, the terrain proved to be mountainous and cratered, with giant volcanoes, of which the highest, Olympus Mons, rises to three times the height of Mount Everest on Earth, and has a base 600km in diameter. A number of scientists expected the Viking spacecraft to find evidence of primitive life-forms, but the experiments all proved negative.

Planetary Atmospheres

Which planets have atmospheres?...Earth's unique features...Venus and its acid rain...Mars and the vestiges of atmospheric water...The giants and their helium rain...Jupiter's spectacular storms...Uranus and Neptune, their curious differences...PERSPECTIVE... The "greenhouse" effect and the terrestrial planets... Seasons and atmospheric engines...Jupiter's mysterious "Great Red Spot"

All the planets apart from Mercury and Pluto have significant atmospheres, although those of Venus, Earth and Mars are much less substantial than the deep envelopes surrounding the giant (Jovian) planets. The Jovian atmospheres are essentially the original gaseous blankets that those planets acquired at the time of their formation, but the terrestrial planets could not hold down atmospheres of this kind. Theirs were produced by outgassing – the release of gases from their interiors by, for example, volcanic activity.

Whether or not a planet retains an atmosphere depends on a number of factors, notably temperature, the chemical composition of the atmosphere, and escape velocity (◀ page 28). If a particle in the outer regions of an atmosphere is moving faster than the planet's escape velocity, it can escape into space. The higher the temperature, the faster the speeds at which the atoms and molecules travel; but at any particular temperature the heavier atoms and molecules move more slowly than the lighter ones. Massive planets have higher escape velocities than less massive ones. The Jovian planets, being both cooler and more massive, have retained much of their hydrogen and helium, while the terrestrial planets have lost these gases.

By a process known as photodissociation (literally "separating by light"), ultraviolet radiation from the Sun breaks down molecules in the upper layers of an atmosphere into their constituent atoms. A molecule of water, for example, could break down into hydrogen and oxygen, and the hydrogen might then escape even though the original water would not have done so.

The mass of an atmosphere is important in several respects. The heavier the atmosphere, the greater the pressure it exerts at ground level – and pressure, together with temperature, determines whether a substance will exist in solid, liquid or gaseous form. The mass of an atmosphere, and its ability to retain heat, also affects its response to changes in the input of energy from the Sun. Both Venus, which has a much more massive atmosphere than the Earth, and Mars, which has a much less massive one, reflect this: the daily temperature variations on Mars are very large, while there is almost a complete absence of any differences between the day and night hemispheres of Venus. The temperature differences between equator and poles (which are responsible for driving global wind systems) on Venus, Earth and Mars are 2 percent, 16 percent and 40 percent respectively.

Atmospheric gains		Atmospheric losses	
Short wave absorbed	22	Short wave scattered to space	4
Short wave scattered	25	Short wave scattered to ground	21
Long wave absorbed	107	Long wave radiated to space	61
Evaporation/latent heat	23	Long wave radiated to ground	97
Conduction/convection	6		
Total	183	Total	183

The greenhouse effect

A planet's atmosphere acts like a blanket, helping to conserve heat and thus to maintain a higher temperature at the surface than would be the case in the absence of an atmosphere. The process is known as the "greenhouse effect" because in some respects it resembles the action of a greenhouse.

Most of the solar energy reaching a planet is in the form of short-wavelength (visible or near-visible) radiation. This penetrates to ground level, where it is absorbed and so heats up the surface. The warmed surface emits infrared radiation, which cannot readily escape into space because it is strongly absorbed in the atmosphere by substances such as water vapor and carbon dioxide. The atmosphere radiates part of this heat back to the ground, so raising the surface temperature further.

The efficiency with which an atmosphere can do this depends upon its chemical composition and its mass. The massive atmosphere of Venus contains a high proportion of carbon dioxide and water vapor, and the greenhouse effect is much greater there than on the Earth. By contrast the Earth's greenhouse is more efficient than that of Mars, where the atmosphere is very thin and has little influence on surface temperatures, despite its high carbon dioxide component.

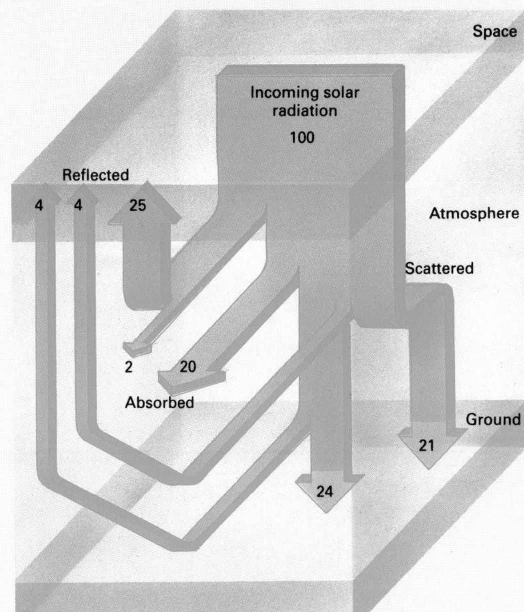

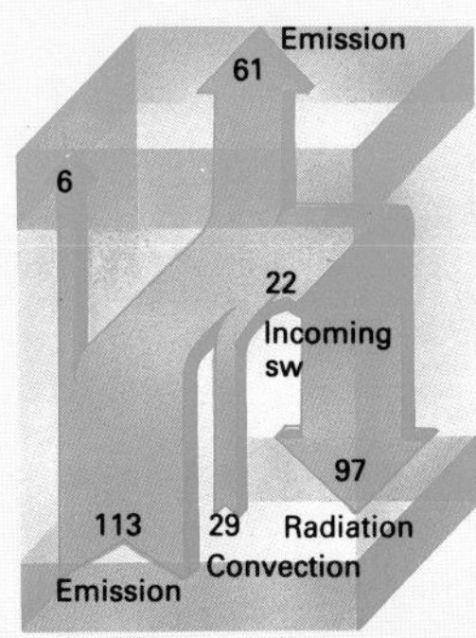

These diagrams show how the atmosphere helps to maintain a balance between incoming and outgoing radiation. The upper diagram represents the distribution of incoming solar radiation, the lower one that of long-wave (infrared) radiation from the Earth. The figures are summarized in the table (left). Gains and losses both total 183 units, so the average surface temperature remains stable.

The Earth

The Earth's atmosphere consists mainly of nitrogen (78 percent by volume) and oxygen (21 percent). Of the other constituents the most abundant are water vapor, argon and carbon dioxide. Average pressure at ground level is 1013 millibars (mb) – equivalent to the weight of a column of mercury 0·76m high, or that of a 10m column of water – and this pressure is termed "one atmosphere".

The greenhouse effect maintains the mean surface temperature at about 290K (17°C), some 35K higher than it would be if there were no atmosphere. It ranges between a maximum of about 313K (40°C) near the equator to a minimum of about 233K (−40°C) at the poles.

It is convenient to divide the atmosphere into a number of layers. The lowest of these is the troposphere, which extends to an altitude of some 12km and contains most of the atmospheric water vapor, clouds and weather phenomena. Because the greenhouse effect is greatest near the ground, the temperature drops with altitude, reaching a minimum of about 217K at the tropopause (the upper boundary of the troposphere). Above this level is the stratosphere, which extends to an altitude of some 50km, and in this layer the temperature increases with height because of the absorption of ultraviolet rays from the Sun (mainly by ozone, which plays a vital role in protecting living creatures from the harmful effects of exposure to these rays).

In the next layer, the mesosphere, the temperature falls again to a minimum of 180K at an altitude of some 85km. Thereafter the rarefied gases of the thermosphere are heated by sunlight to typical values of 1,200K by day and 800K by night. At around 500km this layer merges with the exosphere, from which light atoms such as hydrogen can escape. At heights above about 70km, electrons are knocked out of some of the atoms by solar ultraviolet and X-rays. This produces the layers of electrically-charged electrons and ions which make up the ionosphere – the layer which reflects some radio waves.

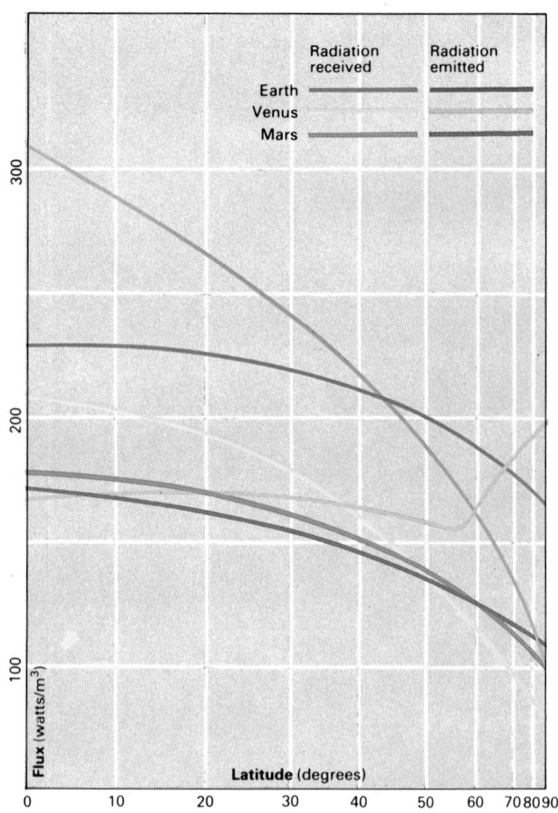

▲ ***At low latitudes the input of solar radiation exceeds the output of infrared; at high latitudes output exceeds input. To balance the sums, heat travels from equatorial to polar regions.***

▼ ***The temperature increase found in the Earth's stratosphere is absent from the atmospheres of Venus and Mars due to a shortage of ozone, which absorbs solar ultraviolet radiation.***

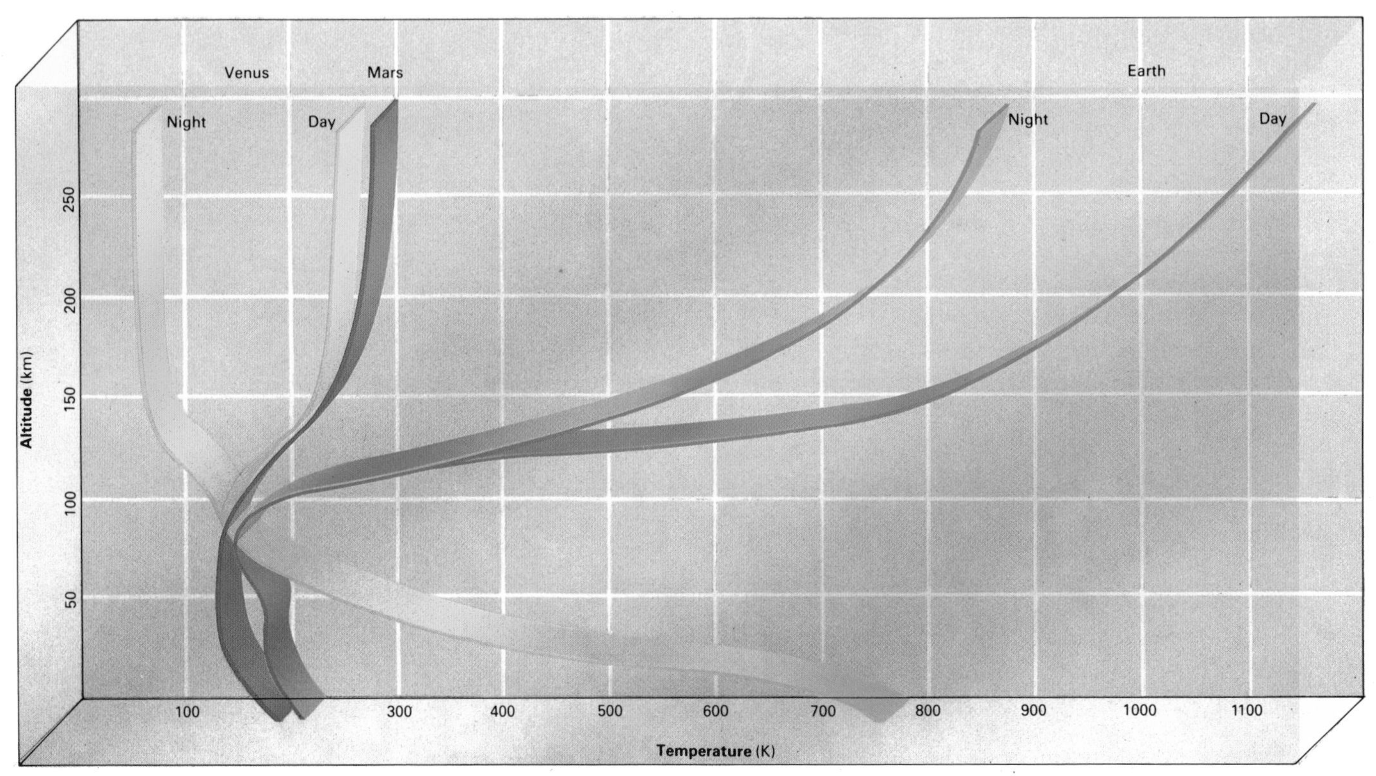

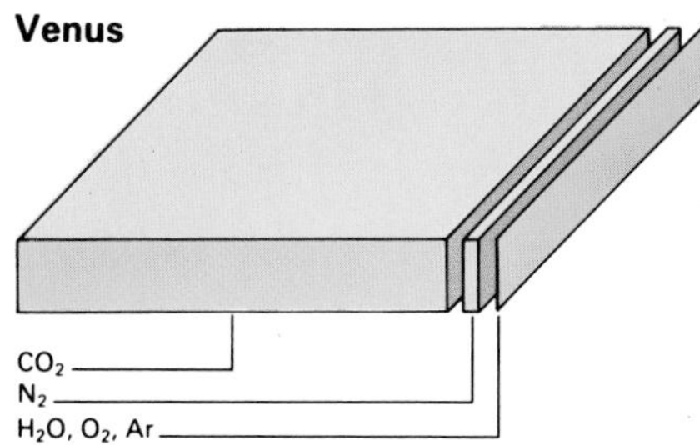

▲ ***The atmosphere of Venus consists almost entirely of carbon dioxide (96 percent).***

▼ ***As on the Earth, the troposphere is heated by infrared radiation from below, while absorption of solar ultraviolet raises the temperature in the thermosphere. The main cloud layer lies between the altitudes of 45 and 60km, with a maximum density at 49-52km. Haze layers lie above and below the clouds.***

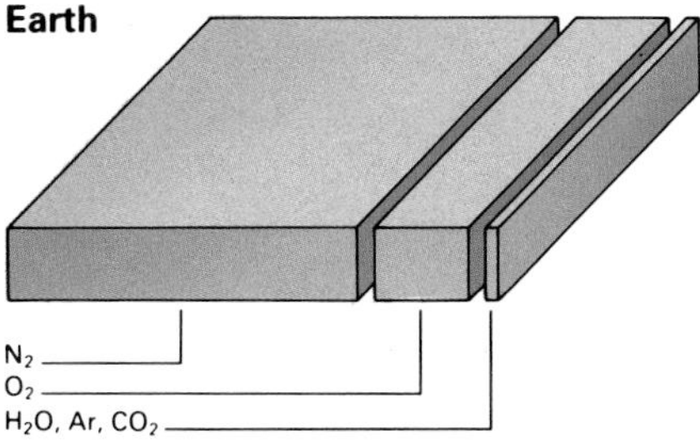

▲ ***Nitrogen and oxygen together make up 97 percent by volume of the Earth's atmosphere.***

▼ ***The lowest atmospheric layer, the troposphere, which contains most of the clouds and water vapor, is heated by infrared radiation from the ground; its temperature falls with increasing height. The stratosphere is heated by incoming ultraviolet light, as is the thermosphere, where the ionized gases of the ionosphere are formed.***

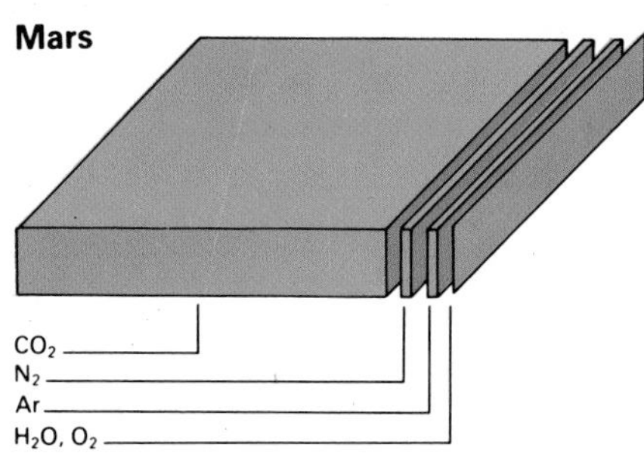

▲ ***The atmosphere of Mars consists largely of carbon dioxide (95 percent), but is very tenuous.***

▼ ***In the thin Martian troposphere, temperature falls with increasing height, while absorption of solar ultraviolet produces only a slight warming in the thermosphere. Clouds of water droplets occur at low levels, where dust is also always present. At a height of tens of kilometers there is a thin haze of water and carbon dioxide crystals.***

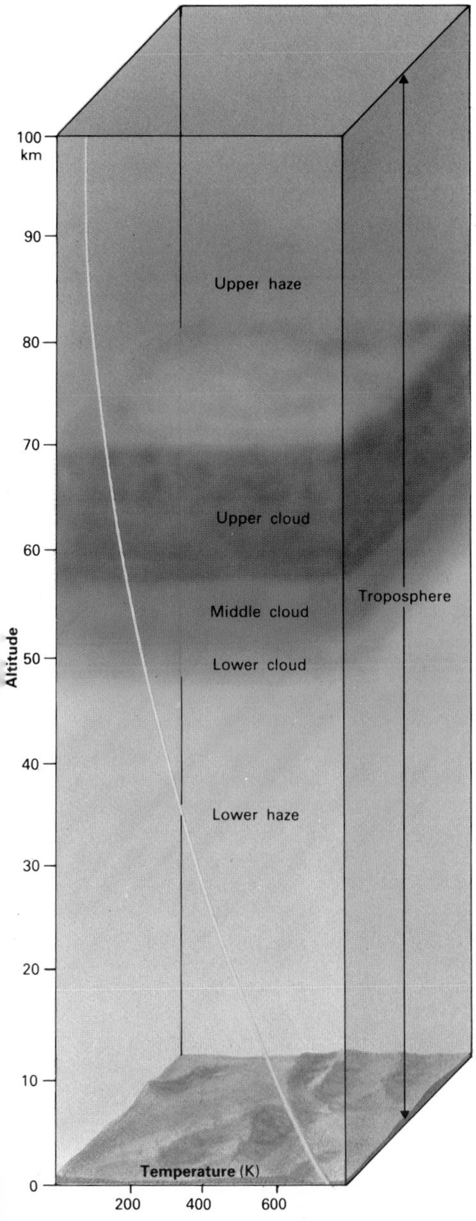

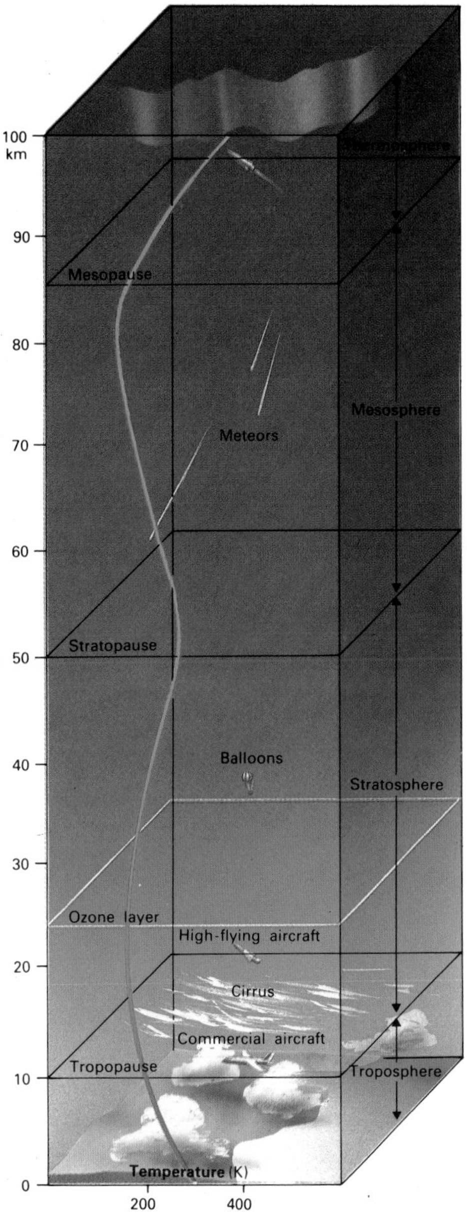

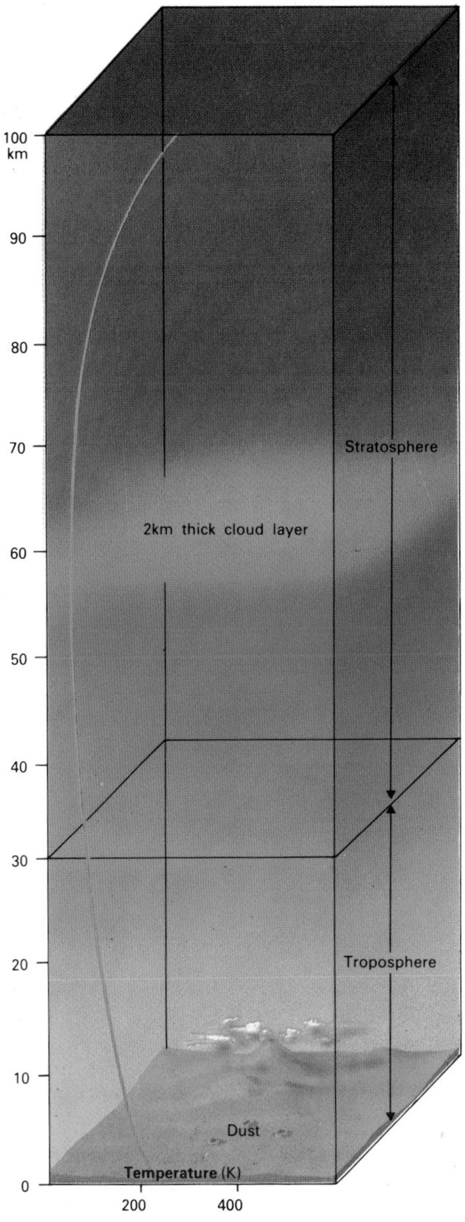

Venus

The dense atmosphere of this hostile world is composed mainly of carbon dioxide (which makes up 96 percent of the total) and exerts a pressure of over 90 atmospheres. The main cloud layer, which consists largely of droplets of sulfuric acid, spans a range of altitudes from about 45km to 60km. A given thickness of Venusian cloud absorbs less light than the same thickness of terrestrial cloud, but the Venusian cloud layer is so much thicker that only about two percent of incoming sunlight reaches the surface of the planet.

The Venusian "greenhouse" traps outgoing infrared radiation so effectively that the surface temperature remains about 750K – nearly 500K higher than it would be in the absence of an atmosphere. Many astronomers believe that at a much earlier period Venus had water on its surface, but that an increase in the luminosity of the Sun raised the temperature enough to start evaporating the oceans, increasing the quantity of water vapor in the atmosphere. Since water vapor is a good absorber of infrared, the greenhouse effect was enhanced, further raising the temperature and speeding up evaporation.

As a result of this "runaway" greenhouse effect, the temperature soared and the oceans evaporated completely. Sunlight must then have broken down the water vapor in the upper atmosphere into hydrogen, which escaped, and oxygen, which combined with surface rocks. The proportion of water vapor in the atmosphere is now only 0·005 percent, but even so it accounts for about 25 percent of the present Venusian greenhouse effect. The other major contributors are carbon dioxide, sulfur dioxide and clouds and haze.

The temperature falls with altitude in the troposphere to a minimum of 180K at 100km. Above this, on the daylight side, solar radiation heats the thermosphere to about 300K, but over the dark hemisphere the temperature of the upper atmosphere drops to about 100K.

The seasons

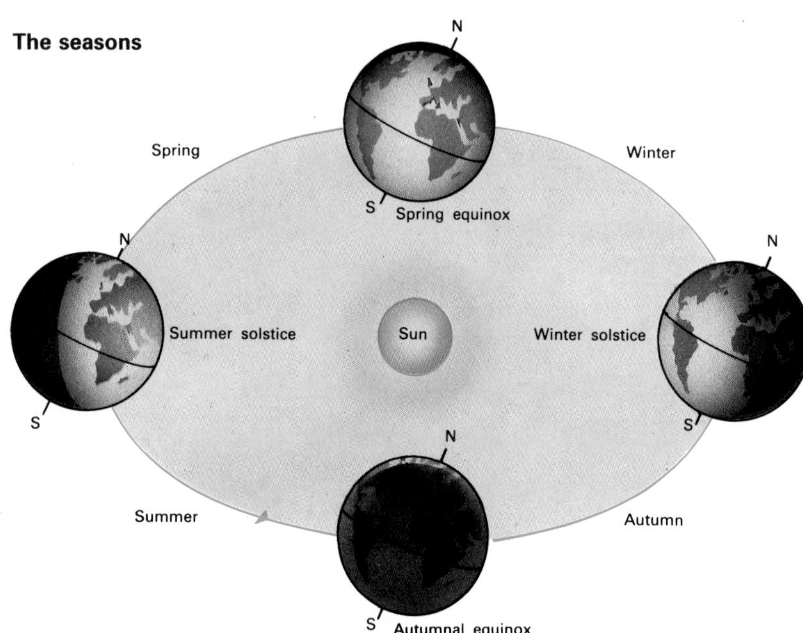

As the Earth orbits the Sun, the tilt of its axis gives rise to the seasons. Spring equinox: the Sun is overhead at the equator. Summer solstice: the Sun is overhead at the Tropic of Cancer (23·4°N). Autumnal equinox: the Sun crosses the equator again. Winter solstice: the Sun is overhead at the Tropic of Capricorn (23·4°S).

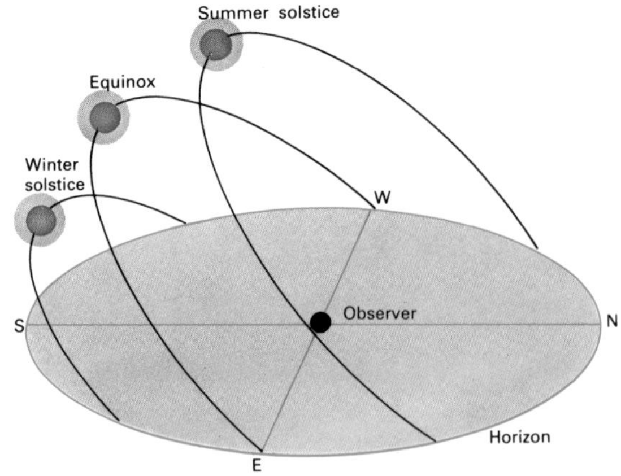

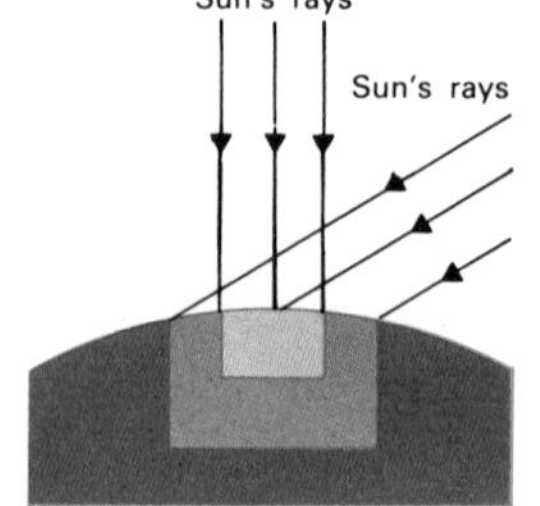

▲ The apparent track of the Sun in midwinter, spring and autumn, and midsummer, shows how the noon elevation and the duration of daylight vary with the seasons.

◄ When the Sun is at a low angle its energy is diluted over a wide area. At a high angle it is concentrated.

Seasons on Earth...

Seasonal effects arise because the tilt of the Earth's axis (23·4°) combined with the motion of the planet round the Sun causes periodic variations in the amount of sunlight reaching the northern and southern hemispheres.

...on Venus...

The Venusian "greenhouse" is so efficient that at the surface and in the lower levels of the atmosphere there are no significant temperature differences between equator and poles, nor between day and night. Because the tilt of the planet's axis is negligible there are no seasonal effects either – the "weather" is uniformly torrid at all times.

...and on Mars

Because the axial tilt of Mars (24°) is almost identical to that of the Earth, the Martian seasons follow the same pattern. However, the orbit of Mars is more elliptical than Earth's, and the input of solar energy is 40 percent greater at perihelion (closest approach to the Sun) than at aphelion (farthest point). This exaggerates the seasonal effects on Mars. A Martian year is almost twice as long as an Earth-year, so each season also lasts much longer than its terrestrial equivalent.

At each pole there is a permanent cap of water ice, but in the winter hemisphere the temperature over the cap is well below the freezing point of carbon dioxide (148K) which, therefore, forms a seasonal layer of frost and ice that extends over and beyond the permanent cap. About 30 percent of the atmospheric carbon dioxide is frozen out in the winter hemisphere, and this causes a drop in atmospheric pressure. With the coming of "spring" and "summer" the cap shrinks and the carbon dioxide returns to the atmosphere.

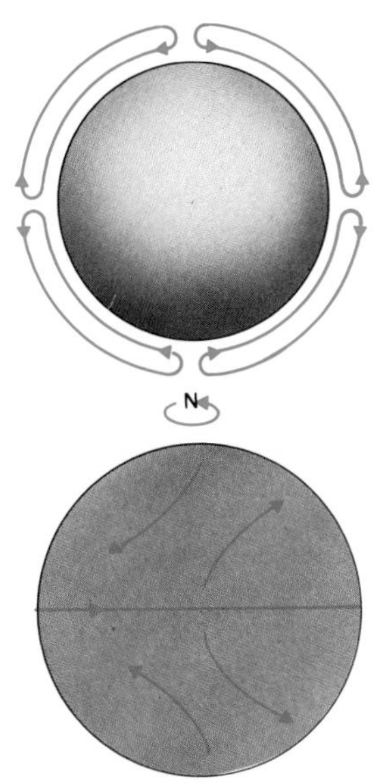

Atmospheric circulation on Earth

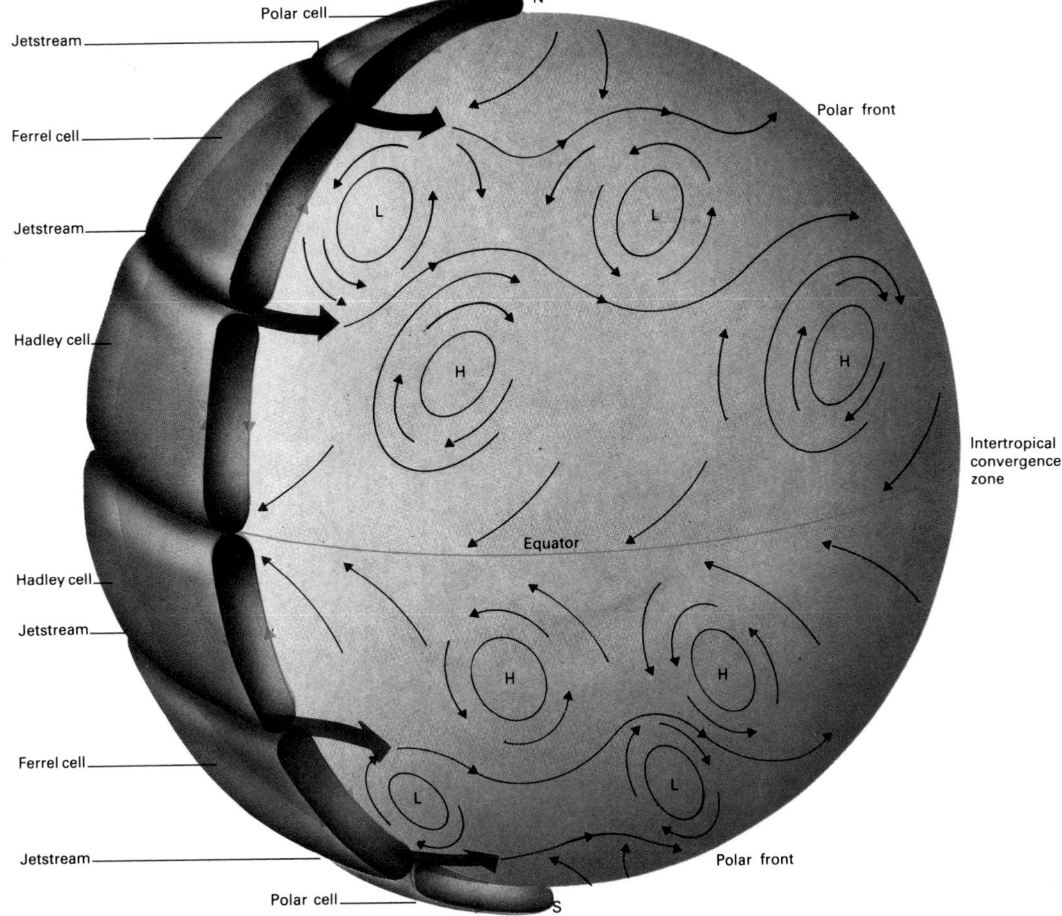

▲ The simplest form of atmospheric circulation is a Hadley cell, in which rising warm air moves polewards and sinking cool air returns to the equator (top). The effect of rotation is to deflect air parcels to the right in the northern hemisphere and to the left in the south.

► The Earth's circulation broadly contains three cells in each hemisphere. It breaks up into turbulent eddies in middle latitudes.

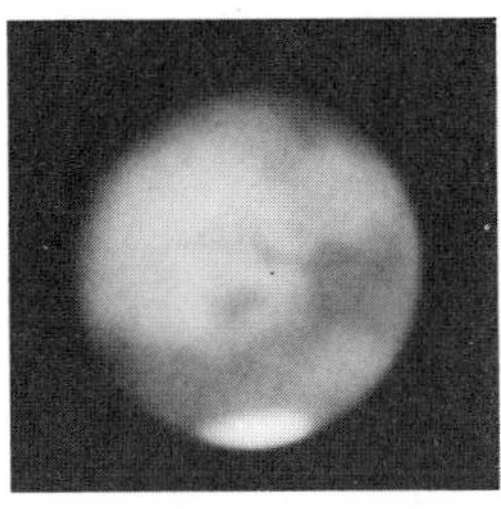

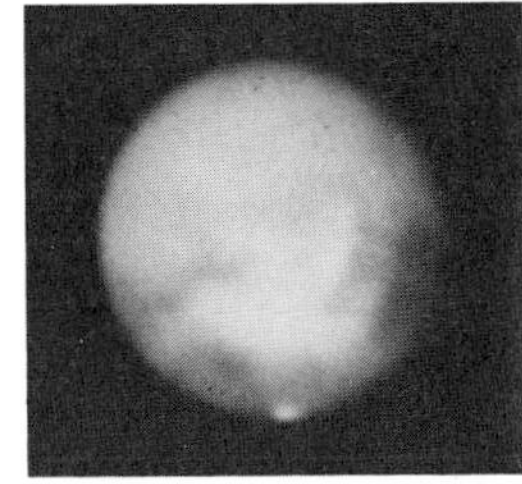

▲ The visible face of Mars can alter dramatically when violent dust storms sweep the planet. These pictures were taken at the Lowell Observatory in August 1971 (left) and October 1973.

▼ The north polar ice cap of Mars dwindles at a variable rate through spring and summer. Vertical scale is latitude of the cap edge; horizontal scale gives longitude of the Sun.

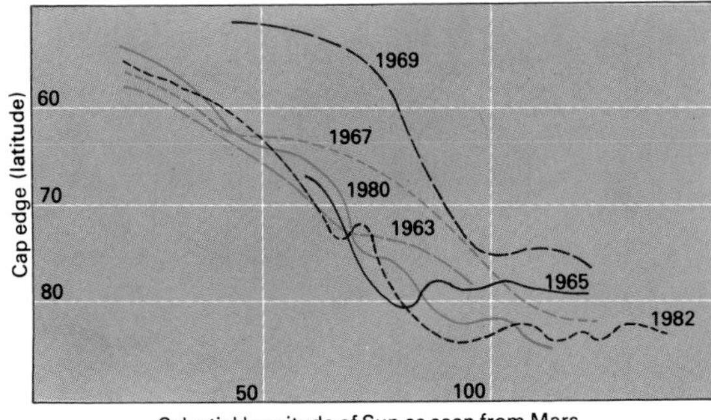

Mars

The rarefied Martian atmosphere consists mainly of carbon dioxide (95 percent) with nitrogen and argon making up most of the rest. There is a small proportion of water vapor – about 0·03 percent – but even if it were possible to liquefy it all, it would produce a surface water layer only a few tens of micrometers thick. The atmospheric pressure is less than 0·01 atmospheres. The surface is so uneven that pressure ranges from 9mb in deep depressions to about 1mb on high mountain tops, but a typical value is 6-7mb.

This tenuous atmosphere raises the surface temperature by less than 10K above the value it would have without an atmosphere. Mean temperature is about 220K, but it ranges from as low as 132K at the winter pole to as high as 300K at the equator. The variation between daytime and night-time temperatures can be as much as 60K.

At one time the atmosphere could well have been sufficiently thick to allow liquid water to exist. The volcanoes must have been a major source of atmospheric water, but the low escape velocity of Mars (5 kilometers per second) allowed gases to leak away quite rapidly.

Atmospheric engines on the terrestrial planets

It is the temperature difference between equator and poles that drives atmospheric circulation. Warm equatorial air flows towards the pole and cooler polar air flows towards the equator. The flow is very complex, particularly on Earth where, in the middle latitudes, it breaks up into turbulent eddies and wedges of cool and warm air intermingle.

Cloud and wind systems on Earth...
Two unique features of the Earth are the presence of oceans of water, and the existence of large quantities of free oxygen in the atmosphere. The oceans act as immense reservoirs of heat which smooth out global temperature variations. Energy is continually exchanged between the oceans and the atmosphere, within which clouds play a major role in the distribution of water vapor. Created by atmospheric motion, large-scale cloud systems reflect the movements of air masses and winds. The intensity of the winds often produces cyclones of up to 240 kilometers per hour.

...on Venus...
Winds at the surface of Venus are sluggish (a few kilometers per hour) but they rise sharply in the cloud deck to around 400 kilometers per hour. The uppermost clouds circle the planet in a period of about four days (60 times faster than the planet itself rotates), and this rapid rotation, combined with the flow of high-altitude air from the equator to the poles, produces characteristic Y- and C-shaped cloud patterns and raised polar "collars".

...and on Mars
The large temperature differences between the Martian equator and poles drive a strong atmospheric circulation. In both hemispheres the contrasts are reduced in summer (when the pole is tilted towards the Sun) and enhanced in winter. In spring the large temperature differences across the boundary of the retreating polar cap can produce particularly strong winds, which raise large quantities of dust high into the atmosphere. The dust absorbs sunlight and this heats the atmosphere, which in turn amplifies the winds and whips up yet more dust. Winds of up to 300 kilometers per hour can arise and spread the dust out to envelop the entire planet, so that it takes months to settle again. Atmospheric dust gives the Martian sky its pinkish hue.

Although cloud cover is generally small, clouds of water, water ice and carbon dioxide occur. Banks of cloud and haze lie over the poles in winter, and mists and hazes form at night in low-lying regions.

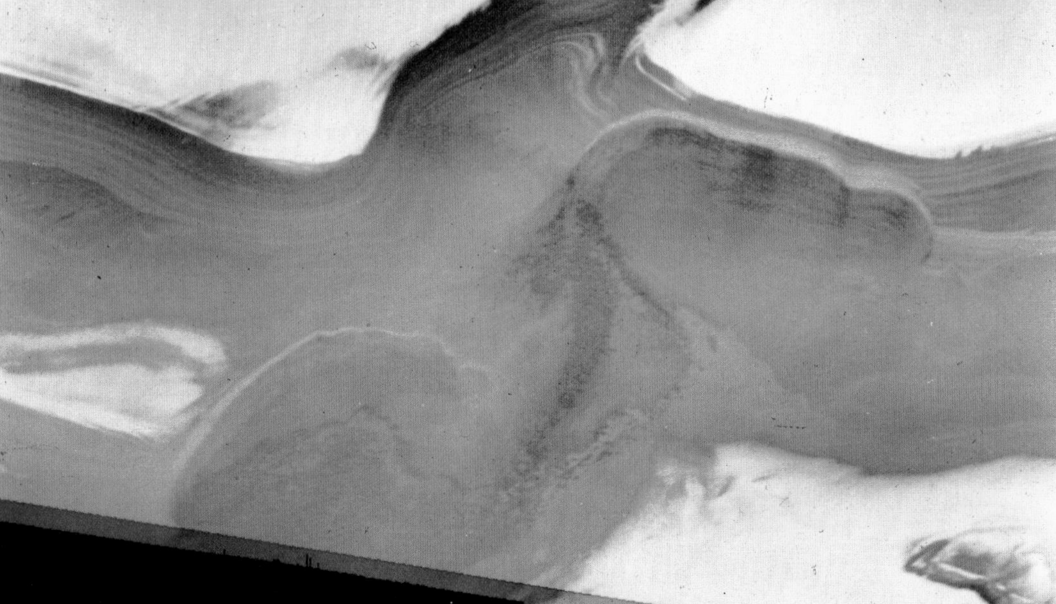

◄ Part of the Martian north polar region, taken by Viking Orbiter 2, reveals the water ice and layered terrain that emerge in midsummer, when the carbon dioxide cap has retreated.

► The irregular orange mass at the top and just left of center is a Martian dust cloud blowing around in Argyre Planitia. It is more than 300km across, and appears to be moving east.

▲ *Cloud vortices such as these over Guadalupe Island are rare on Earth, even in the tropics. These were photographed from Skylab in 1973.*

◀ *The broad sweep of the Earth's weather pattern is frequently interrupted by violent local events – such as Hurricane Gladys (1968), photographed from Apollo 7.*

◀ *Characteristic dark "Y" and "C"-shaped cloud forms encircle Venus. This picture shows the tail of a "Y", the arms of which spanned the planet two days earlier.*

▼ *The north pole is at the center of this infrared photograph of Venus. Blue areas are cold and red areas hot. The Sun was at the six o' clock position.*

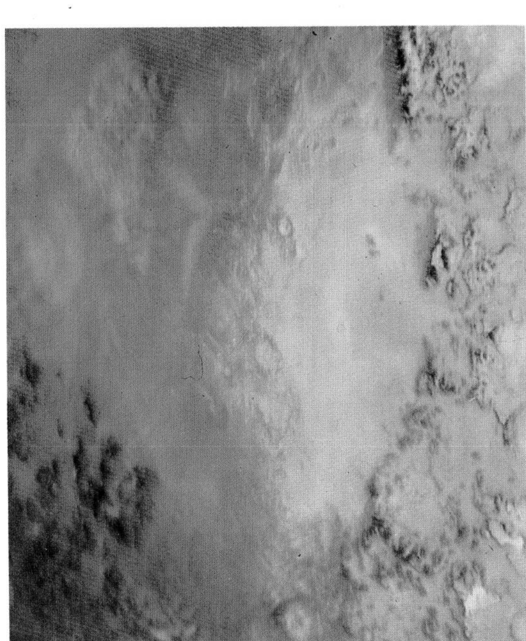

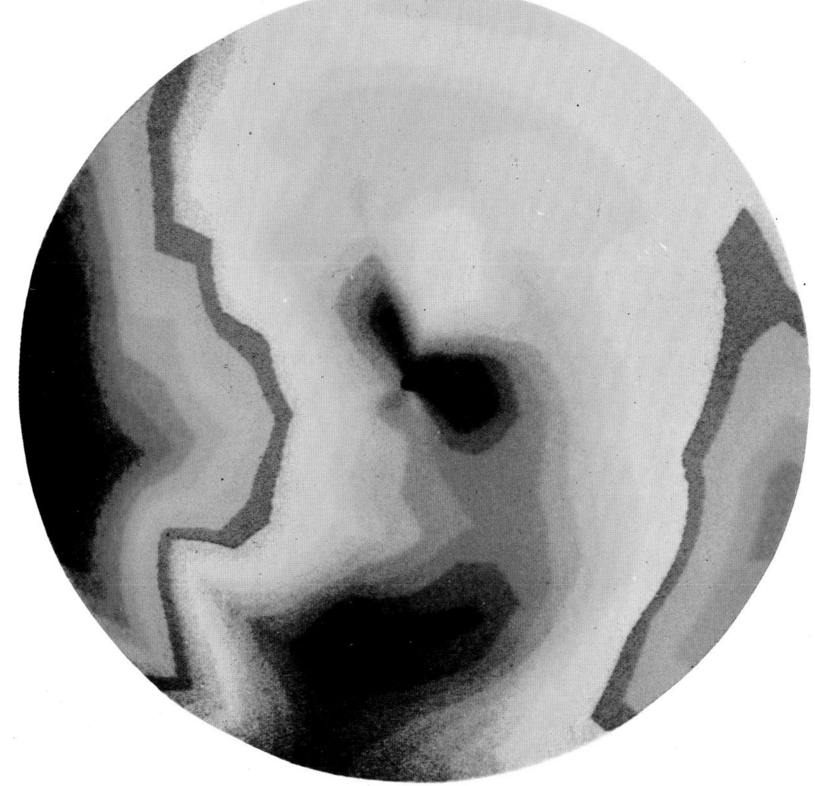

Jupiter, the largest planet, has the shortest "day"

Temperature gradient

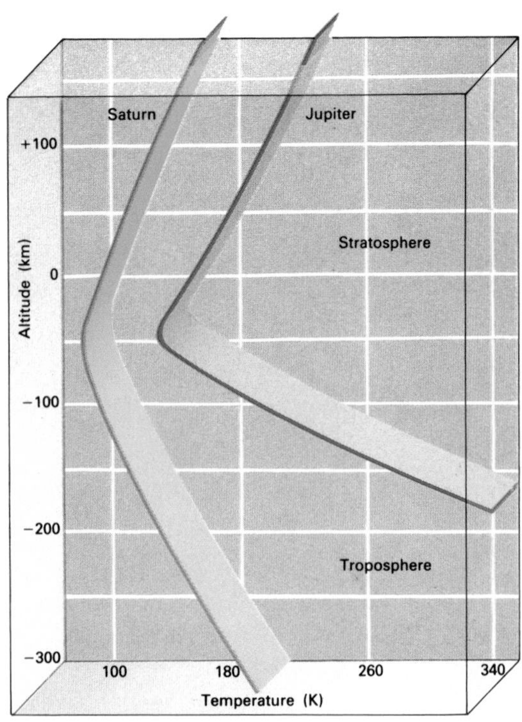

▲ ***The graph shows the vertical variation in temperature through the upper atmospheres of Saturn and Jupiter. Because neither planet has a solid surface from which to measure altitude, heights are referred to the level at which the atmospheric pressure is 100mb (0·1 Earth atmospheres).***

Cloud layers

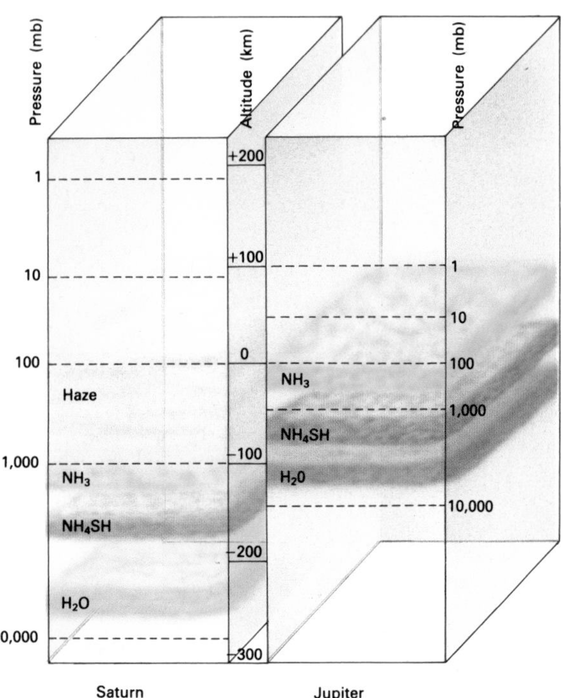

▲ ***From highest to lowest, the three main cloud layers are believed to contain crystals of ammonia (NH_3), ammonium hydrosulfide (NH_4SH) and water (H_2O). Saturn's atmosphere is cooler than Jupiter's and its atmospheric pressure increases more slowly with increasing depth.***

Jupiter

Besides differences of composition, mass and temperature, the Jovian atmosphere contrasts sharply with those of the terrestrial planets in that Jupiter itself has no solid surface with which it can interact, and the planet also emits twice as much heat as it receives from the Sun. Internal heating is a major contributor to the forces which drive the circulation of the Jovian atmosphere. The temperature of the planet is the same all over, the polar regions emitting the same amount of heat as the equatorial regions.

The yellowish flattened disk of Jupiter is crossed by alternating bright zones and darker bands of cloud, the belts and zones being labelled according to their latitudes. There is considerable structure, with waves, plumes and eddies testifying to the turbulence of the clouds. Longer-lasting oval spots also appear, the most famous of these being the Great Red Spot, which measures 30,000-40,000km long by about 14,000km wide and is centered on latitude 22° south.

Since Jupiter has no rigid surface, the planet does not behave in the same way as a solid body: its rotation period varies with latitude, being about five minutes shorter at the equator than at the poles.

Composition and structure

The atmosphere of Jupiter consists mainly of hydrogen and helium, by volume 90 percent and 10 percent respectively, together with hydrogen compounds such as ammonia and methane. Within the troposphere the temperature decreases with altitude at an average rate of 2K per kilometer, reaching a minimum of 105K at the tropopause, where the pressure is about 0·1 atmospheres. The top of the main cloud deck lies about 30km below the tropopause. Seventy kilometers further down, at the base of the cloud layer, the temperature is about 300K and the pressure about 5 atmospheres. At a depth of 1,000km, where the atmosphere meets the liquid interior, the temperature may reach about 2,000K. Above the tropopause there is a stratosphere in which the temperature rises with altitude, mainly because the methane absorbs sunlight. Above this layer the temperature falls at first, then rises to about 1,000K.

Saturn

The composition of Saturn's atmosphere is broadly similar to that of Jupiter except that there is less helium (6 percent compared to 10 percent). Because it is less massive than Jupiter, Saturn has cooled more rapidly, and this has allowed droplets of helium to condense. These fall like rain into the deep interior, depleting the atmosphere of helium and heating the interior as they descend. This probably explains why Saturn emits just over twice as much heat as it receives.

The atmospheric temperature drops with increasing height, reaching a minimum of about 90K at the tropopause. Thereafter methane absorbs sunlight, producing a slight warming in the stratosphere.

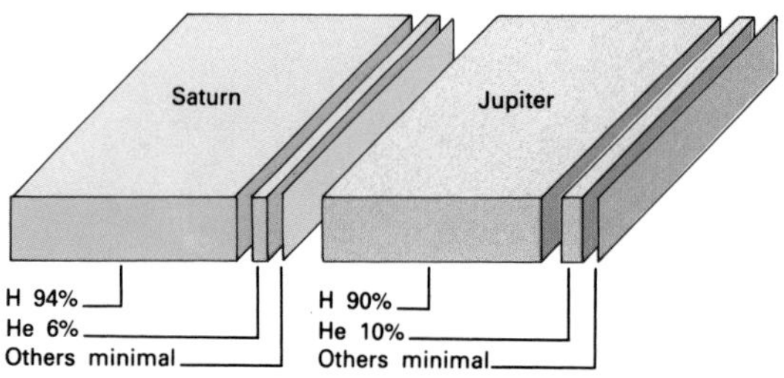

◀ ***The proportions by volume of the main atmospheric constituents of Saturn (left) and Jupiter (right) are compared. Hydrogen and helium are most abundant, but Saturn appears to have significantly less atmospheric helium than Jupiter.***

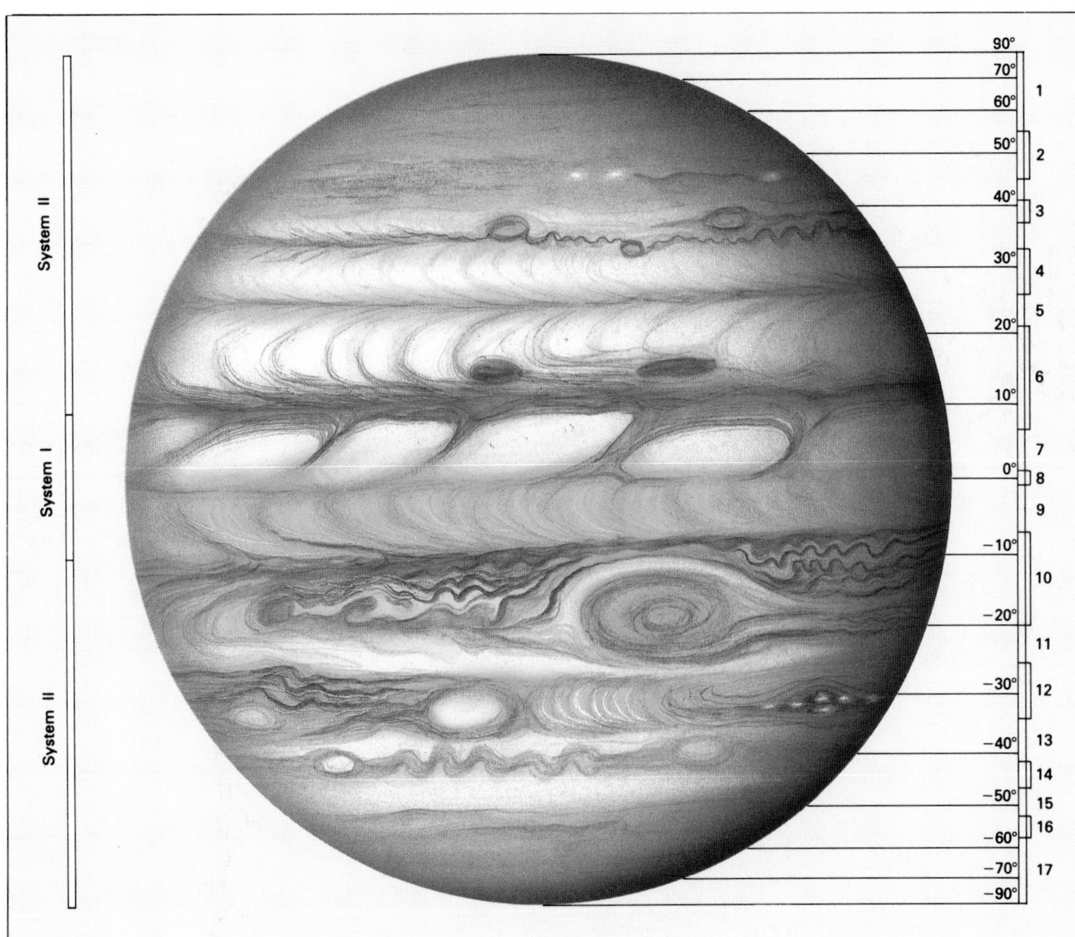

◄*Jupiter's zones and belts. For convenience, observers have assigned three different rotation periods, or systems. System I, a period of 9 hours, 50 minutes and 30·003 seconds, applies to features within about 9° of the equator. System II, a period of 9 hours, 55 minutes and 40·632 seconds, is assigned to the slower-moving clouds farther from the equator. System III is based on measurements of Jupiter's radio emissions, which give a period of 9 hours, 55 minutes, 29·7 seconds.*

1 *North Polar Region*
2 *North North North Temperate Belt*
3 *North North Temperate Belt*
4 *North Temperate Belt*
5 *North Tropical Zone*
6 *North Equatorial Belt*
7 *Equatorial Zone*
8 *Equatorial Band*
9 *Equatorial Zone*
10 *South Equatorial Belt*
11 *South Tropical Zone*
12 *South Temperate Belt*
13 *South Temperate Zones*
14 *South South Temperate Belt*
15 *South South South Temperate Zone*
16 *South South South Temperate Belt*
17 *South Polar Region*

◄ *Saturn's zones and belts. The same system of rotation periods has been assigned to Saturn. The rotation period at the equator – System I – is 10 hours, 15 minutes. The period at higher latitudes – System II – is 10 hours 38 minutes. Like Jupiter, various definite features, such as well marked spots, have rotation periods of their own, so that they drift about in longitude. Saturn's features are far less marked, however, although the two main belts – the North Equatorial and the South Equatorial – can always be distinguished. System III, based on radio emissions, is 10 hours, 39·4 minutes. Like Jupiter, the radio emissions are linked to Saturn's magnetic field, and since this is tied to the interior of the globe, System III is taken to be the "true" rotation period of the planet.*

1 *North Polar Region*
2 *North Temperate Zone*
3 *North Temperate Belt*
4 *North Tropical Zone*
5 *North Equatorial Belt*
6 *Equatorial Zone*
7 *South Equatorial Belt*
8 *South Tropical Zone*
9 *South Temperate Belt*
10 *South Temperate Zone*
11 *South Polar Region*

Jupiter's Red Spot, once believed to be an active volcano, is now known to be a vast, rotating storm

► Jupiter and three of the four Galilean satellites (Io crossing the face of the planet, Europa on the right and Callisto in the lower left hand corner), as seen by Voyager 1.

◄ The Great Red Spot as it appeared when Voyager 1 approached Jupiter in February 1979.

▼ Five months later Voyager 2's similar view shows changes in the white cloud regions, a different white oval, and cloud vortices forming from the wave structures to the left.

Jupiter's distinctive features

Observers have distinguished three main levels of cloud in the Jovian atmosphere. The white zones are the highest and coldest, with temperatures of around 140K. They probably consist of crystals of ammonia ice. The brown clouds of the belts are warmer – about 230K – and lie 30-40km below the ammonia cloud tops. It is likely that they contain ammonium hydrosulfide, a compound of ammonia and hydrogen sulfide which turns brown when exposed to sunlight. The deepest clouds appear bluish and may consist of water ice and dissolved ammonia.

The bright zones are regions where convection is driving warm moist air upwards, and the belts are regions were cool dry gases are descending. As a result of the rapid rotation of the planet, the north-south flow of gases spilling out from the zones is diverted into strong east-west jetstreams with speeds of about 360 kilometers per hour along the boundaries between belts and zones. Astronomers are not sure what sustains the belt-zone pattern. According to one theory, in a rapidly rotating fluid body, concentric cylindrical convection cells will form round the axis of rotation. Where these meet the atmosphere, the belts and zones will appear. On the other hand, computer simulations can reproduce the Jovian belt-zone pattern quite well simply by scaling up the Earth's weather systems and taking account of the very rapid rotation. Perhaps very similar basic processes are at work.

The Great Red Spot seems to be an anticyclonic high-pressure feature rotating anticlockwise in a period of 6 days. The presence of phosphine may account for the color. Scientists believe that convection dredges the phosphine up from a great depth, and that ultraviolet radiation from the Sun breaks it down to release red phosphorus.

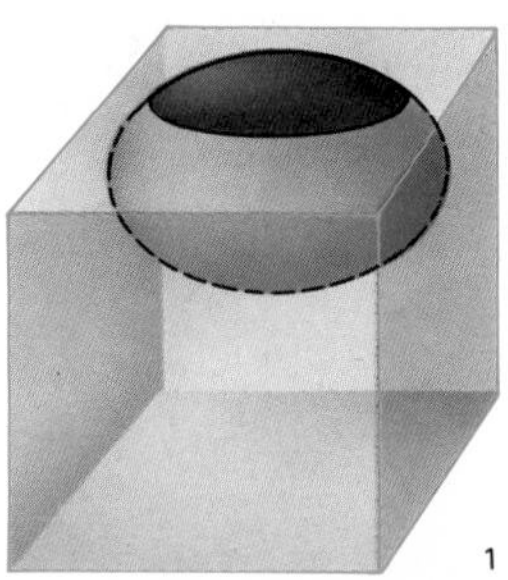

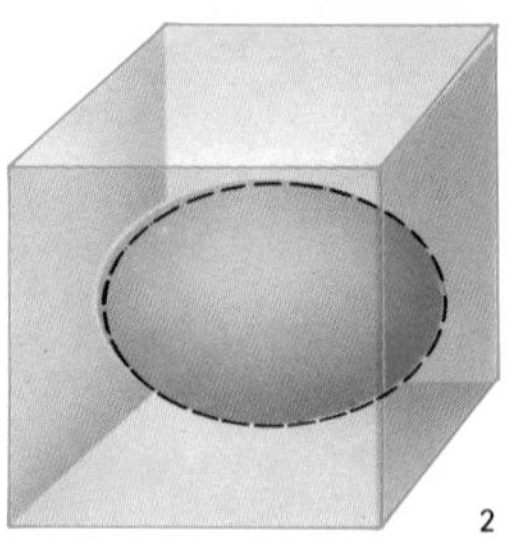

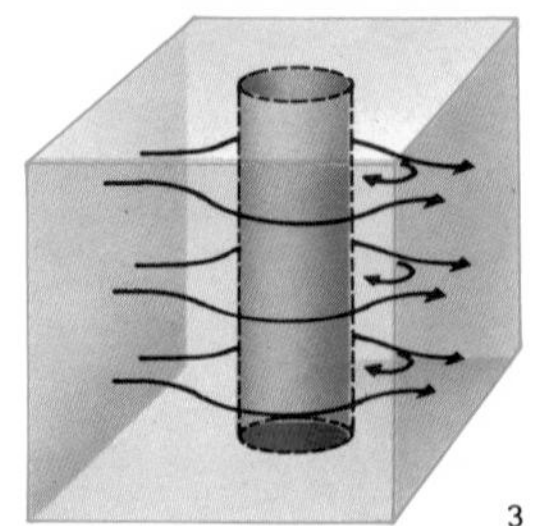

◄ *Early theories of the Red Spot. The "floating raft": a solid or semi-solid body floating in Jupiter's atmosphere (1). When the body sank, the Spot would be covered up and vanish temporarily (2). The "Taylor column": the spot was the top of a column of stagnant gas formed above a surface obstruction interrupting the atmospheric flow (3).*

The Great Red Spot: early theories

Considerable surface detail is visible on Jupiter, even in a small telescope, and from the early days of telescopic astronomy there had never been any serious doubt that the surface was gaseous rather than solid. However, there were many outstanding problems, one of which concerned the nature of the Great Red Spot. Unlike the smaller spots which are short-lived, it has been under observation for several centuries. It sometimes disappears for a few months or a few years, but it always returns.

Early suggestions that it might be a red-hot volcano rising above the clouds were soon rejected, and there was considerable support for a theory proposed by Bertrand Peek (1891-1965), an English amateur astronomer who was one of the best planetary observers of his time. Peek drew an analogy with the behavior of an egg dropped into a tall vase of water. The egg will sink to the bottom, but adding salt to the water increases its density and causes the egg to rise. Peek suggested that the Great Red Spot was a solid or semisolid body floating in Jupiter's outer gases. If the gases became denser, the Spot would rise; if they became less dense, the Spot would sink out of view. This theory was quite plausible, but the close-range results from spacecraft missions have recently disproved it.

Recent discoveries

The Pioneer and Voyager missions have now shown the Great Red Spot to be a whirling storm. Its longevity seems to be due to its exceptional size. It may not be permanent, and there is evidence that it is smaller now than it used to be. It has certainly grown smaller since 1878, in which year it became exceptionally prominent.

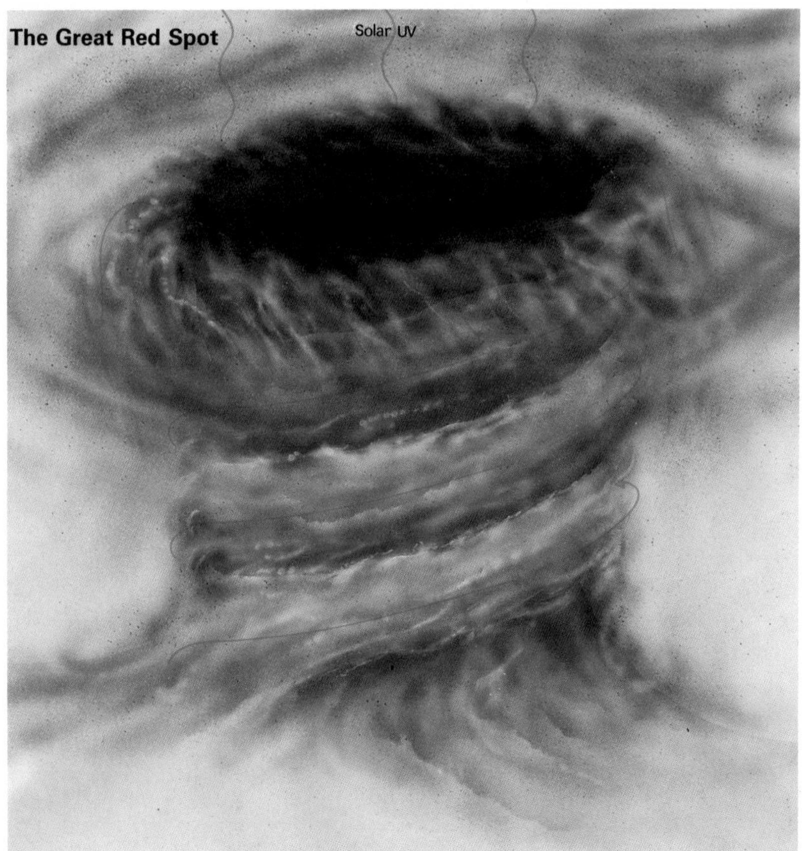

◄ *The Great Red Spot seems to be a high-pressure region, its top standing about 8km above the surrounding cloud layer. The material within it circulates in an anticlockwise direction, which would be normal for a high-pressure system in the southern hemisphere. It seems likely that material flows up from lower atmospheric levels to the top of the Spot, then spreads out and descends at its edges.*

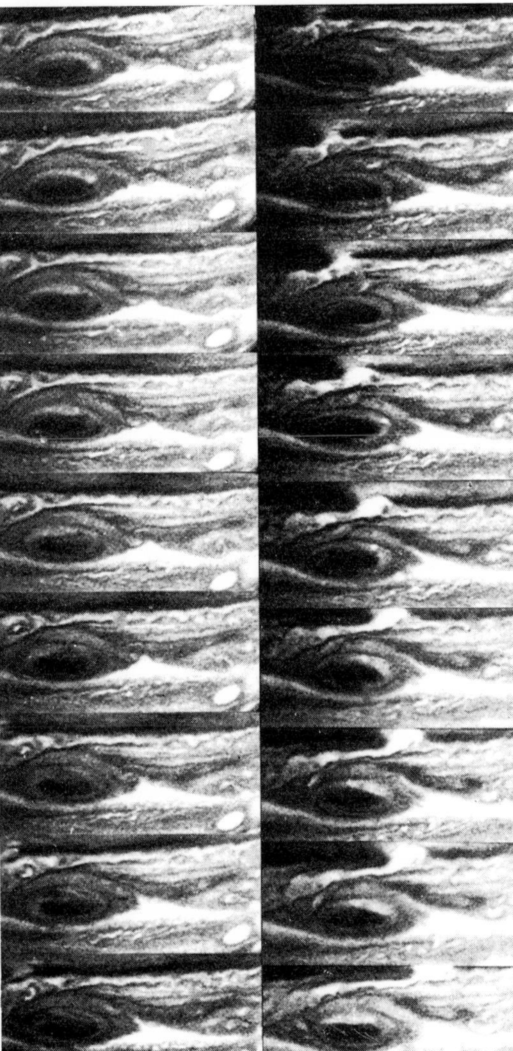

▲ *A time-lapse sequence of photographs reveals the flow of material around the Great Red Spot. Circulating currents to the left and wave-like regions to the right are clearly seen.*

Saturn's muted appearance

The appearance of Saturn itself is blander and more muted than the dramatic panorama presented by Jupiter. The belt-zone pattern is present, but much less conspicuous, and large cloud features are scarce. Circulating spots are there too, but on a smaller scale. Because of the planet's weaker gravitational attraction its atmospheric layers are less compressed than those on Jupiter, and the various cloud layers are more widely spaced in altitude. The clouds which comprise the dark belts lie at a greater depth than those in the Jovian atmosphere and, therefore, may be less conspicuous because they are more heavily obscured by haze. The lower temperature affects the types of chemical reactions which occur, and the rates at which they proceed. Consequently, the more conspicuous coloring agents may be absent from the Saturnian atmosphere.

Saturn has zonal winds which are much stronger than those on Jupiter. The equatorial jet, in particular, blows at a speed of 1,800 kilometers per hour – four to five times faster than the Jovian equivalent. The jets are broader than those on Jupiter and, curiously, do not seem to bear any close relationship to the less well-defined belt-zone boundaries. Their most notable feature, however, is that they are completely symmetrical about the equator, the winds in the southern hemisphere being a mirror-image of those in the north. A possible explanation for this is that the jets may extend down through the planet as a series of concentric cylinders, rather than being confined to the atmospheric layers. The fierce winds, eddies, waves and interacting spots which are present in the Saturnian atmosphere bear witness to the powerful convection of heat from the interior of this intriguing world.

Atmospheric structure of the giants

► *The liquid molecular hydrogen zones of Jupiter and Saturn may consist of concentric cylindrical cells which cause belts, zones and zonal winds where they meet the atmosphere.*

The belts and zones on Saturn are much less striking than those on Jupiter, because there is more overlying "haze"

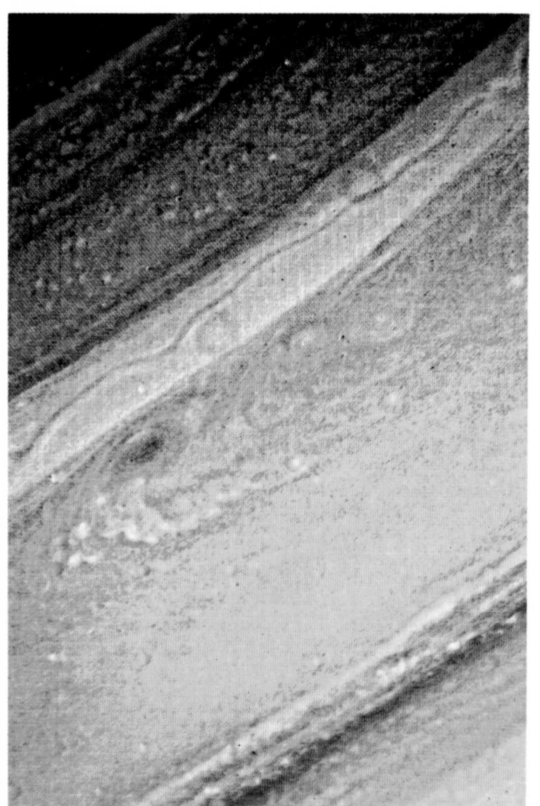

◄ Saturn's northern hemisphere in false color, taken by Voyager 2 in August 1981, the first (far left) from a range of 7·1 million kilometers, the second from 63,000km. Both contain evidence of vigorous activity within the atmosphere, including fast-moving wavelike features and convective zones among the stable ovals.

► Saturn as seen by Voyager 2 from distances of 21 million kilometers (top) and 43 million kilometers. The small white disks in the top picture are three of the planet's moons, and the dark spot is the shadow of one of them (Tethys). Both pictures clearly show the oblateness, or flattening, of the planet. The lower picture is in false color.

▼ This view of Saturn's northern hemisphere shows a convective cloud with a dark ring (light brown zone) and a longitudinal wave (light blue zone).

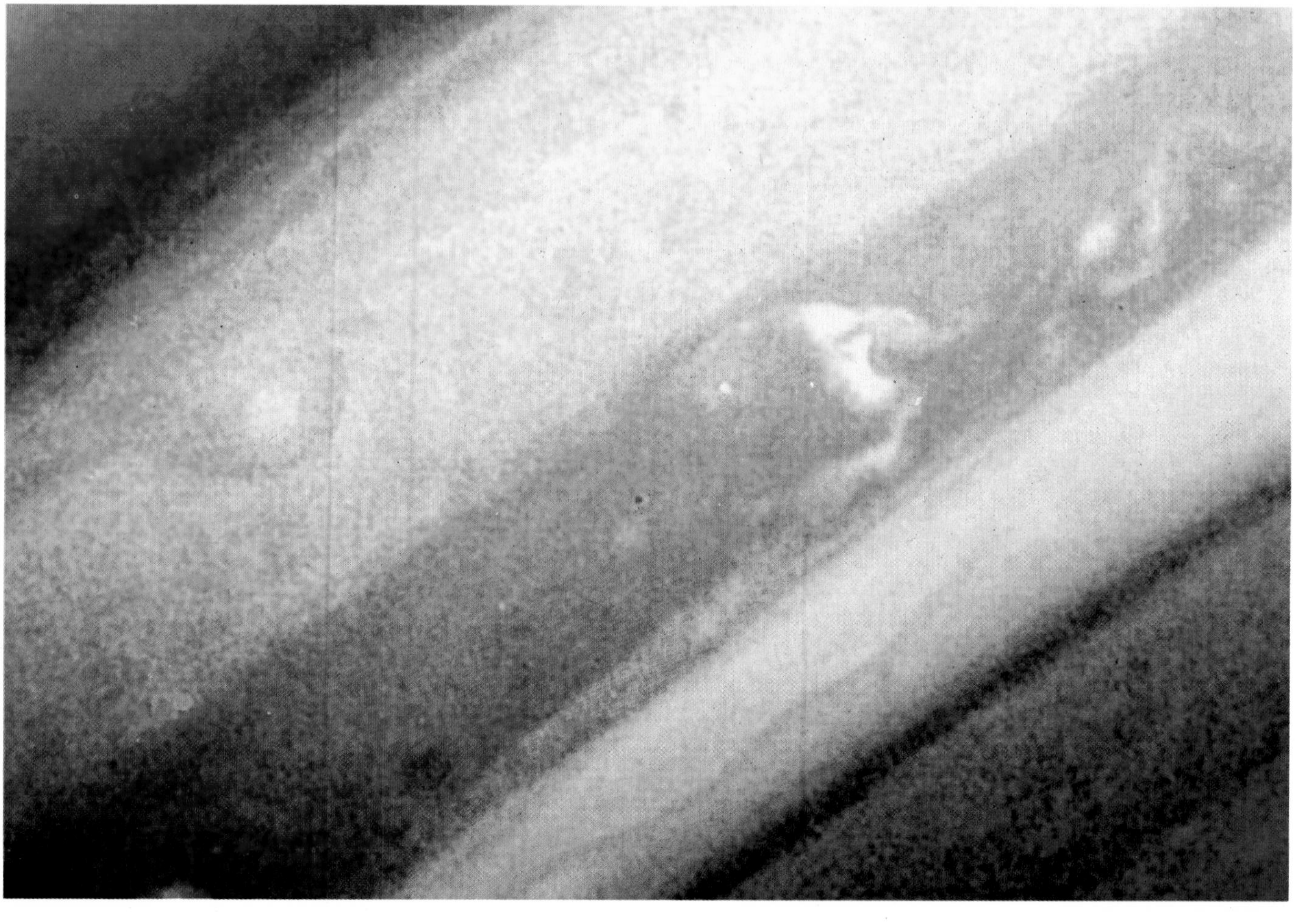

See also
Observing the Planets 21-32
Inside the Planets 33-40
Planetary Landscapes 41-56
Moons and Ring Systems 73-88
The Origin of the Solar System 121-8

The atmospheres of Uranus and Neptune

To an Earth-based observer both Uranus and Neptune appear as small, rather featureless disks, greenish in the case of Uranus and bluish in that of Neptune. Spectroscopic observations have revealed that hydrogen and methane are present in both atmospheres in large quantities. The effective temperature of the two planets is very nearly the same – about 58K – despite the fact that Neptune is half as far again from the Sun as Uranus is, and receives less than half as much sunlight. Neptune emits twice as much heat as it receives from the Sun and, like Jupiter and Saturn, must have an internal heat source which is an important driving force for atmospheric phenomena.

The lack of an internal heat source, together with its curious axial tilt, sets Uranus apart from the other Jovian planets. Because the axis is inclined by 98°, the Sun is almost directly over one of the poles at each solstice. Each pole, in the course of a Uranian year, receives a greater input of solar radiation than does the equator. This, coupled with the lack of a heat source, suggests that atmospheric circulation on Uranus is much weaker than on Jupiter and Saturn, and it does not have the belt-zone structure. At infrared wavelengths Uranus is brighter at the limb than near the center of the disk. This appearance may be due to a high-level layer of methane ice crystals. Near the center of the disk it is possible to see down to a cloud layer which lies at an estimated depth of 500km.

The atmosphere of Neptune is hazier and cloudier than that of Uranus, and infrared images reveal a dark equatorial band due to absorption of infrared by methane. The planet is shrouded in a haze of aerosols (fine particles) and ice crystals – a haze which varies with the level of solar activity. A plausible explanation is that variations in the solar output of particles, X-rays and ultraviolet radiation affect the atmospheric chemistry in an as yet unknown way.

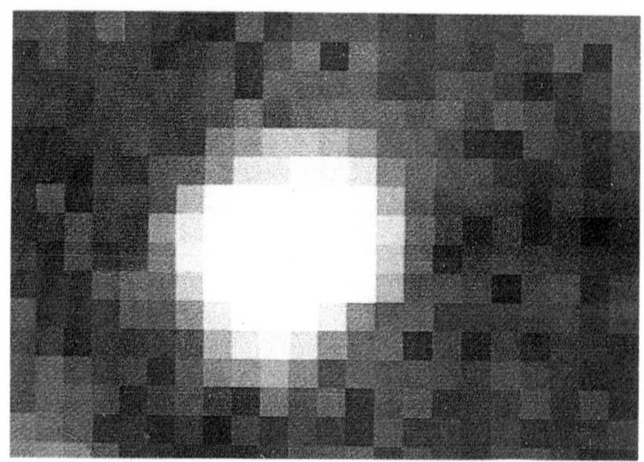

▲ Neptune, photographed from Arizona (top): the disk is brightest in the south. Taken in infrared by D. A. Allen and J. Crawford at Siding Spring, Australia (above), the planet appears extended along its polar axis, possibly by brightening from hazes at high latitudes.

▼ Uranus photographed in 1976 (below) shows no surface markings; the limb brightening results from high-altitude haze. Photographed by D. A. Allen and J. Crawford (right) methane absorption has dimmed the planet, making the rings visible.

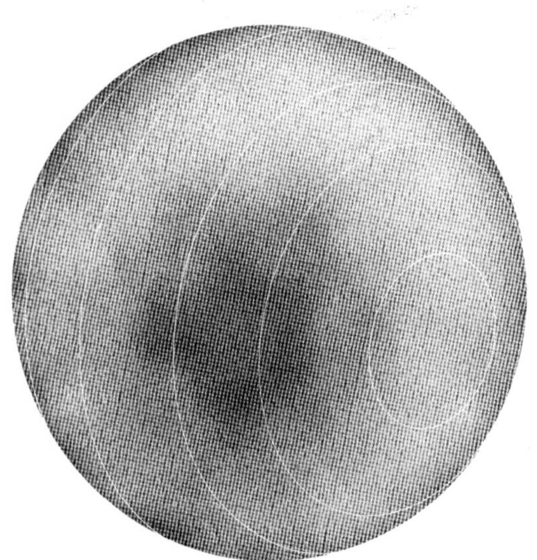

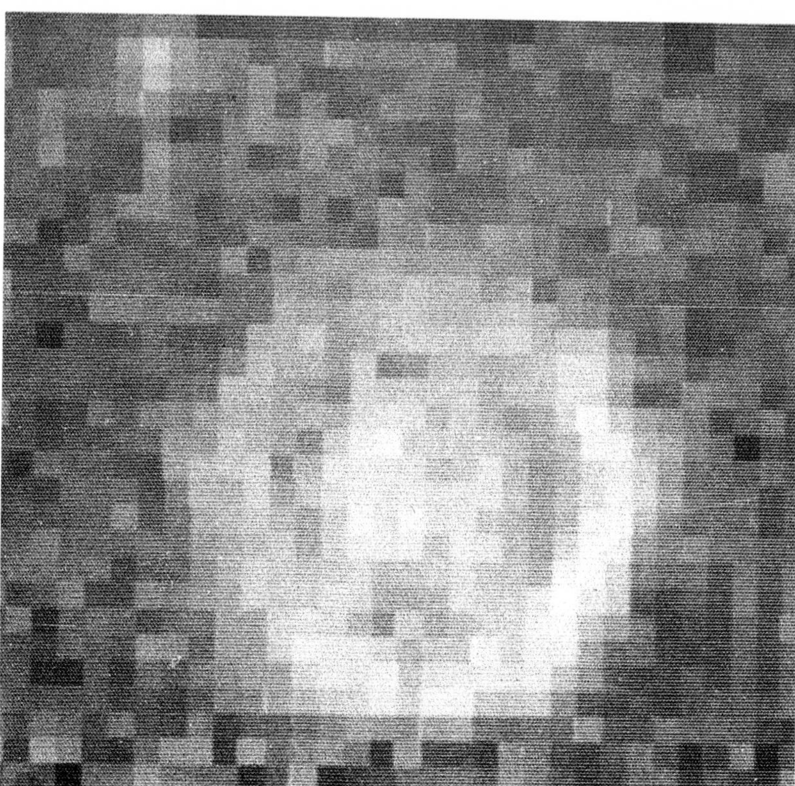

Moons and Ring Systems

6

Comparison of the planets' satellites...The Moon: its tidal effect on Earth...Jupiter's dramatically differing moons, and modest rings...Saturn's giant moon Titan and its smaller companions...Saturn's dramatic complex ring system...Rings and moons of the outer planets...PERSPECTIVE...The tiny Martian moons... Galileo's discovery of Jupiter's four largest moons... Early theories about Saturn's rings

All the planets, with the exception of Mercury and Venus, have natural satellites, or moons, which revolve around them. These range in size from planet-sized bodies, comparable with Mercury, to tiny irregular bodies like asteroids, a few kilometers in diameter. The Earth's Moon has been explored more thoroughly than the others (◀ pages 42-5). It is a rocky, cratered world, the surface displaying both the scars of bombardment by meteorites and the effects of volcanism. Other planetary satellites show a marvellous diversity of structure and form, ranging in composition from rock to ice and in appearance from heavily-cratered to glassy-smooth.

Three of the four giant planets have ring systems made up of small to moderate-sized particles and lumps of matter in orbit above their equators. Saturn's magnificent system is striking even through Earth-based telescopes. Jupiter and Uranus have thin faint rings on a much more modest scale; although not visible in optical telescopes, they can be detected from Earth at infrared wavelengths.

Satellites interact with their planets in various ways, an important one being the generation of tides. The Moon is mainly responsible for the ocean tides which rise and fall twice daily on Earth. A particle of water on the Moon-facing hemisphere is closer to the Moon than the center of the Earth's globe is, so it experiences a slightly stronger gravitational attraction and is accelerated towards the Moon, thus flowing into a bulge in the ocean arising on that side of the Earth. Conversely, a particle on the far side of the Earth experiences a weaker force of attraction than the body of the planet, and "falls behind", away from the direction of the Moon and into a second oceanic bulge. In this way the surface of the ocean assumes an ellipsoidal shape. Although the height of the bulge in mid-ocean is only about one meter, the situation in coastal water is complicated by local factors and a much greater tidal range can occur. The Sun exerts a weaker tide-raising force. When the Sun and Moon are in line (at New Moon and Full Moon) higher "spring" tides are raised. When Sun and Moon are pulling at right angles (First and Last Quarter) lesser "neap" tides arise.

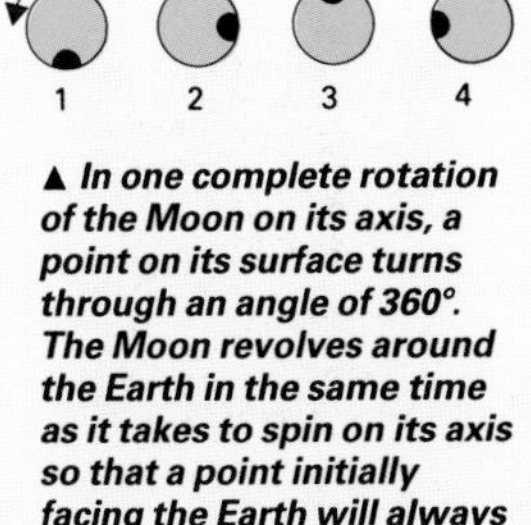

▲ In one complete rotation of the Moon on its axis, a point on its surface turns through an angle of 360°. The Moon revolves around the Earth in the same time as it takes to spin on its axis so that a point initially facing the Earth will always face towards the Earth.

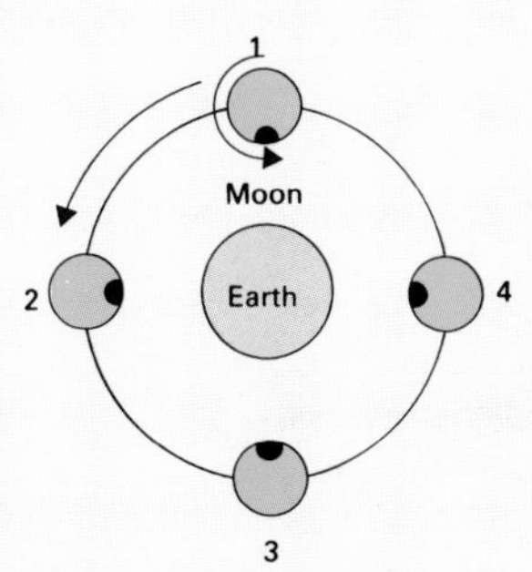

Orbital and rotational effects

It is not only the Earth's oceans that experience tidal effects. The Earth raises tides in the body of the Moon. In the past the effect of the Earth's gravitational attraction on the tidal bulges raised in the Moon was to slow down the Moon's axial rotation until – as now – the Moon rotates on its axis and revolves around the Earth in the same period of time, so keeping the same hemisphere turned always towards the Earth. This phenomenon of synchronous, or "captured", rotation is common to most planetary satellites.

Similarly, the effect of the Moon's gravitational attraction on the Earth's tidal bulges is to slow down the Earth's rotation. The effect is small, between 10 and 15 parts in 100 billion per year, and is equivalent to a lengthening of the Earth-day by about 0·02 seconds per century. Nevertheless, the length of the day will eventually be equal to the period of the Moon's orbital revolution. The energy of rotation lost by the Earth is transferred to the motion of the Moon, and as a result the Moon is gradually receding from the Earth at a rate of some 4·5cm per year.

Satellite-to-planet ratios

Tidal interactions are complex and usually affect the satellite far more than they do the planet. Only where the mass of the satellite is a significant fraction of that of the planet does its effect on the planet become significant. In proportion to the planetary masses, the most massive satellites are the Moon (1/81 that of the Earth) and Charon (1/10 that of Pluto). By contrast, the most massive satellite of Jupiter is less than 1/10,000 of that planet's mass.

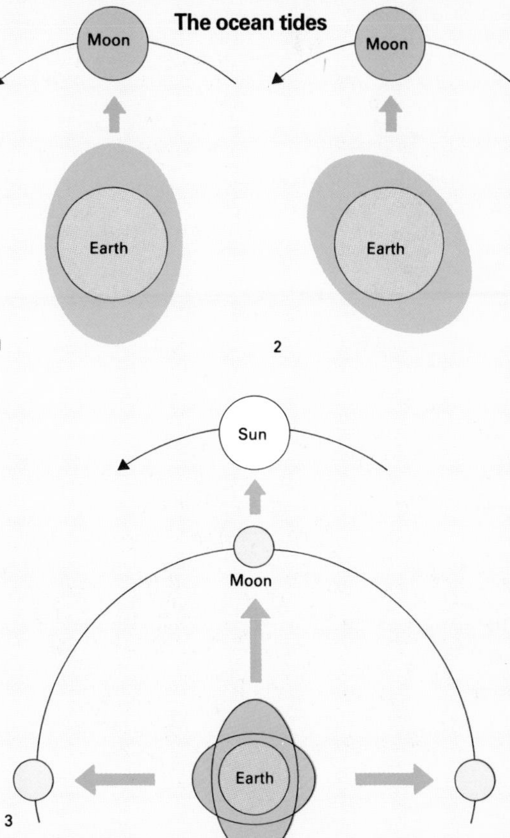

► 1 The Moon's attraction distorts the ocean surface into an ellipsoidal shape. The Earth's rotation carries a point on its surface past the two tidal bulges twice daily. 2 Frictional effects due to the Earth's rotation drag the tidal bulges ahead of the Earth-Moon line. 3 Tidal bulges are highest when Sun and Moon are in line and least when they are at right angles.

Most satellites follow a near-circular orbit close to the plane of their planet's equator

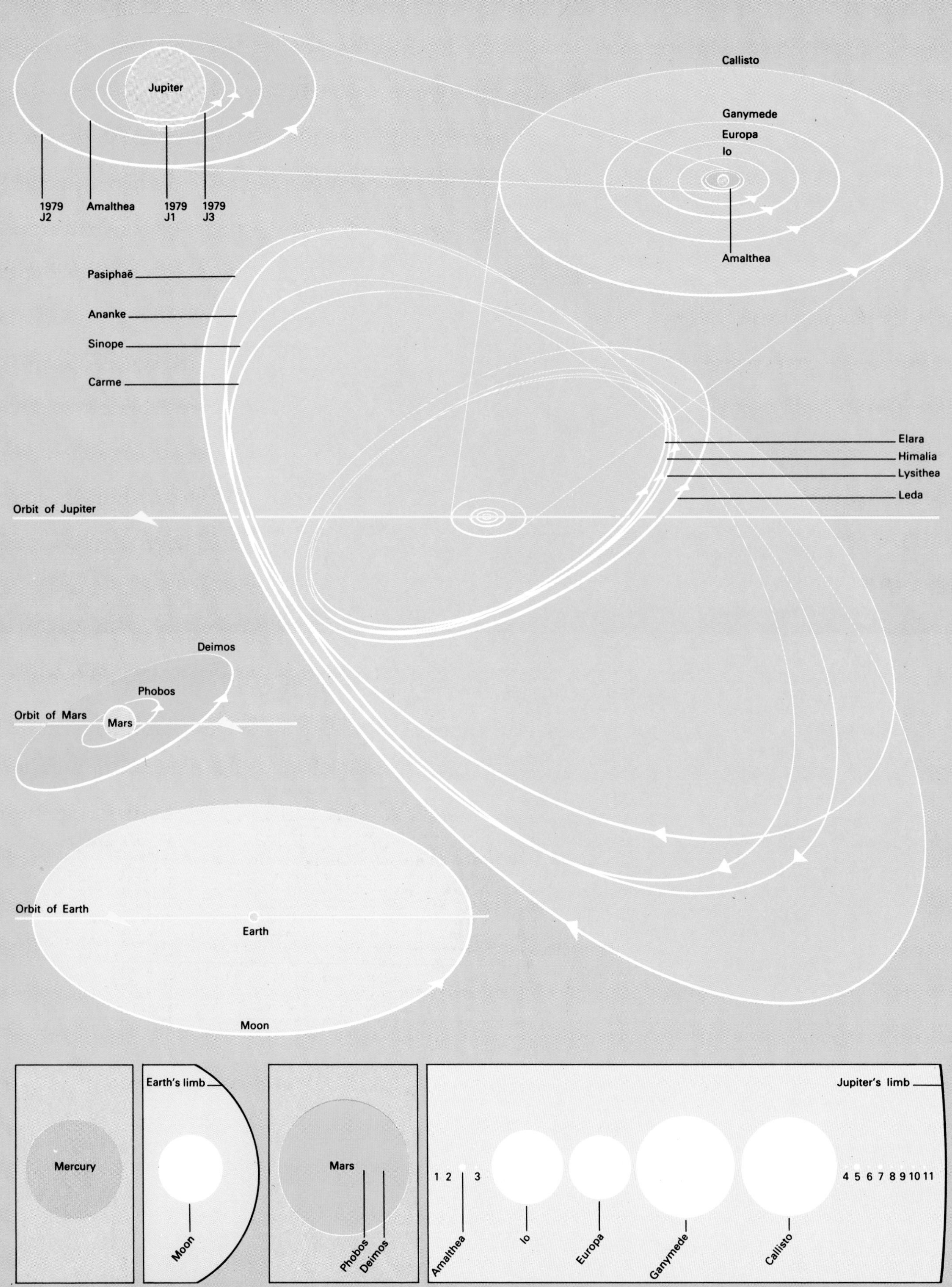

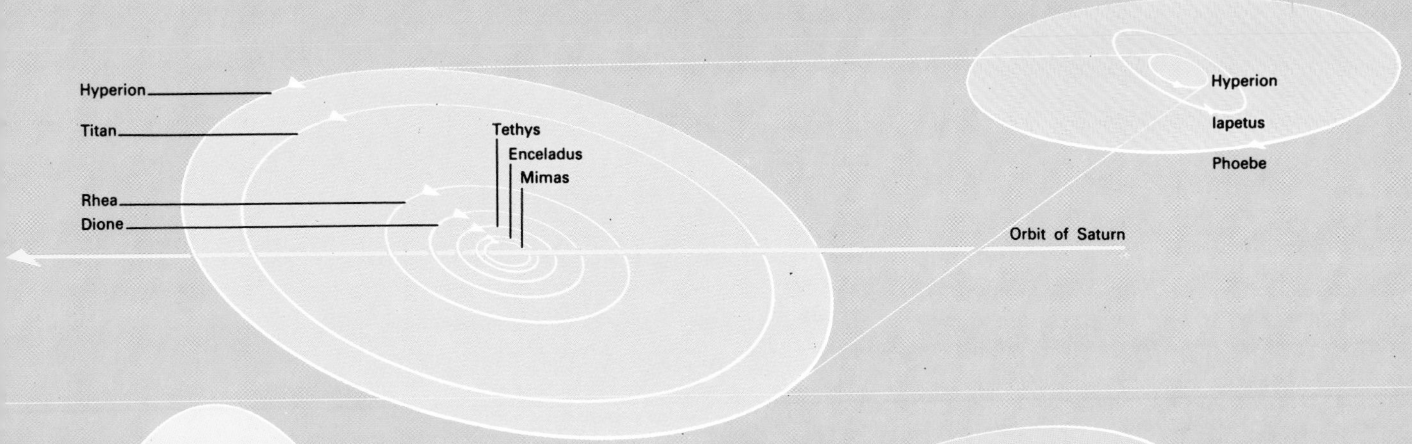

Hyperion
Titan
Tethys
Enceladus
Mimas
Rhea
Dione
Hyperion
Iapetus
Phoebe
Orbit of Saturn

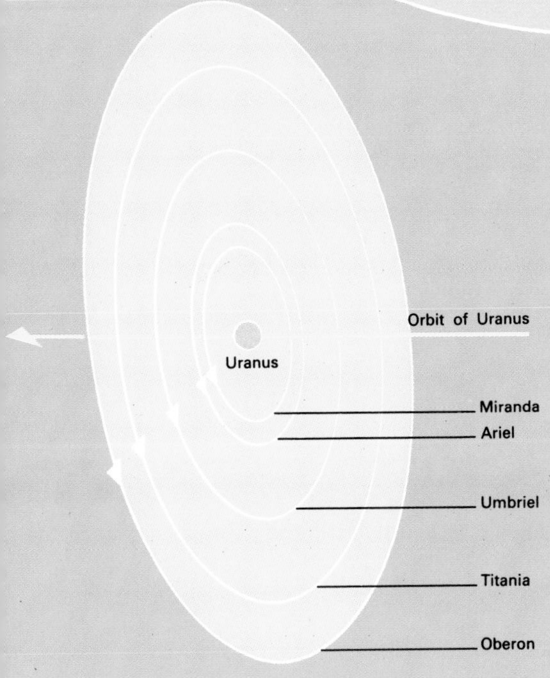

Orbit of Uranus
Uranus
Miranda
Ariel
Umbriel
Titania
Oberon

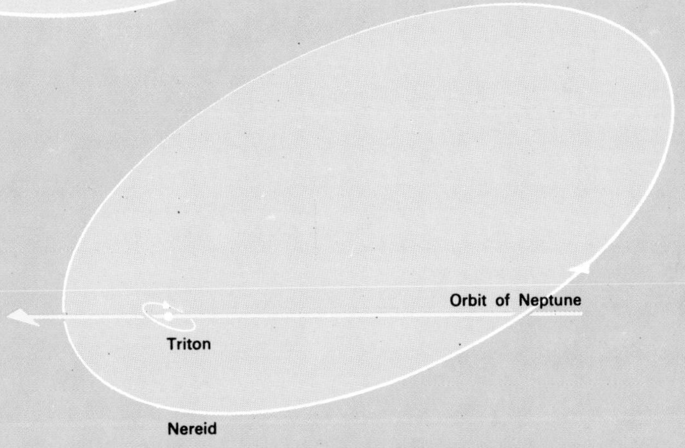

Orbit of Neptune
Triton
Nereid

Charon
Orbit of Pluto
Pluto

The relative spacing and orientation of the orbits of the various planetary satellites are shown. The diameters of satellites are compared with their parent planets (below), but many of the smaller satellites are too tiny to be shown accurately to scale. Of the terrestrial planets, the Earth has one large Moon and Mars has two tiny satellites which follow near-circular orbits above its equator. The largest of Jupiter's 16 moons is Ganymede, with about 1/27 of Jupiter's diameter. The innermost 8 satellites follow near-circular orbits in the plane of the planet's equator, the next 4 have more inclined elliptical orbits, and the outermost 4 have retrograde orbits. Saturn's collection of 21-23 satellites includes one giant moon, Titan, which has about 1/23 of Saturn's diameter. They mostly follow near-circular orbits close to the plane of the planet's equator, but the orbit of Iapetus is significantly inclined, and that of Phoebe is retrograde. Most of the minor satellites are omitted from the diagram as they either share orbits with the largest satellites or lie inside the orbit of Mimas. Uranus has 5 satellites following near-circular orbits in the equatorial plane. The largest, Titania, has about 1/50 of the diameter of Uranus. Neptune's inner satellite, Triton, may be as large as 1/6 of the planet's diameter, and follows a retrograde orbit; the other, Nereid, has an extremely elongated orbit. Pluto's satellite, Charon, has about 1/3 of the planet's diameter.

Jupiter's minor satellites
1 1979 J3
2 1979 J1
3 1979 J2
4 Leda
5 Himalia
6 Lysithea
7 Elara
8 Ananke
9 Carme
10 Pasiphaë
11 Sinope

Saturn's minor satellites
12 1980 S28
13 1980 S27
14 1980 S26
15 1980 S3
16 1980 S1
17 Mimas co-orbital
18 1980 S25
19 1980 S13
20 Tethys co-orbital
21 unnamed
22 unnamed
23 1980 S6
24 Dione co-orbital
25 unnamed

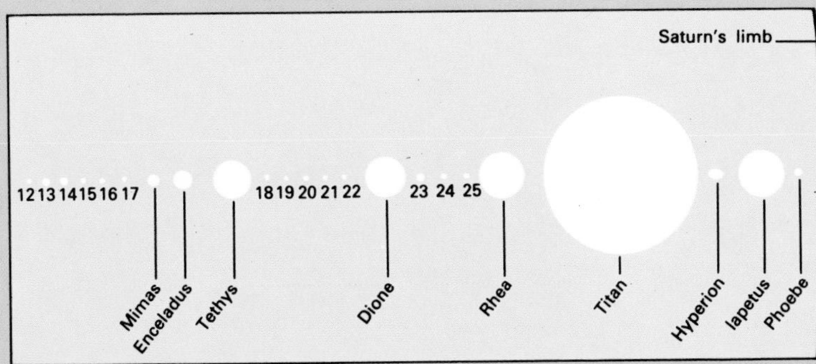

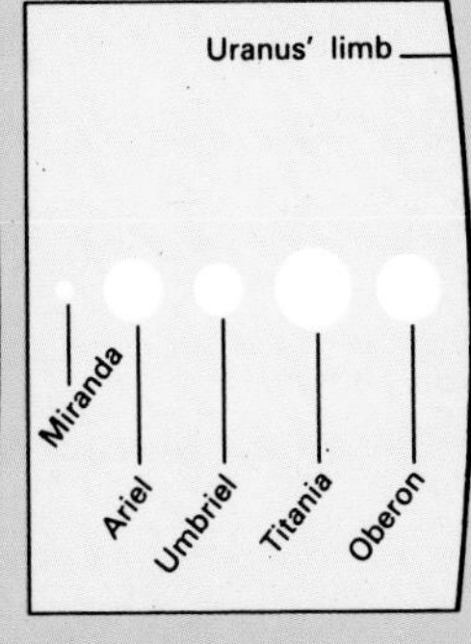

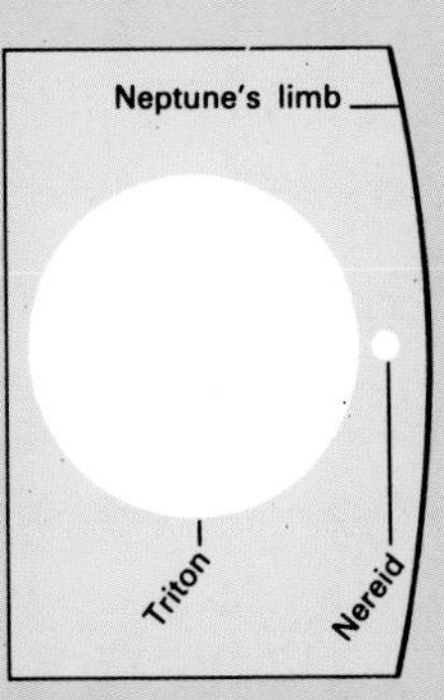

To an observer on Mars, the outer satellite, Deimos, would appear like a large, dim star

A pair of Martian moons

Both Martian moons are tiny. Phobos, the larger of the two, is an irregular ellipsoid with a maximum diameter of about 28km. At a mean distance from the center of the planet of 9,270km, it revolves round Mars in only 7 hours 39 minutes. It is thus the only satellite known to have an orbital period shorter than its planet's axial rotation period, so that it moves from west to east across the Martian sky. Deimos, with a maximum diameter of 16km, lies at a mean distance of 23,400km from Mars and has an orbital period of 30 hours 21 minutes. It moves slowly from east to west across the sky, remaining above the horizon for up to 2·5 Martian days.

The moons are both rocky bodies, apparently similar in nature to a class of carbon-rich asteroid (▸ pages 90-91) found in the outer part of the asteroid belt. It seems likely that Phobos and Deimos are asteroids which, long after their formation, became trapped in the Martian gravitational field.

Both satellites are heavily cratered, and Phobos has one crater, Stickney, which is nearly 10km in diameter. The impact that presumably caused the crater must have come close to shattering the satellite, and a system of striations on its surface are probably another result of the same event.

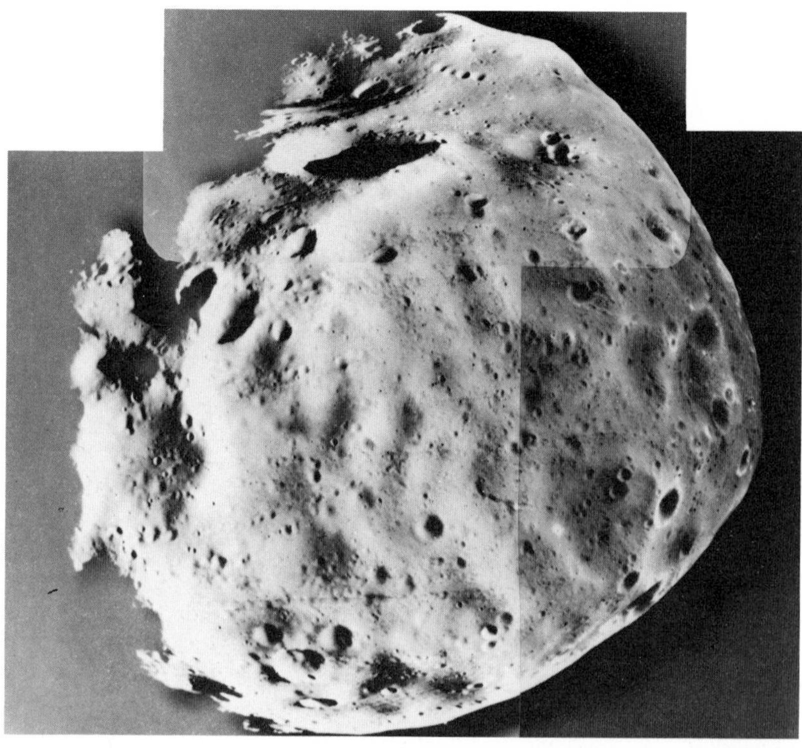

▲ Viking Orbiter 1 took this composite picture of Phobos, the larger Martian moon, from a distance of only 480km.

► Phobos was about to enter the shadow of Mars when Viking took this photograph showing the craters Hall and Stickney.

◄ Deimos resembles Phobos in many respects. However, it lacks grooves and large craters, and its surface is covered with a thinner layer of dust.

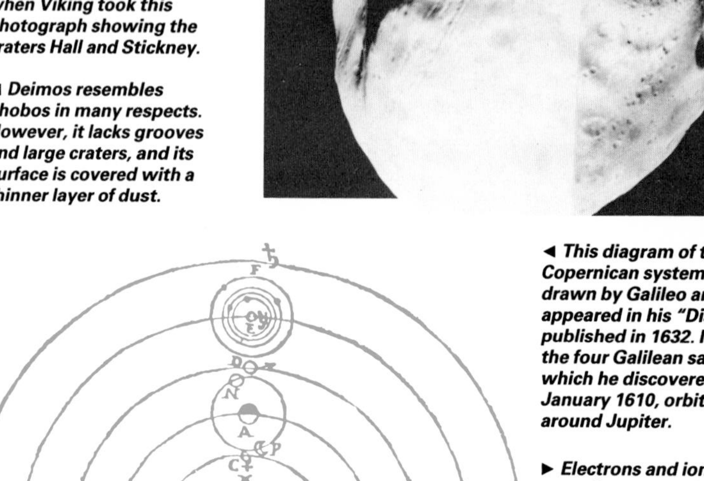

The first planetary satellites

The first satellites to be discovered (apart from the Moon) were the four "Galileans" orbiting Jupiter – Io, Europa, Ganymede and Callisto. At the same time as they were observed by the famous Italian astronomer Galileo Galilei (1564-1642), they were found independently by a German astronomer, Simon Marius (1573-1624). It was Marius who gave the four satellites their names, although they were not officially used until modern times, the satellites merely being referred to as I, II, III and IV. The fifth Jovian satellite was discovered by American astronomer Edward Barnard (1857-1923) in 1892. This was the last to be found visually – all later satellite discoveries have been photographic. The fifth satellite was named Amalthea, although again the name was not used before the Space Age.

Other Jovian satellites were detected later, but all of them were extremely small and faint. The four outermost satellites have retrograde motion, and may well be captured asteroids. Altogether 16 Jovian satellites are now known.

◄ This diagram of the Copernican system was drawn by Galileo and appeared in his "Dialogo" published in 1632. It shows the four Galilean satellites, which he discovered on 7 January 1610, orbiting around Jupiter.

► Electrons and ions – such as sodium, potassium and magnesium – expelled from Io's volcanoes spread out along its orbit to form a doughnut-shaped "plasma torus". The tilt of Jupiter's powerful magnetic field causes the torus to be tilted relative to Io's orbit by about 10°. A stream of electrons and ions flows between Jupiter and Io along a "flux tube" which carries over a million megawatts of electrical power.

Jupiter has 16 satellites, four of which are comparable with, or larger than, the Earth's Moon. These are the "Galilean" satellites: in order of distance from the planet, Io, Europa, Ganymede and Callisto. Apart from Amalthea, an irregularly-shaped body measuring 155km by 270km, and Himalia (170km in diameter), all the other Jovian satellites are considerably less than 100km in diameter.

In terms of their orbits, Jupiter's moons can be divided into three distinct groups. The eight innermost satellites, including the Galileans and Amalthea, lie within 2 million kilometers of the planet's center, and follow near-circular orbits in the plane of Jupiter's equator. The next four, including Himalia, follow more elliptical orbits inclined to the equator by some 26-29°, and at mean distances of between 11 and 12 million kilometers. The outermost four, lying between 20 and 24 million kilometers from the planet, follow highly elliptical retrograde orbits inclined at angles between 147° and 163°. Probably the satellites in the inner group were formed from the same cloud of material as Jupiter itself, while the others are captured asteroids.

Composition of the Galilean satellites

The Galileans differ dramatically from each other. Callisto, the outermost, has a diameter of 4,820km and is similar in size to Mercury. It is much less dense than Mercury, however, suggesting that the satellite contains a substantial proportion of water. It may have a silicate (rocky) core extending to about half its radius, overlain by a mantle of liquid water or soft ice, on top of which lies a rigid crust of rock and ice. Ganymede is about 1·5 times the diameter of the Moon. Slightly larger and a little denser than Callisto, it also probably has a slightly larger silicate core, but is still most likely to contain nearly 50 percent water. An alternative view is that it consists of a fairly uniform mixture of rock and ice throughout. Europa is considerably denser than both Ganymede and Callisto, and probably has a rocky core comprising about 90 percent of its mass.

Finally, Io probably has a crust of sulfur and sulfur dioxide overlying molten silicate and, possibly, a solid core. Its mean density is similar to that of the Moon. Volcanism is evident: it is probably sustained by tidal pumping. Io's orbit is perturbed periodically by Europa and Ganymede, and as a result a tidal bulge rises and falls by as much as 100. The heat released by this motion keeps the volcanoes supplied with molten material. Lying deep within the hostile radiation belts of Jupiter's magnetosphere (➧ page 118), Io is bombarded by charged particles and linked to Jupiter by a "flux tube" of electrons. These can carry an electrical current equivalent to about 2·5 million amperes.

Jupiter's influence on Io

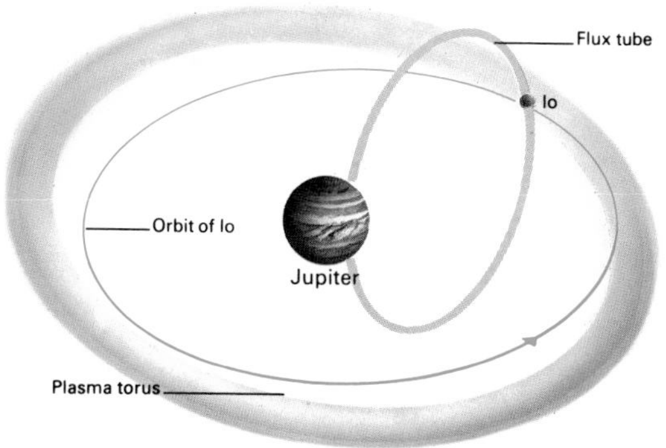

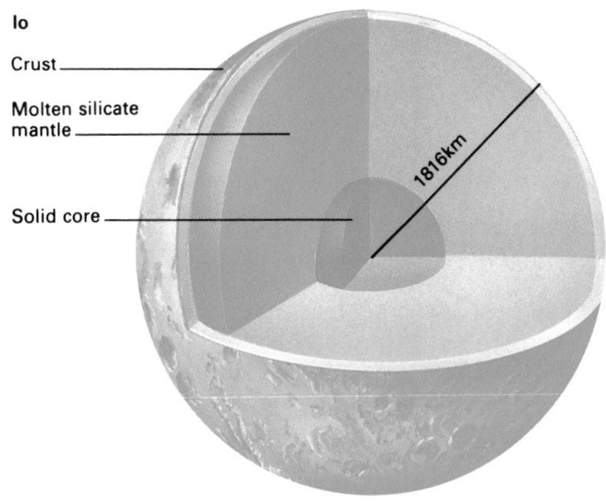

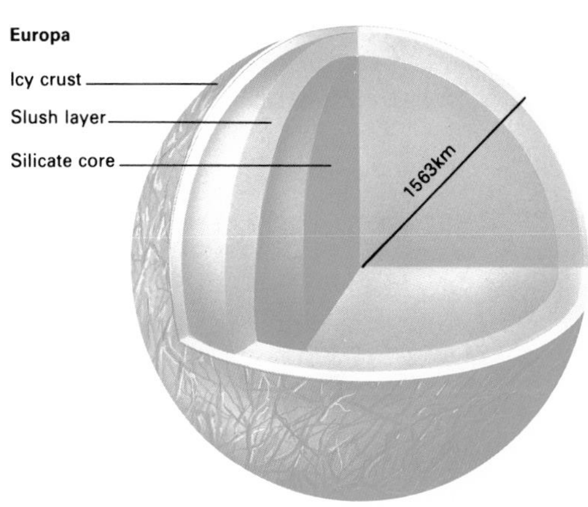

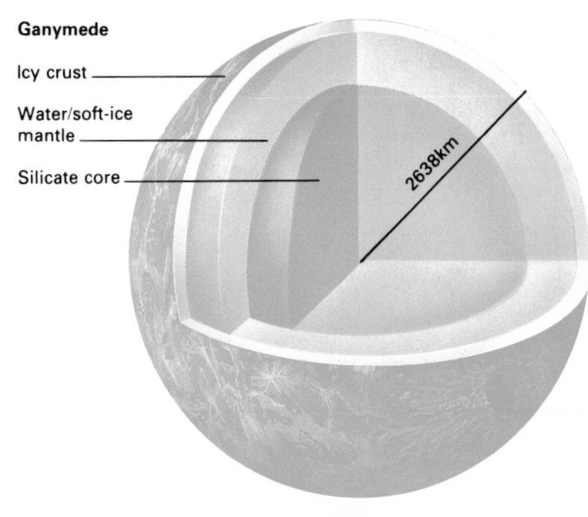

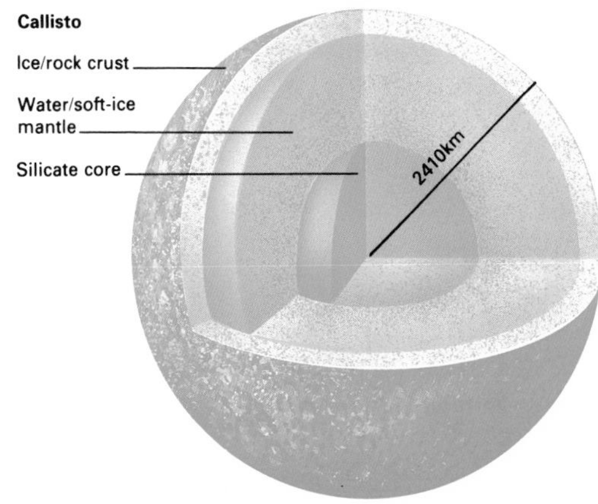

► ***Possible structures of the four Galilean satellites are compared on equal-sized images. Io, the innermost, is the densest (mean density 3,530kg/m³); it may have a solid core. Europa (mean density 3,030kg/m³) is an ice-coated rocky world and is the only one of the four which is smaller than Earth's Moon. Ganymede, the largest, and Callisto, the outermost, are less dense (1,930 and 1,790kg/m³ respectively) and probably comprise roughly half rock and half water or ice.***

Io, Jupiter's innermost large satellite, is the most violently volcanic world known to us

Impurities in the ice give Callisto's surface quite a low albedo (◀ page 41) of 0·2. It is by far the most heavily cratered body in the Jovian system. The albedo of Ganymede is higher (0·4) and the surface is strikingly different from that of Callisto. Like the Moon, it has two types of terrain, bright and dark, but in Ganymede's case the dark areas are the heavily cratered ancient surfaces, presumably composed of dirty ice, whereas the brighter areas seem to consist of ice which melted later and flowed over the older terrain before freezing again. The differences between Callisto and Ganymede are probably due to greater heating and convection in the latter, resulting partly from the decay of radioactive elements.

Europa is the most reflective of the Galileans (albedo 0·64) and it has such a smooth surface that it has been compared to a billiard ball. There are no craters, but the surface is covered with a pattern of bright and dark stripes and a network of cracks, grooves and short narrow ridges. Vertical relief is generally less than a hundred meters, or at most a few hundred. Any craters that may have existed at one time must have been obliterated by ice flowing over the surface.

By far the most intriguing of the four is the innermost one, Io. Its orange-red surface is pockmarked with volcanic caldera and violently active volcanoes. Voyager 1 recorded eight volcanoes erupting actively, and six of these were still erupting during the later flyby of Voyager 2. Plumes of matter are expelled to heights of 250km or more at speeds of up to 1km/s (3,600km/h) – 20 times faster than ejecta from Earth's Mount Etna. The absence of volcanic craters suggests that the surface is no more than a million years old and that this strange world is resurfaced by volcanism so rapidly that it must have been turned inside out and reprocessed several times since its formation.

▲ The surface of Io is subject to colossal tidal and electrical forces, and is continuously shaken by volcanic activity.

▼ The surface of Europa, the smallest of the Galilean satellites, is remarkably smooth.

► The four Galilean satellites are shown in equal-sized images so that their appearances can be compared. Io (top left) was photographed from 862,000km, Callisto (top right) Ganymede (bottom left) and Europa (bottom right) from 1·2 million km.

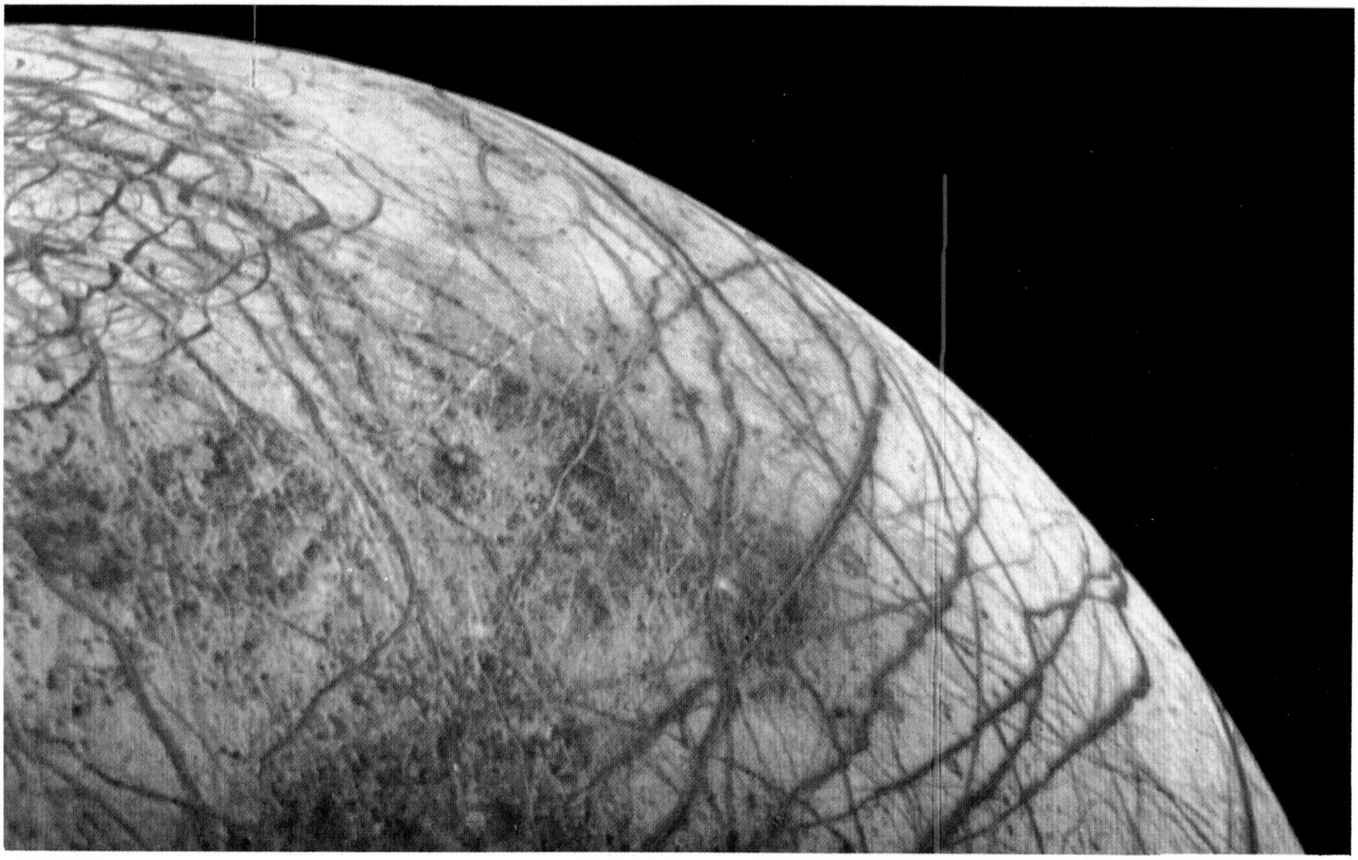

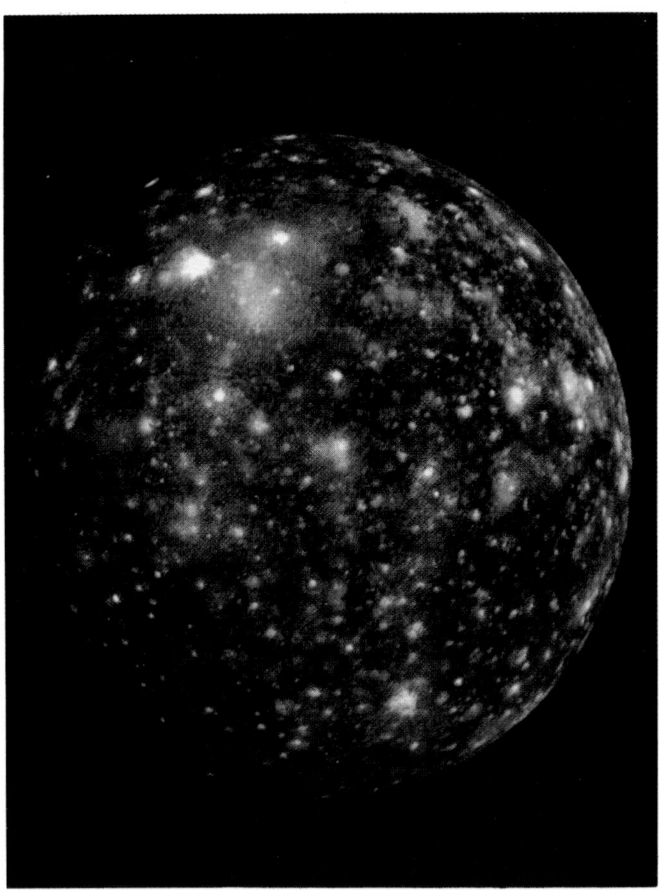

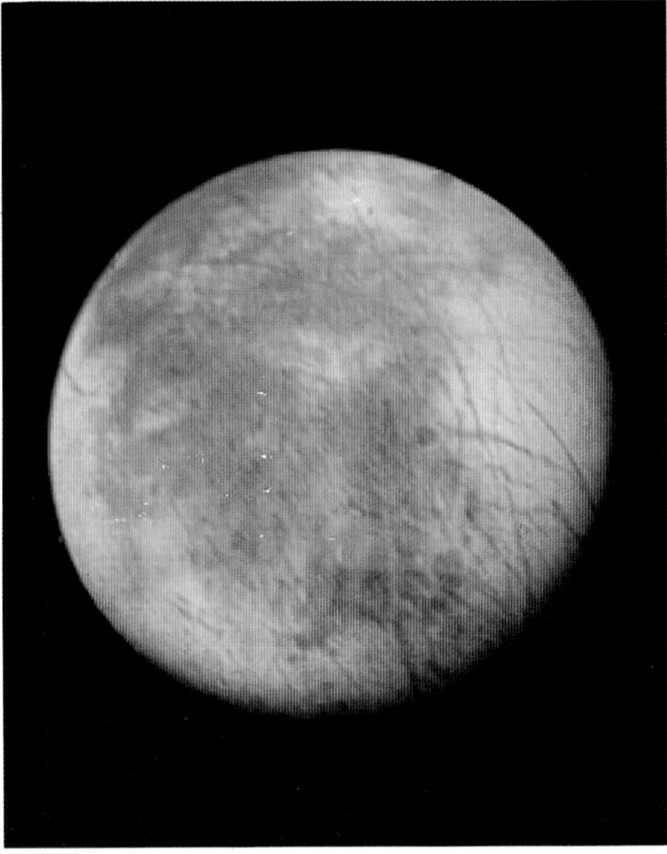

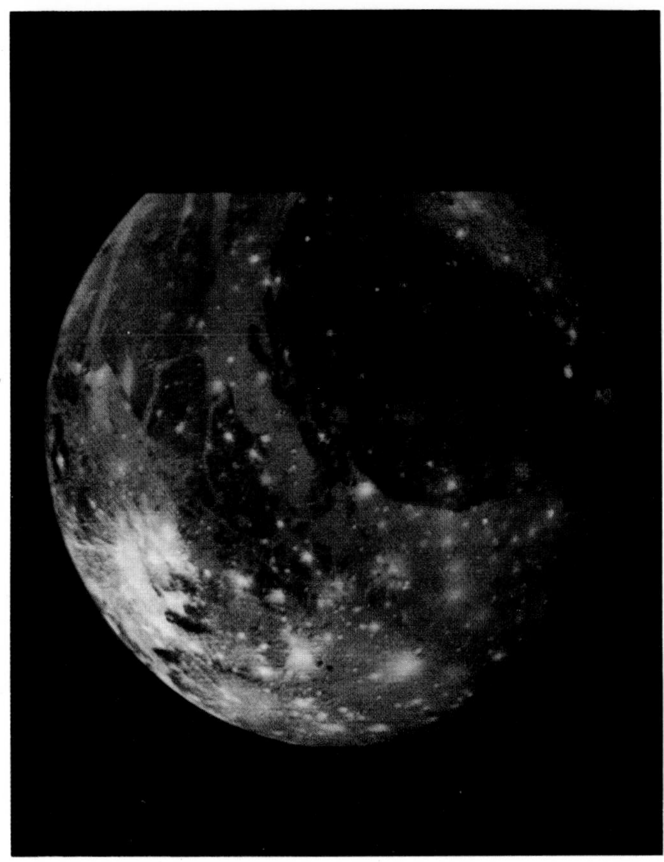

Unlike Jupiter, with its four major satellites, Saturn has one large moon and several of moderate size

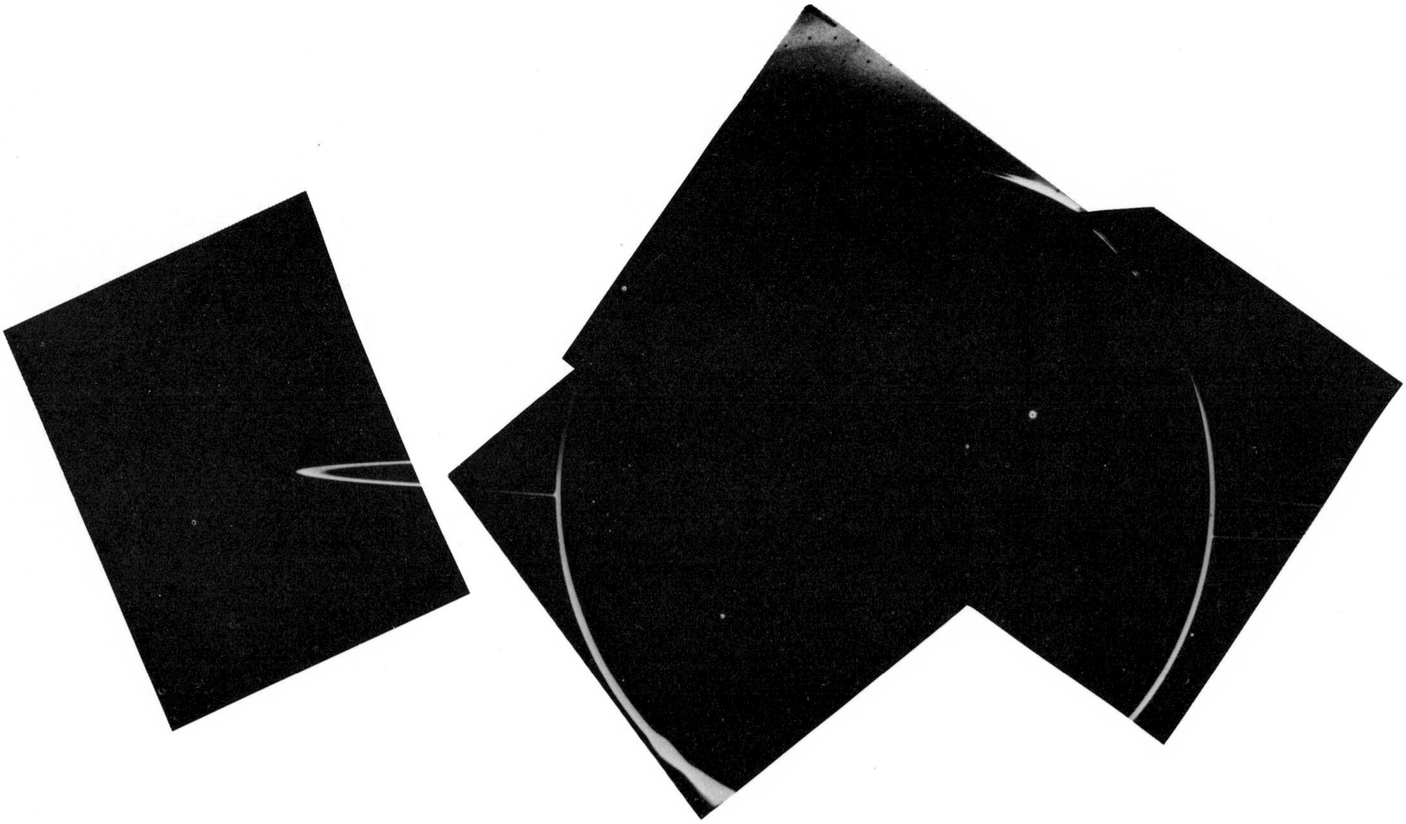

Jupiter's rings

Discovered by Voyager 1 and examined in more detail by Voyager 2, Jupiter's rings are modest by planetary ring system standards. The main ring has quite a sharply-defined outer edge at a radius of some 128,000km, and extends inwards in a rather featureless sheet about 6,000km wide. There is a brighter, narrower zone towards the outer edge of the sheet, and a more tenuous sheet of particles seems to extend inside the main ring down to Jupiter's atmosphere. The particles are tiny, only a few micrometers in diameter; at this size radiation effects will cause them to spiral slowly into the Jovian atmosphere. If the rings are a permanent feature, their material must be continually replenished, possibly by the erosion of material from the surfaces of Jupiter's innermost satellites.

▲ Following the discovery of Jupiter's rings by Voyager 1, Voyager 2 was programmed to take additional pictures giving better resolution. In this one the ring showed up particularly brightly when Voyager 2 passed behind Jupiter with respect to the Sun. The planet's shadow obscures part of the ring in the direction of the camera. The micrometer-sized particles in Jupiter's ring contrast with the Saturnian ring particles, which are estimated to be several centimeters in size.

▼ This composite image of Jupiter's ring was made by Voyager some 26 hours after flying past the planet, at a distance of 1,550,000km. The forward scattering of sunlight reveals a radial distribution of very small particles extending inward from the ring toward Jupiter. There is an indication of structure within the ring, but the spacecraft motion during these long exposures blurred the highest resolution detail. The ring has a characteristically orange color.

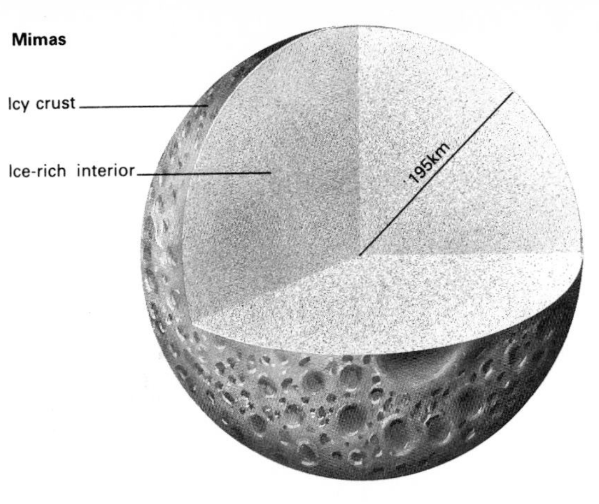

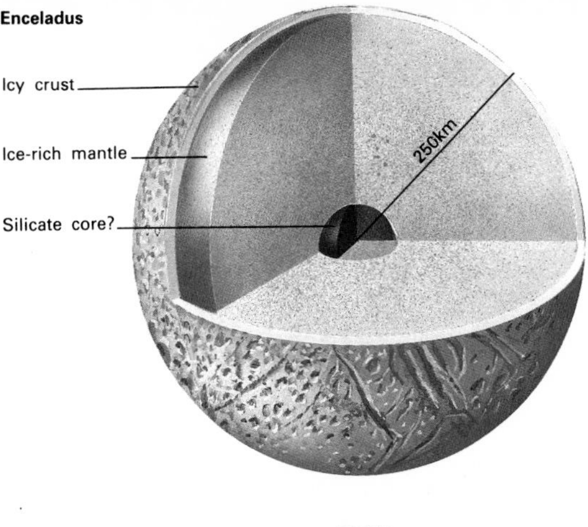

▶ Possible interiors of some of Saturn's larger satellites are compared on equal-sized images. They are all icy, with differing proportions of rocky material, but their internal structures are very uncertain. Mimas (mean density 1,400kg/m³) is probably composed mainly of water-ice and a proportion (about one-third) of denser materials. It is probably solid throughout, and it has a heavily cratered surface. Enceladus (mean density 1,200kg/m³) is also predominantly icy but may have a small rocky core. Much of its surface is covered by smooth plains and fissures. Rhea, larger and denser (1,300kg/m³) than Enceladus, may have a larger core. Titan is by far the largest and densest (1,900kg/m³) of Saturn's satellites. It probably consists of at least 50 percent rock, most of it in the central core. Iapetus (1,100kg/m³) may be composed of ices of various kinds.

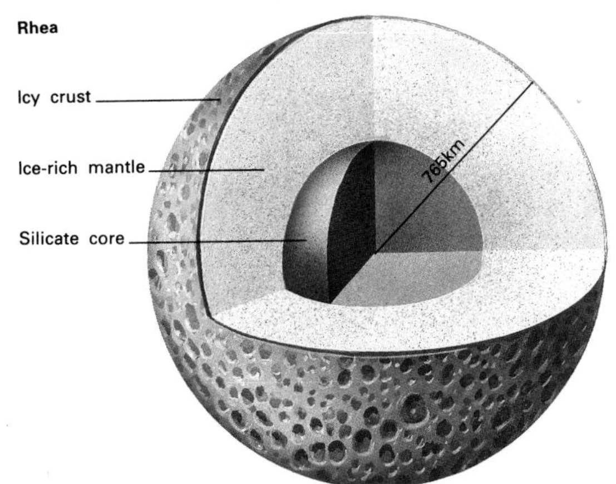

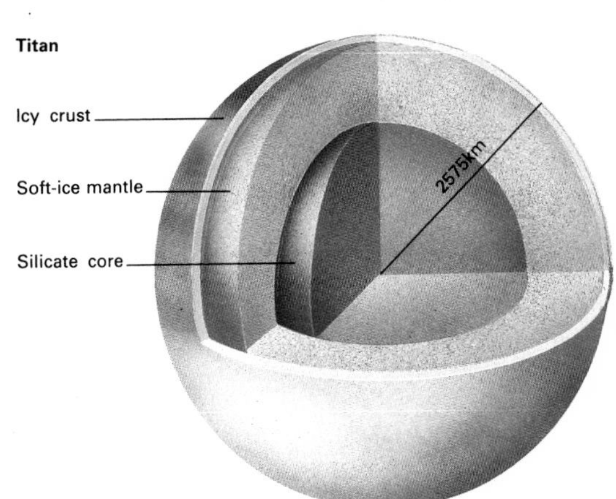

Satellites of Saturn

Saturn has a retinue of 21 or 23 satellites. Difficulties of positive identification mean that the exact figure is open to question: in any case, there are doubtless other small satellites yet to be found. Their distances range from 137,760km to 12,950,000km from the planet's center. There is one giant satellite, Titan, 5,120km in diameter – only fractionally smaller than Ganymede. Of the remainder, four (Rhea, Iapetus, Dione and Tethys) are of moderate size, ranging in diameter from 1,050km to 1,530km, seven are small (160-500km), two are about 100km in diameter, and the rest are tiny (less than 35km across). All follow direct orbits apart from Phoebe, the outermost satellite, which pursues a retrograde path inclined at an angle of 150° to Saturn's equatorial plane.

About half of Titan's mass probably consists of a silicate core, most of the rest being a water- and methane-ice envelope. The most intriguing feature of this satellite is its atmosphere, which consists, by volume, of about 85-95 percent nitrogen, 5-10 percent argon and about 1 percent methane, together with minor constituents such as hydrogen cyanide (one of the building blocks of nucleic acids fundamental to the construction of living matter). Titan is the only planetary satellite to have a permanent, substantial atmosphere.

At ground level on Titan the temperature is about 95K and the atmospheric pressure between 1,500 and 1,600mb, about 50 percent greater than atmospheric pressure at the Earth's surface. The temperature declines with height to a minimum of 70K at the tropopause (the top of the lowest atmospheric layer) some 40km above the surface, and thereafter increases to about 160K. The satellite's surface is hidden from view by a deep layer of orange-red smog made up of aerosols (fine particles and droplets), and extending downwards from an altitude of 200km. A layer of methane clouds, from which methane rain probably falls, is thought to lie just below the tropopause (40km altitude).

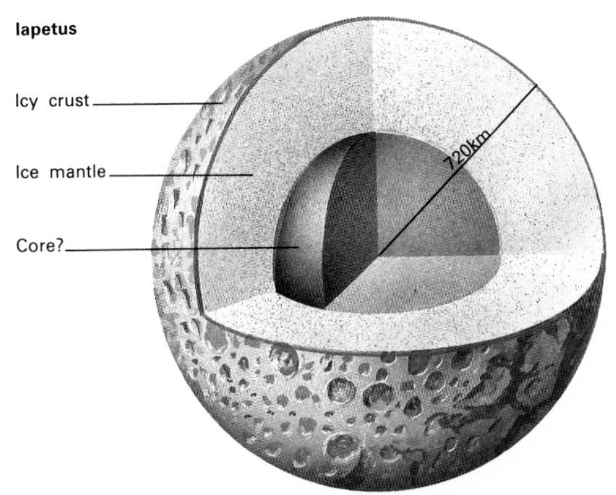

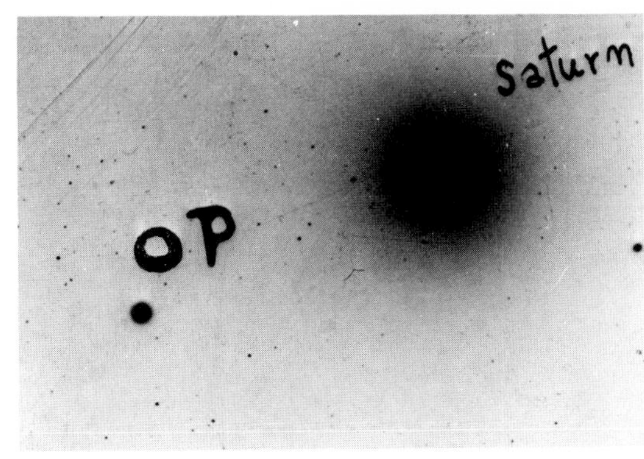

▲ The discovery photograph of Phoebe. In 1898 W. H. Pickering, a foremost American planetary observer, used photography for the first time in the search for new satellites around Saturn. He exposed four plates for two hours each. Between them they revealed more than 400,000 stars. The plates were compared and an object, clearly not a star, was found to have moved. Saturn's ninth satellite had been found.

On Titan, there may be cliffs of solid methane and rivers of liquid methane

At the time of the Voyager flybys the northern hemisphere of Titan was darker and redder than the southern hemisphere, and between 1980 and 1981 the dark hood at the north pole developed into a dark collar. These are probably seasonal effects arising from the fact that Titan's orbit lies in Saturn's equatorial plane and is, therefore, inclined to the ecliptic (the plane of the Earth's orbit) by some 27°. Alternate polar regions are thus illuminated by solar radiation.

The nature of Titan's surface remains a matter of speculation. The temperature is close to the triple point of methane, that is, the temperature at which it can exist in solid, liquid or gaseous form. The surface may be completely covered with methane ice or an ocean of liquid methane. Alternatively, there may be methane lakes and methane cliffs, over which a thin drizzle of liquid nitrogen falls.

The moderate-sized satellites in Saturn's system are icy worlds, some of which may have significant rocky cores. Their icy surfaces are peppered with craters, but there are significant differences in the extent to which partial melting and resurfacing has occurred. Iapetus displays a striking contrast of albedo, the trailing hemisphere being about 10 times more reflective than the very dark leading hemisphere. The reflectivity and reddish color of the dark hemisphere is similar to that of carbonaceous chondritic meteorites (▶ pages 90-1). The source of the dark material may be methane-ice that welled up from the interior and then decomposed leaving deposits of carbon "soot".

The small satellites also have interesting features: Mimas boasts a 130km crater, one-third of the satellite's size, and Enceladus is the most reflective body known in the Solar System. Of the multitude of tiny moonlets, three are intimately involved in "marshaling" the planet's ring particles (▶ pages 84-5), two share the same orbit as each other, and several others share orbits with the larger satellites Mimas, Tethys and Dione.

▲ The surface of Mimas is heavily cratered. As well as many small craters as little as 2km across, there is a giant called Herschel, 130km across, with a conspicuous central peak and raised rim. The cratering is evidence of the bombardment that the Solar System underwent four billion years or so ago, shortly after its formation.

► Enceladus, although the eighth farthest satellite from the planet, is the most reflective. Some parts of the surface are smooth, whereas other regions show craters up to 35km across. The surface is also crossed by linear grooves several hundred kilometers long, which may be faults resulting from deformation of the crust.

▲ Dione is one of Saturn's larger satellites, with a diameter of 1,120km. The trailing hemisphere ("trailing" in the sense of its motion around Saturn), on the left of this photograph, shows contrasting light and dark areas, the former being probably due to surface water-ice. The leading hemisphere is more uniform.

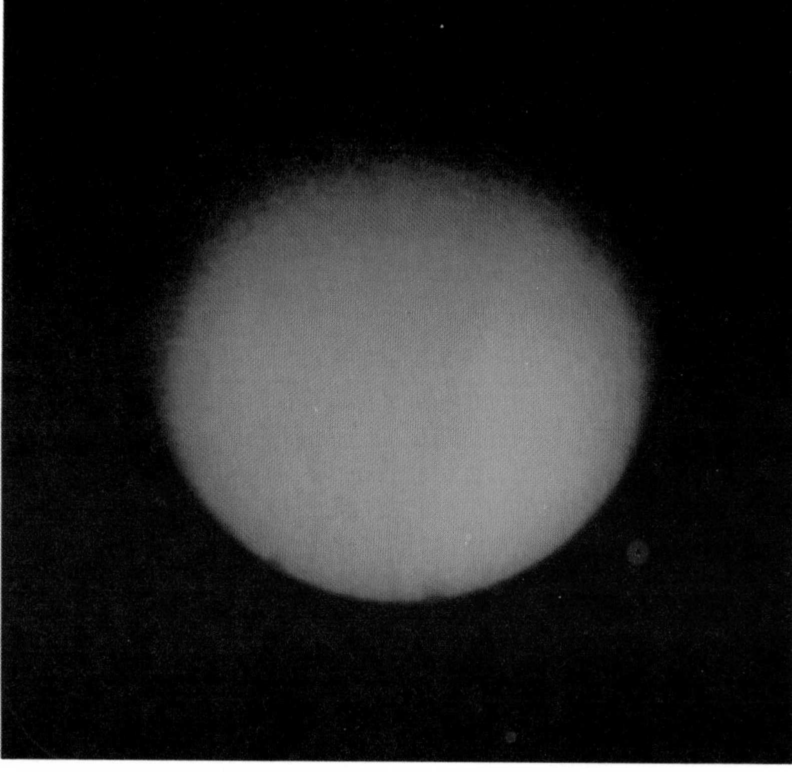

► Titan is the second largest satellite in the Solar System (Jupiter's Ganymede being the largest) and the only body apart from the Earth to have a dense, nitrogen-rich atmosphere. Its northern hemisphere is darker and redder than the southern hemisphere, and both Voyagers found that there was darkening at the north pole.

The thickness of Saturn's rings may be as little as one kilometer, and is certainly not much more

▲ **The Dutch astronomer Christiaan Huygens.**

Early ideas about rings

One of the objects observed by Galileo in 1610, with his newly-made telescope, was Saturn. Galileo realized at once that there was something unusual about the planet's shape, but he could not make out exactly what it was. Finally he concluded that "Saturn is not one alone, but is composed of three, which almost touch one another...the middle one is about three times the size of the lateral ones."

Two years later he was surprised by the fact that the two attendant globes had disappeared. In fact, this was because the ring system had become edgewise-on to the Earth, and at such times not even large telescopes will show the rings clearly.

The discovery that the strange appearance of the planet was due to a flat ring – or, more accurately, a system of rings – was made by the Dutch astronomer Christiaan Huygens (1629-1695) in 1655, but it was not for some time that his interpretation was accepted. Sir Christopher Wren (1632-1723) in England worked out a theory involving an elliptical corona around Saturn, but never published his ideas because as soon as he heard Huygens' interpretation he accepted it.

In 1675 the Italian astronomer Giovanni Cassini (1625-1712), working in Paris, found a dark gap in the main ring. This is still known as the Cassini Division. It was, naturally enough, assumed to be empty, and this belief continued until the first probes visited the planet. Pioneer 11, having passed by Jupiter, was rerouted to encounter Saturn in 1979, and the plans included a course straight through the Cassini Division. Fortunately this was not actually attempted. If it had been, Pioneer would have stood little chance of survival, since astronomers now know that there are several ringlets within the Cassini Division.

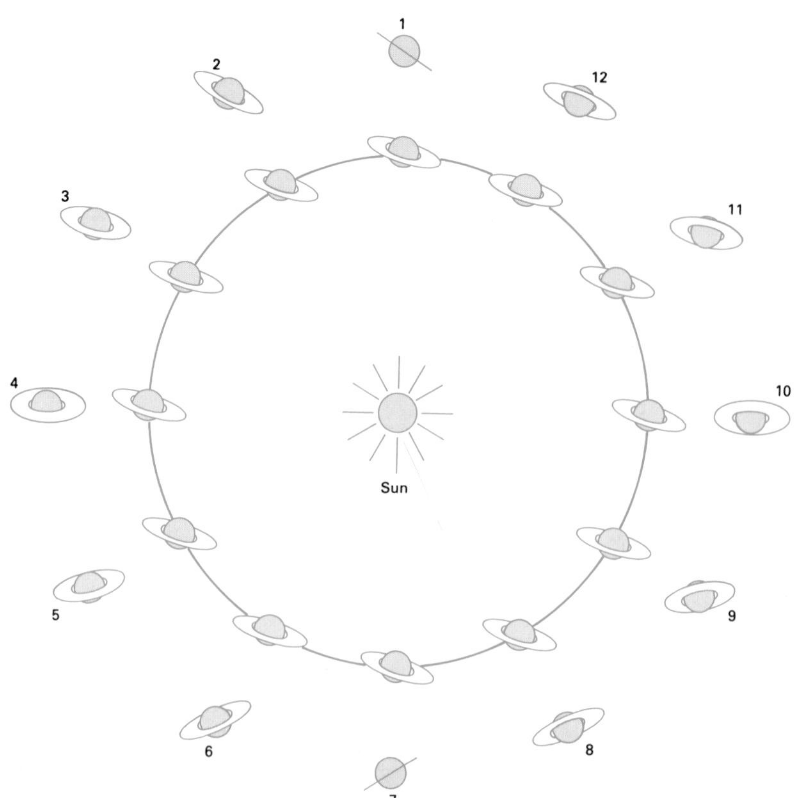

▲ ***Due to the tilt of Saturn's axis, the aspect of the rings seen from Earth varies as the planet moves around the Sun. The observed aspect is shown adjacent to the successive numbered positions. At 1 and 7, the rings are edge-on; at 4 and 10 they are seen at their widest angles. The rings were edge-on in 1980 (1), the northern face most fully displayed in 1987 (4) and they will be edge-on again in 1995 (7).***

Saturn's rings

Seen through Earth-based telescopes, Saturn's ring system was for a long time believed to consist of three principal components, rings A, B and C. The outermost of these rings, ring A, had an overall diameter of 272,000km and was separated from ring B, the brightest ring, by a 4,000km gap, the Cassini Division. The innermost of the three, ring C (the Crêpe Ring), was very faint and quite transparent when seen against the disk of the planet. The Voyager spacecraft have revealed that the system is in fact astonishingly complex, each of the main rings being subdivided into thousands of ringlets, some separated by gaps, others being no more than ripples in the main ring structures. There may be over 100,000 such features altogether.

An extremely tenuous D-ring extends in a thin sheet inwards from ring C, possibly down to the cloud-tops of the planet itself. The Cassini Division is not a truly empty gap, but contains at least five bands of ringlets. Farther out, at a distance of 140,000km from the center of the planet, is the peculiar F-ring. Less than 150km wide, and deviating from circularity by up to 400km, it is made up of a number of intertwining strands, each about 10km wide. The particles making up the F-ring are marshaled by gravity into this narrow band by two "shepherd" satellites – 1980 S26 and 1980 S27 – which move, respectively, just outside and just inside the ring. Particles on the outer fringe of the ring are slowed down as they pass S26, and drop inwards, while particles on the inner fringe are accelerated by S27 as it overtakes them, and they move a little way outwards.

◀ *These two photographs, taken 15 minutes apart by Voyager 1, reveal dark spokelike features revolving with the rings themselves. Scientists believe that magnetism is involved in their formation, but the exact mechanism is not understood.*

▼ *Two satellites (S26 and S27) "shepherd" Saturn's thin F-ring. The inner satellite laps the outer one every 25 days. Their combined influence pulls the ring into intertwining strands, as well as keeping it out of circularity by up to 400km.*

▶ *The multiple fine ringlets and divisions in the ring system may be produced by small "shepherd" satellites. As a moonlet moves through the ring, the successively numbered positions show how a faster-moving inner particle is slowed down as it overtakes the moonlet and, as a result, drops down into an orbit closer to the planet: a slower-moving outer particle is accelerated as the moonlet passes by and is elevated into a higher orbit farther from the planet. In this way particles could be swept out of certain orbits and concentrated into bands.*

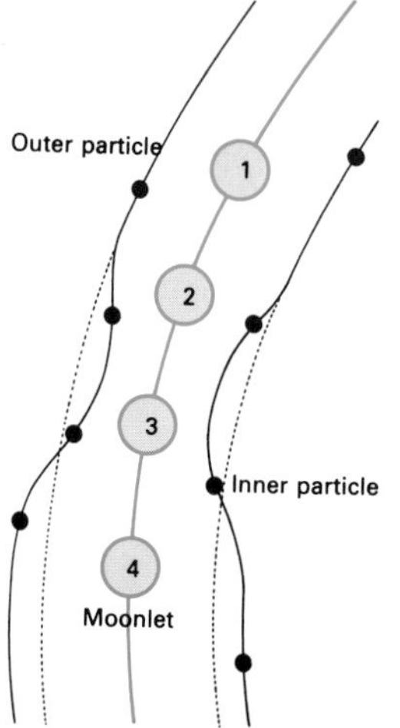

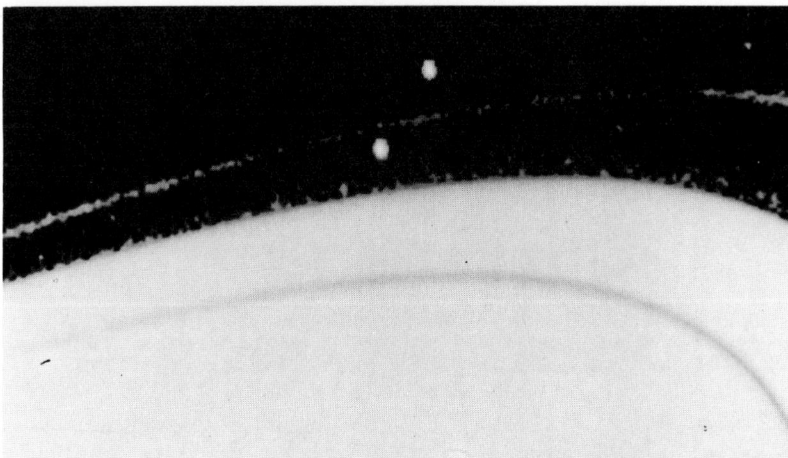

Farther out still from Saturn is an extremely thin G-ring, beyond which a very tenuous sheet of particles, ring E, extends from a distance of 210,000km to nearly 500,000km. The inner rim of this sheet lies beyond Mimas, and Enceladus orbits within it.

The classical theory of the rings suggested that the divisions were due to orbital resonances, rather like vibrations, caused by the gravitational effects of the satellites. For example, a particle at the distance of the Cassini Division would have half the orbital period of Mimas and so at regular and frequent intervals would be perturbed away from that orbit by the gravitational attraction of Mimas. In this way, it was thought, the Cassini Division would be "swept clean" of particles. It now seems unlikely, however, that a resonance theory of this type can account for all the ring features. It may be that density waves spread out, like ripples on a pond, from resonance locations. Another intriguing ring phenomenon is the occurrence of transient dark radial features that form and disperse within ring B, revolving with the ring rather like the spokes of a wheel. These may be due to electrically charged particles elevated away from the ring plane by the planet's magnetic field.

The origins of the ring system remain a matter of debate. It is possible that they represent fragments of a satellite torn apart by gravitational forces exerted by Saturn, but it seems more likely that most of the ring particles are material dating from the formation of the planet that failed to coagulate into a satellite. Certainly Saturn's bright rings are quite different from the darker systems of Jupiter and Uranus.

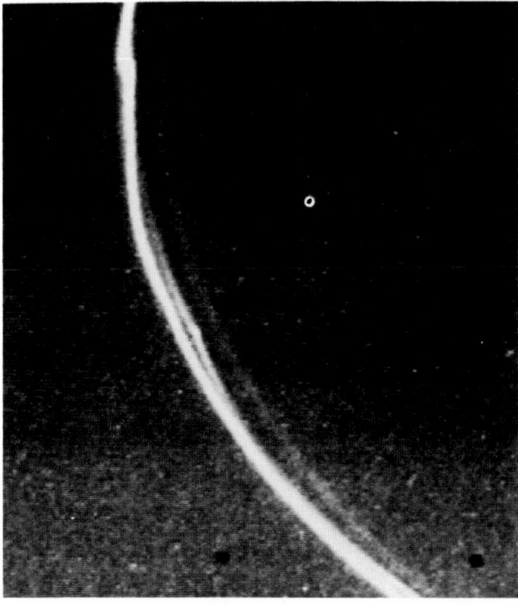

▲ *This detail from the puzzling F-ring shows two distinct bright strands, which appear to be braided. In fact, Voyager pictures have shown that there are at least ten strands in all within this very narrow ring, though it is not clear whether the braiding phenomenon affects them all. The braiding may not be a permanent feature of the F-ring, or it may even be confined to one or more small areas.*

Saturn's rings are so unlike those of Jupiter and Uranus that they may have a completely different origin

▲ ***Voyager 2 took this view of Saturn from a range of 3·4 million kilometers. The rings seem to be less than 500m thick; the planet's disk is clearly visible through the ring material.***

▼ ***This image is constructed from data on ring density obtained by Voyager 2, and shows the brightest strand of the F-ring as seen from inside the curve and slightly above the ring plane.***

Rings of the outer Solar System

In 1787 the German-born astronomer William Herschel (1738-1822) reported seeing two rings around the planet Uranus. Such rings do in fact exist, but Herschel could not possibly have seen them with the equipment he was using – they were due simply to distortions in his new reflecting telescope. The rings of Uranus were in reality discovered in 1977, while measurements were being made of the occultation of a star by the planet. The star was seen to flicker several times just before the occultation, and again several times after the event, as if it had been temporarily obscured at intervals by concentric rings around the planet. Subsequent observations have revealed the presence of nine thin and rather dark rings, and the ring system as a whole has been observed from the Earth at infrared wavelengths. The rings are at distances from the center of Uranus of between 42,000km and 52,000km and all are slightly elliptical. They are all extremely narrow, only about 10km wide, apart from the outermost one, the epsilon ring, which varies in width from 20 to 100km. It is the most elliptical of all the rings, its distance from the planet varying by 800km, and it may, in fact, consist of two adjacent rings of slightly different ellipticities. The ring particles may be maintained in their orbits by as yet unseen shepherd satellites.

▲ ***This computer-generated image reveals the Encke division in the A-ring; the central ringlet shows clearly in this picture. Called the "Encke Doodle", it is probably one of two.***

▼ ***Saturn's C-ring is the main element of this false-color picture; the B-ring can also be seen at top left. The different colors denote different compositions of the ring particles.***

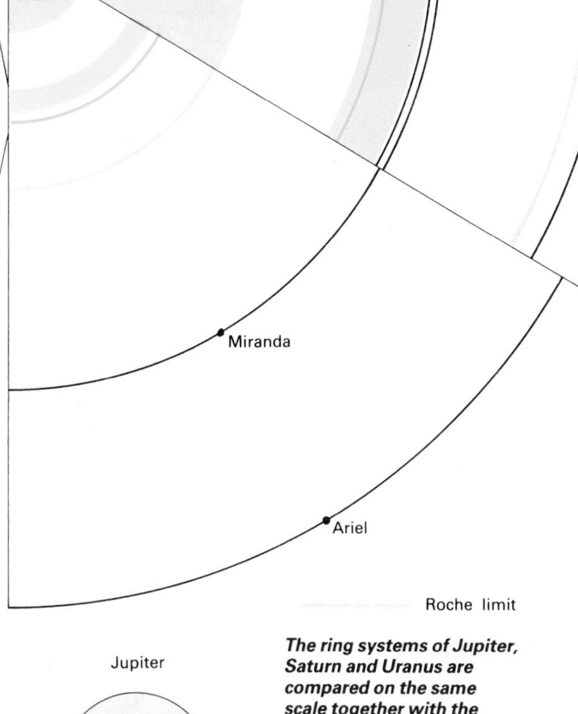

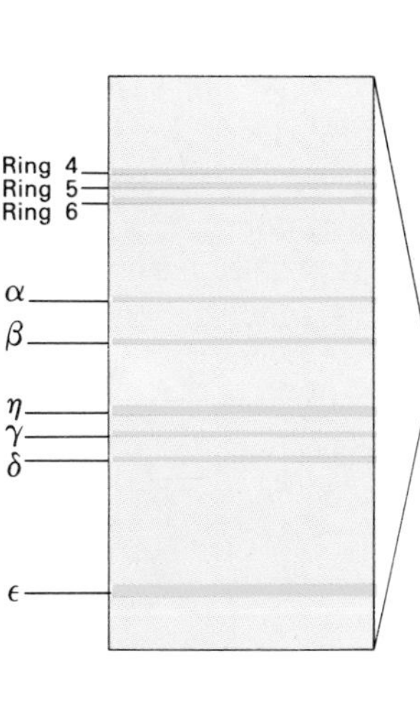

Jupiter

Saturn

Uranus

The ring systems of Jupiter, Saturn and Uranus are compared on the same scale together with the orbits of some of the inner satellites. The gray line denotes the Roche limit within which a satellite having the same density as its parent planet would be torn apart by tidal forces. Saturn's rings contain particles ranging from micrometers to tens of meters in diameter, many of which are composed of, or coated with, ice. The much darker Jovian and Uranian rings contain tiny rocky particles. Compared with Saturn's ring system (simplified here), the tenuous Jovian ring has little structure. The Uranian system contains nine very narrow ringlets. Due to the different axial inclinations of the planets the rings are tilted relative to the planetary orbits by the angles shown (left).

See also
Planets and Orbits 9-20
Inside the Planets 33-40
Planetary Landscapes 41-56
Planetary Atmospheres 57-72
The Origin of the Solar System 121-8

The satellites of Uranus

Uranus is known to have five satellites, but information about them is incomplete. In order of distance from the planet, they are Miranda, Ariel, Umbriel, Titania and Oberon. Spectroscopic measurements suggest that they have icy surfaces, and it seems likely that they are similar in nature to the small and moderate-sized Saturnian satellites. Assuming a mean albedo of 0·5, their diameters range from about 320km for Miranda to about 1,040km for Titania.

The satellites of Neptune

Neptune has two very different satellites, each intriguing in its own way. The inner satellite, Triton, at a mean distance of 353,000km, moves in a circular retrograde orbit inclined at 160° to Neptune's equator. It is a large satellite, possibly comparable in size with Mercury, but there is uncertainty about its exact diameter, estimates ranging from 3,000km to 6,000km. Opinions about the nature of its surface also differ considerably, but one possibility is that there are continents of methane-ice and oceans of liquid nitrogen. Both elements have been detected spectroscopically. There seems to be a methane atmosphere with a pressure of about 0·0001 atmospheres. The satellite's peculiar orbit produces major seasonal changes in the amount of solar radiation reaching different parts of its surface. A cycle of melting and refreezing will result, causing release of gases (mostly methane and nitrogen) which may produce as much as a thousandfold change in the total volume of the atmosphere.

Another odd feature of Triton is that, because it pursues a retrograde path, the tidal interaction with Neptune is causing it to spiral inwards at such a rate that the satellite may break up or collide with the planet within the next 10 to 100 million years. Neptune's other satellite, Nereid, is much smaller (500-800km in diameter) and has the most elliptical orbit of any known satellite. Its distance from the planet ranges between 1,400,000km and 9,700,000km.

Charon – Pluto's satellite

Charon lies at a mean distance of 20,000km from the center of Pluto and revolves around the planet in the same period of time as the planet rotates on its axis. With a diameter of about 800km, Charon is about one-third the size of Pluto, and is the largest of all satellites in relation to the size of its planet (the Moon is about one-quarter of the Earth's diameter, and all other satellites are much smaller in proportion to their primaries). Like Pluto, Charon is probably an icy world, and if so it has about one-tenth of Pluto's mass.

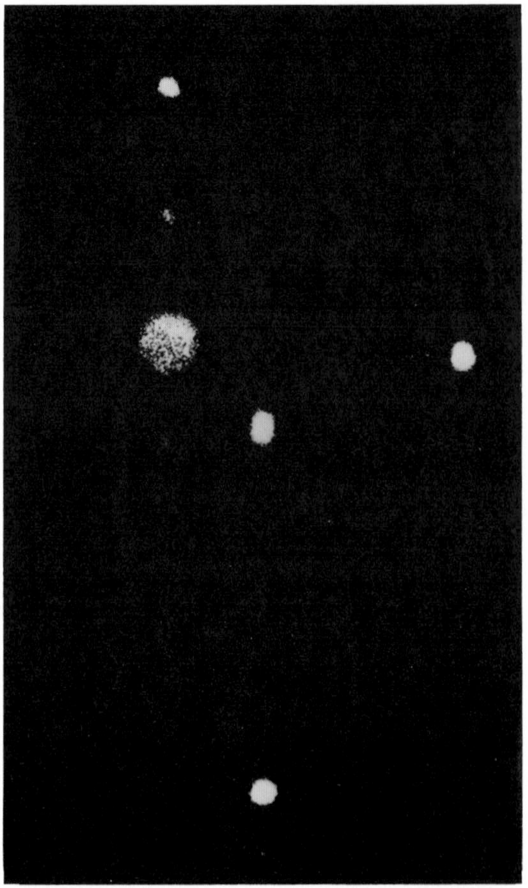

▲ Uranus and family: the planet has five moons, all of which are visible in this photograph. From top to bottom, the objects in the picture are: Umbriel, Miranda (very faint), Uranus, Oberon (over to the right), Ariel and Titania. The axial tilt of the planet is 98°, but the orbits of all five satellites are in the plane of its equator, so that seen from this angle they would appear to have more vertical than lateral movement. The event that tilted Uranus may also have produced the satellites.

▼ Neptune is difficult to photograph because of the enormous distances involved, but this picture has succeeded in capturing the planet with both its satellites (arrowed). Almost lost in the contrast glare of Neptune is Triton, the larger of the two moons. Its orbit is retrograde, and some astronomers expect it to spiral into the planet within 10-100 million years. The smaller moon, Nereid, has a highly eccentric orbit. Voyager 2 should make more information available about Neptune in 1989.

► Pluto and Charon, a picture obtained by a special technique known as speckle interferometry. Pluto is to the left (the apparent bars projecting from it are instrumental). The two bodies are so alike that it may be better to regard them as twin planets – or, more correctly, twin asteroids, since their combined masses are much less than that of our Moon. Although far from being generally accepted, one theory is that they may once have been members of the Neptunian system.

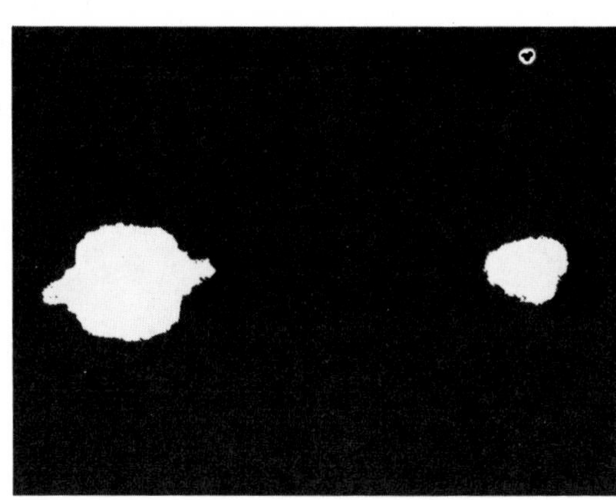

Asteroids and Comets

The Solar System's debris, asteroids and comets... The asteroid belt and Jupiter's powerful influence... Meteorites, their devastating impact...Comets, the spectacular wanderers...The death of comets and meteoroids...PERSPECTIVE...The "Celestial Police" and their search for asteroids...Examining stones from space...How to name an asteroid...Comets, harbingers of doom...Halley and his comet

The asteroids, or minor planets, are pieces of rock, most of which pursue orbits lying between those of Mars and Jupiter. The main part of the asteroid belt lies between 2·2 and 3·3AU (astronomical units) from the Sun. Known asteroids range in diameter from 1,000km (for the largest, Ceres) to less than 1km, but only about 200 are known to be larger than 100km in diameter. Just over 2,700 have been named.

Although the larger asteroids are approximately spherical, many smaller ones are elongated or irregular. Some may be double or multiple bodies, loosely held together by gravity, while others may have satellites. Herculina, for example, 220km across, has a suspected satellite with a diameter of 50km at a distance of approximately 1,000km. Deimos and Phobos, the irregularly shaped satellites of Mars, are only 28km and 16·5km respectively at their widest points. It is likely that they formed within the asteroid belt, much nearer to Jupiter, and were captured by Mars during its formation.

▼ Johann Schröter's observatory at Lilienthal, near Bremen, was the venue for the inaugural meeting of a group of observers who aimed to discover a planet predicted between the orbits of Mars and Jupiter. The largest telescope of continental Europe, built by William Herschel, was housed in Schröter's observatory – destroyed in 1813 when the invading French burned Lilienthal.

► Johann Elert Bode was one of the first scientists to bring astronomy within the grasp of people who had no training in science. His popular "Introduction to Astronomy" went into nine editions; his other works included a Celestial Atlas. Bode popularized, but did not devise, the law which sometimes bears his name. "Titius-Bode's Law" encouraged observation as a basis of mathematically guided prediction.

A strange law and the Celestial Police

The story of the minor planets began in the early 1770s with the announcement, by Johann Titius (1729-1796) of Wittenberg, of a strange law linking the distances of the various planets from the Sun. In 1772 this theory was popularized by the German astronomer Johann Bode (1747-1826) and it is known as Titius-Bode's Law. The relationship of the distances can be demonstrated by taking the numbers 0, 3, 6, 12, 24 and 48, each of which, apart from the first, is double its predecessor. Adding 4 to each, and taking the Earth's distance from the Sun as 10 units, the figures give the distances of the planets out to Saturn with reasonable accuracy (Uranus, Neptune and Pluto were then unknown).

Planet	Bode distance	Actual distance
Mercury	4	3·9
Venus	7	7·2
Earth	10	10
Mars	16	15·2
—	28	—
Jupiter	52	52·0
Saturn	100	95·4

Bode did not think this was a coincidence and his conviction was reinforced in 1781, when the German-born English astronomer William Herschel (1738-1822) discovered Uranus. The Bode distance was 196, while the actual distance was 191·8.

The search for a planet to fill the gap

The only gap in the system corresponded to the Bode number 28, and it seemed likely that there should be a planet in that region. Obviously it would be faint, as otherwise it would have been discovered long before. In 1800, a group of observers calling themselves the "Celestial Police" met at Lilienthal, where the German astronomer Johann Schröter (1745-1816) had his observatory, and worked out a plan for a systematic hunt.

In the event, the "Police" were forestalled. At the Sicilian observatory of Palermo, the Italian astronomer Giuseppi Piazzi (1746-1826) came across a star-like object which moved across the sky from night to night. The object was named Ceres, in honor of the patron goddess of Sicily. But the "Police" were not satisfied – Ceres seemed surprisingly small – and so the search continued. Within the next eight years they had detected three more small planets, Pallas, Juno, and Vesta.

In 1832 a new search was begun by a German amateur, Karl Hencke (1793-1866). In 1845 he was rewarded with the discovery of asteroid number 5, Astraea, and a second discovery in 1847. From 1891 onwards, because of the use of photography, asteroids were discovered in large numbers. Their trails have appeared on photographic plates exposed for quite different reasons, and they have even been called the "vermin of the skies".

It was not until the discovery of the planet Neptune in 1846 at an actual distance of 306 on the Bode scale, that the law was shown to be clearly invalid.

In handling a meteorite, you are probably touching an object which came from the asteroid belt

◀ ***A large meteorite can produce an impact crater such as this one in Arizona. It is over 1,200m in diameter and 175m deep, has a slightly raised rim and a regular bowl shape. Geologists believe that the main mass of the impacting body was totally evaporated, although part may be buried under the south wall. The site bears the name Meteor Crater.***

The perturbing effect of Jupiter's powerful gravitational influence sweeps certain areas of the Solar System clear of any orbiting bodies. As a result there are gaps, named Kirkwood gaps after their discoverer, where no asteroids occur. In addition two families of asteroids, the Trojans, share Jupiter's orbit, lying some 60° ahead of and 60° behind the planet (the angles measured at the Sun).

The albedos of asteroids range from below 0·02 (darker than a blackboard) to about 0·4, and observations of their colors and spectra reveal that there are several distinct types of asteroid. About 75 percent (C-types) are dark, carbon-rich bodies apparenty similar in nature to carbonaceous chondritic meteorites (see below). About 15 percent (S-types) are reddish in color, have moderate albedos, and seem to have a high content of iron and magnesium silicates, while a smaller group (M-types) seem to be composed almost entirely of a mixture of iron and nickel. The S-types are found mainly in the inner part of the asteroid belt, while the C-types dominate the outer part. This distribution probably reflects the conditions prevalent at the time of the formation of the Solar System (▶ pages 122-3). The asteroids are probably debris left over from that event, some being planetesimals which never assembled into a single planet, others being fragments of larger asteroids that were broken up by collisions.

Meteorites

Interplanetary space contains a host of meteoroids, debris ranging in size from microscopic particles through sand-grain size to bodies tens or even hundreds of meters in diameter. Small meteoroids, which can enter the Earth's atmosphere at speeds of up to 70 kilometers per second, are destroyed (▶ page 96). Larger bodies, which survive their fiery passage through the atmosphere, are called meteorites. Meteorites of less than 100 tonnes are severely decelerated by the atmosphere and reach the ground – in one piece or in fragments – relatively gently, but more massive meteorites are hardly braked at all and can excavate craters. Meteorites are probably fragments of asteroidal collisions. This view is supported by the similarity between spectra of meteoritic types and those of the different classes of asteroid, and by the fact that the orbits of some meteorites can be traced to the asteroid belt.

Irons or stones

Traditionally, three types of meteorite have been recognized – stones (aerolites), irons (siderites) and stony-irons (siderolites). The stony meteorites are made up of silicates and contain chondrules, rounded droplets of rocky material often 1-2mm in diameter. Carbonaceous chondrites are an interesting subset of this class. They are rich in carbon compounds and about 5 percent of their mass is in the form of complex organic material. Iron meteorites contain mainly iron with about 5-10 percent nickel. The stony-irons are the least abundant of the main types and contain rock and metal in roughly equal proportions.

The irons, stony-irons and some of the stones show evidence of having been melted in the past, possibly in the interiors of larger bodies, but many of the stones contain material that has not been significantly modified since the formation of the planets. Most meteorites contain mineral crystals of similar age – about 4·6 billion years old – which is taken to be the age of the Solar System.

Tektites

Tektites are curious pebble-sized objects, the nature and origin of which is an intriguing puzzle. Although resembling volcanic glass in some respects, they differ significantly in composition from terrestrial volcanic glasses. Their distribution on Earth is also most peculiar. They are mostly confined to four large fields, in North America, Australasia, the Ivory Coast of Africa and Czechoslovakia (the Moldavite field). They differ considerably from other meteorites and, in particular, the absence of cosmic ray tracks within them suggests that they cannot have spent much time in space. Their origin is a mystery, but the most likely possibilities are either that they originated on Earth – from violent volcanic eruptions, or from impacts that hurled them through the atmosphere and back to Earth – or they were expelled from the Moon by impacts or volcanism in recent geological times.

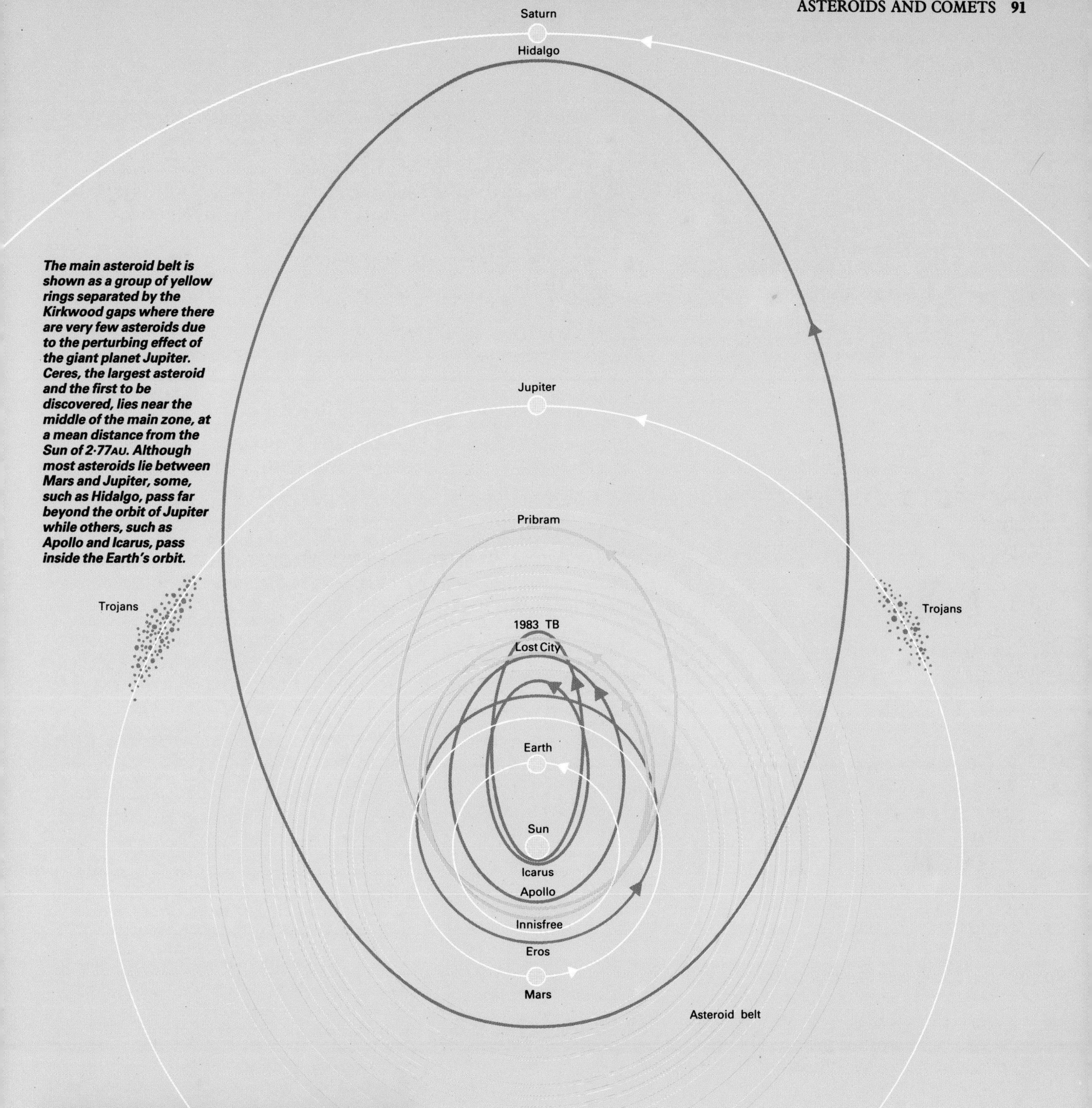

Saturn
Hidalgo
Jupiter
Pribram
Trojans
Trojans
1983 TB
Lost City
Earth
Sun
Icarus
Apollo
Innisfree
Eros
Mars
Asteroid belt
Kirkwood gaps

The main asteroid belt is shown as a group of yellow rings separated by the Kirkwood gaps where there are very few asteroids due to the perturbing effect of the giant planet Jupiter. Ceres, the largest asteroid and the first to be discovered, lies near the middle of the main zone, at a mean distance from the Sun of 2·77AU. Although most asteroids lie between Mars and Jupiter, some, such as Hidalgo, pass far beyond the orbit of Jupiter while others, such as Apollo and Icarus, pass inside the Earth's orbit.

Asteroid 1983 TB, discovered by the IRAS satellite, passes closer to the Sun than any other known asteroid. Its distance from the Sun ranges between 0·1AU and 2·5AU so that it travels from inside the orbit of Icarus to beyond that of Mars. Also shown are the pre-impact orbits of three meteorites (Pribram, Lost City and Innisfree) which indicate that these meteorites came from the asteroid belt.

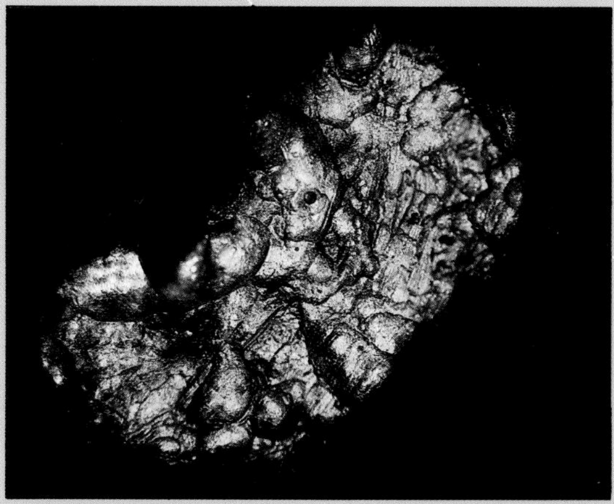

◄ Two types of objects. The iron meteorite (left) is comparatively common because of its durability. The rarer tektite (right) is quite different, and may not be of meteoritic origin.

On several occasions the Earth has passed through a comet's tail without suffering the slightest damage

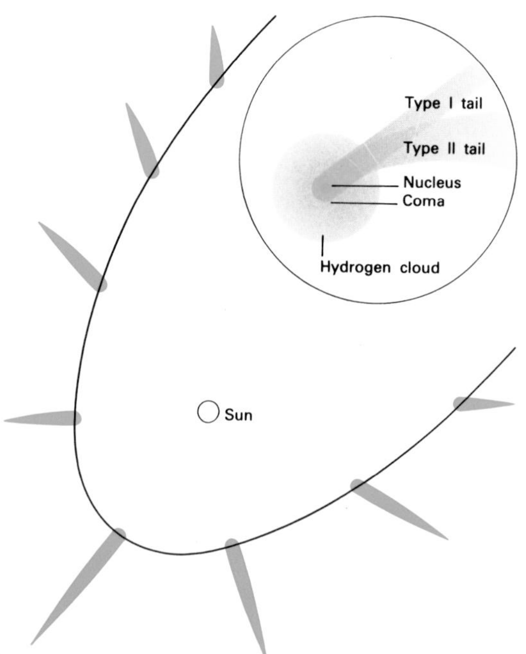

▲ **A comet's tail always points away from the Sun, following the head as it approaches and preceding it as the comet departs after perihelion. The icy nucleus is surrounded by the coma, a cloud of gas and dust, and some comets are enveloped in a huge hydrogen cloud. Type I plasma tails are fairly straight while Type II (dust) tails are curved.**

Comets

A comet consists of a compact nucleus surrounded by a cloud of gas and dust called the "coma", from which emerges a tail, or tails. In some larger comets the coma is surrounded by an extensive hydrogen cloud, millions of kilometers in diameter. Most comets follow long, highly elliptical orbits. As a comet approaches the Sun, the coma and tail begin to develop and become more conspicuous. The tail always points away from the Sun, so that it lags behind the head on the inward journey and precedes the head as the comet recedes.

There are two types of tail. Type I tails point almost directly away from the Sun and show considerable structure, whereas Type II tails curve gracefully but show little structure. Type I tails are typically as much as 10 times longer than Type II tails, and extend to tens or, exceptionally, hundreds of millions of kilometers in length.

Ultraviolet radiation from the Sun knocks electrons out of some of the atoms and molecules in the coma to produce electrically charged ions. These are swept away from the head of the comet by a stream of charged particles known as the solar wind (➧ pages 114-5). This gives rise to a Type I tail of gas which shines by the process of fluorescence, whereby ions absorb solar ultraviolet radiation and are thereby excited to emit visible light.

▼ **Austin's Comet of 1982, photographed by R. Arbour. The comet, a bright telescopic object, has a period so long that it is classed as non-periodic.**

Naming the asteroids

The discoverer of a minor planet has the right to choose a name for it – and some recent names are not too serious. R. S. Dugan, discoverer of minor planet 518, named it Halawe, after an Arabian sweet of which he was particularly fond. No. 694 is Ekard, "Drake" spelled backwards – the orbit was computed by students at Drake University. No. 1625 is The NORC, in honour of the Naval Ordnance Research Calculator at Dahlgren, Virginia. No. 1581 is Abanderada ("one who carries a banner") and No. 724 is Hapag, after the Hamburg-Amerika shipping line. The 250th asteroid was discovered by the Italian astronomer Palisa, who sold his right of naming for £50 to Baron Albert von Rothschild, who named the asteroid Bettina after his wife.

There are also the Trojan asteroids, which move in the same orbit as Jupiter. They were named after the heroes of the Trojan War, in Greek mythology, a conflict between the Greeks and Trojans culminating in a ten-year siege of Troy.

Chiron – the odd one out

One of the most unexpected discoveries was that of Chiron, by Charles Kowal from Palomar, in 1977. (Chiron should not be confused with Charon, the satellite of Pluto.) Chiron is large by asteroidal standards, with a diameter of several hundred kilometers, but it is far beyond the main swarm or even the Trojans, and its orbit lies mainly between those of Saturn and Uranus. Its revolution period is 50 years. Calculations show that in 1664 BC it approached Saturn to within 16,000,000km, which is not so very much greater than the distance between Saturn and its outermost satellite, Phoebe. Phoebe has a retrograde orbit, and may be a captured asteroid, in which case it and Chiron might be of the same nature. It has also been suggested that Chiron is in an unstable orbit, so that eventually it may be thrown out of the Solar System altogether.

Type II tails are made up of micrometer-sized dust particles driven from the cometary head by the pressure of light itself. These particles move away from the head at slower speeds than the ions and so get "left behind" as the head moves along its orbit, forming a curved tail which shines by reflecting sunlight.

According to the widely-accepted icy conglomerate or "dirty ice" model, the nucleus of a comet is a lump of ice, no more than a few kilometers in diameter, within which are embedded particles of silicate dust and larger rocky fragments. As the icy nucleus approaches to within about 3AU from the Sun, solar heating begins to vaporize its skin, releasing gases and dusty particles into the developing coma which can extend its diameter to between 100,000 and 1,000,000km.

Each time a comet passes the Sun it loses some of its mass. Some calculations indicate that a layer 1-3 meters thick may be lost from the nucleus each time, so that a comet cannot survive more than a few thousand perihelion passages before being completely disrupted. Many astronomers believe that fresh comets are perturbed from time to time into the inner Solar System from a vast reservoir of icy conglomerates – known as the Oort cloud – which extends some 40,000-50,000AU from the Sun and contains some hundreds of billions of potential comets.

▲ *"The figure of a fearful comet", from Dr Ambroise Paré's "Surgery", gives a subjective impression of the comet of 1664-5. Swords, daggers, coffins and men's heads were supposed to be omens of plague, and plague did indeed break out in London in June of 1665, within 3 months of the comet's visit.*

▲ *At the Tunguska event in Siberia in 1908, trees were blown down and animals killed. Possibly a fragment from a cometary nucleus caused the explosion.*

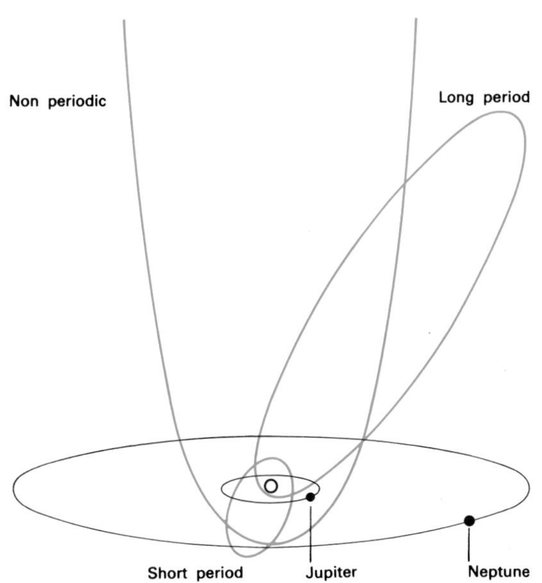

► *Periodic comets – those for which reasonably accurate orbits have been computed and which have been seen at more than one return – are divided into "short-period" (period of up to 20 years), "medium-period" and "long-period" (over 60 years). Many short-period comets have aphelion points close to the orbit of Jupiter; long-period comets travel beyond the orbit of Neptune. Comets of very long period are called "non-periodic" because it is impossible to compute their orbits accurately.*

Blaming the comets

Occasionally cometary nuclei must collide with the Earth, and it has been suggested that a great explosion in Siberia in 1908 – the Tunguska event – was due to part of the icy nucleus of a comet which exploded in the atmosphere and did not produce a crater as a meteorite might have done.

An intriguing and controversial suggestion has been made by British astronomers Sir Fred Hoyle (b. 1915) and Chandra Wickramasinghe (▶ page 128) that comets contain bacteria and viruses and that the incidence of epidemic diseases on Earth may be related to close encounters with comets. This view is hotly disputed, however, by most astronomers and biologists.

Malice in the skies

A famous description of one non-periodic comet was given in 1528 by a French doctor, Ambroise Paré: "This comet was so frightful, and it produced such great terror that some died of fear and others fell sick. It appeared to be of extreme length, and was the colour of blood. At the summit of it was seen the figure of a bent arm, holding in its hand a great sword as if about to strike. At the end of this point there were three stars. On both sides of the rays of this comet were seen a great number of axes, knives, and blood-colored swords, among which were a large number of hideous human faces, with beards and bristling hair."

No doubt the good doctor's description is wildly exaggerated, and he may even have been describing an aurora rather than a comet, but it illustrates the alarm with which comets were regarded in those days.

Halley's Comet

▲ Edmond Halley examined the records of comets that had appeared in 1531, 1607 and 1682, and deduced from his study of their orbits that it had been the same comet on each occasion. He predicted that it would return again in 1758, but regrettably did not live to see his prediction fulfilled. The comet now bears his name. Halley also showed that comets followed solar orbits that were consistent with Newton's theory of universal gravitation.

▲ Halley's Comet was first recorded in the year 240 BC, since when it has been observed returning to the inner Solar System at intervals of 76 years. This photograph was taken at its 1910 return, on 25 May from Helwan, Egypt.

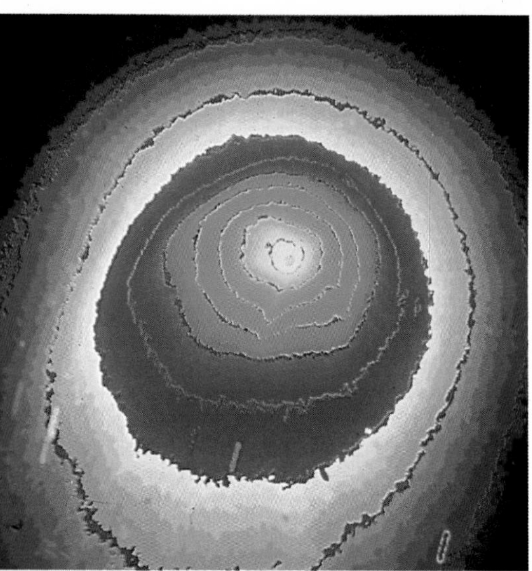

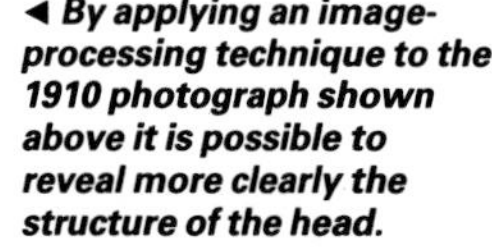

◀ By applying an image-processing technique to the 1910 photograph shown above it is possible to reveal more clearly the structure of the head.

An unusually bright comet

Practically all spectacular comets have very long orbital periods (and are known as non-periodics), far too long to be calculated with any precision. Those with periods of less than a few hundred years (periodics) tend to be faint objects without conspicuous tails. The one exception is Halley's Comet, which returns to perihelion at intervals of 75-76 years and is usually a conspicuous object on these occasions. It was last seen in 1910 and next returns to perihelion in February 1986, when it will not be so spectacular.

Comets have always been regarded as unlucky, and the great fear of them had not even been dissipated by the 1910 return of Halley's Comet, when an enterprising salesman made a large sum of money by selling anti-comet pills, though he did not explain just what they were meant to do.

History and Halley's Comet

Halley's Comet has made an impact throughout history. It appeared in AD 79, when the Roman emperor Vespasian commented that the comet "menaced rather the king of the Parthians; for he is hairy, while I am bald". (Comets had often been referred to as "hairy stars".) It is perhaps worth mentioning that Vespasian died in the same year.

The return of 1066 took place as the Normans were preparing to invade England, and a scene in the Bayeux Tapestry, said by some to have been designed by the Conqueror's wife, shows King Harold tottering on his throne, while his courtiers look on aghast as the comet blazes overhead. In 1546 the current pope Calixtus III publicly preached against the comet as an agent of the Devil.

Spacecraft to Halley's Comet

The 1986 return of Halley's Comet will be the most important in history – not because the comet will be favorably placed (indeed, the conditions are just about as poor as they could possibly be), but because of the striking technological advances that have been made since the last return, that of 1910. In particular, several space missions to the comet will be despatched.

Plans for a US probe have been cancelled, but there are to be two Russian probes, Vega 1 and Vega 2, which will go to Halley's Comet by way of Venus, dropping probes into the atmosphere of that planet as they pass by. The Japanese have two small probes, and the European Space Agency has one, Giotto, which will – it is hoped – penetrate the coma and send back details of the nucleus itself. All these probes will rendezvous with the comet after perihelion passage in February 1986, so that the comet will be receding from the Sun. It is unlikely that Giotto will survive the encounter, but with luck it may last for long enough to send back pictures from inside the coma.

▲ This episode from the famous Bayeux Tapestry depicts King Harold quaking on his throne while his courtiers tremble at the sight of Halley's Comet.

► Halley's Comet follows an elliptical orbit ranging from within the orbit of Venus to out beyond that of Neptune. Its motion is retrograde.

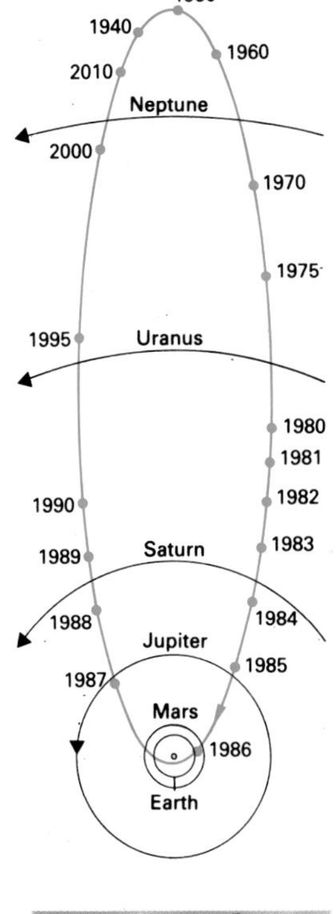

▲ The European spacecraft Giotto is due to pass the nucleus of Halley's Comet in 1986. It is named after the Italian artist Giotto di Bondoni (1267-1337) whose "Adoration of the Magi" (left) was probably inspired by the apparition of Halley's Comet in 1301.

See also
Planets and Orbits 9-20
Inside the Planets 33-40
Moons and Ring Systems 73-88
The Sun's Influence 113-20
The Origin of the Solar System 121-8

The death of comets

An orbiting satellite has photographed a comet disintegrating on impact with the Sun, but such a violent end must be rare. What happens to a comet when it has lost all its gases? Possibly the nucleus remains in orbit round the Sun in the guise of an asteroid. This theory could account for the so-called "Earth-grazers", such as Hermes, which are fundamentally different from the larger asteroids that are farther away from the Sun, and keep strictly to the zone between the orbits of Mars and Jupiter.

Two British astronomers, Victor Clube and Bill Napier, share this view. They believe that comets are formed from interstellar material, and that every time the Sun passes through one of the spiral arms of the Galaxy it collects a new swarm of comets. Inevitably some of these collide with the Earth and, according to Clube and Napier, one such collision occurred 65,000,000 years ago, causing so great a climatic change that the dinosaurs were unable to adapt to the new conditions, and became extinct.

Meteors and meteor showers

When a tiny meteoroid enters the atmosphere at high speed, its destruction is marked by a brief streak of light which lasts for at most a few seconds. These phenomena are called meteors. Most of the millions of meteors daily entering the atmosphere are too faint to be seen, but radar beams pick up their trails of ionized gas by day or night.

Sporadic meteors can appear at any time from random directions, while shower meteors appear fairly regularly at particular times of the year. Meteor showers occur when the Earth crosses the path of a stream of meteoroids strung out along a particular orbit round the Sun. Since the meteoroids are following parallel tracks, they seem to radiate from a particular point in the sky, known as the radiant. Each shower has its own particular radiant point.

Shower meteors are thought to consist of cometary debris – dusty particles spread out along the orbits of dead, or dim, elderly comets. A significant number of correlations between the orbits of meteor streams and of known comets have been established.

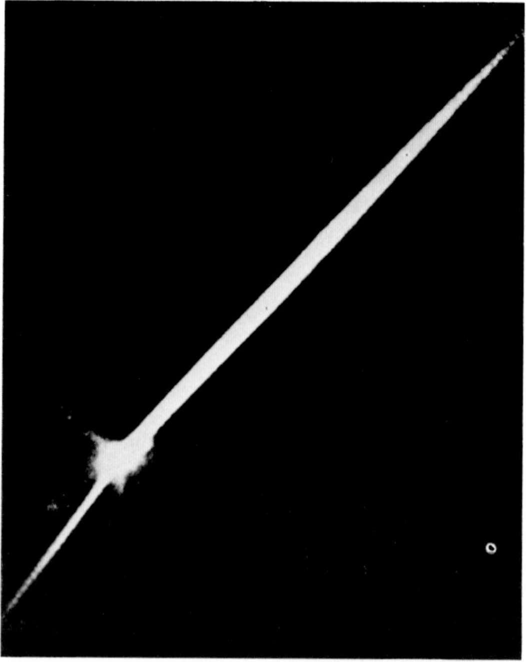

▲ The Bielid meteor shower is now very feeble, but in 1895 it produced this meteor which exploded in the Earth's atmosphere.

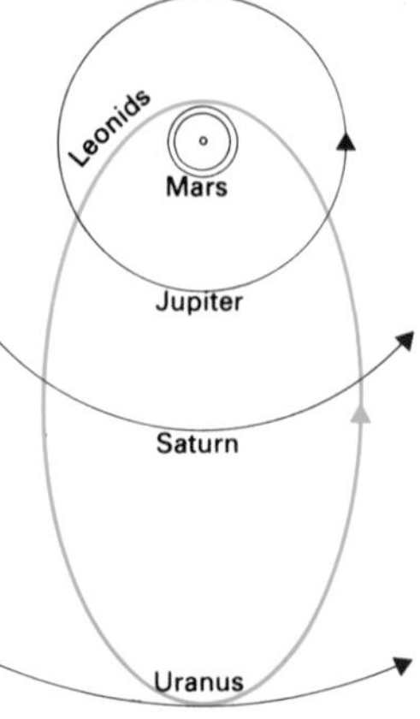

◄ The Leonid meteor shower occurs when the Earth crosses the orbit of a stream of meteoroids associated with the comet Tempel-Tuttle (left). Seen from Arizona, the Leonids produced a spectacular shower in 1966 (below); as indicated, all the individual meteors appear to radiate from one particular position in the sky because they are moving in parallel paths.

Shooting stars to watch for

Meteors, or shooting stars, are associated with certain comets. The Perseids in August are the most reliable shower, and occasionally the November Leonids are brilliant. They are associated with the periodic comet Tempel-Tuttle, which has a period of about 33 years. There were magnificent displays in 1799, 1833, 1866 and again in 1966 (the displays of 1899 and 1933 failed to materialize because the orbit of the main swarm had been perturbed by the giant planets and the Earth did not pass through it).

Comets lose material each time they pass the Sun, and by cosmic standards comets are short-lived. They have been observed in their death-throes. Biela's periodic comet broke in two during its return of 1845. At the comet's next appearance, in 1852, the pair had separated, and they were never seen again, although the debris of the defunct comet was seen in 1872 as a meteor shower coming from the direction in which the comet should have appeared. A few Bielid meteors are still seen every year, although the shower has now become very feeble.

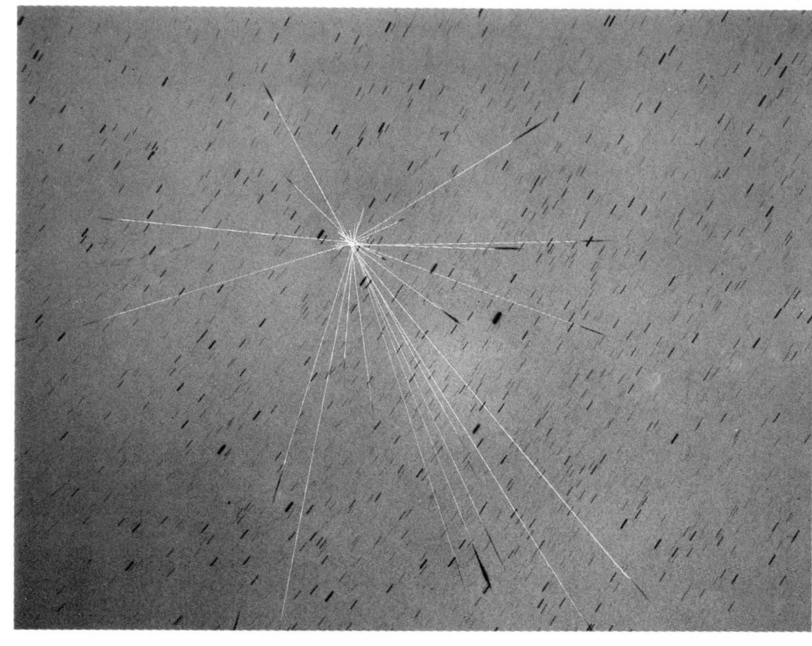

8 Inside the Sun

Our Sun as an ordinary star...Its central core: a nuclear powerhouse...The outer layers...The Sun's visible spectrum, its stellar fingerprint...Particles with no mass, a key to the Sun's interior...Is the Sun shrinking?... PERSPECTIVE...The discovery of infrared, "the invisible light"...Great pioneers of spectrum analysis... "Catching" particles from the Sun

The Sun is a large brilliant body at the center of the Solar System. Its diameter is 1·3 million kilometers, or more than 100 times the diameter of the Earth. This is small compared with some stars, however, the largest of which have diameters of up to 400 million kilometers.

The volume of the Sun is 1,392,000 times greater than that of the Earth, but its mass is only 330,000 times greater. It follows, therefore, that the mean density of the Sun is only 1,410 kilograms per cubic meter (that is, 1·41 times the density of water), which is about a quarter of the mean density of the Earth. This low mean density suggests that the Sun must be composed mainly of the lightest chemical elements. In fact, it is believed to be composed of about 73 percent hydrogen, 25 percent helium and 2 percent heavier elements.

The "visible" surface of the Sun is a thin gaseous layer from which practically all the Sun's light is emitted; it is called the photosphere (▶ page 105). The temperature of the photosphere is about 5,800K (about 5,530°C). Most of the Sun's prodigious outpouring of energy emerges in the form of near-ultraviolet, visible and infrared radiation but the Sun also emits lesser quantities of all kinds of radiation from gamma and X-rays to radio waves (◀ page 26). It also emits atomic and nuclear particles. The total amount of energy per second emitted from the Sun is known as the solar luminosity and is expressed in the power units "watts" in the same way as the power of a light bulb or electric fire is described. The solar luminosity is $3{\cdot}86 \times 10^{26}$ watts and, although the Earth intercepts only about 4 ten-billionths of this output, the amount of energy per second falling vertically on to one square meter at the top of the Earth's atmosphere is still 1,370 watts. This quantity is known as the solar constant.

The nature of light

The white light of the Sun is split up into a band of colors called a spectrum. This visible spectrum looks exactly like a rainbow, which is no coincidence because raindrops act as prisms and split up sunlight to produce a rainbow. Isaac Newton (1643-1727), who discovered the compound nature of light, preferred to think that light consisted of tiny particles which he called "corpuscles". By thinking of light traveling in waves, however, the explanation for the spectrum is clear. Light waves possess a range of frequencies – the number of vibrations passing every second – which correspond to the distance between the crest of one wave and the next – the wavelength; different frequencies produce different colors. The ordering of the seven colors of the rainbow – violet, indigo, blue, green, yellow, orange and red – reflects the range of frequencies. Violet is the wave motion with the highest frequency and therefore the shortest wavelength. Moving towards red the frequency falls and the wavelength rises. On hitting a prism, rays of different wavelength are bent by different amounts. The shorter the wavelength the greater the bending. In this way the colors of the visible spectrum can be distinguished.

◀ In 1666 Newton illustrated what happens when a controlled ray of light falls on a glass prism. By positioning a screen beyond the prism the ray could be seen to break down into a brightly colored band. By passing single colors through a hole in the screen on to a second prism, Newton proved that these colors were themselves primary.

▲ It was not until 100 years after Newton's discovery that it was recognized that the Sun's spectrum covered only visible wavelengths. In 1800 William Herschel discovered, by moving a thermometer with a blackened bulb along the solar spectrum, that beyond the red there are radiations which, like those of the visible spectrum, have a heating effect.

▶ *Anatomy of the Sun. A journey into the center of the Sun would take a traveler from the visible "surface" down through a convective zone where hot gases rise to the top and radiate their heat, to the region known as the radiative zone and thence to the hot, dense central core.*

At the center of the Sun the temperature is believed to be about 15 million K and the density about 160,000 kilograms per cubic meter (160 times the density of water). The central core, extending to about a quarter of the solar radius, is the nuclear powerhouse within which energy is generated and released primarily as highly energetic photons (quantum of radiant energy) – gamma rays and X-rays.

Surrounding the core, and extending out to about 80 percent of the solar radius, is the radiative zone, through which energy is transferred in the form of radiation. Within this region photons cannot travel far without colliding, or being absorbed and reemitted, each time in a random direction. An individual photon experiences so many events of this kind that it takes from 10,000 to 1,000,000 years to reach the solar surface. As photons move out through the radiative zone, their energies decline, and gamma and X-rays are thus transformed mostly into visible and near-visible radiation.

In the outer 15-20 percent of the solar globe, large-scale convection takes place. Hot gases rise towards the surface and cooler gases descend, and in this way heat is carried bodily to the photosphere from which it is radiated into space. This layer is known as the convective zone.

The outer layers

Above the photosphere is the chromosphere, a more rarefied layer a few thousand kilometers thick, within which the density of gases drops from about 0·00001kg per cubic meter to about 10^{-11}kg per cubic meter. The temperature drops from the photospheric value of nearly 6,000K to a minimum of about 4,300K at an altitude of a few hundred kilometers, but thereafter it increases rapidly to the transition zone which separates the chromosphere from the Sun's outer atmosphere, the corona. The corona extends to a distance of several solar radii and has a temperature of between 1 million and 5 million K. The density of coronal material is extremely low, and despite the high temperature the quantity of heat contained in the corona is only a minute fraction of that contained in the cooler but much denser photosphere. Temperature describes the speed at which individual atoms and electrons are moving around, and within the rarefied corona these speeds are very high indeed.

The amount of visible light coming from the corona is less than one-millionth of that emitted by the photosphere, and the corona is, therefore, far too faint to be seen without the aid of specialized instruments (▶ pages 110-11), except during a total eclipse, when the glare of the photosphere is obscured, for at most a few minutes, by the disk of the Moon.

A stream of atomic particles, mainly protons and electrons (▶ page 100), flows out from the corona into interplanetary space. This is known as the solar wind (▶ pages 114-15).

The Sun in cross-section
1 Filament
2 Faculae
3 Sunspot
4 Core
5 Radiative zone
6 Convective zone
7 Photosphere
8 Chromosphere
9 Corona
10 Prominence
11 Solar wind

1

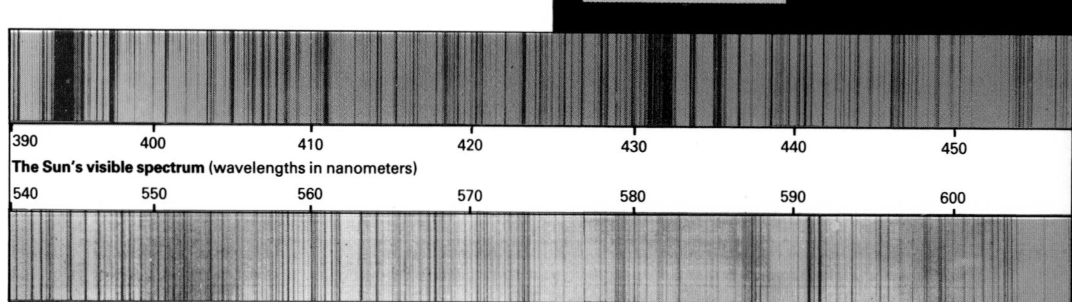

▶ ***The spectrum of the Sun, showing the full range of visible wavelengths from 390 to 690nm. The Sun's spectrum is continuous with dark absorption lines which originate in the solar atmosphere. Each line indicates a particular chemical element. The more prominent are calcium (393, 397nm) and hydrogen (656nm)***

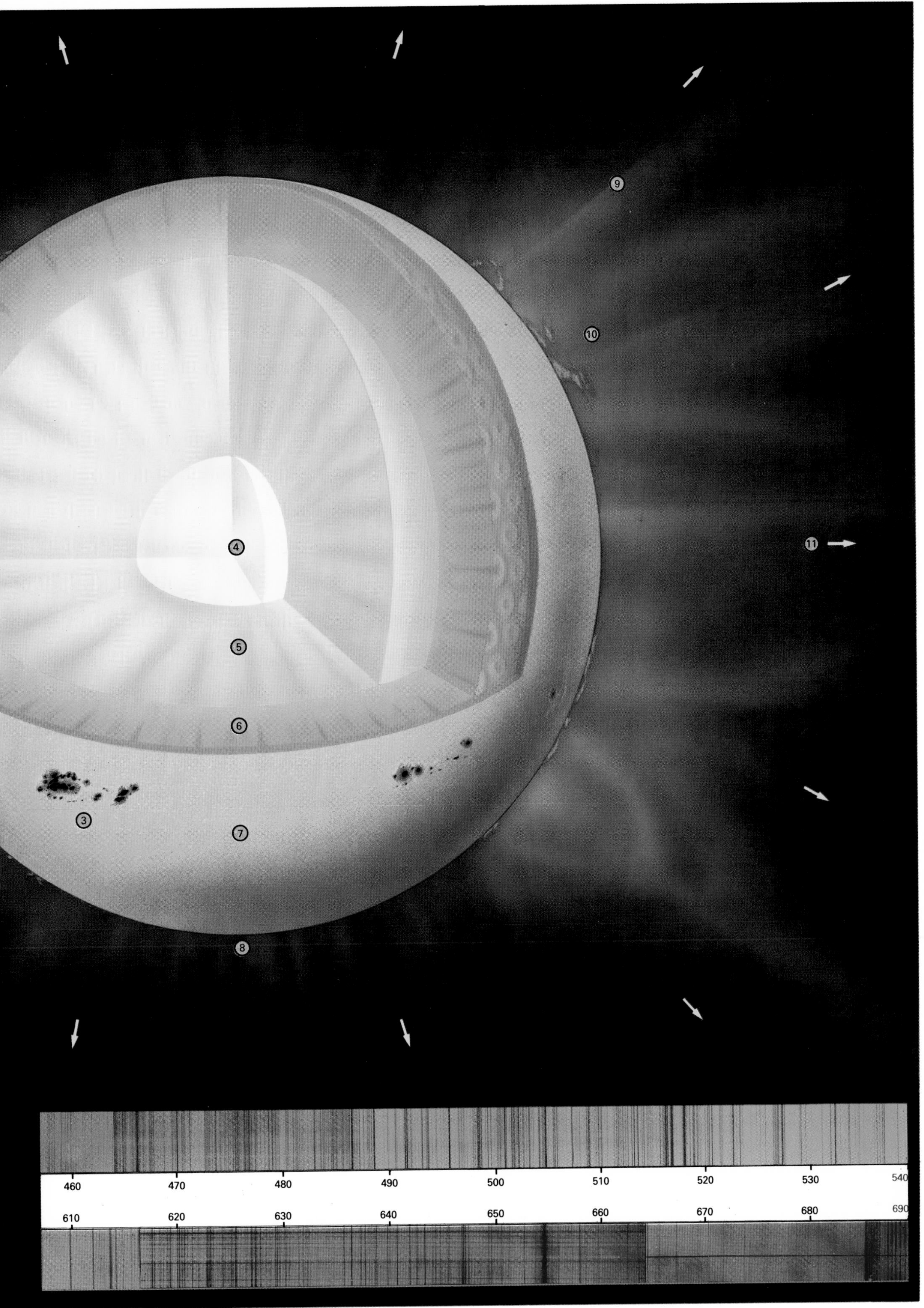
9
10
11
4
5
6
3
7
8
460
470
480
490
500
510
520
530
540
610
620
630
640
650
660
670
680
690

The origin of spectroscopy

In 1802 the English chemist William Wollaston (1766-1828) studied the spectrum of the Sun, using a slit to produce a band of light instead of a series of overlapping disks, as Herschel had done. Wollaston observed a series of dark lines crossing the rainbow of colors, but he thought that they merely marked the boundaries between neighboring bands of color. The true interpretation was first revealed by the work of German physicist Joseph von Fraunhofer (1787-1826), followed by that of his fellow German scientists Gustav Kirchhoff (1824-1887) and Robert Bunsen (1811-1899).

Fraunhofer produced a solar spectrum in the same way as Wollaston had done, but realized that the lines did not mark any particular boundaries of colors. Moreover, they were permanent and never changed either in position or in intensity. Fraunhofer produced the first "map" of the Sun's spectrum, showing several hundred lines. Given time, he might have produced a satisfactory theory to explain them, but he died before his 40th birthday in 1826. The explanation finally came in 1859, as a result of work carried out by Kirchhoff and his colleague Bunsen.

The photosphere emits light of all wavelengths ("white light"). When sunlight is passed through a glass prism, the constituent wavelengths are spread out, and the light split into a rainbow band of colors from red to blue and violet. Superimposed upon this continuous spectrum are thousands of dark absorption lines. These are produced in the lower reaches of the solar atmosphere where atoms of different chemical elements each absorb light at their own particular set of wavelengths. Absorption lines in the spectrum thus identify particular chemical elements, rather like a set of fingerprints.

The way in which these lines are produced depends upon the physics of the atom. The simplest atom, hydrogen, consists of a central proton (a heavy particle of positive electrical charge), around which there revolves a single electron (a light particle of negative electrical charge). According to quantum theory, which explains the emission and absorption of quantities of energy by a hot body, this electron can exist only in one of a number of possible orbits, each of which corresponds to a different "energy level" of the atom. If an electron absorbs a quantity of energy exactly equal to the difference in energy between its initial level and a higher one, it will jump up to a higher one. Conversely, if an electron drops down from a higher level to a lower one, it will emit a quantity of energy equal to the energy difference between the two levels.

Light can be regarded either as a wave motion or as a stream of particles called photons. Each photon carries a certain amount of energy, and the higher the energy of the photon, the shorter the wavelength of the light – that is, wavelength is inversely proportional to energy. White light consists of a mixture of photons of all kinds of energy levels. If an electron absorbs a photon of sufficient energy to lift the electron to a higher orbit, this produces an absorption line in the solar spectrum. The wavelength of each absorption line corresponds to the difference in energy between two levels within an atom of a particular chemical element. Each element produces its own distinctive pattern of absorption lines corresponding to the various possible "jumps", or transitions, between its different energy levels. When an electron drops down from a higher to a lower orbit, a photon of the appropriate energy and wavelength is emitted, and this can produce an emission line spectrum.

Which lines are present in the spectrum, and whether they are conspicuous or faint, broad or narrow, single or multiple, depends upon factors such as chemical composition, temperature, pressure, rate of rotation and the presence or absence of magnetic fields.

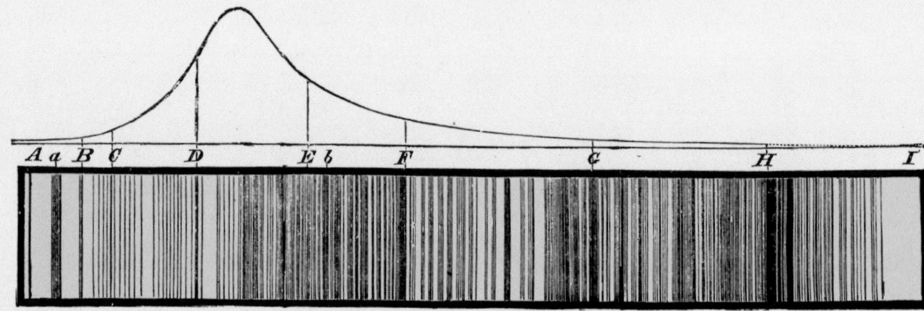

◀ *Josef von Fraunhofer made a grating of parallel lines which produced spectra by diffracting light. He used the grating to study the solar spectrum and in 1814 produced a map of it showing 324 dark lines, now known as Fraunhofer lines (above). These "absorption lines" proved to be the key to a vast store of information about the Sun and stars.*

◄ Gustav Kirchhoff extended the work of Josef von Fraunhofer, and discovered experimentally the connection between dark absorption lines and bright emission lines.

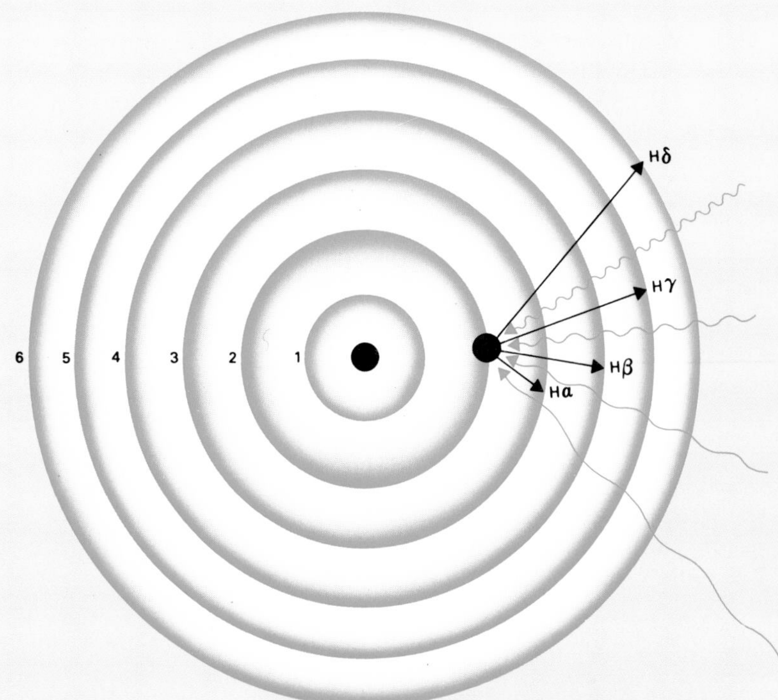

▲ When an electron, initially in permitted orbit number 2, absorbs light of the correct energy, it will jump up to a higher orbit. This process causes absorption lines in the visible spectrum. For example, red light of wavelength 656·3nm is absorbed when an electron jumps up from level 2 to level 3 (line Hα), and green light of 486·1nm is absorbed in the jump from level 2 to level 4 (line Hβ).

Discoveries by Kirchhoff

Kirchhoff found that incandescent solids, liquids or gases at high pressure will produce a rainbow or "continuous" spectrum, whereas gases at low pressure yield an "emission" spectrum, made up of disconnected bright lines. Each line is the "trademark" of an element or group of elements, and cannot be imitated. Two bright yellow lines, for example, in a particular part of the spectrum, can only indicate the presence of sodium.

Kirchhoff knew that the Sun's photosphere yields a rainbow spectrum. Above the solar surface there are low-pressure gases which would usually yield bright lines, but against a bright background the lines are "reversed", and appear dark. Their position and intensities remain unchanged, however, so they can still be identified: two prominent dark lines in the yellow part of the Sun's spectrum also indicate the presence of sodium.

Kirchhoff's explanation marked the beginning of astronomical spectroscopy – the study of spectra. One element was found in the Sun before it was detected on Earth. Some of the solar lines did not correspond to any known element, and it was therefore deduced that they must be the signature of an element not yet discovered on Earth. The element was named helium. Not until later was helium finally isolated in the laboratory.

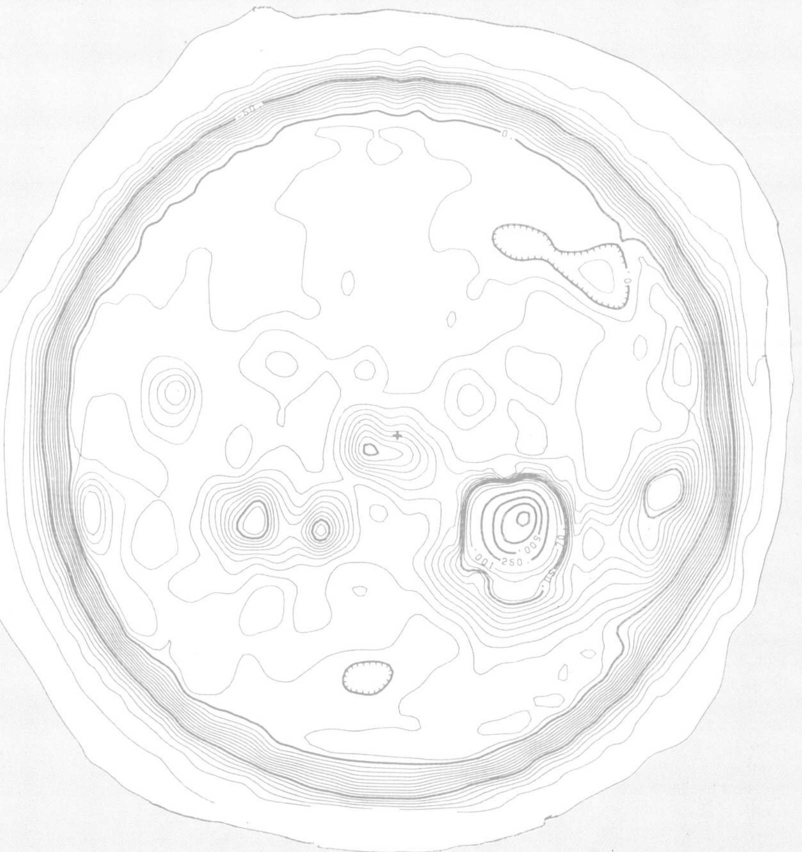

► This contour map shows the Sun at a wavelength of 2·8cm – that is, in the radio region of the spectrum. Emissions at radio wavelengths increase over active regions, and radio bursts may last for hours or for less than a second. The map was made with the 100m telescope at the Max Planck Institute in Bonn, West Germany.

One of the world's most unusual observatories is a mile underground, in a gold mine near Deadwood Gulch!

The Sun generates energy in its core by the process of fusion, whereby light elements are welded together to form heavier ones with an associated release of energy. In the case of the Sun, and in similar "middle-aged" stars, the dominant process is thought to be the proton-proton reaction, in which nuclei of hydrogen atoms (that is, protons) are fused together to form nuclei of helium atoms (the nucleus of a helium atom consists of two protons and two electrically neutral neutrons). In each reaction four protons are welded into one helium nucleus, but two of these are converted into neutrons.

In each reaction a small quantity of mass (*m*) is converted into energy (*E*) in accordance with Einstein's famous relationship $E=mc^2$. This demonstrates that mass and energy are interchangeable and that since *c* denotes the speed of light (a large number), enormous quantities of energy are released by the destruction of relatively small quantities of matter. In each individual reaction only 0·7 percent of the mass involved is converted into energy, but so many reactions occur that the Sun uses up more than 4 million tonnes of matter per second in order to maintain its luminosity.

Despite this prodigious loss of mass the Sun has enough fuel to keep it shining at its present rate for a further 5 or 6 billion years. It is thought to be just under 5 billion years old at present, and to be about halfway through the stable period of its existence.

Peculiar particles from the Sun

One of the most intriguing puzzles of present-day solar research is the neutrino problem. Neutrinos are peculiar particles which carry a certain amount of energy but have zero electrical charge and zero mass (recently it has been suggested that the neutrino does, in fact, have a tiny mass, but this is as yet unproven), and which can penetrate great masses of material. The nuclear reactions in the Sun are thought to produce vast numbers of neutrinos, most of which, traveling at the speed of light, shoot right through the solar globe into space. Some of these reach the Earth just over 8 minutes after being produced in the solar core, and measurements of the number arriving (the "neutrino flux") are made at the observatory in Brookhaven. There is a similar observatory in the Soviet Union.

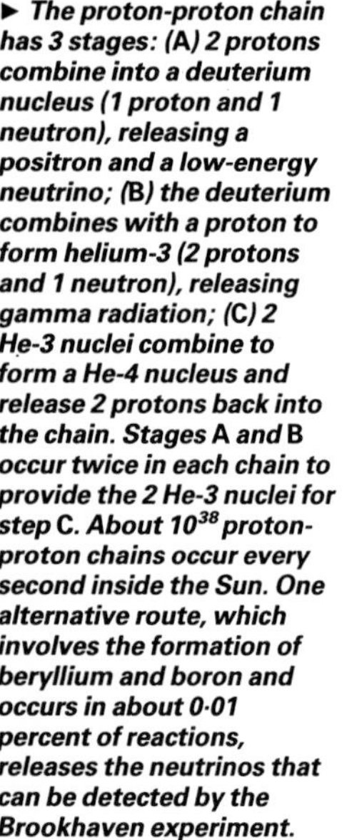

► The proton-proton chain has 3 stages: (A) 2 protons combine into a deuterium nucleus (1 proton and 1 neutron), releasing a positron and a low-energy neutrino; (B) the deuterium combines with a proton to form helium-3 (2 protons and 1 neutron), releasing gamma radiation; (C) 2 He-3 nuclei combine to form a He-4 nucleus and release 2 protons back into the chain. Stages A and B occur twice in each chain to provide the 2 He-3 nuclei for step C. About 10^{38} proton-proton chains occur every second inside the Sun. One alternative route, which involves the formation of beryllium and boron and occurs in about 0·01 percent of reactions, releases the neutrinos that can be detected by the Brookhaven experiment.

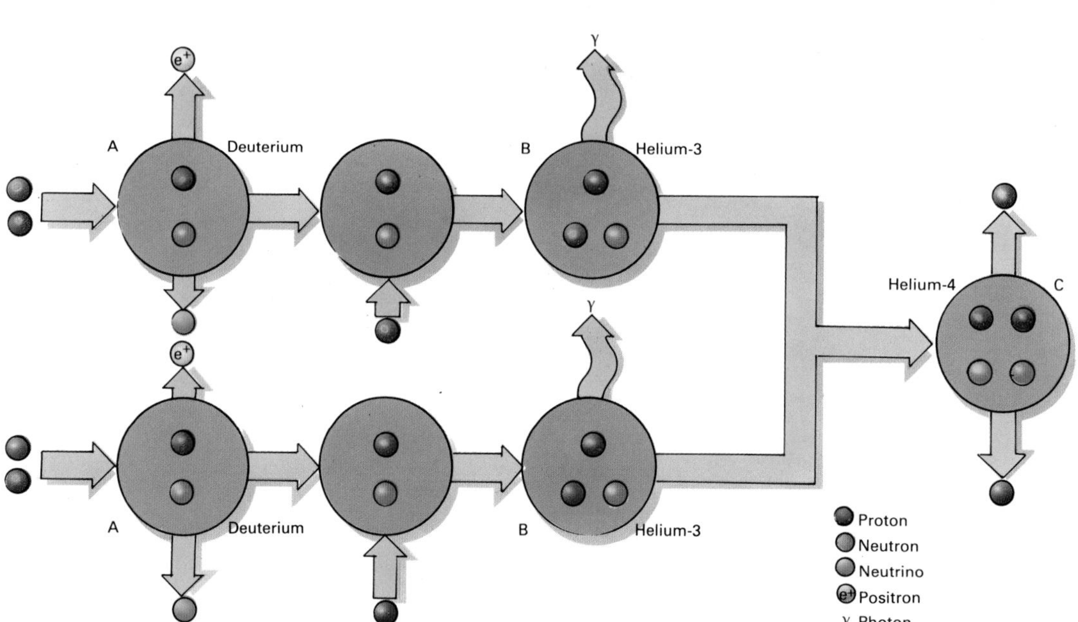

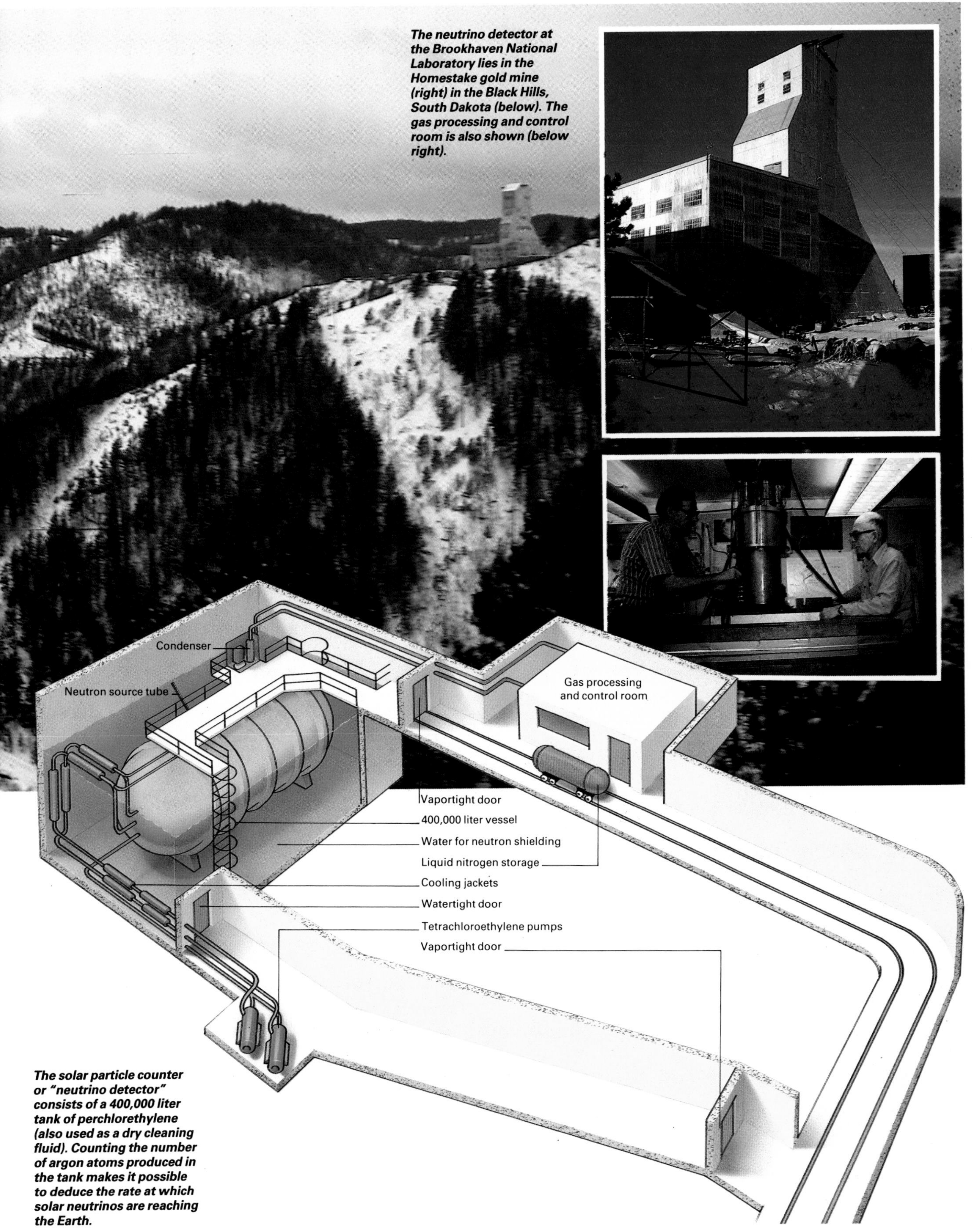

The neutrino detector at the Brookhaven National Laboratory lies in the Homestake gold mine (right) in the Black Hills, South Dakota (below). The gas processing and control room is also shown (below right).

The solar particle counter or "neutrino detector" consists of a 400,000 liter tank of perchlorethylene (also used as a dry cleaning fluid). Counting the number of argon atoms produced in the tank makes it possible to deduce the rate at which solar neutrinos are reaching the Earth.

See also
The Surface of the Sun 105-12
The Sun's Influence 113-20
The Origin of the Solar System 121-28
The Basic Properties of Stars 145-52
Birth, Life and Death of Stars 153-60

Although neutrinos are extremely difficult to "catch", detectors have been in operation since 1964. At no stage, however, have these instruments detected the number of neutrinos predicted in theory. The average detection rate, in fact, is about one-third of the theoretical value, which seems to suggest that there is something wrong with present-day understanding of the solar interior or of the reactions which are taking place there. If the core were about 10 percent cooler than is thought, the scarcity of neutrinos could be explained, but this would raise other major problems. Perhaps the current rate of energy generation in the core is lower than normal. It has even been suggested that the Sun has a small iron core or contains a small black hole, both of which could, in principle, account for the low neutrino flux. Yet another possibility is that neutrinos, if they do have tiny masses, can decay into three different types, only one of which can be picked up by existing detectors (thus the neutrino flux would be one-third of the expected value). As yet the problem remains unsolved.

The vibrating and shrinking Sun

Just as a bell vibrates at its natural frequency when struck, so the Sun is vibrating at a number of "frequencies" or periods ranging from a few minutes to a few hours. The amplitudes of these vibrations are small, but they can be used to study the solar interior. Some of these oscillations fit in well with the standard model of the Sun, but others do not. Once again, the problem remains to be resolved.

An analysis of daily measurements of the solar diameter made from 1836 to 1954 led American solar physicist John Eddy to suggest that the Sun is shrinking at the surprisingly high rate of 0·1 percent per century. Such a trend could not be a long-term one, for it would imply that the Sun was twice its present size just 100,000 years ago, and would shrink to the size of a pinhead in 100,000 years' time. Recent measurements indicate that the Sun shows small variations in diameter, perhaps over a number of different periods ranging from decades to centuries. The best-established variation seems to be an 80-year cycle with an amplitude of 0·02 percent of the solar diameter.

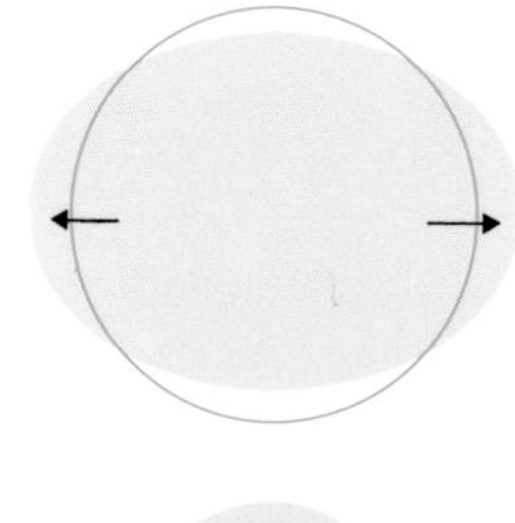
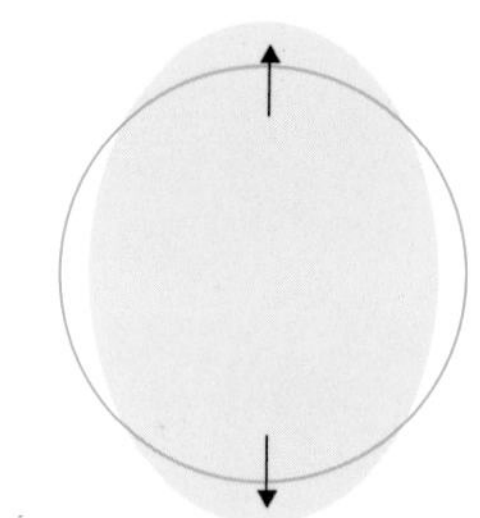
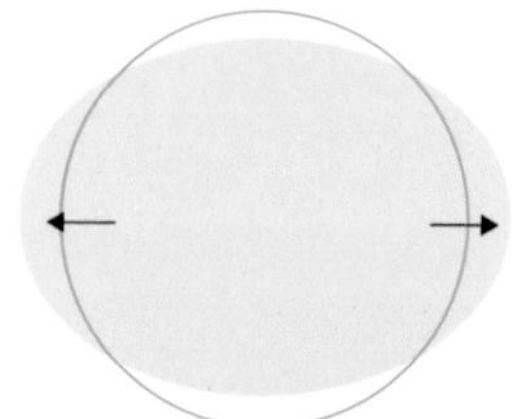

◀ The solar globe vibrates rather like a jelly at a number of different frequencies. The 5 minute oscillation is only a surface ripple, but longer-period oscillations represent reverberations of the entire body of the Sun. The basic oscillation causes the Sun to bulge alternately, by a few kilometers, at the equator and at the poles. Analysis of these vibrations gives information about the internal structure of the Sun. One of the observed periods (of 52 minutes) is consistent with established ideas about the interior, but another (2 hours 40 minutes), requires the Sun to be much more uniformly dense inside than is thought to be the case.

▼ Methods of determining the Sun's diameter include timing eclipses, timing transits of Mercury, and measuring the Moon's shadow during eclipses (2). Results for transits (disks) and eclipses (circles) show periodic fluctuations (1).

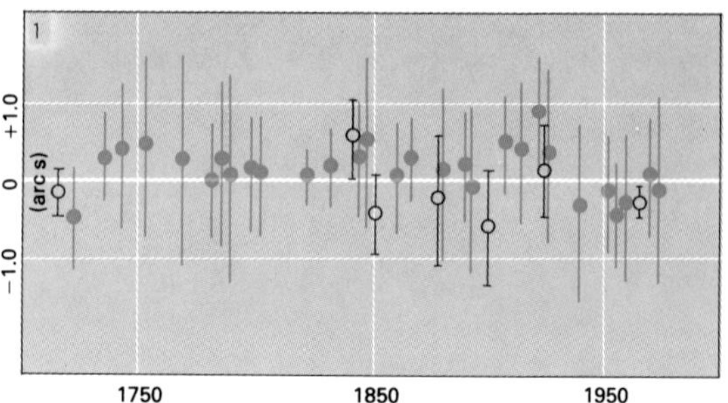

▲ Measurements of the size of the Moon's shadow during the eclipse seen from Java in 1983 were marginally smaller than the "standard diameter".

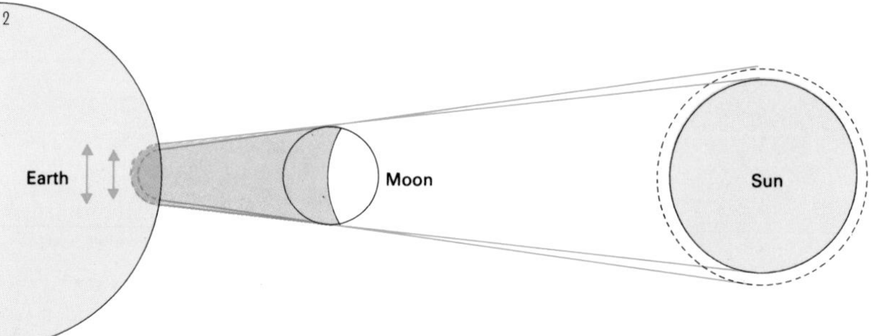

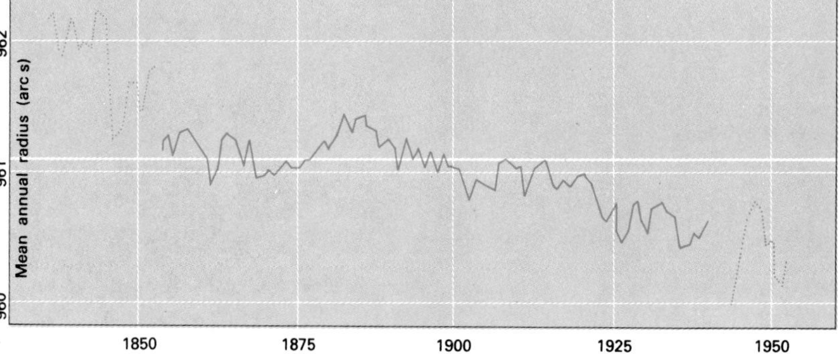

▲ Possible change in the Sun's east-west diameter is based on records of the time taken for the Sun to cross the meridian (solid lines denote better results).

The Sun's Outer Layers

9

Structure of the Sun's photosphere...The formation and cyclic variation of sunspots...An explanation of sunspots...Prominences and flares...The activity of the corona...PERSPECTIVE...Modern solar telescopes... Observing at specific wavelengths...Viewing the corona at total eclipse...Historical reactions to solar eclipses

Almost all of the Sun's light comes from a region called the photosphere (literally "sphere of light"). The solar disk appears less bright towards the edge (the "limb") than in the center. This phenomenon, known as "limb darkening", arises because the Sun is gaseous and partly transparent. When looking at the center of the disk an observer sees vertically down to the deeper layers of the photosphere, which are hot and bright. When looking towards the edge of the disk he is looking at an angle into the cooler and less bright outer layers.

The photosphere consists of millions of bright granules which form, shift about and vanish with an average lifetime of about eight minutes. The individual granules are about 1,000km in diameter, and each one is a cell within which hot gases rise from the interior, then spread out, cool and descend once more into the darker lanes between the granules. There is also a larger-scale network of supergranular cells, each of which contains hundreds of individual granules. They represent patterns of circulation that extend deeper into the Sun.

Sunspots

The most easily seen phenomena on the photosphere are sunspots. A typical sunspot consists of a dark central region, the "umbra", surrounded by a less dark region, the "penumbra", made up of light and dark threads extending out from the center like the spokes of a wheel. Sunspots are cooler than the rest of the photosphere and so appear dark by contrast. The temperature in a sunspot umbra is typically just over 4,000K, compared with about 5,500K in the penumbra and 6,000K in the granules of the photosphere.

Sunspots range in size from tiny pores, about the same size as the granules, to complex structures covering billions of square kilometers. About five percent of spots are large enough to be seen with the naked eye when conditions are favorable. A typical group consists of a pair of spots (or more complex regions) having opposite magnetic polarity, one spot behaving as a north magnetic pole and the other as a south magnetic pole. The magnetic field in the umbra may have a strength of between 1,000 and 4,000 gauss (0·1 to 0·4 Tesla), which is nearly ten thousand times stronger than the magnetic field at the surface of the Earth, and the pattern of magnetic field lines between a sunspot pair is similar to that around a normal bar magnet. Most spot groups last for about two weeks on average, but the largest ones can survive considerably longer. Spot groups are often accompanied by brighter gaseous patches, or "faculae", which lie higher in the photosphere and which often last longer than the spots themselves.

Sunspot observations have shown that the Sun rotates faster at the equator than at the poles: the rotation period is 25 days at the equator, and about 35 days near the poles. Sunspots appear to rotate round the Sun about four or five percent faster than the adjacent photosphere.

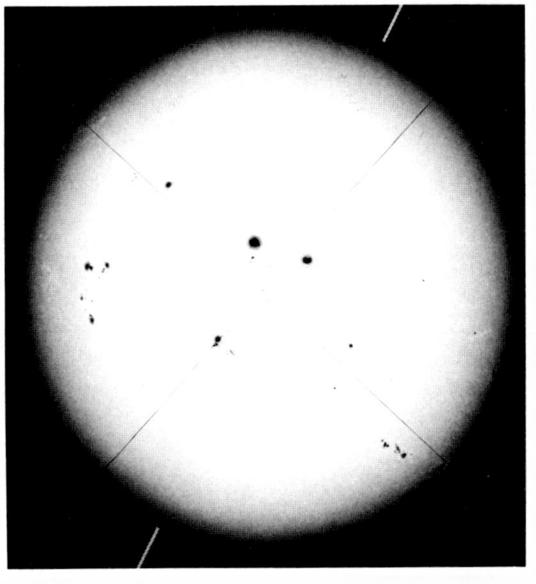

▲ This photograph of the solar disk was taken close to sunspot maximum in April 1970, and shows many spots. Limb darkening – the fading of the disk towards the edge – is also evident.

◄ The solar surface rotates faster at the equator than at the poles. Therefore, after one complete rotation, spots closer to the equator will have moved ahead of those at higher latitudes.

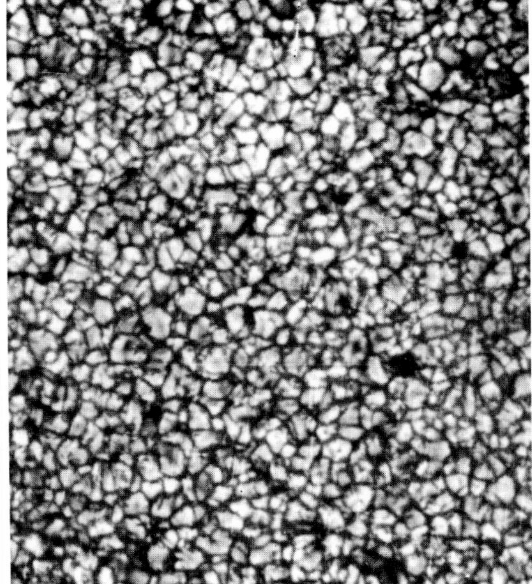

▼ The "granules" of the photosphere are evidence of seething convective activity near the surface of the Sun. Hot gases rise to the top, then sink back down in the darker lanes between the granules.

At present the sunspot cycle is about 11 years long, but is it a permanent feature of the Sun's activity?

The numbers of sunspots vary in a cycle lasting about 11·1 years. At the start of this cycle, activity is at a minimum and few if any sunspots are present. For the next four or five years activity increases towards a maximum, when the disk is heavily spotted, and for the following five or six years it declines again to the next minimum. The level of activity varies considerably between successive cycles, and there is evidence that the varying level of activity at successive maxima reflects a longer-term cycle lasting about 80 years. This cycle may be linked to fluctuations in the diameter of the Sun (◂ page 104). Historical records show that there have been several major gaps in the sunspot cycle, in particular the "Maunder minimum" which lasted from 1645 to 1715, during which sunspot and related activity seems to have been almost completely absent.

The latitudes at which sunspots appear vary throughout the cycle. At the start, spots occur at latitudes 30° to 40° north and south of the Sun's equator; at maximum most spots are at around 15° latitude, and by the end of the cycle they are quite close to the equator.

Magnetic polarity reverses in each successive cycle. In one cycle all the leading spots in groups in the northern hemisphere will have north magnetic poles, and all the leading spots in the southern hemisphere will have south magnetic poles. In the next cycle the opposite pattern will hold in both hemispheres.

A possible explanation of sunspots

Although there is no theory which completely accounts for all aspects of sunspots and the sunspot cycle, the Babcock-Leighton model gives a good explanation of most of the phenomena. If at some starting date the lines of force of the solar magnetic field lay along lines of longitude from the north to the south poles of the Sun, in time differential rotation would stretch and wind these lines round and round the Sun, concentrating the field where lines were tightly bunched together. Where kinks occurred in the field a tube of force would erupt through the solar surface, giving rise to a pair of spots where it emerged from and reentered the photosphere. A spot of north magnetic polarity would arise where the magnetic lines of force emerged and a spot of south polarity where they reentered the solar surface. The winding-up operation would move the bands of sunspot activity progressively closer to the equator, after which the lines would unwind to give a situation similar to the starting conditions but with the overall polarity reversed. Then a new cycle would begin.

▲ The heliostat at Kitt Peak National Observatory.

► A Kitt Peak solar magnetogram showing regions of negative (dark blue) and positive (yellow) polarity.

Modern solar telescopes

The most advanced solar telescope in existence is that at the Kitt Peak National Observatory in Arizona. This has an upper mirror, or "heliostat", 2m in diameter, placed at the top of a 2,089m tower. The heliostat is rotatable, allowing it to follow the Sun across the sky and to reflect its light continuously down a sloping tunnel. At the bottom of the tunnel is a 1·5m mirror which reflects the image back up the tube to yet a third mirror, which in turn reflects the light down into the observation room. The advantage of this arrangement is that heavy equipment need not be moved at all.

The Kitt Peak telescope can produce a solar image 76cm in diameter, containing a large amount of detail. It can also be used at night for studying the spectra of stars.

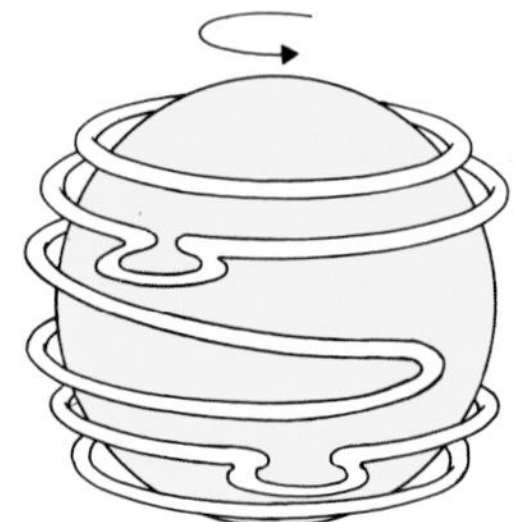

◄ Because of the Sun's differential rotation, magnetic field lines are stretched and twisted below the solar surface. Where concentrated bundles of them develop kinks, loops of magnetic force penetrate the photosphere and cause pairs or groups of sunspots to form.

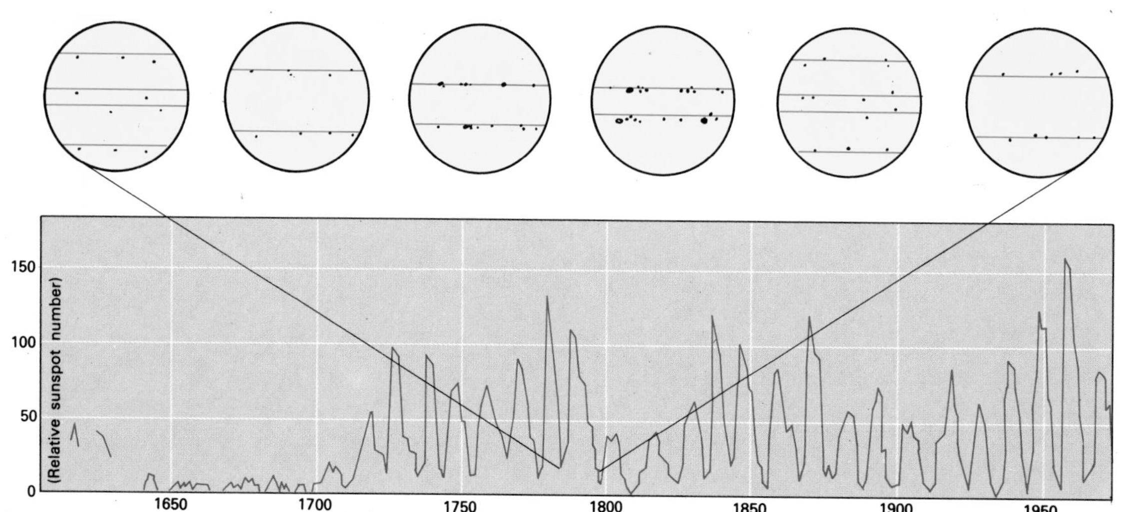

◄ This series of solar disks shows schematically the changing appearance of the Sun during a typical 11-year cycle. At first a few spots of the previous cycle may be seen near the equator; spots of the new cycle appear at higher latitudes. As the cycle continues spots tend to occur in two bands moving progressively closer to the equator.

◄ The variation in mean annual sunspot numbers over a period of several centuries clearly shows the 11-year cycle. The gap in activity (1645-1715) is the "Maunder minimum".

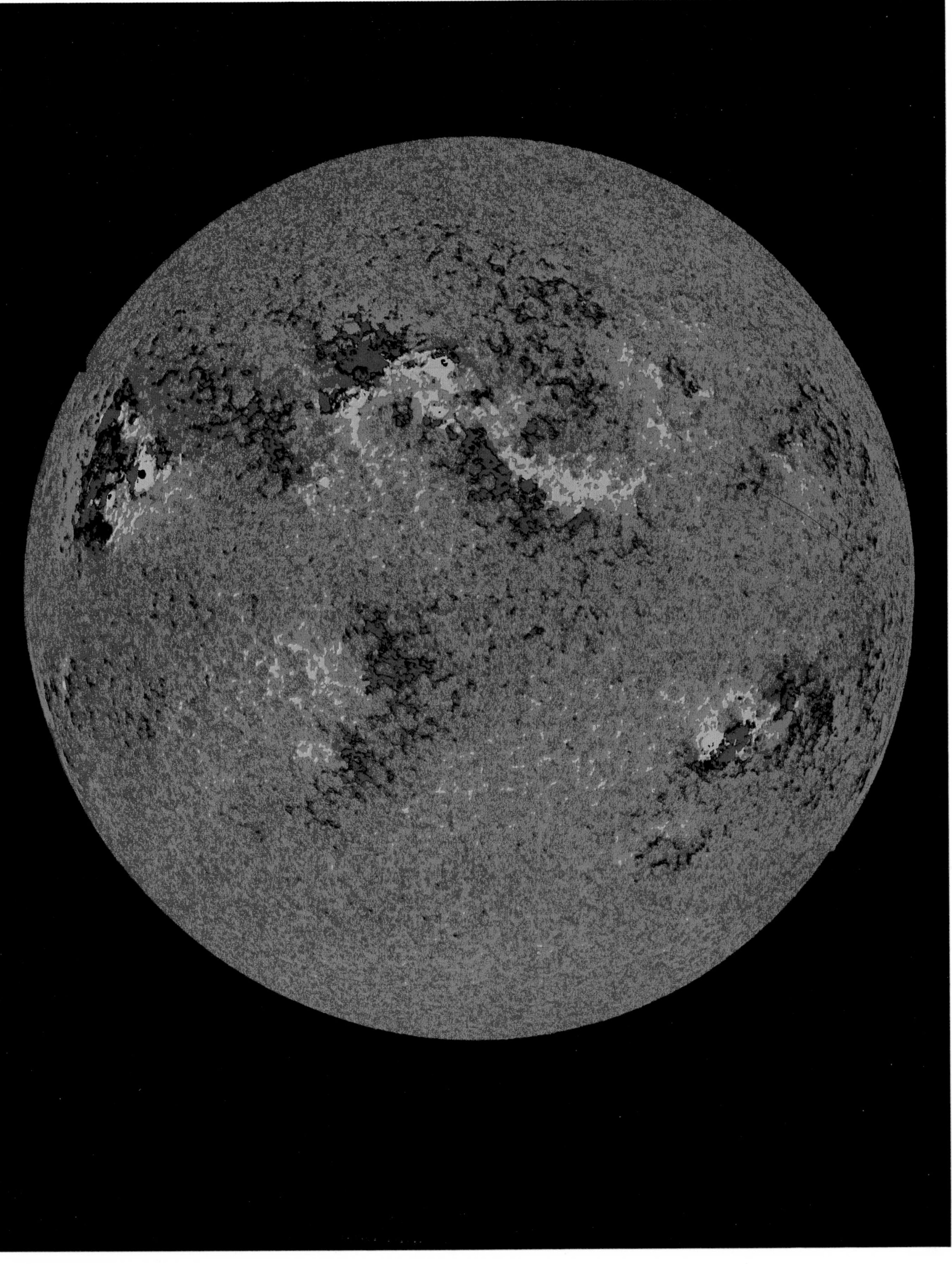

Spicules

Outside the Sun's photosphere is a gaseous layer, the chromosphere, which varies in thickness from 2,000km to about 10,000km. It consists of hundreds of thousands of hot gaseous tongues or "spicules", with temperatures of between 10,000K and 20,000K. They rise to heights of some 10,000km before falling back, and have an average lifetime of five to ten minutes. On a larger scale the chromosphere shows a cellular structure known as the "chromospheric network", the lines of which are marked by bunches of spicules concentrated at the boundaries of the supergranules in the underlying photosphere.

Prominences

The most beautiful of solar phenomena are the prominences. These are clouds, tubes and tongues of gases at lower temperatures but higher densities than the surrounding material of the upper chromosphere and corona. If they occur at the edge of the solar disk they have a flame-like appearance. They are easiest to see in wavelengths at which hydrogen and helium emit (or absorb) light, although they can be seen directly during a total eclipse of the Sun. When examined in wavelengths at which they absorb light, they appear against the solar disk as long, dark filaments.

Some prominences are quiescent – they hang in the corona, suspended by lines of force of the local magnetic field, for months or even for a year or more. Others are active, short-lived phenomena showing dramatic changes in minutes. The most violently active hurl tongues or loops of material to heights of several hundred thousand kilometers, and may even expel material from the Sun altogether.

Flares

Flares are the most violent of solar events. Energy stored in the warped and twisted magnetic fields associated with complex sunspot groups is suddenly, explosively released – nobody knows exactly how – and atomic particles are blasted forth, shock waves surge across the solar surface and through its atmosphere, and all kinds of electromagnetic radiation from gamma rays to radio waves are emitted. A flare reaches peak brightness within a few minutes and then declines, the total duration varying from a few minutes to a few hours.

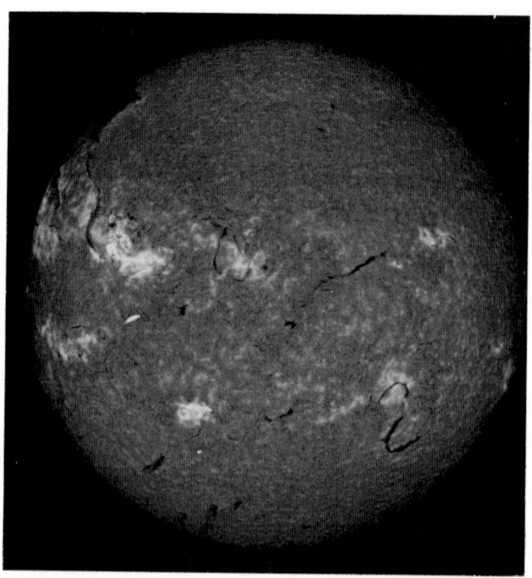

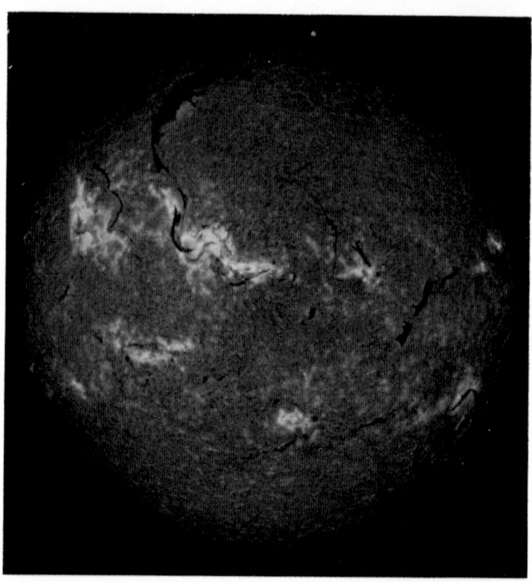

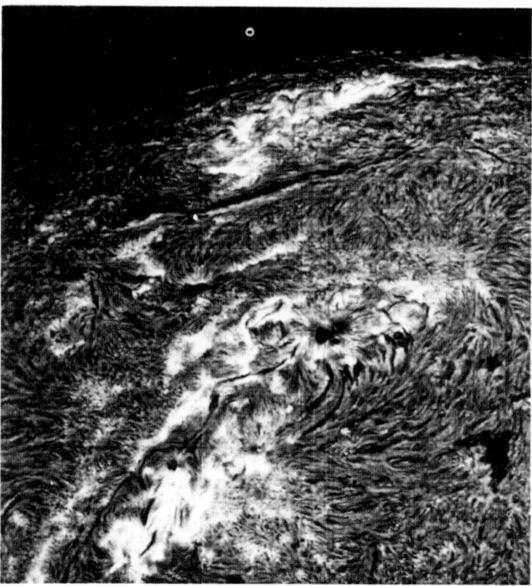

◄ *Spectroheliograms are of great value in all studies of solar phenomena. This one was taken on 4 May 1958, close to an energetic solar maximum, in the red light of hydrogen. The structural details are clearly shown.*

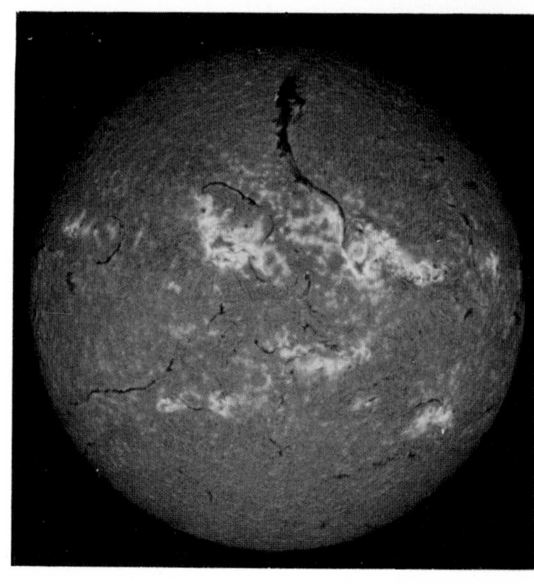

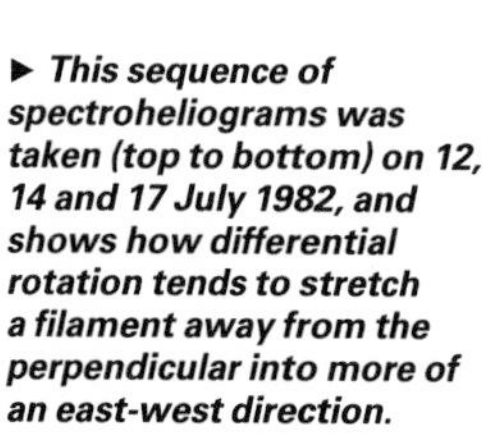

► *This sequence of spectroheliograms was taken (top to bottom) on 12, 14 and 17 July 1982, and shows how differential rotation tends to stretch a filament away from the perpendicular into more of an east-west direction.*

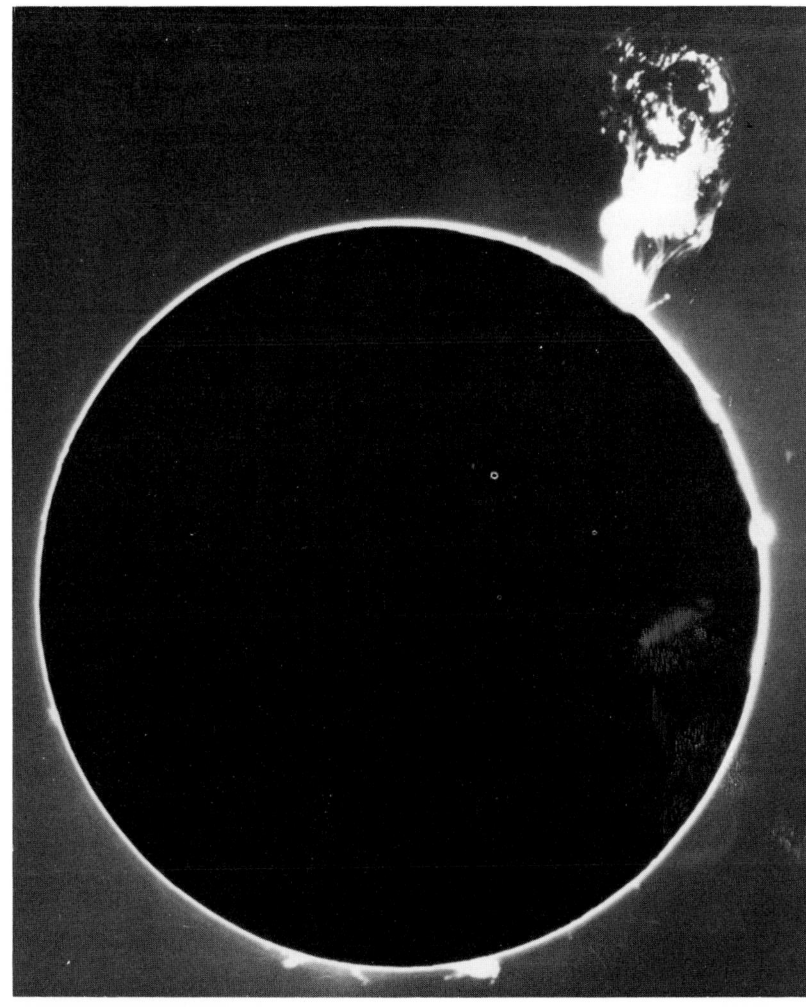

Observing the Sun's outer layers

One of the most important developments in the field of solar research was the invention of the spectroheliograph in 1891 by the American observer George Ellery Hale (1868-1938). Hale was an outstanding student of the Sun, and it was he who made the important discovery that sunspots are associated with powerful magnetic fields.

The spectroheliograph makes it possible to photograph the Sun in the light emitted by one element only – generally hydrogen or calcium. It contains a series of prisms which produce an extended solar spectrum. A fine slit masks off all but a narrow strip of the Sun, while a second slit isolates the light of one wavelength, blocking all the rest. As the Sun's image drifts across the field of view, the photographic plate is driven immediately behind it, so that an image of the Sun in the light of this wavelength only is built up. Later Hale devised the visual equivalent of the spectroheliograph – the instrument called the spectrohelioscope.

The French astronomer Bernard Lyot (1897-1952) approached the problem from a different angle, developing a special filter to cut out all the unwanted wavelengths. Observing at hydrogen wavelengths makes it possible to see prominences against the bright disk of the Sun as dark filaments, sometimes known as "flocculi".

The corona

The discovery that prominences could be kept under constant observation was a major step forward in solar research, but there remained a much greater problem relating to the corona. Even when the Sun is totally eclipsed, the corona produces only about as much light as the full Moon, so that under normal conditions it is hopelessly lost against the brightness of the sky. Bernard Lyot attempted to overcome this difficulty in 1930. He constructed a coronagraph, which is essentially a telescope fitted with a series of internal screens to block out the light from the photosphere. Using this, he succeeded in observing the inner corona.

▼ Some prominences are quiescent, but this loop is a violent and dramatic event. The structure highlights the magnetic field above a sunspot group.

▲ Flares, such as the one which triggered this violent eruption, shower energetic particles and X-rays into the Earth's magnetic field and atmosphere.

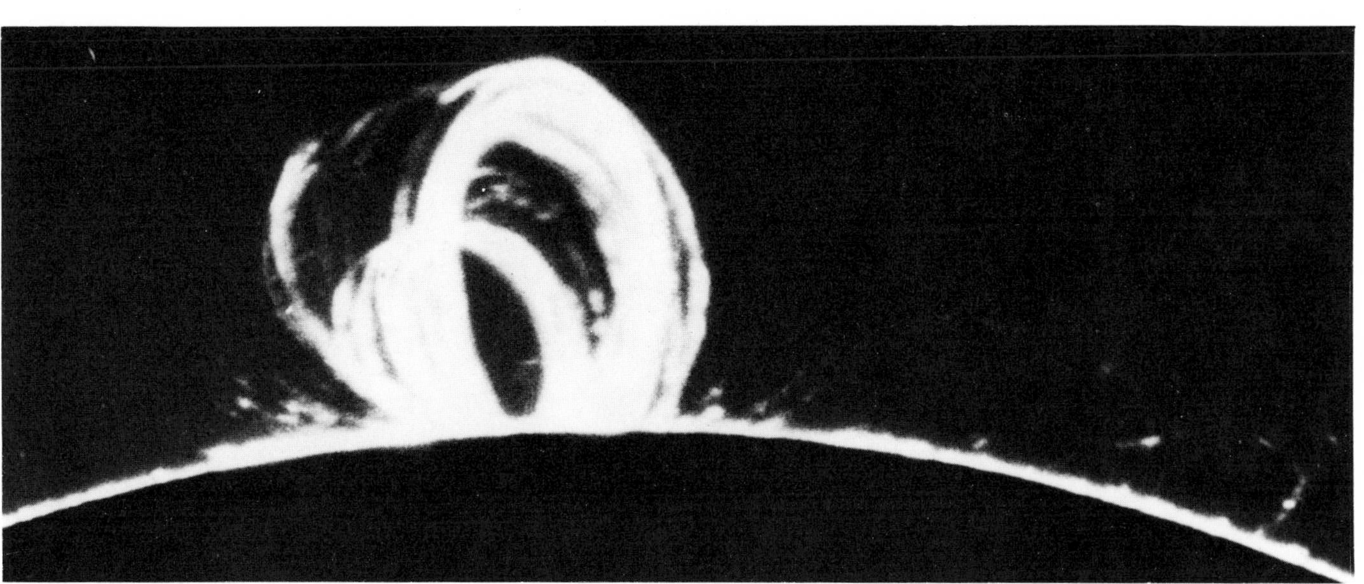

Though the temperature of the corona is over a million degrees, it is not "hot" in the usual sense

The appearance of the corona varies markedly. Usually when the Sun is more active – around sunspot maximum – the corona is more conspicuous and more evenly spread around the disk, while at minimum it is less even and shows long streamers over equatorial regions. The detailed structure of the corona has been revealed by X-ray and extreme ultraviolet observations. The corona is far hotter than the photosphere, and because of its high temperature it emits X-rays strongly, while the photosphere does not emit X-rays in significant quantities. Thus at X-ray wavelengths the photosphere is dark while the corona shows clearly in front of the solar surface. Such observations show that the corona is lumpy and uneven in structure, and consists of active regions, quiet regions and coronal holes.

Features of the corona

Active regions, which emit X-rays strongly, consist of hotter and denser gases held within closed loops of the Sun's magnetic field. Quiet regions are less conspicuous areas of lower temperature and density, held by weaker magnetic fields.

Coronal holes, which show up as dark patches in X-ray and ultraviolet photographs, are cooler low-density regions where the magnetic field lines are open (that is, they spread out into interplanetary space rather than curving back to the Sun). Electrically charged particles such as protons and electrons flow unhindered out of these holes into the solar wind (➧ page 113). The energy that maintains the corona at its very high temperature may come from shock waves originating in the bubbling "noisy" photosphere, and from other processes such as magnetic heating (energy released by electrons flowing through the coronal magnetic fields).

Occasionally the corona is wracked by colossal disturbances known as "coronal transients", which are loops or arcs of material hurled outwards at speeds of up to 1,000 kilometers per second. These disturbances, which are often triggered by eruptive prominences, extend beyond the corona into interplanetary space.

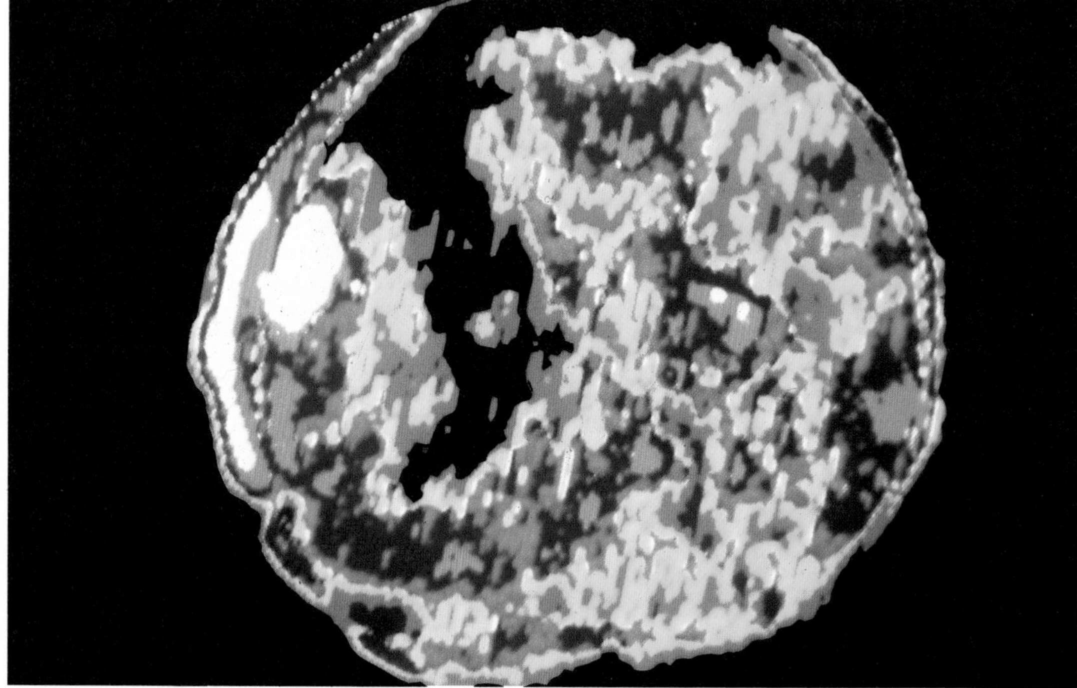

▲ In a conventional photograph of the 1980 solar eclipse (top) the inner corona is overexposed and all delicate detail washed out. The detail is revealed in the color-contour photograph below it.

◄ The false-color enhancement of an extreme ultraviolet image shows the black S-shaped area of a coronal hole.

► Three months earlier the same coronal hole was visible in the X-ray image.

►► Coronal streamers and a rising prominence loop are seen in this color-contour image. The different colors correspond to values in the density of the corona.

The color-coded photograph above it shows coronal emissions extending for millions of kilometers. Here the different colors distinguish levels of brightness.

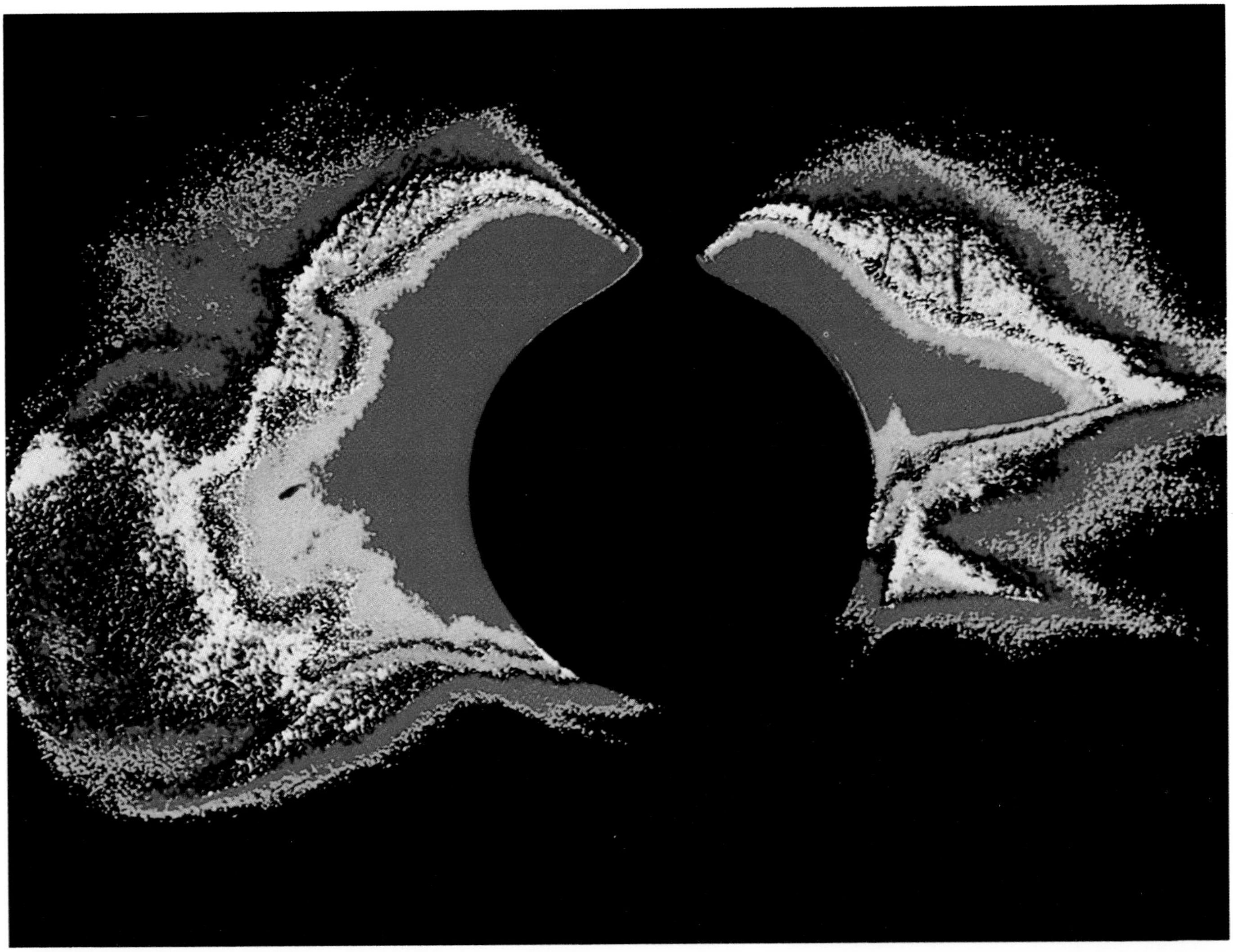

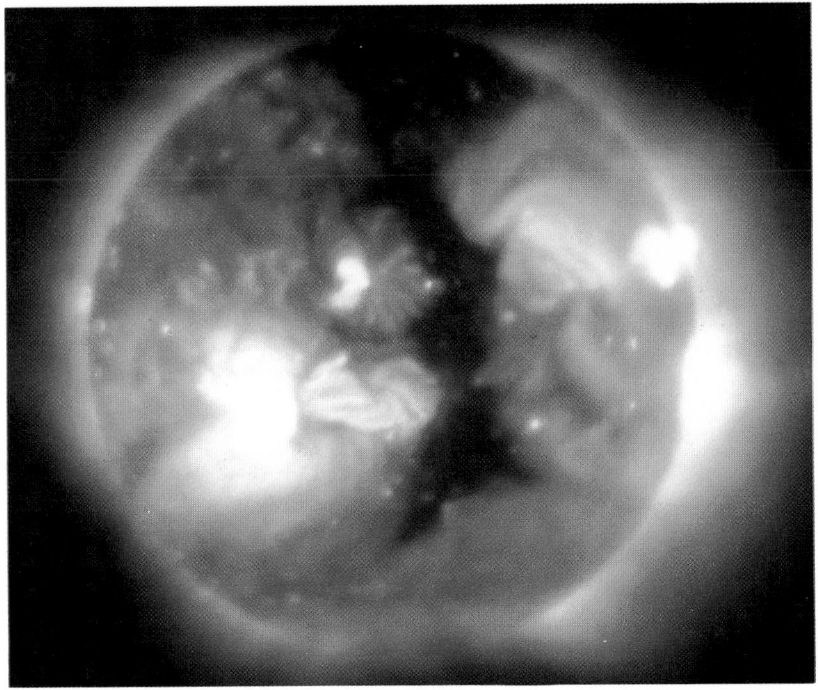

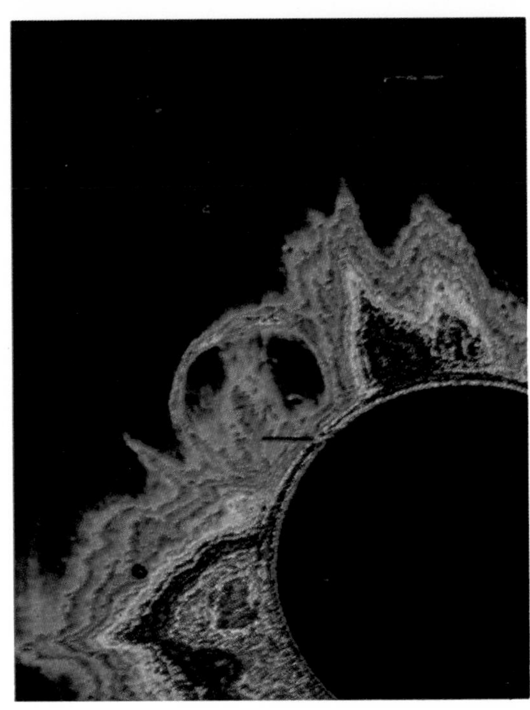

See also
Inside the Sun 97-104
The Sun's Influence 113-20
Observing the Universe 137-44
The Basic Properties of Stars 145-52
Birth, Life and Death of Stars 153-60

History of coronal observation

During total eclipses the corona appears as a lovely pearly "mist" stretching outward from the Sun. The earliest reference to it probably occurs in the writings of the Greek biographer Plutarch (AD 46-120), but it was definitely recorded at the eclipse of 968. The German astronomer Christopher Clavius (1537-1612) saw it on 9 April 1597, but thought it was the uncovered edge of the Sun. Johannes Kepler disproved this, and attributed the corona to an extensive lunar atmosphere. However, at the eclipse of 16 June 1806, the Spanish astronomer Don José de Ferrer pointed out that in this case the Moon's atmosphere would have to be fifty times as high as that of the Earth, and this seemed improbable. Nevertheless, there was still doubt as to whether the corona belonged to the Sun or the Moon – a problem which was not finally solved until the eclipse of 1842.

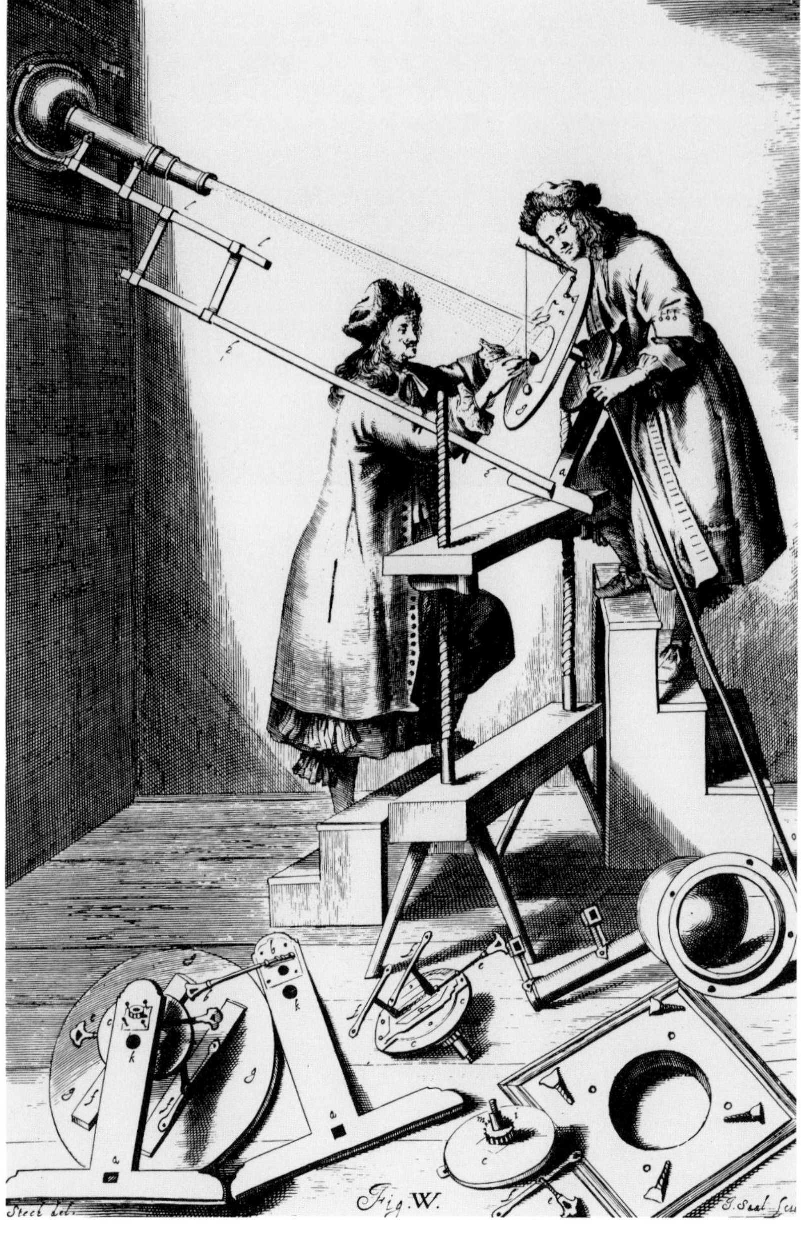

◄ Johannes Hevelius, builder of some admirable contraptions for astronomical observation, demonstrates here the safe way to observe a solar eclipse – by projection.

▲ An elderly astronomer photographs the solar eclipse very early in the morning of 19 June 1936, from the almost deserted streets of Hornsey, North London.

Historic eclipses of the Sun

The Moon's shadow is conical and its tip only just reaches the Earth, so that a total eclipses of the Sun is seen only from a narrow track. Moreover, totality is always brief, and cannot last for more than 7 minutes and 31 seconds. During the eclipse of 30 June 1973 a specially equipped Concorde aircraft kept pace with the Moon's shadow, so that the astronomers on board saw a total eclipse lasting for 72 minutes. The Moon's shadow travels at over 3,000 kilometers per hour, so Concorde is the only commercial aircraft that can match it.

The ancient Chinese believed that at the onset of an eclipse a hungry dragon was trying to devour the Sun. Their remedy was to shout, scream, beat drums and gongs and generally make as much noise as possible in order to scare the dragon away – a procedure which always worked, naturally.

Even without detailed knowledge of how the Sun, Moon and Earth really move it is possible to forecast eclipses, at least roughly. The three bodies return to the same relative positions after an interval of 6,585·321 solar days, or about 18 years and 11 days, so that an eclipse is likely to be followed by another one 18 years and 11 days later. This period, called the Saros, was known to the first of the great Greek philosophers, Thales of Miletus (c. 624-537 BC), who successfully predicted an eclipse for 25 May 585 BC.

Contemporary sources relate that the eclipse of AD 840 so terrified Louis, Emperor of Bavaria, that he died of fright – after which his three sons embarked on a ruinous war over the succession.

In 1870 the French astronomer Jules Janssen was extremely anxious to observe a total eclipse due on 22 December, but he was in Paris, outside the band of totality, and the city was besieged by German forces. Janssen daringly built a balloon and made his escape in that, but he missed the eclipse anyway because the sky was overcast.

The Sun's Influence

How the planets are affected by charged particles from the Sun...The Sun's magnetic field, the controlling factor...How ice ages, thunderstorms and aurorae are linked with solar activity...Planetary magnetic fields and solar particles – what happens when they interact?... PERSPECTIVE...Great pioneers in solar particle research...Predicting solar activity...Exploring out to the limit of the Sun's influence

The solar wind is a stream of charged particles flowing out from the Sun's corona (outer atmosphere) at typical velocities of 400-500 kilometers per second and which, on average, take 4-5 days to reach the vicinity of the Earth. The wind consists mainly of elementary particles such as protons and electrons in roughly equal numbers, together with heavier particles such as alpha particles (helium nuclei). At the Earth's distance from the Sun (1AU) the average number of particles per cubic meter is about 5,000,000, which is approximately equivalent to 1 ten-million-million-millionth of the density of air at sea level. High-speed streams of particles emanate from what are known as coronal holes, and violent events in the Sun's atmosphere, such as flares and eruptive prominences, send bursts of high-speed particles surging through the slower-moving wind.

The Sun loses more than a million tonnes of matter per second into the solar wind. This may appear to be an alarming statistic, but in fact it is an infinitesimal quantity compared to the mass of the Sun, and there is no danger of its fading away by this means before it runs out of nuclear fuel in 5 or 6 billion years' time.

The Sun's magnetic field

Close to the Sun, the solar magnetic field ensures that the particles of the solar wind are dragged round with the solar rotation so that they "co-rotate" with the corona. Farther out, where the magnetic field becomes weaker, the motion of the charged solar wind particles controls the direction of the magnetic lines of force in interplanetary space. Although individually the particles move more or less radially away from the Sun, taken collectively they form a spiral pattern rather like that produced by a spinning garden sprinkler, and the interplanetary field conforms to this pattern.

If the Sun had a simple dipole magnetic field (that is, like a bar magnet), there would be a flat "neutral sheet", a plane extending out from the solar equator, on either side of which, due to the opposite magnetic polarities in opposite hemispheres of the Sun, there would be oppositely-directed lines of force. In reality, the Sun's magnetic field is more complex and the equatorial plane of the photosphere (solar surface) usually comprises a number of alternating sectors of opposite polarity. This causes the neutral sheet to be buckled, so that as the Sun rotates the neutral sheet rotates in interplanetary space rather like a warped gramophone record whose surface wobbles up and down relative to the plane of the Earth's orbit. Each time the warped boundary passes the Earth the direction of the field lines changes, giving the impression of alternating sectors in the interplanetary field. The solar wind speed is usually low near a sector boundary and high away from it, so that each time a boundary passes by, sharp changes in solar wind speed are detected.

▲ Sir Oliver Joseph Lodge, seen here at work in his home near Salisbury in 1928, suggested that the Sun might be a source of continuous torrents of particles, and that these might be responsible for atmospheric phenomena such as aurorae and magnetic storms.

◄ Eugene N. Parker developed and refined the theory, and coined the term "solar wind".

Early solar theory

In 1900 an important paper was published by the British physicist Sir Oliver Lodge (1851-1940), in which he suggested that the Sun might be the source of a cloud of particles spreading out in all directions. He was correct, but the modern theory of what is now called the solar wind dates only from 1958 and research by Eugene N. Parker involving the nature of the Sun's corona.

Parker realized that the corona is at a very high temperature, amounting to well over one million degrees. This is much higher than the temperature of the bright solar surface, which is below 6,000 degrees, and yet the corona is not hot in the accepted sense of the word. Temperature depends upon the speed at which the various atoms and molecules move around – the greater the speeds, the higher the temperatures. In the corona, the speeds are very great, but the particles are few in number, so that the temperature is high, but actual heat is negligible.

Parker also pointed out that because of its high temperature, the corona must be in a state of continuous expansion, in which case it is replenished from below. It has no definite limit, and X-ray observations carried out by the Skylab astronauts showed that there are "holes" in it, from which flow particles of the solar wind. The wind is "gusty", being much stronger at times of solar maxima, when sunspot groups and violent, short-lived outbreaks known as solar flares are common.

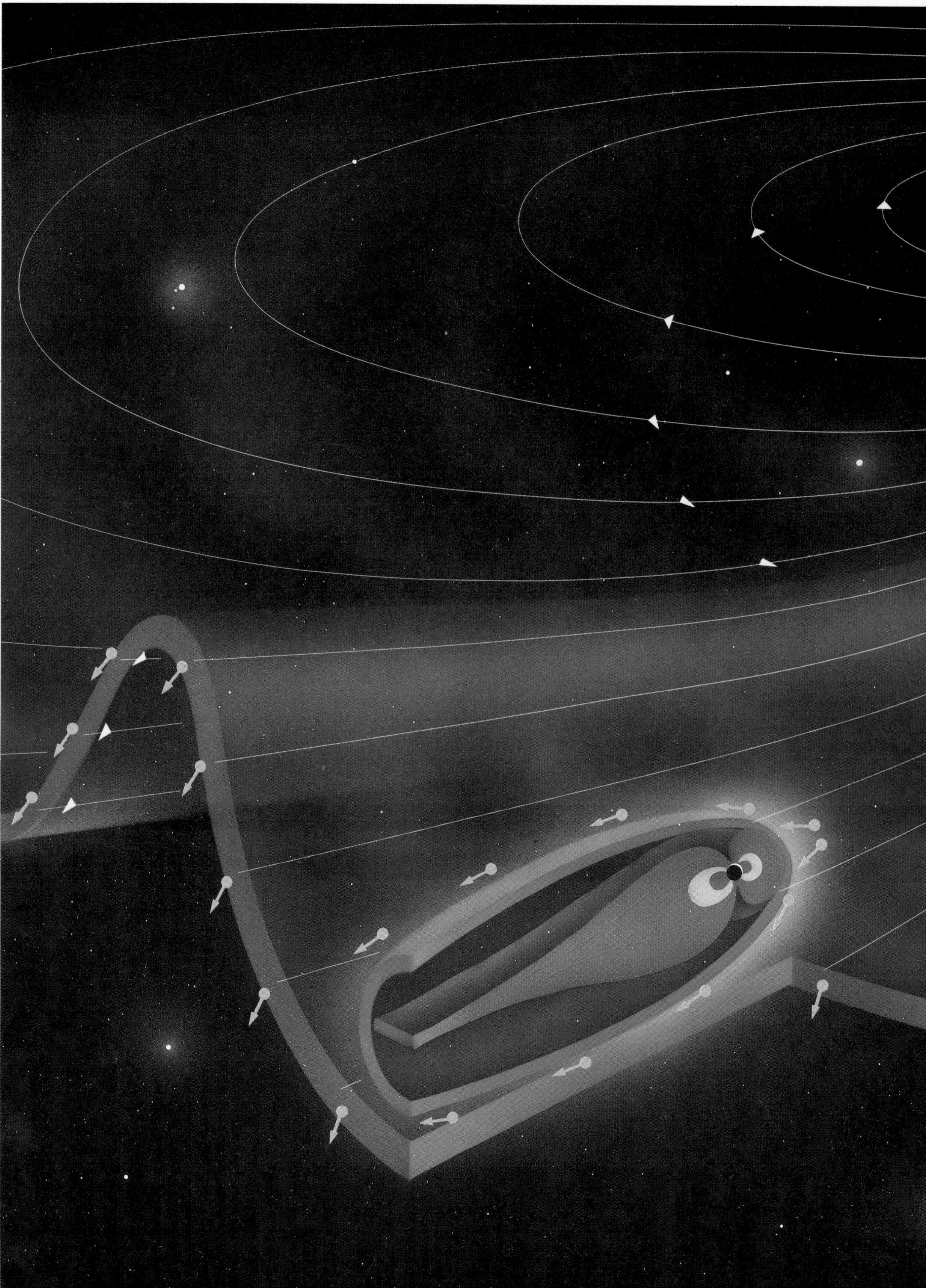

◄ ***Lines of magnetic force, outwards from and inwards to the Sun, follow a spiral pattern in interplanetary space. Solar wind particles flow around the Earth's magnetosphere, but high-speed particles from violent solar storms can penetrate this shield, producing spectacular displays of aurorae.***

The magnetosphere is the region of space within which the Earth's magnetic field dominates the interplanetary field. On the sunward side the pressure exerted by the solar wind squeezes the magnetopause (the boundary of the magnetosphere) in to a distance of about 10 Earth radii, whereas on the other side it draws it out into a long tail (the magnetotail), which extends to about 1,000 Earth radii, or about 6,000,000km. The magnetosphere is shaped like an elongated teardrop, and is similar in some respects to the tail of a comet.

The magnetosphere acts as a shield against the solar wind whose charged particles cannot readily cross the magnetopause. Relative to the solar wind, the magnetosphere acts like an obstruction plowing through the wind at some 400-500km. Just as a supersonic aircraft piles up a shock wave ahead of it, so the Earth produces a shock wave, known as the bow shock, in the solar wind, some 3-4 Earth radii on the sunward side of the magnetopause.

The solar wind is "gusty", and variations in the wind strength produce changes in the magnetosphere. A sudden increase will squeeze the magnetosphere, causing fluctuations (magnetic storms) in the magnetic field at ground level. Particularly violent magnetic storms are caused by bursts of particles emitted by major solar flares. These particles usually arrive at Earth within a couple of days of the event.

A small proportion of solar wind particles penetrates the magnetosphere, and these, together with ionized particles expelled from the Earth's upper atmosphere, become trapped by the Earth's magnetic field. They are strongly concentrated in two zones – the Van Allen radiation belts – centered at about 1·5 and 5 Earth radii, although there is no sharp distinction between one belt and the next. Disturbances produced by solar magnetic storms shake particles out of the magnetosphere into the upper atmosphere. These enter mainly through a narrow oval band some 10° to 20° in radius and centered on each magnetic pole. The excitation and ionization of atoms and molecules that they cause in the upper atmosphere generates the light, predominantly red and green in color and taking the form of rays, curtains and arcs, which makes up the aurorae. The frequency of auroral displays is normally greatest at times of sunspot maximum.

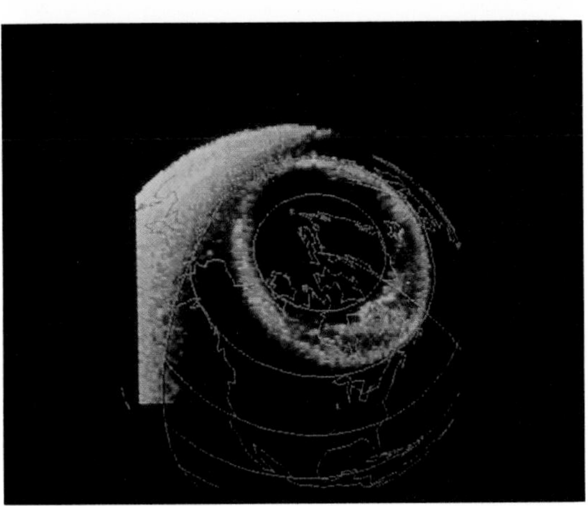

▲ ***Viewed from above, an aurora is seen forming through a narrow oval band centered on the north pole; the direct result of a solar magnetic storm.***

► ***Viewed from the ground auroral bands present a brilliant display. Here they are seen at Fairbanks, Alaska, a not uncommon sight at this latitude.***

Between 1645 and 1715 there were almost no sunspots – and in England the River Thames often froze over

There is growing evidence to suggest that there are many subtle links between changes in solar activity and atmospheric and climatic phenomena on Earth. There seems to be a definite correlation, for example, between the passing by of a sector boundary in the interplanetary magnetic field and the amount of vorticity (rotational motion) in the Earth's atmosphere. This reaches a peak about 3-4 days after such an event, and results in an increase in the storminess of the atmosphere. There also appears to be a link between fluctuations in solar activity, the atmospheric electrical circuit, and the amount of thunderstorm activity. There are very possibly aspects of the Earth's atmospheric engine which are so finely balanced that small changes in energy input, resulting from fluctuations in solar activity, are sufficient to trigger significant changes.

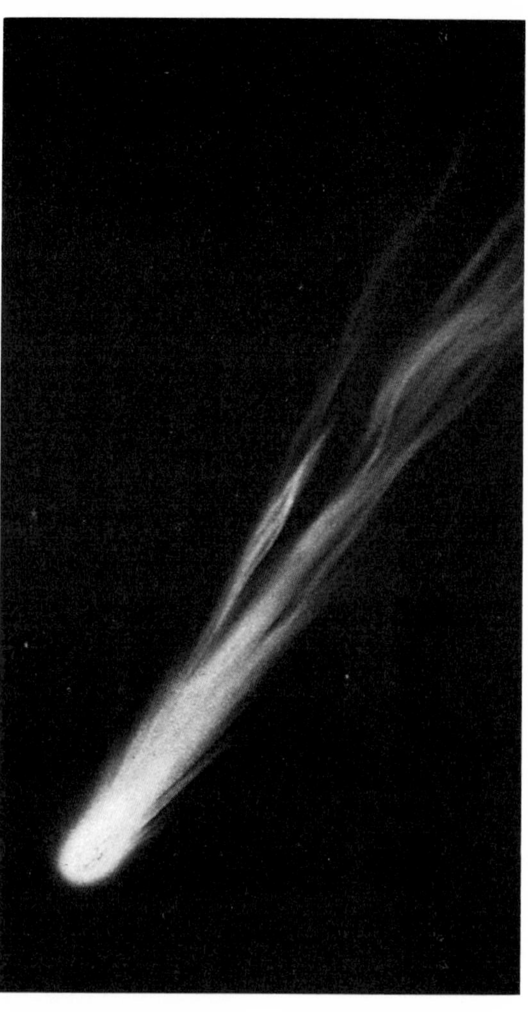

▲ *The solar wind is not visible in the conventional sense of the word, but, like atmospheric winds on Earth, it reveals itself in its effects on other substances. For example, ionized (electrically charged) gases in the tails of comets are swept away from the Sun by the wind, which is why the tails always point away from the Sun.*

Sunspots and ice ages

The Maunder minimum in the sunspot cycle (◆ pages 106-107) between 1645 and 1715 coincided with a series of severe winters in Europe, which became known as the Little Ice Age. There is other evidence, admittedly tenuous, for there having been other prolonged periods of high or low solar activity corresponding, respectively, to periods of unusually warm or exceptionally cold climatic conditions.

Recent observations have shown that the solar constant (◆ page 97) undergoes short-term fluctuations of about 0·1 percent, and it is estimated that a long-term change of only about 1 percent would be enough to precipitate another ice age, such as that which ended about 10,000 years ago. Such changes may be responsible for triggering ice ages, but the general view is that periodic changes in the Earth's orbital eccentricity, the direction of the Earth's axis and the time of year in which the Earth reaches perihelion (closest approach to the Sun) alter the input of solar energy into polar regions enough to bring about ice ages. These phenomena are known as Milankovič cycles.

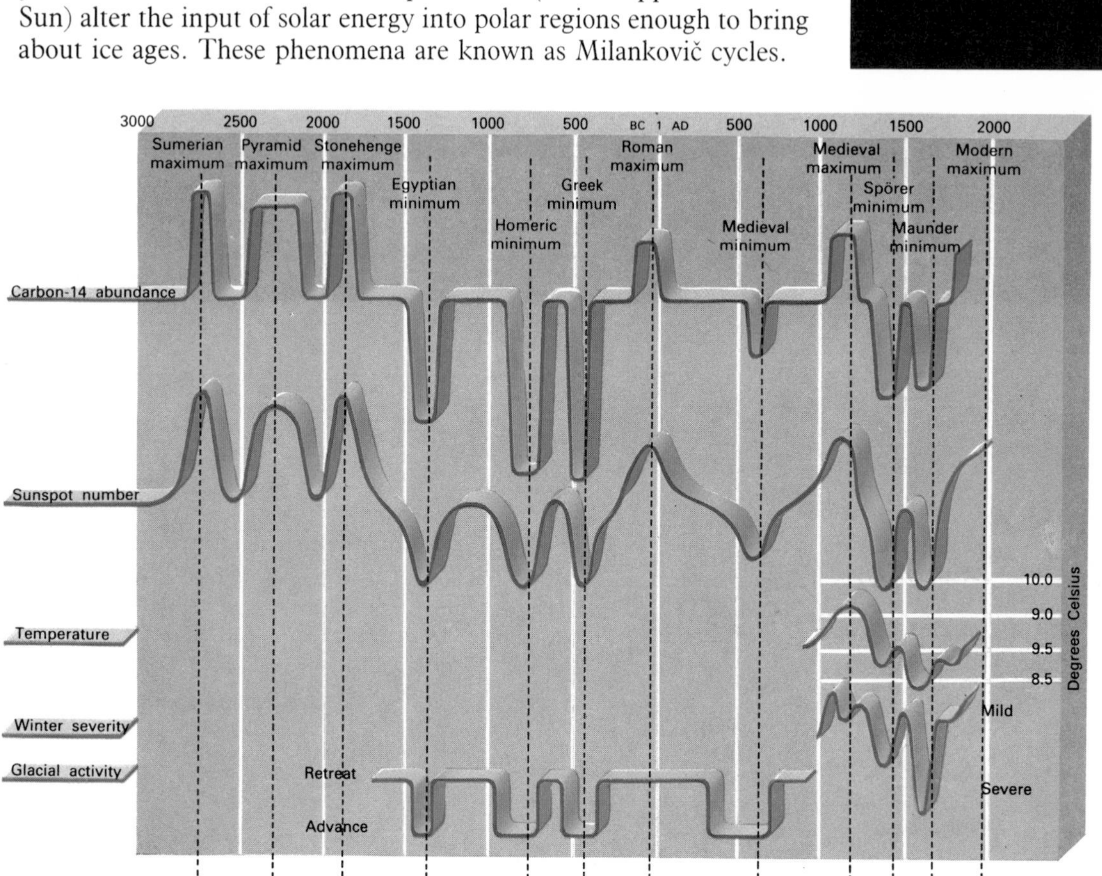

◄ *When solar activity is high the Sun's magnetic field shields the Earth from cosmic rays which produce carbon-14 in the atmosphere. Studies of the carbon-14 content of tree rings in long-lived Bristle-cone pines give an indication (top graph) of variations in solar activity over the past 5,000 years. The second graph gives an estimate of corresponding fluctuations in sunspot numbers. Comparison with the other three graphs suggests there may be a link between long-term variations in solar activity and certain climatic factors such as the severity of winters and the retreat or advance of glaciers.*

▲ **During the winters of 1683-1689 in the "Little Ice Age", the River Thames froze so solid that the citizens of London held "Frost Fairs" on the ice, erecting stalls, pitching tents and lighting bonfires.**

Ice ages – the search for an explanation

Life depends entirely on the Sun. Fortunately for mankind, the Sun is a steady, well-behaved star, whose energy output has not altered much for a major part of the Earth's independent existence. If the energy received from the Sun increased by a mere 1 or 2 percent, the polar ice-caps would melt and many of the continents would be flooded. If the energy were reduced by a comparable amount, we would be plunged into a new Ice Age.

Why do Ice Ages occur? There have been many theories. One, the Milankovič theory, depends upon changes in the Earth's orbit and axial inclination. It has been suggested that the Solar System may have passed through an interstellar cloud which blocked part of the Sun's radiation; alternatively, Sir Fred Hoyle and others believe that the cause was a gigantic meteoritic impact, which had profound effects on our climate. Other authorities maintain that the cause is to be found in actual changes in the amount of radiation sent out by the Sun. Whatever the answer, the so-called solar constant is misnamed inasmuch as it is not quite constant! For example, it increased slightly during the late 1970s.

Sifting the evidence

Ice Ages have occurred periodically throughout Earth's history. The most recent ended only 10,000 years ago, and was made up of four "cold waves" separated by warmer spells or "interglacials", and we may really be in the midst of an interglacial at the present time. Unfortunately our knowledge of the state of the Earth during the Ice Age is limited, because records from the southern hemisphere are much less complete than those in the north. However, it now seems definite that the whole of the Earth was colder than it is today, so that the Ice Age was global rather than local.

Studies of carbon-14 in tree rings link the 17th century "Little Ice Age" to low solar activity. Carbon-14 is produced by cosmic rays in the Earth's upper atmosphere, and is affected by the state of activity in the Sun: when it is "quiet", more cosmic rays reach the surface. Carbon-14 is absorbed into tree rings during the growth cycle, and during the Maunder minimum of 1645-1715 the content was exceptionally high, indicating that solar activity was low. There is some evidence of a similar period between 1450 and 1550.

It is impossible to predict whether there will be any marked alteration in solar output in the foreseeable future. However, any changes are not likely to be marked for many centuries, and it will be at least five billion years before the Sun uses up its store of available hydrogen "fuel" and changes its whole structure – with effects that are certain to be disastrous for the planet Earth.

► *Jupiter's invisible cosmic shield. The "magnetic bubble" or magnetosphere surrounding Jupiter is the most extensive in the Solar System. The fluctuation in strength of charged particles from the Sun can cause considerable variation in the overall size of the "bubble".*

The magnetospheres of the terrestrial planets

Mercury's magnetic field is weak compared to that of the Earth. It still has a teardrop shape, the bow shock being located just 2 planetary radii (about 3,500km) from the planet's center. The field appears to be too weak to trap and retain charged particles, and there are no Van Allen-type radiation belts. Although Venus lacks any appreciable magnetic field, the solar wind interacts directly with the planet's ionosphere, which prevents solar wind particles from reaching the planet's surface and acts like a magnetopause. A bow shock occurs "upstream" of the planet. Mars, too, despite its lack of any significant field, produces a shock and tail in the solar wind.

The giant planets

Jupiter has the most extensive magnetosphere of all the planets, partly as a result of its strong magnetic field, but also because the solar wind strength is much weaker out there. Jupiter's "magnetic bubble" is about 100 times larger than the Earth's, the magnetopause on the sunward side occurring at about 70 Jupiter radii from the planet, and the bow shock some 15 Jupiter radii further upstream. The size of the Jovian magnetosphere varies considerably with fluctuations in the strength of the solar wind, and the magnetopause on the sunward side moves in and out from about 50 to 100 Jupiter radii (some 3·5 to 7 million kilometers). The magnetotail extends beyond the orbit of Saturn (more than 700 million kilometers).

The Jovian magnetosphere is filled with a curious mixture of ions, which is significantly different from the composition of the solar wind and includes, apart from hydrogen and helium, large quantities of carbon and sulfur. A large proportion of these particles, particularly in the innermost part of the magnetosphere, are probably supplied by Jupiter's satellites, in particular Io (◀ pages 76-79) which sprays out material from its volcanoes into a large torus (doughnut-shaped volume) some 2 Jupiter radii in extent, around its orbit. Most of the particles in the inner region are confined to a disk which rotates with the planet. Jupiter's magnetic equator is tilted to its rotational equator by about 10° so this disk, together with Io's plasma torus, flaps up and down as the planet rotates. It is thought that some of Jupiter's radio emissions are triggered by this flapping motion.

Saturn's magnetosphere is smaller than Jupiter's but about 20 times larger than the Earth's. Within a tenuous torus of plasma extending to about 800,000km from the planet the fast-moving particles have a temperature of 400-600 million K, the highest known temperature in the Solar System.

At present astronomers have no real knowledge of the nature or extent of the magnetospheres of Uranus and Neptune.

Jupiter's magnetosphere

1 Bow shock
2 Magnetosheath
3 Magnetopause
4 Magnetic field lines
5 Orbit of Callisto
6 Solar wind
7 Plasma sheet
8 Magnetospheric wind
9 Ecliptic

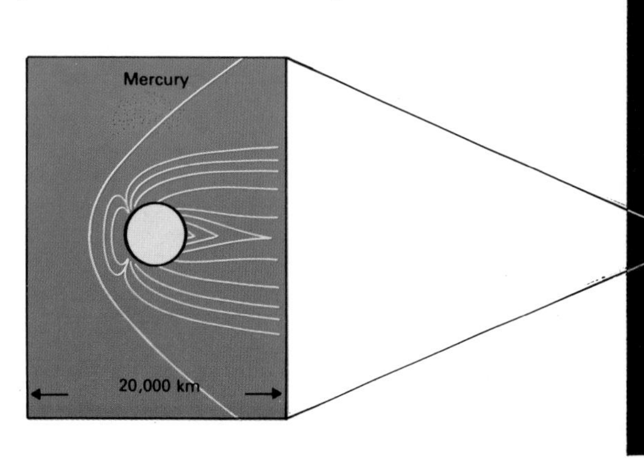

► ***The overall dimensions of the magnetospheres of Mercury, Earth, Jupiter and Saturn are compared in order of increasing size.***

5
6
7
8
9
Saturn
8,000,000 km
Jupiter
60,000,000 km

See also
Inside the Planets 33-40
Asteroids and Comets 89-96
Inside the Sun 97-104
The Sun's Outer Layers 105-12
The Basic Properties of Stars 145-52

The heliosphere – the Sun's whole empire

The solar wind blows out beyond the orbit of Neptune, but eventually both the wind and the interplanetary magnetic field are halted by the small but finite pressure exerted by interstellar gas and an interstellar magnetic field. The boundary of the solar "magnetic bubble" is known as the heliopause and the region within which the solar field dominates over the interstellar field is the heliosphere. The Solar System moves through the interstellar gases at a speed of about 20 kilometers per second, so there is an "interstellar wind" apparently blowing past the heliosphere and distorting it into a teardrop shape. Where exactly the heliopause lies is not known for certain, but it is probably some 50-100AU (astronomical units), or 7·5 to 15 billion kilometers, from the Sun.

Since the heliosphere is "inflated" by the solar wind and the wind strength varies with the solar cycle, the heliosphere is believed to expand and contract with the solar cycle and to "wobble" in extent as a result of individual major solar storms.

The heliosphere partly shields the inner Solar System from cosmic rays – high-speed atomic nuclei approaching from the depths of space – and the extent to which these penetrate the inner System and reach planets such as Earth also varies in line with the solar cycle.

Exploring the heliosphere

The outer reaches of the heliosphere are presently being explored by four of NASA's most successful planetary spacecraft, Pioneers 10 and 11, and Voyagers 1 and 2. Pioneer 10, launched on 3 March 1972, made the first-ever close flyby of Jupiter. On 13 June 1983 it passed beyond the orbit of Pluto, and on that date became the first man-made object to pass beyond all the known planets. Most of its instruments are still returning data. Among other discoveries, Pioneer 10 has shown that, contrary to previous expectations, solar wind particles do not slow down significantly as they get farther from the Sun. It has also been able to detect interstellar hydrogen and helium flowing into the Solar System. Pioneer 10 is heading down the long tail of the heliosphere but Pioneer 11, is heading in the opposite direction, towards the "bow shock". Since the heliopause should be squeezed in on that side, Pioneer 11 may have a better chance of crossing the boundary before contact is finally lost, but it remains to be seen whether any of the four spacecraft will be able to relay information about the boundary of the Sun's "magnetic bubble", or about conditions beyond the the solar wind.

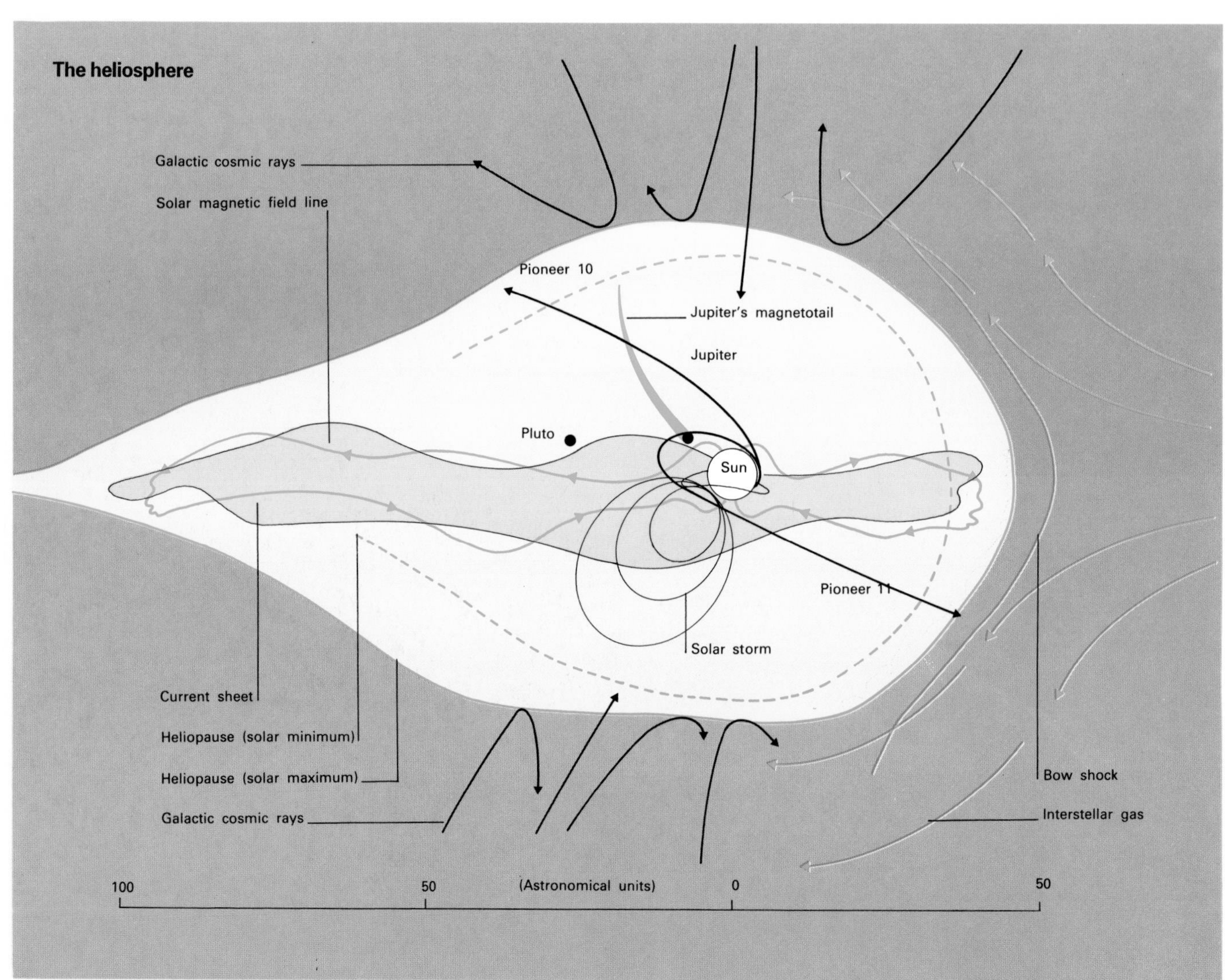

The Origin of the Solar System

How and why the Solar System came into existence...The two principal theories – conflicting or complementary...Why the Earth does not have a poisonous Venusian-like atmosphere...Is there life elsewhere on other planets?...Origin and evolution of life...PERSPECTIVE...The early theorists, from hoaxers to philosophers...Bug-eyed monsters, the facts and fiction...Impacts from space and the vulnerability of life

Many scientists have attempted to explain how and why the Solar System came into existence, but so far none of them has been able to answer all of the most difficult questions. Some have suggested that the Sun was born first, and that as it moved through an interstellar cloud it collected the material that eventually became the planets. Perhaps this material was blasted out of some former companion star, or sucked out of the Sun or a passing star during a close encounter. The most popular view, however, is that the Sun and planets were born at the same time out of a single cloud of gas and dust – a nebula. This is called the "nebular" hypothesis.

It seems likely that the birth of the Solar System began with the collapse of a cloud of gas and dust of between one and two solar masses. Clouds of this size do not usually collapse under their own gravity unless they are squeezed by some sort of "trigger" event. Analysis of meteorites – rocks falling out of space – suggests that short-lived radioactive materials (notably aluminium-26 which decays to magnesium-26) were added to the Solar System material just before the meteorites and planets were formed. Since these materials are created inside exploding stars, known as "supernovae", it may be that such an explosion was the trigger for the collapse of the cloud that then became the Solar System.

The fate of a collapsing cloud

If it rotates very rapidly, a collapsing cloud will form a binary system (two stars revolving round each other). Otherwise it will form a central condensation which evolves into a star, and a surrounding disk of gas and dust. The straightforward collapse of a cloud of gas would form a star rotating far faster than the Sun does today, but various processes acted to slow down the rotation of the embryonic Sun, or "protosun", by transferring angular momentum (rotational motion) from it to the surrounding nebula. Such processes could include friction between materials in the disk, the magnetic coupling of the protostar to the surrounding nebula and to the interstellar magnetic field, and the effects of a superpowerful solar wind.

The inner part of the disk-shaped nebula would be hotter than the outer parts, but nobody really knows what the exact temperatures would have been. Given a high enough temperature at the center, all the grains from the protosolar cloud would have vaporized and would then have solidified once more as the nebula cooled down. If the temperature was not so high, all the original metal and rocky grains would have survived and just the icy ones would have vaporized. Either way, the temperature gradient in the nebula would have produced a difference in chemical composition between the grains which solidified or remained solid in the inner Solar System, and those which solidified or remained solid in the outer parts of the system.

James Jeans made important advances in the study of stellar evolution, from which he developed his theory for the formation of the Solar System. The idea of a cigar-shaped filament of solar material accounted for the size of the giant planets, as illustrated (right). Jeans was knighted in 1958 at the age of 51.

Variations on an old theme

In the past some ingenious theories have been proposed to explain the formation of the Solar System. The British astronomer, writer and broadcaster Sir James Hopwood Jeans (1877-1946) worked out a theory according to which the planets condensed out of a long, cigar-shaped filament drawn out of the Sun by a passing star. On the face of it this seemed plausible, since the largest planets, Jupiter and Saturn, would have been in the thickest part of the "cigar", but detailed mathematical analysis of the proposition showed that it could not be made to work.

Another modification was put forward by the eccentric New Zealand scientist Alexander William Bickerton (1842-1929), who had an excellent knowledge of chemistry and physics – indeed at one time the great Lord Rutherford was one of his pupils – combined with a total ignorance of higher mathematics such as calculus. Bickerton believed that a wandering star struck the Sun a glancing blow, and he used his theory of "partial impact" to explain a wide-ranging sequence of phenomena. Regrettably for Bickerton, his theory could not survive the rigorous tests to which it was subjected by the mathematicians.

A further variation on the encounter theory was developed in the 1960s by Professor M. M. Woolfson of the University of York, England. He suggested that the Sun made a close encounter with a young protostar of very low temperature and density and of a size comparable with the Solar System. In the encounter cool material was drawn from the protostar into orbit round the Sun, eventually building up the planets and satellites. Although it overcame many of the objections to the Jeans theory, Woolfson's theory did not find favor among most astronomers.

► *Two routes to the origin of the planets. Possibly, small particles began to stick together forming progressively larger lumps which assembled into planets. Alternatively, each planet may have formed from an individual blob of gas within which heavier elements settled to form a core.*

At temperatures below about 1,400K solid grains of iron and nickel would condense (or remain solid), and a little farther out from the center of the nebula, silicate grains would also form. By the outer fringes of what is now the asteroid belt, water-ice would occur and a little farther out other ices, such as ammonia and methane, would have condensed. These differences seem to be reflected in the differing compositions of the planets, satellites and asteroids.

The "snowball" effect

As the nebula surrounding the infant Sun continued to cool, it seems likely that these micrometer-sized grains collided and hung together, building up centimeter-sized particles which rapidly settled into a thin disk. More collisions occurred and built up meter-sized bodies, which, under gravitational attraction, began to fall together to form kilometer-sized bodies or "planetesimals". Collisions between these fragmented some and built up others. Large planetesimals then collided to form the planetary cores, which subsequently mopped up the remaining planetesimals. In this way the planets were probably assembled within about 100 million years.

The giant planets in the cooler outer parts of the nebula acquired massive envelopes of hydrogen and helium. The inner planets were unable to do so, however, because the higher temperature would aid the escape of light gases, and large gaseous envelopes would be disrupted by solar tidal forces. After the planets had formed, the solar wind probably blew the remaining gases right out of the system.

Gaseous protoplanets – an alternative theory

One alternative view of planetary formation suggests that the solar nebula broke up into large, massive blobs, each one of which collapsed separately to form a planetesimal, or even to form a whole planet directly. According to this view, heavy elements sank to the center of the protoplanets to form the planetary cores. Solar tidal forces probably tore away the gaseous envelopes of the inner planets, while the outer planets retained their original envelopes of light gases.

The process whereby planets were assembled from planetesimals – the accretion process – is considered the most likely mechanism, but both processes may have played a part, with the giant planets forming from gaseous protoplanets and the terrestrial planets by accretion.

5

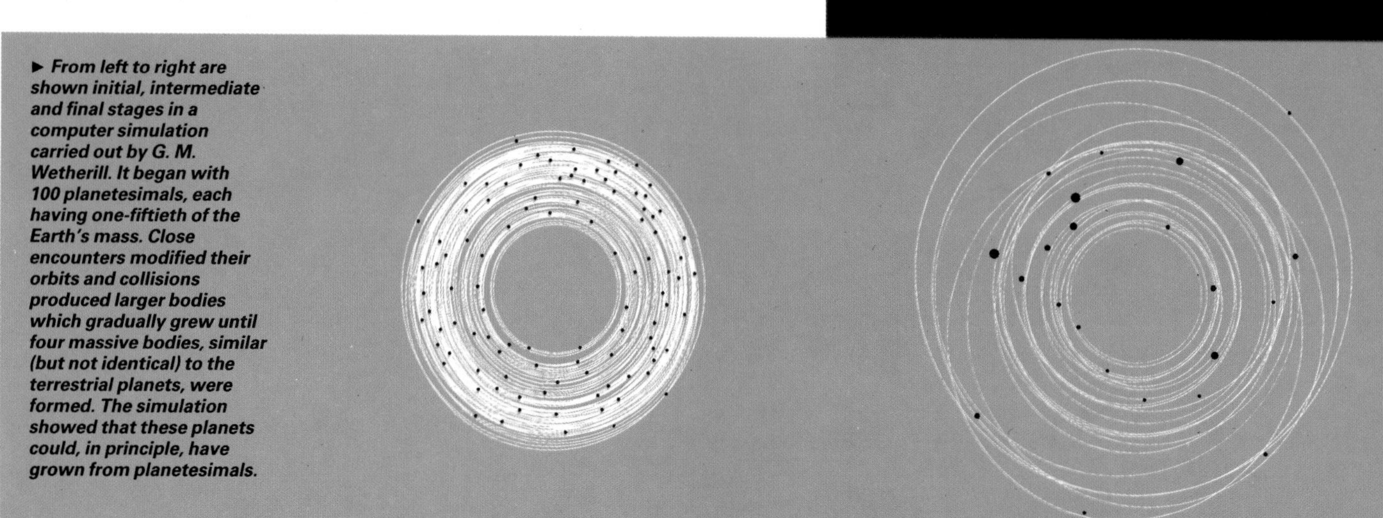

► *From left to right are shown initial, intermediate and final stages in a computer simulation carried out by G. M. Wetherill. It began with 100 planetesimals, each having one-fiftieth of the Earth's mass. Close encounters modified their orbits and collisions produced larger bodies which gradually grew until four massive bodies, similar (but not identical) to the terrestrial planets, were formed. The simulation showed that these planets could, in principle, have grown from planetesimals.*

Accretion theory

1 Micron-sized grains condense and stick together
2 Centimeter-sized particles build up meter-sized objects
3 Mutual gravity pulls together meter-sized objects
4 Collisions of kilometer-sized objects
5 New planets bombarded by remaining debris

Protoplanet theory

A Solar nebula breaks up into large blobs
B Blobs collapse under their own gravity
C Heavy elements settle to form cores
D Solar wind disperses the remaining gases

From accretion to the emergence of life

Accretion round other stars?

Astronomers believe they may have found evidence for the accretion process happening around other stars. For example, the object known as MWC 349 appears to be a very young, hot and highly luminous star surrounded by a glowing disk which is estimated to contain enough material for a planetary system, and which seems to be evolving rapidly. During 1983 the infrared satellite IRAS discovered many stars which seem to be surrounded by dusty shells or disks. Most surprising of all was the discovery that the bright naked-eye star Vega, three times the mass of the Sun and nearly sixty times more luminous, is surrounded by a disk of infrared-emitting particles at a temperature of about 85K and extending to a radius of some 85AU (comparable with our planetary system). Indications are that the particles may be millimeter-sized or larger – much larger than normal dust-grains – in which case they must have formed by accretion. However, other interpretations are possible, and it remains to be seen whether or not any of these dusty shells are actually planetary systems in the making, or already contain asteroidal or planetary bodies.

Later stages in accretion

If the planetesimal theory is correct, moons and ring systems would have formed from disks of matter surrounding the more massive planets in much the same way as the planets themselves were formed. Asteroids are leftover and fragmented planetesimals, some of which were later captured by planets to form part of their satellite systems. The Moon was probably formed separately, but close to the Earth. Comets probably represent icy planetesimals dating back to the early era of planetary formation.

Energy released by the impact of planetesimals, combined with heat released by the decay of short-lived radioactive elements, melted the interiors of the planets and of some of the satellites, so that heavy elements sank to the center while lighter elements floated to the surface. The cratering record, so widespread throughout the Solar System, testifies to the mopping up of most remaining planetesimals.

The atmospheres of the terrestrial planets probably leaked out from the hot interiors through volcanoes and fissures, and have themselves evolved from their original chemical composition. The primitive atmosphere of the Earth was probably hydrogen-rich, containing hydrogen compounds such as water, ammonia, methane and hydrogen cyanide in addition to other gases such as carbon monoxide and carbon dioxide. Hydrogen would soon be lost to space, and any free oxygen would rapidly react with other gases or with surface rocks to remove itself from the atmosphere. Water condensed and formed the oceans about 3·6 billion years ago, and most of the carbon dioxide in the Earth's atmosphere was then combined with water to form carbonate rocks. In this way the Earth avoided having a massive carbon dioxide atmosphere like that of Venus (◆ page 60).

The emergence and development of life on Earth played a fundamental part in modifying the atmosphere. Green plants, which use sunlight to form glucose out of water and carbon dioxide (by photosynthesis) release oxygen into the atmosphere, and since their appearance on Earth, over two billion years ago, the proportion of free oxygen in the atmosphere has settled to a fairly steady 21 percent. The industrial activity of mankind which, among other side-effects, releases carbon dioxide into the atmosphere, has now become a major factor in modifying the atmosphere and climate.

Early Speculatio

▲ *The great French philosopher René Descartes.*

▼ ***Swedenborg thought that at some time a crust had formed round the Sun, that this crust had split into an inner and an outer layer, and that the planets and satellites had formed from the outer layer as illustrated.***

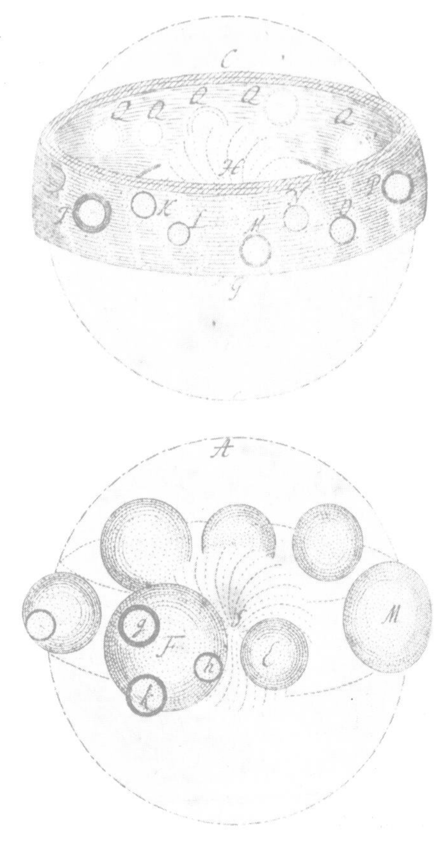

Groping towards an explanation

As observation continued to provide new facts, speculation constantly endeavored to assimilate these into viable theories. Two of these sought to explain planetary motion in terms of vortices. The French philosopher René Descartes (1596-1650) developed a theory according to which the universe is filled with vortices; the planets are held in an ethereal vortex which swirls around the Sun, so that they are carried along like corks in a whirlpool.

Isaac Newton had no patience with this scheme, but apparently he did not himself speculate about the origin of the planets. The first person to do so after the publication of Newton's laws was a Swede, Emanuel Swedenborg, who graduated from Uppsala University in 1709 and wrote voluminously on astronomy, economics, chemistry, mechanics and algebra. According to Swedenborg, the Sun was simply a huge vortex; it had developed a solid crust which expanded and broke up to produce the planets.

Any mathematical analysis will show that this is totally unsound. Moreover, Swedenborg believed that he was in constant touch with beings from other planets, and from the spirit world. Once, for a joke, he claimed to have seen an archbishop's recently deceased cards partner "shuffling his cards in the company of the Evil One, and waiting for your worship to make up a threesome".

► **Pierre de Laplace, author of "Système du Monde" in which he described his so-called nebular hypothesis for the formation of the planets.**

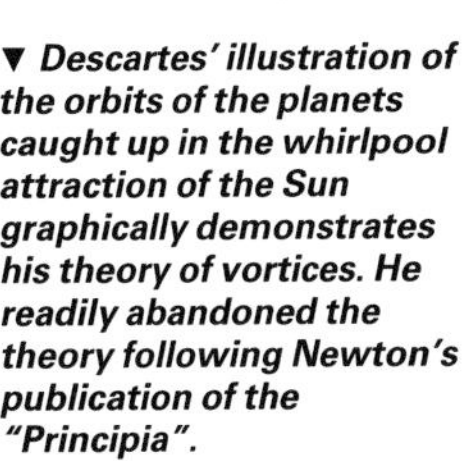

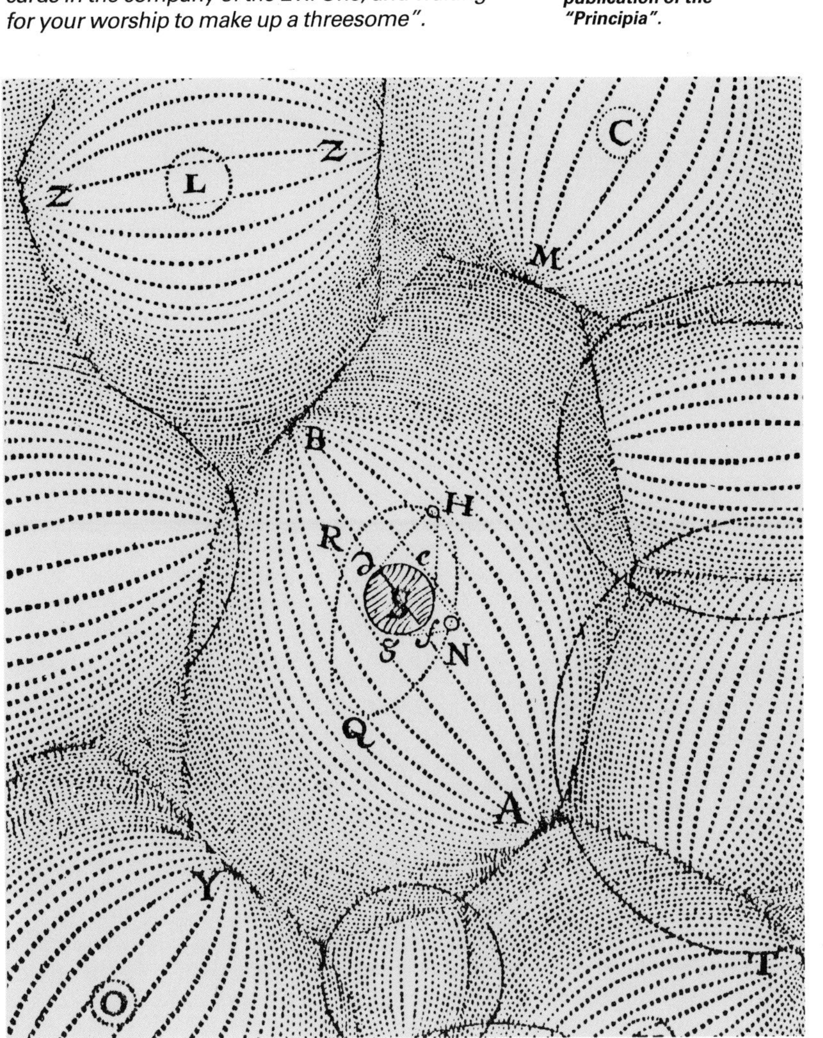

▼ **Descartes' illustration of the orbits of the planets caught up in the whirlpool attraction of the Sun graphically demonstrates his theory of vortices. He readily abandoned the theory following Newton's publication of the "Principia".**

Getting closer to the truth

In 1795 the French mathematician Pierre Simon de Laplace (1749-1827) produced the most properly scientific ideas to date concerning the origin of the Solar System. According to Laplace, the Sun began as a huge cloud of material, spinning round. Under the influence of gravity it began to shrink, and as it did so it threw off its outermost section, forming a ring which slowly condensed into a planet. The shrinkage continued, and as it did so fresh rings were thrown off until the planetary system had developed to its present form. The Sun itself was the last remnant of the original cloud. Outwardly Laplace's "Nebular Hypothesis" seemed plausible, but in its original form it was untenable on purely mathematical grounds.

It was followed by a number of "catastrophic" theories involving a wandering star. In 1900 the American astronomers Thomas Chamberlin (1843-1928) and Forest Ray Moulton (1872-1952) suggested that the gravitational force of a passing star had pulled matter out of the Sun, and that this had subsequently formed the planetesimals. Mathematicians showed, however, that due to their high temperatures the solar gases would disperse before they could combine into planets.

Sir Fred Hoyle, in the late 1940s, dispensed with a wandering star, but suggested instead that the Sun used to be a double or binary system; the companion star evolved more quickly than the Sun and exploded as a supernova, planet-forming material being scattered in the process while the tremendous recoil sent the erstwhile companion outward. Although this is impossible to disprove, most modern theories mark something of a return to Laplace's Nebular Hypothesis.

Titan is the only other world in the Solar System with a thick, nitrogen-rich atmosphere, but is too cold to support our kind of life

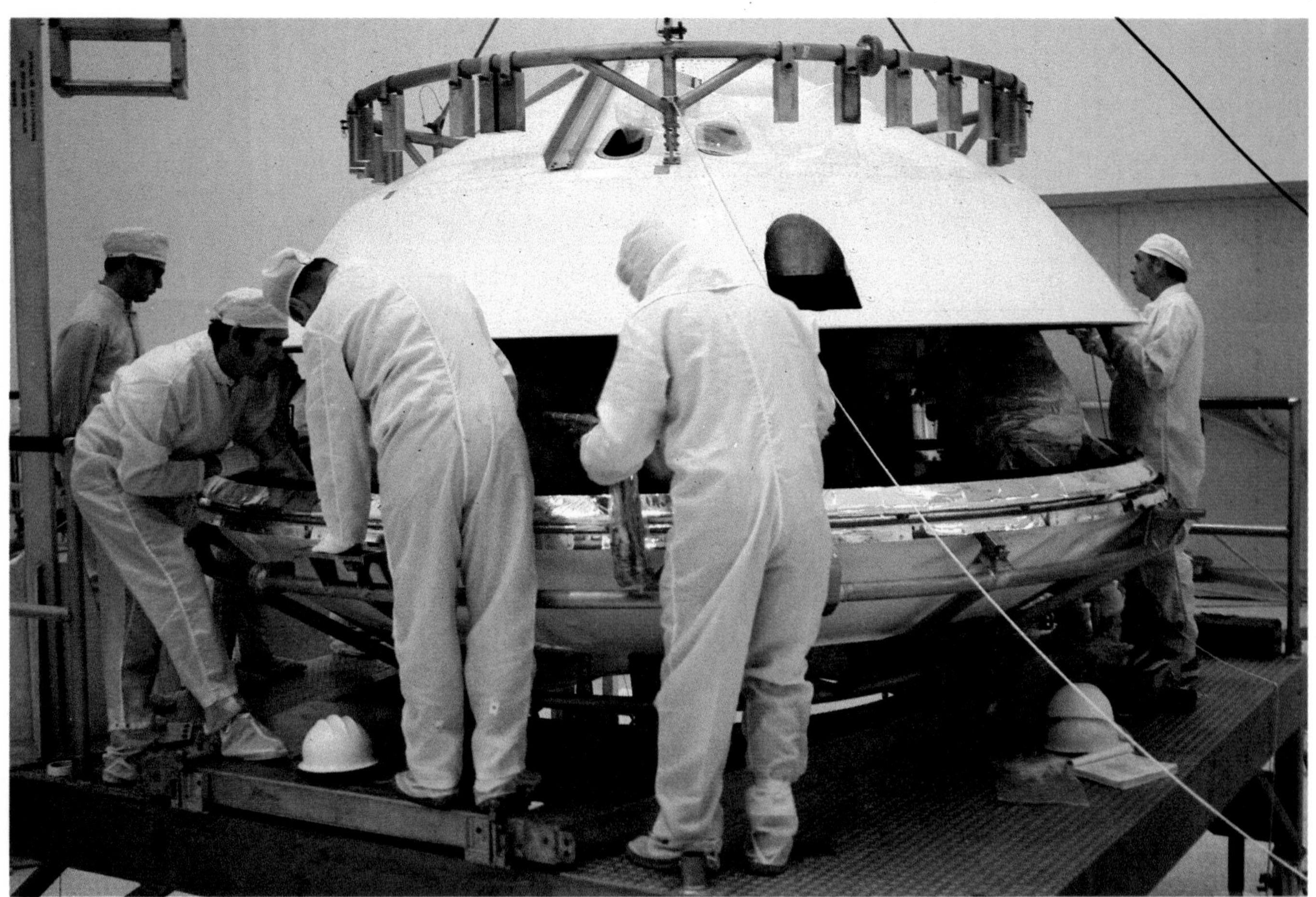

▲ **One of the Viking Lander Capsules is unpacked after delivery from the manufacturers to the Kennedy Space Center. The Lander was "baked" before launching to remove all traces of terrestrial life. The purpose of this was twofold – to avoid contamination of the Martian surface, and to prevent the life-detecting equipment being misled by stowaway organisms from Earth.**

The presence of liquid water seems to be an essential prerequisite for the existence of life as we know it, and the lack of liquid water on the other planets in the Solar System suggests that they are unlikely to support even the most basic life-forms. The two Viking spacecraft which landed on Mars in 1976 carried a variety of experiments designed to seek for signs of life. Analysis of Martian soil samples failed to reveal any organic molecules. Other tests designed to measure biological activity, such as the reaction of organisms to nutrients, at first seemed to give positive results, but these were not repeatable and are now widely believed to be due merely to some quirk of Martian soil chemistry. These results appear to have eliminated the possibility of even basic bacterial life on that planet, although there is an outside chance that some kind of organisms may exist near the Martian ice caps where there is more water.

Some theorists have speculated that suitably equable conditions may prevail at some level within Jupiter's atmosphere, and that a form of life may exist there for which ammonia takes on the role played here by water.

Farther afield Titan, one of Saturn's satellites, has an atmosphere which is similar in composition to the primitive terrestrial atmosphere. Although it is far too cold to support life now, in the distant future the Sun will swell up to become a red giant, and Titan may become as warm as the Earth is now. However, this state of affairs will be short-lived and the emergence of life on Titan seem to be extremely unlikely. The balance of evidence strongly suggests that, within the Solar System at least, life is unique to the Earth.

Searching in vain for life elsewhere

No subject can have produced a greater volume of eccentric literature than that of life beyond the Earth. Originally, of course, the idea of intelligent beings on the Moon seemed quite logical. The great German astronomer Johannes Kepler (1571-1630) wrote a novel, published posthumously, in which obliging demons took his hero to the Moon. There he found the local inhabitants to be highly civilized; any Moonchild showing signs of depravity was exiled to Earth, where his depravity would go unnoticed.

Some even more far-fetched ideas followed, including William Herschel's picture of a habitable Sun. But by the middle of the 20th century it was clear that there were only two planets besides Earth that might be possible sites for life: Mars and Venus. Unmanned spacecraft investigated both planets in the 1970s, but the Mariner and Viking probes revealed that the atmosphere on Mars is extremely thin and that there is no evidence of life of any sort, while Venus has proved to be one of the most hostile of all the worlds in the Sun's family. Where else is there to look?

◄ *These book jackets are of allegedly factual events: far left, Kenneth Arnold sights the first flying saucer; near left, writer Frank R. Paul meets a friendly Venusian.*

▼ *In 1938 an American radio station broadcast a dramatized version of H. G. Wells' novel "The War of the Worlds". The production (by Orson Welles) was so realistic that listeners mistook it for a genuine news bulletin reporting an invasion of the Earth by loathsome Martians with fighting-machines and heat rays. Widespread panic broke out; people claimed to have seen the fighting-machines, felt the heat rays and watched the slaughter of innocent people.*

Science fact and science fiction

Titan, the largest of Saturn's satellites, has a dense nitrogen atmosphere, but is bitterly cold. All the other satellites are without appreciable atmospheres (with the possible exception of Triton in Neptune's system), and although some theorists have suggested that living organisms could survive in a sub-crustal sea on Europa, one of Jupiter's satellites, or inside the outer gases of Jupiter and Saturn, there is no evidence to support this.

Less serious suggestions concerning "BEMs" (bug-eyed monsters) that could live on airless worlds or in the corrosive carbon dioxide atmosphere of Venus are not only unscientific in themselves: if such creatures exist, then they must lie outside the accepted parameters of modern science.

The idea that the Earth was visited by aliens in near-historical times has been popularized by the Swiss writer Erich von Däniken, who has produced a mass of "evidence" – ancient maps showing Antarctica, cave-drawings of astronauts, saucer landing-strips in Peru, and so on. All this is an intriguing diversion, but it is hardly science.

The flying saucer story began in 1947, when an amateur pilot named Kenneth Arnold observed some disk-shaped objects over Washington. Other sightings followed, and before long there were accounts of direct contact with aliens: for example, one George Adamski related how he met a "Venusian" in the Californian desert.

See also
Planets and Orbits 9-20
Asteroids and Comets 89-96
Birth, Life and Death of Stars 153-60
The Evolving Universe 229-44
Communications and Travel 245-8

The origin and evolution of life is an emotive and controversial subject. Some scientists believe that the number of steps in the chain which led to the formation of life is so enormous that life must have originated in the vast "laboratory" of interstellar space, and that basic living organisms – bacteria – were dumped on this planet, possibly by comets. However, most biologists firmly believe that life originated here on Earth.

Life as we know it depends on the ability of the element carbon to form long complex chains and so to assemble elaborate structures. Organic molecules are composed primarily of the elements carbon, hydrogen, oxygen and nitrogen – which are among the most abundant elements in the universe. The basic building blocks of life, complex molecules known as amino acids, are easy to synthesize from the elements that scientists believe were present in the primitive atmosphere, using energy sources such as ultraviolet radiation, lightning, volcanic eruptions and meteoritic impacts, all of which were abundant early in the planet's history.

Biologists believe that a primordial organic "soup" formed in the oceans, and that more complex self-replicating molecules developed where the organic matter became concentrated, for example by evaporation or freezing. Such molecules provided the basis for the production of simple cells which reproduced by the simple act of repeatedly dividing. They probably appeared on Earth at least 3·3 billion, possibly as much as 4 billion, years ago. Mutation (caused, for example, by radiation) modified the nature of some of the self-replicating organisms, and while most mutants died, some prospered because they were better suited to their environment. In this way, over billions of years, life evolved to yield the diverse range of about half a million different species which now inhabit the Earth. Although many of the steps are poorly understood, the overall process of evolution is established beyond reasonable doubt.

Impacts and life

Catastrophic impacts by asteroids or cometary nuclei, despite being rare events, must have exerted an influence on the atmosphere, the climate and on life. Over the past few hundred million years, a kilometer-sized body must have struck the Earth, on average, once every million years or so, and an object of 10 kilometers once every 50-100 million years. The impact of a 10km asteroid would produce a blast comparable to the detonation of 100 million one-megaton nuclear bombs, and would throw tens or even hundreds of millions of millions of tonnes of dust into the atmosphere; if such an impact occurred in an ocean it would raise an equivalent amount of water. The dust would severely diminish the amount of sunlight reaching the Earth's surface, lowering temperatures and disrupting food chains on a global scale. In the case of an oceanic impact, the water vapor may have increased the greenhouse effect and raised the temperature to the detriment of some species of life. Short-term chemical changes in the atmosphere, the disruption of the ozone layer, and tidal waves would all affect terrestrial life. Although the atmospheric and climatic effects may have been short-lived, the long-term effects on life and its evolution could have been substantial.

A sudden major extinction of species, including the dinosaurs which had dominated the planet for 130 million years, occurred 65 million years ago. It is possible that this event was the result of a massive impact. If the impact occurred on land, it may yet prove possible to identify the eroded remains of the resulting crater; but if the impact occurred in the ocean, the crater has probably been filled in or eradicated by plate tectonics by now.

► Professor A. H. Delsemme of the University of Toledo in America is an expert in the field of cometary analysis. Working with a spectrograph, he has calculated that comets contain the same proportions of hydrogen, carbon, nitrogen and oxygen as living organisms on Earth. Comets and living organisms share this chemical balance with each other and with no other bodies in the known universe. If this is not a bizarre coincidence, then comets seem to be implicated somehow in the emergence of life on Earth.

Overview of the Universe

Astronomical distances...The light year...Our Sun compared with other stars...Stars, nebulae and galaxies introduced...Grasping the vastness of the universe... PERSPECTIVE...Early speculations on the scale of the universe...The first attempts at measuring stellar distances...The true status of our Solar System

Like the Sun, the stars are self-luminous gaseous bodies generating energy in nuclear reactions. Their distances from the Earth and from each other are vast. The star nearest to the Solar System, Proxima Centauri (a dim red star in the constellation of Centaurus), lies at a distance of over 40 million million kilometers, some 270,000 times greater than the distance between the Earth and the Sun.

When considering astronomical distances it is convenient to think not in conventional units like meters or kilometers, but in terms of how long it takes a ray of light to travel across these distances. Moving at about 300,000 kilometers per second, light takes 1·3 seconds to travel from the Moon to the Earth, 8·3 minutes to reach the Earth from the Sun and about 5·4 hours to cross the distance between the Sun and the planet Pluto. However, a ray of light takes 4·3 *years* to reach us from Proxima Centauri. This, more graphically than huge numbers, sets the scale of stellar distances.

The distance that light travels in one year – 9·46 million million kilometers – is called a "light year", and this provides a convenient unit for describing distances in the universe. Proxima Centauri lies at a distance of 4·3 light years, and in our locality the average distance between stars is from three to four light years.

The Earth and the center of the universe

Civilizations such as China and Egypt carried out useful observations, notably of comets and eclipses, but had no idea of the real nature of the universe. To the Egyptians, the universe was shaped like a rectangular box, with a flat ceiling supported by pillars at the four cardinal points; these pillars were connected by a ledge along which ran the celestial river Ur-nes, in which the boats carrying the Sun, Moon and other gods sailed. The Egyptian Pharaoh Akhenaton (reigned c. 1379-1362 BC) even founded a new cult of Sun-worship, which was soon rejected after his death.

Astronomy as a true science began with the Greeks; up until the first of the great Greek philosophers, Thales (c. 636-546 BC), mythology had been used to explain the nature of the physical world. With the realization that the Earth is indeed a globe and not flat, one or two of the Greeks – notably Aristarchus of Samos (c. 320-250 BC) – even relegated it to the status of a planet moving round the Sun. Unfortunately Aristarchus found few supporters, and the later Greeks reverted to the idea of a central, motionless Earth. Ptolemy of Alexandra (fl. AD 2nd century), last of the major figures of the Greek school of astronomy, maintained that the Earth could not be spinning, as in this case there would be a constant gale as the world rotated beneath its atmosphere. On the other hand Ptolemy was an excellent observer and mathematician and his theory of the universe persisted until the 16th century. In it the Solar System was thought to be the entire universe with the Earth as its center and the distant stars located just beyond the farthest planet. It was not until 1530 that it was first suggested that the stars are at very great distances compared to the planets.

◄ *An astonished observer, having poked his head through the sphere of stars, looks out towards the driving mechanism of the cosmos. This medieval representation of the universe illustrates the long-held view that the stars were fixed to a sphere which rotated round the Earth once a day, driven by a prime mover.*

The Scale of the Universe

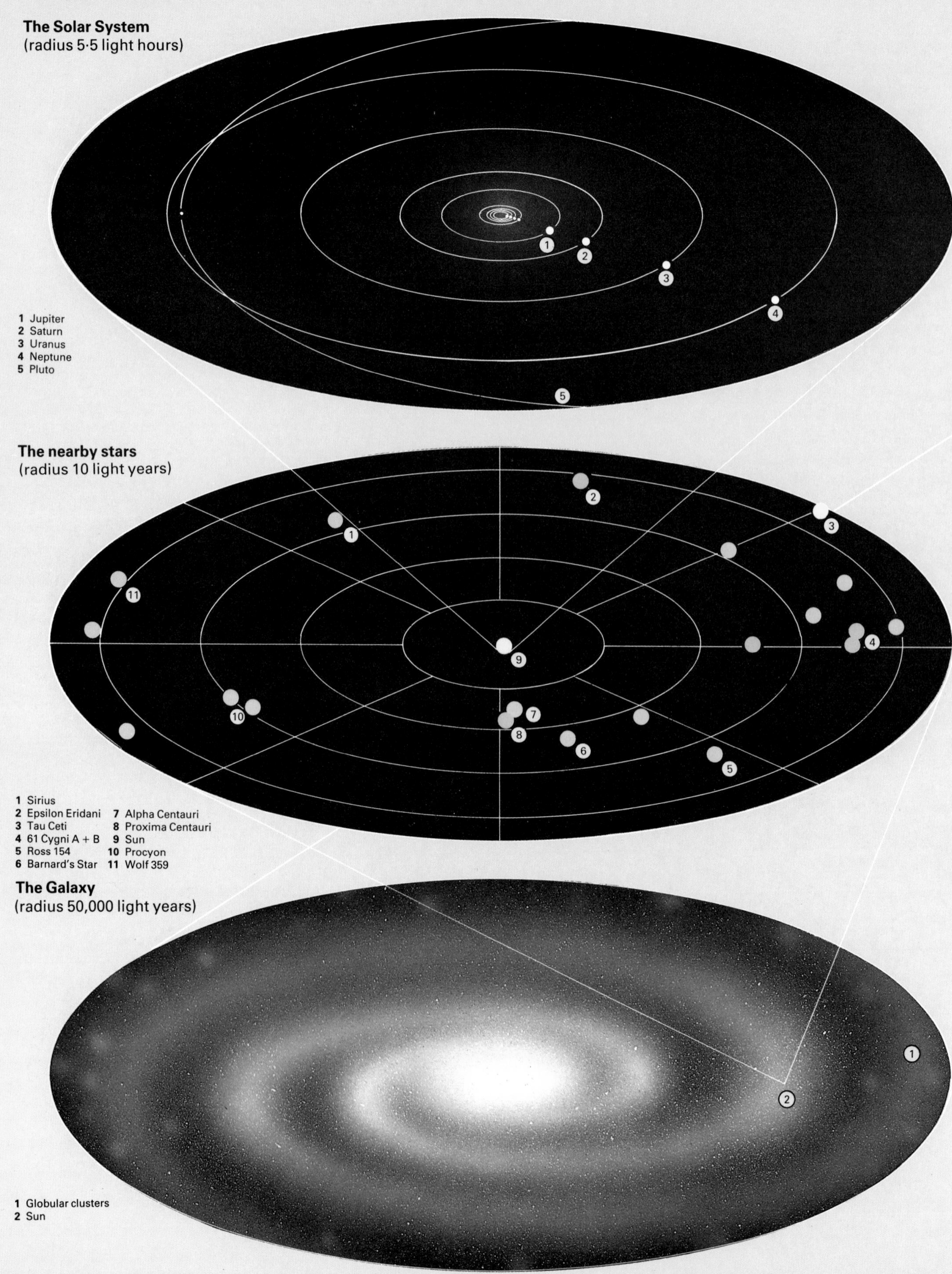

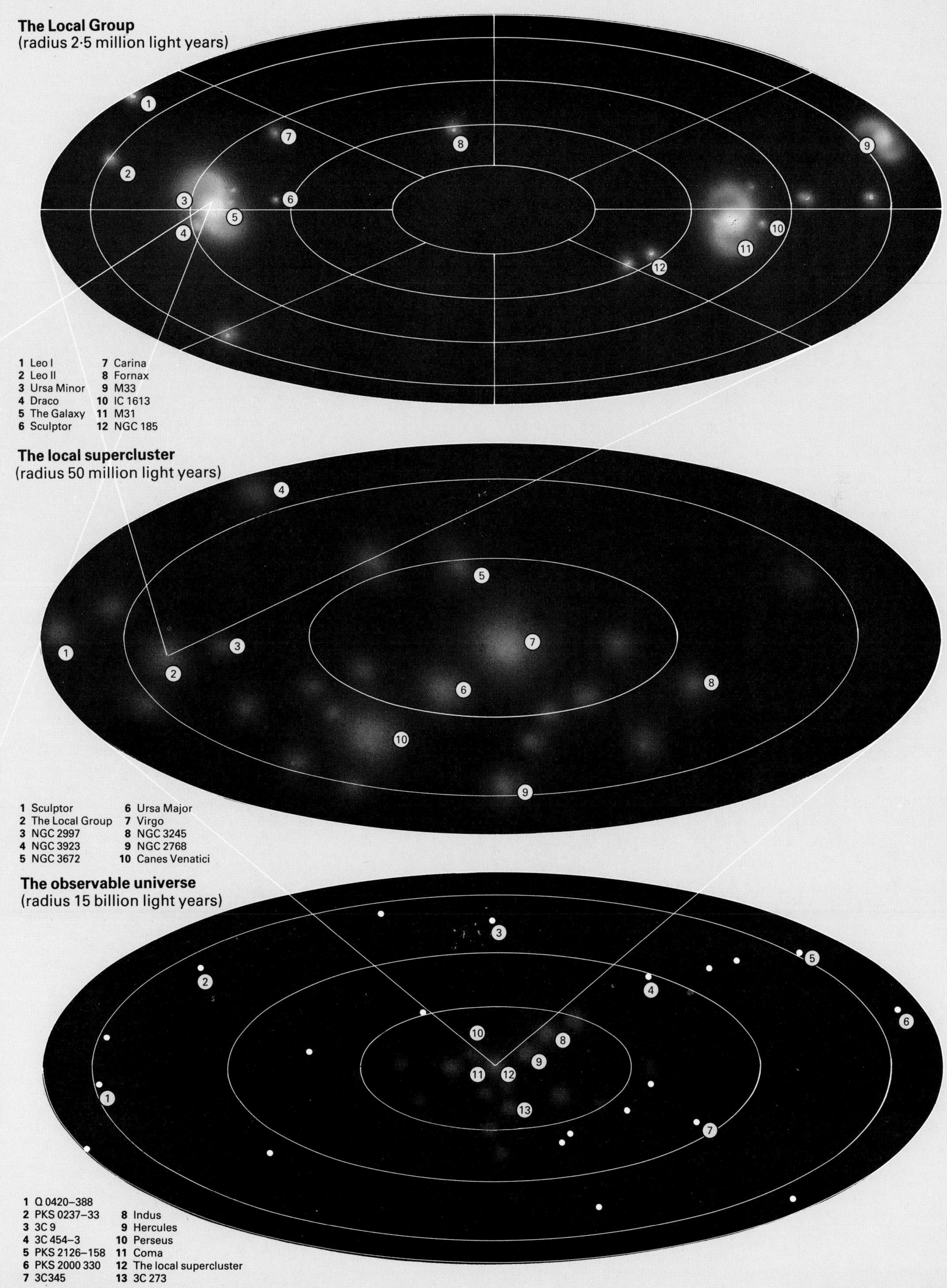
The Local Group
(radius 2·5 million light years)
1 Leo I
2 Leo II
3 Ursa Minor
4 Draco
5 The Galaxy
6 Sculptor
7 Carina
8 Fornax
9 M33
10 IC 1613
11 M31
12 NGC 185
The local supercluster
(radius 50 million light years)
1 Sculptor
2 The Local Group
3 NGC 2997
4 NGC 3923
5 NGC 3672
6 Ursa Major
7 Virgo
8 NGC 3245
9 NGC 2768
10 Canes Venatici
The observable universe
(radius 15 billion light years)
1 Q 0420–388
2 PKS 0237–33
3 3C 9
4 3C 454–3
5 PKS 2126–158
6 PKS 2000 330
7 3C345
8 Indus
9 Hercules
10 Perseus
11 Coma
12 The local supercluster
13 3C 273

See also
Planets and Orbits 9-20
The Basic Properties of Stars 145-52
The Milky Way Galaxy 189-96
Galaxies 197-204
The Evolving Universe 229-44

Although the space between the stars (interstellar space) is a near-perfect vacuum by terrestrial standards, it contains a certain amount of matter in the form of thinly spread gases and tiny particles of dust. Some of the gas clouds are visible directly as luminous patches in the sky. Known as "nebulae", from the Latin word for "clouds", most are revealed by their emission of radio or infrared radiation (◀ page 26). New stars are being born continuously within clouds such as these.

Galactic neighbors

The Sun is a member of a system of stars, gases and dust known as the Galaxy, or as the "Milky Way" galaxy. The diameter of the Galaxy is about 100,000 light years (▶ page 194), and it contains about 100 billion (10^{11}) stars. Beyond the edges of our own star system astronomers can see billions of other galaxies. Many of these are smaller than the Milky Way system, but some are significantly larger. Our Galaxy is a member of a small group of about thirty galaxies known as the Local Group. Most galaxies are members of clusters, some of which contain several thousand members, and there even seem to be clusters of clusters (superclusters) in the hierarchy of the universe.

The Galaxy has two smaller satellite galaxies, the Large and Small Magellanic Clouds – named after the Portuguese navigator Ferdinand Magellan (*c.* 1480-1521) – which lie at distances of 160,000 and 190,000 light years respectively.

Looking back in time

The nearest large galaxy comparable with our own is the Andromeda spiral, or M31. At a distance of 2,200,000 light years, it is just visible without telescopic aid on a good, clear, dark night, and it is by far the most distant object visible to the naked eye. The light which is now arriving from M31 was emitted 2,200,000 years ago, and depicts that galaxy not as it is now, but as it used to be all that time ago. The farther astronomers probe into deep space, the more "out of date" their new information becomes, but there is a compensating advantage in that they are able to study distant parts of the universe as they were billions of years ago, and in this way they can attempt to trace out the evolution of the universe itself.

In addition to "ordinary" galaxies, there exists also a wide variety of violently active galaxies which emit far more energy than normal systems. These include radio galaxies and quasars (▶ page 205) which are so brilliant that it is possible to detect them at distances of well over 10 billion light years. The universe also contains exceedingly tenuous intergalactic matter and radiation, and recent observations seem to indicate that the visible matter in galaxies may be considerably outweighed by non-luminous forms of matter (▶ page 223). The entire universe seems to be expanding and evolving (▶ page 229).

Astronomers can look out to distances greater than 10 billion light years and can probe back through aeons of time to an era when the universe was much younger than it now is. Compared to this broad panorama, interplanetary distances within the Solar System seem microscopic indeed.

The true vastness of the universe

It was almost impossible for early peoples to form any real idea of the distance-scale of the universe, but at least it was known that the Sun is a long way away; Ptolemy gave its distance as 8,000,000km. Obviously the stars were more remote still, but it was not then known that they were suns in their own right; it was widely believed that they were lamps attached to an invisible crystal vault.

With the revelation that the Earth is a planet moving round the Sun – a change in outlook which was complete by 1687, with the publication of Newton's great work – the status of the Sun among the stars became questioned. Thomas Wright (1711-1786) suggested that the universe might extend to infinity. The great Danish astronomer Tycho Brahe (1546-1601) had already shown that the stars must be much more remote than the Sun. But actual distance-measures were very difficult; even William Herschel (1738-1822) arguably the greatest of all observers, failed in his efforts – though he did make the discovery that many double stars are binary systems instead of being mere line-of-sight effects.

Success was finally achieved by F.W. Bessel (1784-1846) in 1838, when he measured the distance of the faint star 61 Cygni as around 11 light years. He used the parallax method (▶ page 145). Other measures followed; the parallax method worked well enough for relatively close stars, but not for those which were beyond a few hundred light years, and less direct methods were found.

The next step was to decide upon the status of our star-system or Galaxy. Did it comprise the entire universe, or was it a mere unit? This fundamental problem was not cleared up until 1923, when Edwin Hubble (1889-1953) observed short-period variable stars in "spiral nebulae" and proved that these objects are independent galaxies. Today we know that the universe is vaster than ancient peoples could have credited; we see the most remote known objects not as they are now, but as they used to be before the Earth existed.

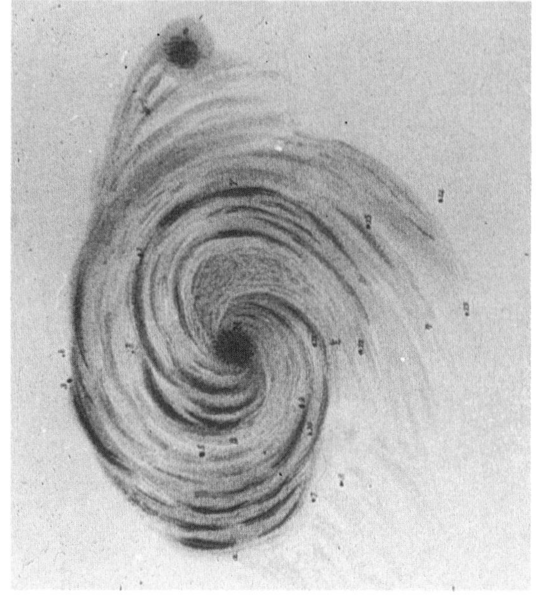

▶ When the third Earl of Rosse (1800-1867) drew this spiral galaxy in 1845, his was the only telescope of the time powerful enough to show the spiral forms of the objects we now know to be galaxies. Their nature was then a mystery, and it was not until 1923 that it was finally shown that the spirals and other so called "starry nebulae" are external to our Galaxy; most of them many millions of light years away.

The Moving Sky

Introducing the motions of the stars – the "celestial sphere"...Defining the exact position of a star...Patterns in the sky – the constellations...The significance of the pole star...Famous constellations...PERSPECTIVE... Astronomy and astrology – worlds apart

To an Earth-based observer the sky looks like a hemispherical dome bounded by the horizon. The point vertically overhead is the zenith, and the imaginary circle passing through the zenith and the north and south points of the horizon is the observer's meridian. Stars rise on the eastern horizon and move across the sky in a westerly direction, reaching their greatest elevation as they cross the meridian. They are then at upper transit, or culmination. As the night progresses they sink lower in the sky, eventually setting below the western horizon.

This motion is not real but apparent, arising from the rotation of the Earth from west to east. However, for defining the positions of stars it is convenient to imagine that they are fixed to a huge sphere – the celestial sphere – which rotates round the Earth once a day. If the Earth's axis is extended into space, it meets this imaginary sphere at two points, the north and south celestial poles. Likewise, if the plane of the Earth's equator is extended into space it meets the sphere in a circle known as the celestial equator. Observers define the positions of the stars by reference to the celestial equator and poles.

To an observer at the north pole of the Earth, the north celestial pole is vertically overhead, and the stars appear to trace out circles centered on the pole and parallel to the horizon. The celestial equator coincides with the horizon, and the entire southern half of the celestial sphere is permanently hidden below the horizon. An observer situated on the equator has a very different view. The celestial equator passes vertically overhead and the poles coincide with the north and south points of his horizon. Although only half of the celestial sphere is visible at any particular instant, every part of the sphere comes into view at some time or other.

For an observer located anywhere between the equator and one of the poles, it is possible to divide the sky into three regions. Stars close to the near pole never set; stars close to the distant pole never rise, and those in the middle part of the sphere rise and set in the normal way.

Apparent motion of the stars

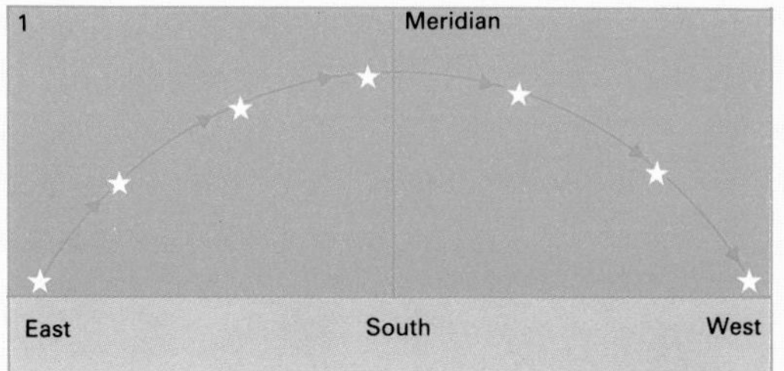

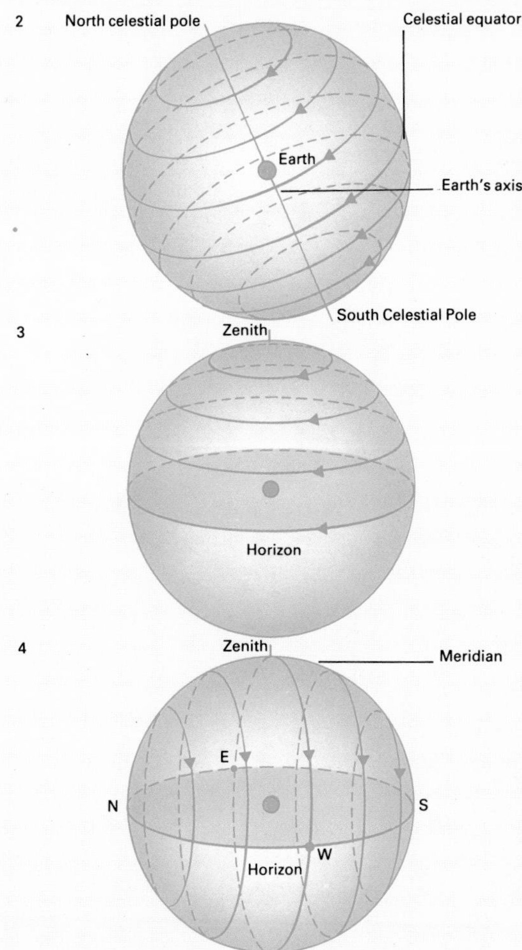

▲ The apparent motion of a star across the sky is shown (1). The Earth's spin causes the apparent east to west rotation of the celestial sphere (2). Viewed from the poles (3) stars appear to move parallel to the horizon, while at the equator (4) they rise and set vertically.

◄ A group of students from the University of California, absorbed in practical studies in 1895, reflect the popularity of amateur astronomy.

If the stars were visible in daylight, the Sun would be seen progressively to change its position relative to the background stars, tracing out a complete circle – the ecliptic – in the course of a year. Because the terrestrial equator is inclined to the plane of the Earth's orbit by an angle of about 23½°, the ecliptic is inclined to the celestial equator by the same angle, and crosses it at two points – the equinoxes. The vernal equinox is the point where the Sun passes from south to north of the equator around 21 March each year; at the autumnal equinox it passes back from north to south, some six months later.

Celestial positions

Positions on the Earth's surface are described in terms of latitude and longitude, latitude being the angular distance north or south of the equator and longitude being the angular distance east or west of the Greenwich meridian. Astronomers use a similar system on the celestial sphere. Declination is the angle, measured north (+) or south (–), between the celestial equator and the star in question. Right ascension is the angle, measured in an anticlockwise direction (eastwards) from the vernal equinox. Right ascension takes any value between 0° and 360°, but is expressed in time units, from 0 to 24 hours (this is for historical reasons connected with the use of the Earth's rotation and measurements of stars to regulate clocks). One hour is equivalent to 15° since that is the angle through which the Earth spins in one hour. Thus 1° is equivalent to four minutes, that being the angle through which the Earth rotates in four minutes of time.

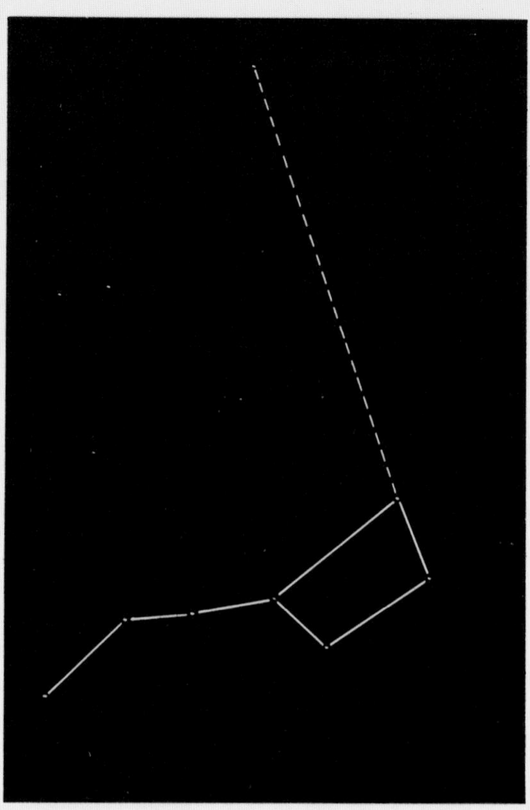

◀ *Several hours' worth of starry paths around the Pole Star are etched onto a single photograph. The celestial sphere rotates once every 24 hours.*

▲ *Polaris, the Pole Star, is at the end of the "handle" of the Little Dipper or Bear.*

▼ *Orion – one of the easiest of all constellations to find.*

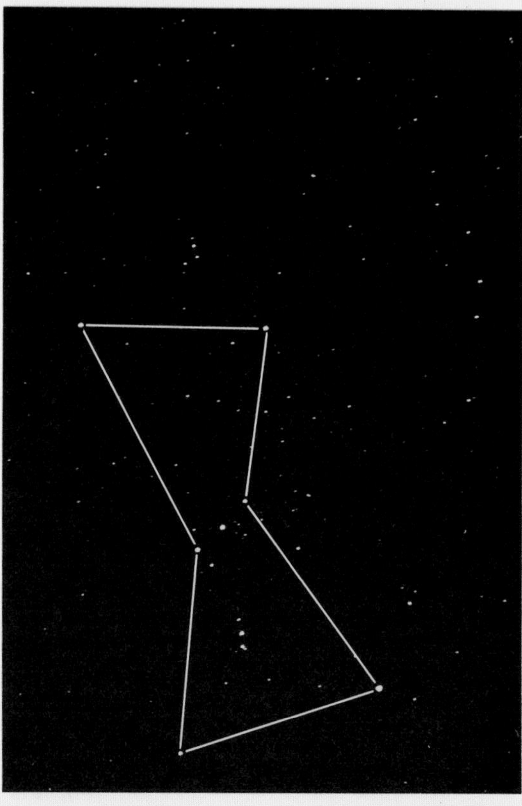

The celestial sphere

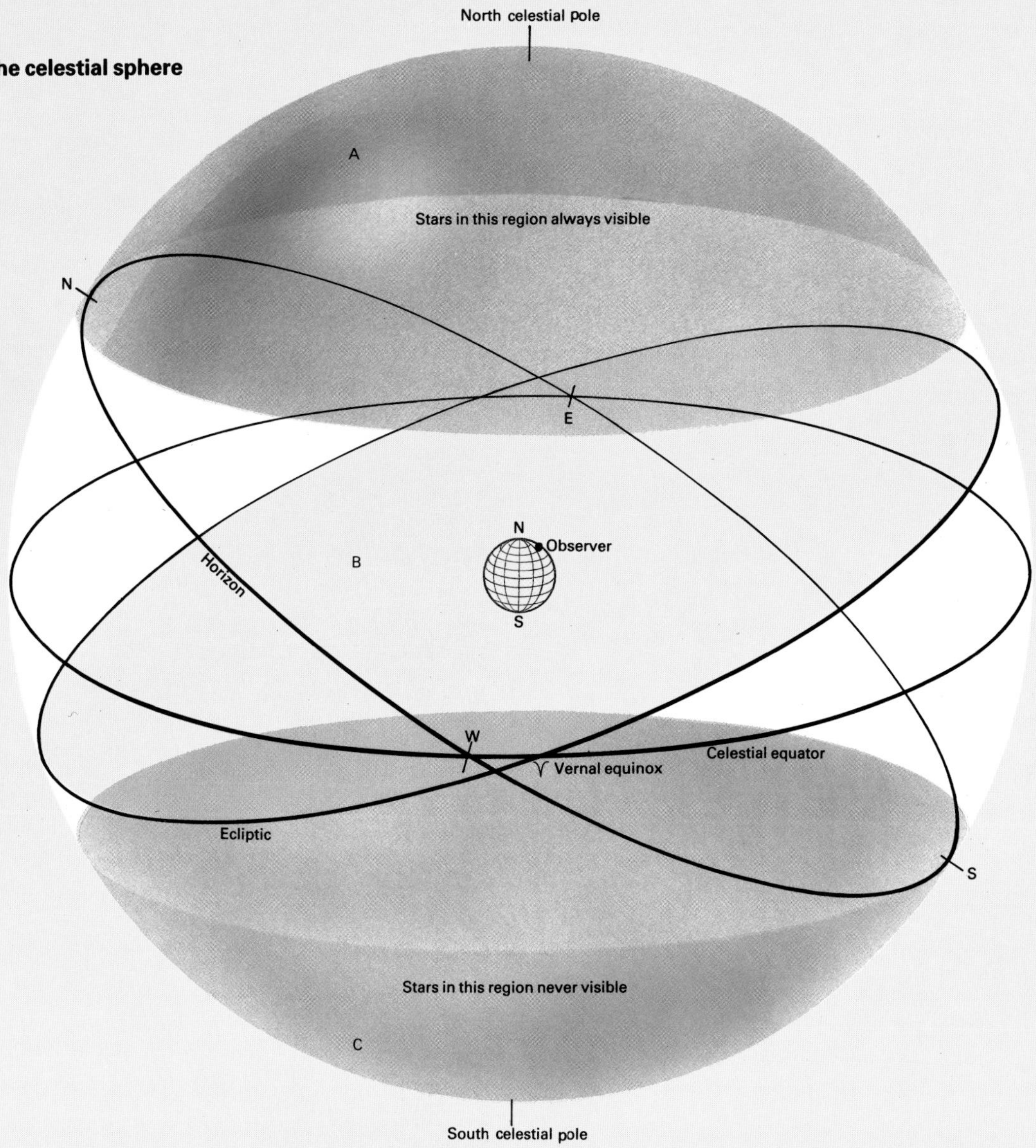

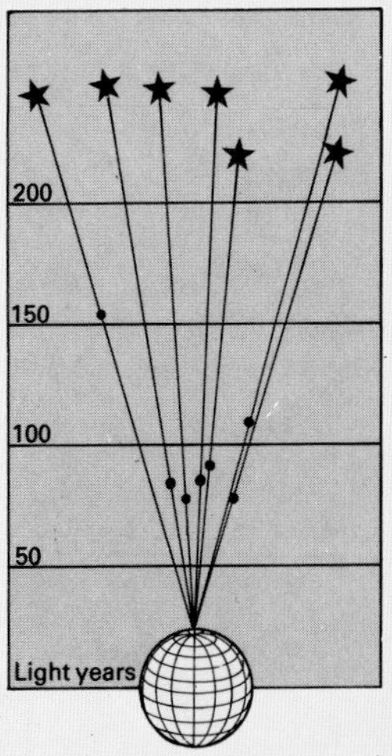

▲ *The seven principal stars of the Big Dipper or Plough lie in similar directions but at quite different distances.*

◄ *The celestial sphere as seen by an observer at latitude 60° north (the observer is on the inside, looking out). Scales of right ascension and declination run along and perpendicular to, the celestial equator. Stars seem to circle round the celestial poles. For this observer, stars in region A are circumpolar – they never set – stars in region B rise and set, and in region C they never rise.*

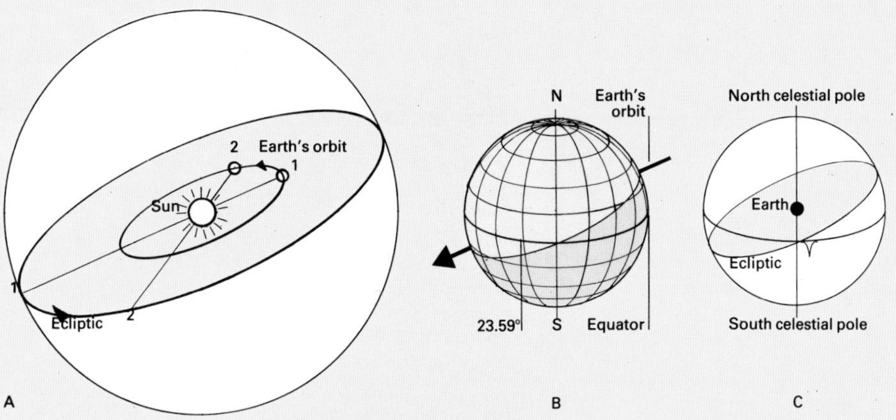

▲ *As the Earth moves along its orbit so the Sun appears to trace out a circle on the celestial sphere – the ecliptic – which lies in the same plane as the Earth's orbit (A). Because the Earth's equator is tilted to the plane of its orbit by an angle of about 23½° (B) the celestial equator and ecliptic are inclined to each other by the same angle (C). One of the two points where these circles cross is the vernal equinox (♈) which provides a reference point for position measurement.*

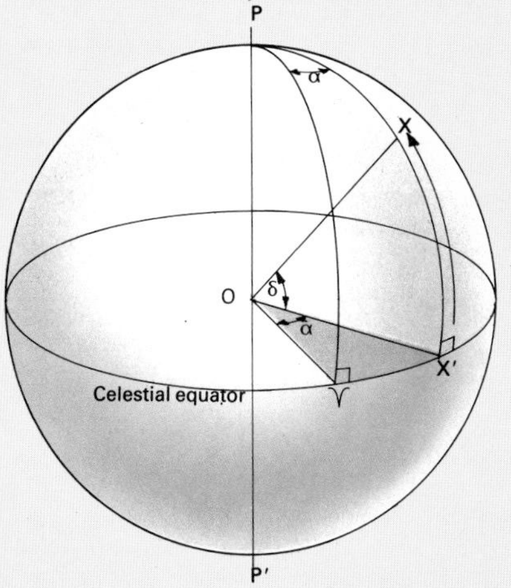

▲ *For an observer, O, looking at a star, X, the declination of X (δ) is the angle X′0X measured perpendicular to the celestial equator. Values of declination range between 0° for a star on the celestial equator to ±90° for stars at the north (+) and south (–) celestial poles. The right ascension of X (α) is the angle ♈PX (= angle ♈0X′) – the angle between the vernal equinox (♈) and the star measured eastwards from (♈) and expressed in time units (hours, minutes, seconds).*

See also
Planets and Orbits 9-20
Observing the Planets 21-32
Observing the Universe 137-44
Binary and Variable Stars 161-8
Galaxies 197-204

The constellations

Some groups of stars seem to make up identifiable shapes, and these groups are known as constellations. The constellations have no physical significance. They comprise stars which happen to lie in similar directions as seen from the Earth, but which in most cases lie at very different distances from the Sun and from each other. There are 88 constellations spanning the entire celestial sphere, many of them named after mythological figures such as Andromeda, the daughter of Cepheus and Cassiope, and Orion, the Hunter. In most cases the resemblance between the shape of the constellation and the creature or object after which it is named is extremely tenuous.

The brightest stars in each constellation are identified by Greek letters, α being the brightest, while the constellations themselves have Latinized names. Thus Sirius is also known as α Canis Majoris – the brightest star in Canis Majoris (the Large Dog). The best-known constellation in the northern hemisphere is Ursa Major, the Great Bear, within which is the very familiar group of seven stars known as the Big Dipper or Plough (a grouping within a constellation is called an "asterism"). The two stars at the front of the "saucepan" shape are known as "the pointers" since a line through them leads to Polaris, a star which lies within 1° of the north celestial pole and which, therefore, is known as the Pole Star. The most conspicuous of all constellations and one which, because it straddles the celestial equator can be seen from anywhere on Earth, is Orion. These two easily identified constellations provide useful signposts to locating any of the others.

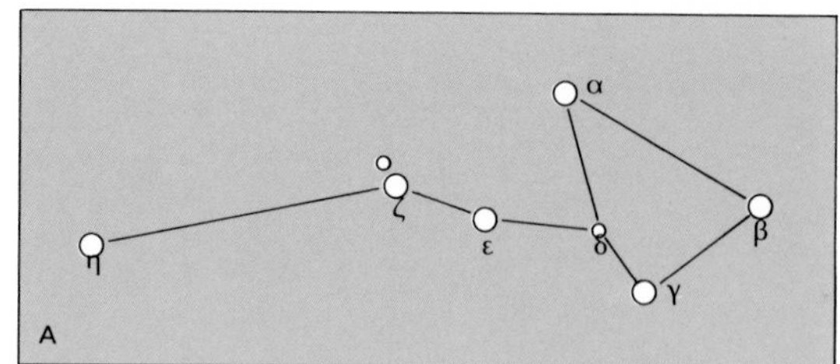

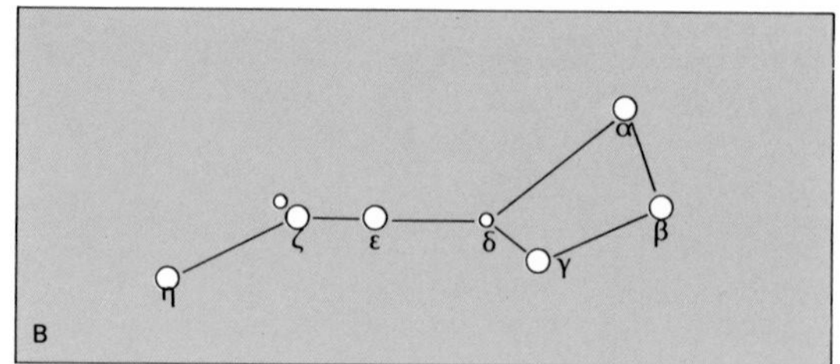

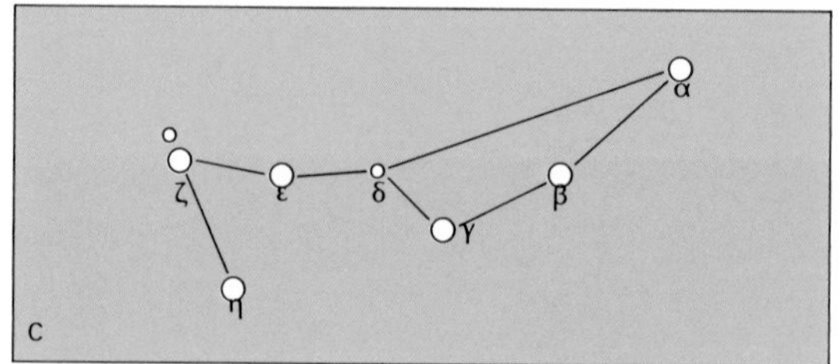

◄ This tapestry from Bali depicts astrological star signs and symbols – just one interpretation of the patterns that the stars can create in the night sky.

▲ Individual stellar motions change constellation shapes. The Big Dipper or Plough is shown as it was 100,000 years ago (A), as it is now (B), and as it will be in 100,000 years' time (C).

Astronomy and astrology

In the primitive world it was natural to regard the celestial bodies as gods, or at least as the dwelling-places of gods. Later this view became modified, but most people still believed that the positions of the Sun, Moon and planets at the moment of birth could have a profound influence on a child's destiny. The first person to draw a distinction between the sciences of astronomy and astrology was Isidorus (c. AD 570-636), a Carthaginian who became Bishop of Seville.

The constellation patterns cannot, however, have any significance for human destiny. Two stars which appear close together on the celestial sphere, and are part of the same constellation, may be widely separated in space. For example, Betelgeux and Rigel, the two brightest stars in Orion, are 500 and 900 light years away respectively, so that Rigel is nearly as remote from Betelgeux as the Earth is – they simply happen to lie in much the same direction as seen from Earth. Viewed from a planet between the two, Betelgeux and Rigel would be on opposite sides of the sky.

Yet astrology persisted for a long time, and until the 17th century it was still regarded as a true science. Even Newton believed that the heavens may provide signs, and today there are still people who seek to prove that human personality is associated with the positions of stars and planets.

Observing the Universe

Introducing the observable universe...Computer-controlled telescopes...From photographic plate to silicon chip...The world's greatest observatories... Radio astronomy – the largest telescopes...Infrared astronomy – the most sensitive...Observing above the Earth's atmosphere...PERSPECTIVE...Problems of accuracy for early observers...The Herschel family of astronomers

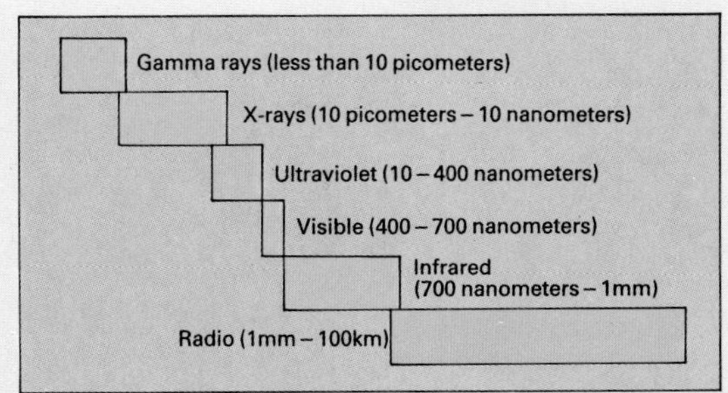

▲ ***The wavelengths of the electromagnetic spectrum.***

Practically all information about the universe arrives in the form of electromagnetic radiation (◆ pages 26-7). In many respects light and other forms of electromagnetic radiation behave like waves, with the distance between successive wavecrests being the wavelength. Human eyes respond only to wavelengths between about 400nm and 700nm, but there is a vast range of wavelengths. This is known as the electromagnetic spectrum, and it is divided (from the shortest to the longest) into gamma rays, X-rays, ultraviolet, infrared and radio waves. In other respects, electromagnetic radiation behaves like a stream of tiny particles – photons – which have zero mass but which carry a quantity of energy inversely proportional to wavelength: the shorter the wavelength, the higher the energy. Thus gamma-ray and X-ray photons are highly energetic, while infrared and radio photons are of low energy.

The atmosphere prevents most wavelengths from reaching ground level. However, satellites orbiting above the atmosphere can view the entire spectrum, and since the dawn of the Space Age, satellite-borne instruments have opened many new "windows" on the universe and brought about a major revolution in our perception of it. Ground-based astronomy has also benefited from developments in electronics and computer technology, and is in a healthy state of expansion.

Instruments of the early observers

Naked-eye astronomy has obvious limitations. Using accurate measuring instruments it is possible to fix the positions of the stars with reasonable precision, and to follow the movements of the planets, but it is impossible to find out anything about the physical nature of celestial bodies. Yet the observers of pre-telescopic times learned a remarkable amount. Tycho Brahe's star catalog, compiled between 1576 and 1596 by naked-eye work alone, was a masterpiece. It is ironic that Tycho, who would have made good use of telescopes, missed them by less than ten years.

Galileo was a skilful observer. Historians have recently shown that while plotting the positions of the satellites of Jupiter, in January 1610, he also recorded the planet Neptune, and thus anticipated the "discovery" of that planet by over 230 years. Kepler, on the other hand, was no observer at all – largely because of poor eyesight – even though he did invent an improved type of eyepiece. Since Galileo's first refractor, reflecting telescopes, introduced in 1671, have now become far more important in astronomy.

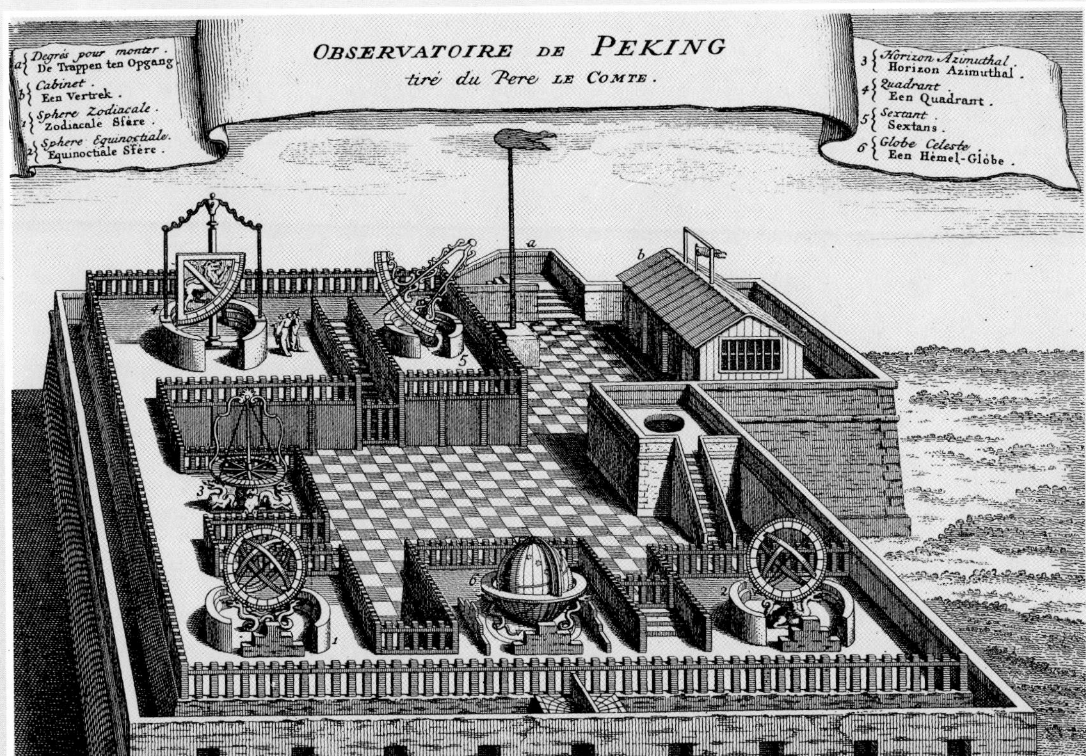

◄ **Peking Observatory, in the year 1747. Although there are no telescopes, various sophisticated measuring instruments are illustrated. These were originally introduced in the 17th century from western Europe. The figures near the quadrant (top right) give an impression of scale.**

A Family of Astronomers

The Herschels

Few families have made such an impact on astronomy as the Herschels. The first was Wilhelm, always known as William because he spent almost all his life in England. He was born in Hanover in 1738 of a musical family. His father was a bandmaster in the army, and William also joined as a musician. He visited England once during his military service and learned to speak English.

On leaving the army he returned to England, at first making a living by copying music. However, he was a talented organist, and eventually became organist at the Octagon Chapel in the city of Bath, then one of England's most fashionable centers. Herschel became fascinated by astronomy after reading some books on the subject. He borrowed a small telescope, but found it inadequate, and decided to make a reflector. By the mid-1770s he had already begun observing on a systematic basis.

In 1781 William Herschel discovered the planet Uranus, and became famous almost overnight: King George III invited him to become King's Astronomer, and in this capacity he continued his surveys of the sky. With the help of his sister Caroline, who had left Hanover to become his assistant, he discovered thousands of double stars, clusters, nebulae and other objects.

◄ William Herschel was a professional musician until he discovered the planet Uranus from his garden in Bath. Never a wealthy man, he could not afford to buy a telescope and had to assemble his own using home-made mirrors. After his discovery of Uranus he received a small salary as royal astronomer. His finances were eventually improved through marriage in 1788.

▼ Caroline Herschel, William's sister, was also a talented musician. She left a life of drudgery at the family home in Hanover in order to become William's assistant, and was willing to endure long hours of discomfort in the service of astronomy. She was an able observer in her own right, and discovered six comets. After William's death Caroline returned to Hanover, a move she regretted. She died in 1848 at the advanced age of 98.

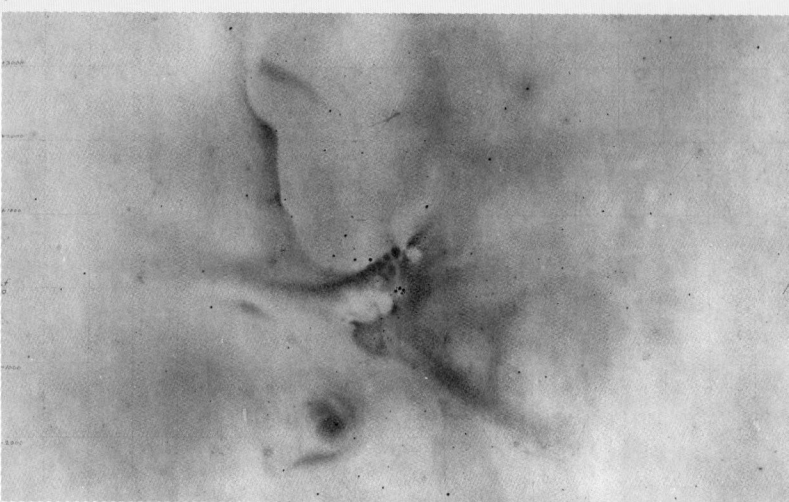

William's son John

When William Herschel died in 1822 he had explored the northern sky. His son John (1792-1871) extended the survey to the southern stars, and in the 1830s he spent some years at the Cape of Good Hope, erecting an observatory at Feldhausen. His work was triumphantly successful: he produced the first really detailed survey of the southern stars, and observed Halley's Comet at its return of 1835.

John Herschel's career as an active astronomer more or less ended with his return from Feldhausen in 1838, but it took him many years to reduce all his observations, and he remained active in associated fields – for example, his book "Outlines of Astronomy" was regarded as the standard work for many decades.

Another member of the family, Alexander, achieved a considerable reputation through his studies of meteors and meteoritic phenomena.

◄ John Herschel took over where his father had left off, transporting one of William's telescopes to Feldhausen, South Africa, where he built an observatory in order to survey the southern sky. Among the objects he observed from the Cape of Good Hope with his "20-feet reflector" was Halley's Comet in 1835, although this proved to be an unwelcome distraction. "To tell the truth, I am glad he is gone!" wrote John to his aunt in Hanover. Father and son were both knighted in recognition of their services to science. John's drawing of the Orion nebula is reproduced above left.

William Herschel completed his "40-feet reflecting telescope" in 1787; this print was published in 1819. The telescope had a 1.3m mirror – the "40 feet" refer to the focal length of 12m. With this telescope Herschel discovered Mimas and Enceladus, two of Saturn's moons, but it was extremely cumbersome and awkward to use. The inset figure (above) shows the new reflector named after William Herschel.

The performance of modern telescopes is limited by the poor optical quality of the atmosphere itself

Optical astronomy

Optical telescopes use lenses or mirrors to collect visible light and form images of distant objects. Observers can then view these images directly, photograph them, or amplify and analyze them electronically. The larger the telescope's aperture (the clear diameter of the main lens or mirror) the more light it collects, enabling it to "reach" fainter and more distant objects. The light-gathering power of the instrument is proportional to the area of the lens or mirror (that is, to the square of its diameter) so that, for example, a 5m aperture will collect 25 times more light than a 1m aperture. In principle, the larger the aperture the better the telescope's resolving power (its ability to reveal fine details), but in practice atmospheric turbulence blurs the images to such an extent that large Earth-based telescopes cannot achieve their theoretical resolving powers.

Most of the new generation of telescopes are reflectors of about 4m aperture, but designs for much larger instruments are being considered. To minimize the cost and engineering difficulties associated with large one-piece mirrors, these will probably use mirrors made up of a number of segments, computer-controlled to maintain their alignment, or will take the form of multi-mirror telescopes such as the one currently operating on Mount Hopkins in the USA. (This combines the light received by six 1·8m mirrors to give the light grasp of a single 4·5m telescope.)

For observing deep space, photographic emulsion is much more effective than the human eye, for it provides a permanent record – possibly of millions of star images on one plate – and, unlike the eye, it integrates (in other words, within practical limits, light accumulates in the emulsion to reveal sources too faint to be seen directly by the eye). Yet even photographic emulsion does not collect light with an efficiency of more than a few percent, and for many purposes modern electronic devices are much more effective.

The "Leviathan of Parsonstown"

The story of the third Earl of Rosse (1800-1867), an Irish aristocrat, is one of the most remarkable in the history of astronomy. Having studied science at university, he became attracted to astronomy and, while still a young man, determined to make a large telescope. He duly constructed a 91·5cm reflector, based on the Herschel reflectors. It was completed in 1836, and proved to be very successful.

Lord Rosse was not satisfied, however, and decided to build a telescope with a 183cm mirror – far larger than anything previously attempted. Techniques of glass-casting in those days were inadequate for the task, so the mirror, like Herschel's, had to be of speculum metal – an alloy of copper and tin. After one unsuccessful attempt Lord Rosse succeeded in casting the huge disk, which he subsequently shaped into the correct optical curve.

There followed the problem of mounting the telescope. To make it fully maneuverable would have been impossible. Instead Lord Rosse mounted the tube of the telescope between two massive stone walls, pivoting it at the bottom. This meant that the view of the sky was restricted to a strip either side of the meridian (north-south line), and any object could be kept in view for only a limited period before the tube was stopped by one of the walls. The telescope was cumbersome and it never had a finder; yet its light-grasp was excellent, and in 1845, soon after it was completed, Lord Rosse discovered the spiral nature of some of the objects now called galaxies.

Until the age of the great refractors, this "Leviathan of Parsonstown" was in a class of its own, and it was not until 1917 that a larger reflector – the 245cm Mount Wilson – was made.

▲ *The third Earl of Rosse.*

► ***The Earl of Rosse's huge telescope at Birr Castle, Parsonstown, was completed in 1845 and remained the largest in the world throughout its working life. It was taken out of use in 1909.***

► *Several major telescopes stand on Kitt Peak, Arizona, including the world's largest solar telescope and a modern 4·01m reflector. Selecting the site for an observatory can present many problems. It needs to be at a high altitude to minimize atmospheric disturbance, in a location subject to good weather conditions, and well away from the "light pollution" of big cities. Kitt Peak satisfied these conditions, but there were problems of another kind: it is in the reserve of the Papago Indians, and Babuquivari, a prominent landmark from the top of Kitt Peak, is a sacred mountain to them – they regard it as the center of the universe, with the gods living in nearby caves. Before negotiations were concluded the Observatory had to make a firm guarantee that the caves will never be disturbed.*

◄ *The 2·54m telescope at Mount Wilson, California, was commissioned by the founder of the observatory, George E. Hale, in 1917, eight years after the Earl of Rosse's telescope was dismantled. It was the world's largest for 30 years, until the 5·08m telescope at Mount Palomar was built. It is still one of America's major telescopes.*

▲ *Modern astronomers may use computers to store the data provided by telescopes, as here in the control console of the Anglo-Australian telescope at Siding Spring. Using image-enhancement techniques they can isolate minuscule variations in the wavelength or intensity of radiation, and present them in vivid color.*

Modern electronic devices

The photomultiplier consists of a light-sensitive surface – the photocathode – which emits electrons (electrically charged particles) when struck by light. A high voltage accelerates these down a tube in which they collide with further sets of electrodes, releasing more electrons at each collision and so forming a cascade of electrons. The output is an electrical current proportional to the incoming light. Measuring this current reveals accurately the brightness of faint sources.

In an image intensifier, the electrons emitted by a photocathode are accelerated to high speed and focused magnetically onto a light-emitting screen. This produces an image far brighter than the one formed at the focus of the telescope.

The image photon counting system (IPCS) takes the process a stage farther: an image tube is viewed by a television camera, and a computer system removes unwanted background "noise" (stray specks of light) to yield an image up to nearly a million times brighter than the incoming light.

The charge-coupled device (CCD) is a silicon chip divided into an array of tiny squares (picture elements, or "pixels"). The CCD is exposed for a time at the focus of a telescope and an electrical charge builds up on each pixel in proportion to each part of the telescopic image. The charge is then read off from each pixel; a computer stores and processes the data and produces an enhanced image. Charge-coupled devices collect light very rapidly and, with an efficiency of about 70 percent, are approaching the maximum possible sensitivity for any light-sensitive device.

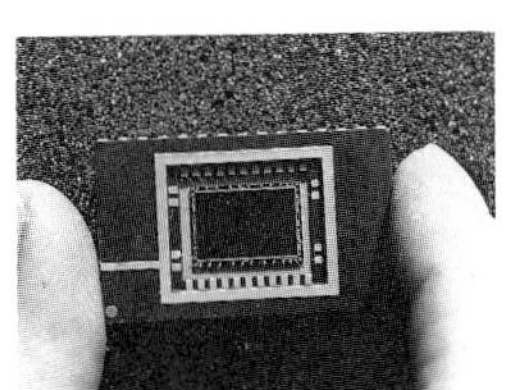

▲ *This silicon chip is divided into an array of 385 by 576 tiny squares – a total of 221,760 – and each square, or "pixel", records the light falling on it and transmits the data to a computer for analysis. This is the "eye" of a charge-coupled device (CCD).*

Some of the most significant information arriving from space is carried by waves which are only just outside the range of human vision

◀ ***This new telescope pivots in a spherical reinforced concrete housing, and is the prototype for a new interferometer under construction in southeast France.***

▲ ***Mauna Kea, an extinct volcano on the island of Hawaii, is one of the best observing sites in the world, despite the arid, cold hostility of its environment. The two nearest structures are infrared telescopes.***

▼ ***The multi-mirror telescope (MMT) at Mount Hopkins, Arizona, uses a laser beam to align its six mirrors. The whole telescope is cheaper than a conventional reflector of comparable light grasp.***

The interferometer

1

A B

2

A B

◀ *Radio waves arrive at the Earth rather like waves rolling onto a beach. If "crests" arrive together at two radio dishes (A,B), the combined outputs from A and B add together (1). If a crest arrives at B when a trough arrives at A, the two outputs cancel out (2).*

A B X

Constructive interference

λ =wavelength

A B Y

Destructive interference

Beyond the visible spectrum: radio astronomy

Radio telescopes vary widely in design. The simplest consists of a dipole (length of wire), while the most elaborate are fully steerable dishes which collect radio waves and focus them onto radio receivers.

The resolving power of a telescope depends on the ratio of the wavelength of incoming radiation to the aperture of the instrument. For a given aperture, the longer the wavelength, the poorer the resolving power. Radio telescopes are often used at wavelengths a million times greater than those of visible light, and to attain the same resolution as their optical counterparts would require apertures a million times larger. Even at its shortest operational wavelength (about 2cm) the world's largest steerable radio antenna – the 100m dish at Effelsburg in Germany – has a resolution no better than the unaided human eye (about one minute of arc).

This problem can be overcome by the technique called interferometry. In its simplest form a radio interferometer consists of two telescopes separated by a known distance (the "baseline"). As a radio source moves across the sky due to the Earth's rotation, the crests of the radio waves arriving at the two antennas are alternately in step and out of step. If the outputs are combined the incoming waves will interfere with each other, compounding or cancelling each other out. Analysis of the interference pattern reveals the detailed structure of a source with greater precision than the output from a single antenna can do. The greater the baseline the better the resolution, although the resolution is limited to the direction of the baseline.

◀ ***The technique behind the Very Large Array was devised by Sir Martin Ryle at Cambridge, England. With the moveable dishes it is possible to build up detailed radio "pictures" of the sky.***

▼ ***The world's largest radio telescope dish is built into a natural hollow near Arecibo, Puerto Rico, in the Caribbean. Although it is not steerable, its sheer size (305m) enables it to detect more radiation than any other single dish.***

The largest telescopes of all

Very long baseline interferometry (VLBI) uses two radio telescopes separated by thousands of kilometers (up to the diameter of the Earth). Each looks at the same source, and the output is recorded on magnetic tape against a timing standard provided by very precise atomic clocks. By combining the two sets of data it is sometimes possible to obtain a resolution as good as 0·001 arcsec – a great improvement on any single dish.

Aperture synthesis is a more elaborate technique which utilizes the Earth's rotation to allow an array of fixed and movable antennas to build up an image. This technique gives the same resolution as a single dish would yield if its aperture were equal to the maximum separation of the antennas. The largest instrument of this kind, the Very Large Array (VLA) at Socorro, New Mexico, uses 27 antennas on a Y-shaped railway track to synthesize a single dish 27km in diameter.

See also
Planets and Orbits 9-20
Observing the Planets 21-32
The Moving Sky 133-6
Space, Time and Gravitation 225-8
Communication and Travel 245-8

Beyond the visible spectrum: infrared astronomy

Most of the infrared radiation that reaches the Earth is absorbed in the atmosphere, particularly by carbon dioxide and water vapor. Wavelengths between about 30 micrometers and 300 micrometers cannot penetrate the atmosphere at all, but Earth-based telescopes on high mountain sites which are above most of the water vapor can study some of the shorter wavebands.

Infrared detectors are devices with electrical properties which change when they absorb radiation. To minimize background "noise" they have to be cooled to very low temperatures. Some operate at the temperature of liquid nitrogen (77K), while others have to be cooled to as low as 2K (−271°C) by the use of liquid helium. Objects at room temperature "shine" brightly with infrared radiation, especially at wavelengths between 5 and 20 micrometers, so that the atmosphere and even the telescope itself are powerful sources of radiation. To overcome this, observers have developed a technique whereby the detector looks alternately at sky plus source and at sky alone, and by comparing the results they are able to separate the feeble astronomical signal from the background noise.

The first satellite to be devoted entirely to infrared astronomy was IRAS (▶ page 168) which, during a 10-month period of operation in 1983, discovered about 250,000 cosmic infrared sources.

Infrared radiation with wavelengths of 1mm down to about 0·3mm can be observed from very dry high-altitude sites, and this is a promising development area in ground-based astronomy.

Shorter-than-visible radiation

Since radiation of all wavelengths shorter than about 310nm is absorbed in the atmosphere, astronomers can observe them only by the use of rockets and satellites. Several satellites, such as the IUE (International Ultraviolet Explorer), have already examined the universe at ultraviolet wavelengths, using telescopes and detectors broadly similar to those used in optical astronomy. But some of the most spectacular results have come from X-ray astronomy – the study of sources which emit at wavelengths from 10 micrometers down to 0·01 micrometers. X-rays are detected by devices such as the proportional counter, in which a gas (usually argon) absorbs the X-rays, causing them to release electrons. These are multiplied thousands of times to yield an electrical current proportional to the energy of the incoming rays.

Gamma rays extend downwards in wavelength from 0·01 micrometers and are the most energetic form of electromagnetic radiation. Scientists use scintillation counters and spark chambers to detect them. A scintillation counter contains crystals which convert gamma rays into visible light, and this can then be amplified by photomultipliers. A spark chamber consists of a target plate followed by a pile of plates held at alternately zero and high (typically 10,000 volt) potential. A gamma ray striking the target plate releases an electron and its antiparticle, a positron. As these plow through the stack, they cause sparks to leap from plate to plate, and analysis of their tracks reveals the direction and energy of the incoming gamma rays. So far, the resolutions obtained in gamma-ray astronomy are no better than one degree. However, scientists are currently developing new techniques which suggest that this may be a spectacular growth area in the 1990s.

The next biggest single step forward in observational astronomy will be in the visible waveband with the launching – it is hoped, in 1986 – of the Edwin Hubble Space Telescope.

The Space Telescope

▲ The Space Telescope, in orbit above the atmosphere and equipped with the latest electronic gadgetry, will see sources 50 times fainter than any telescope on Earth.

▼ The European Space Agency's X-ray satellite Exosat carries two X-ray telescopes, among other detectors. This photograph shows an engineering model under test in France.

X-ray telescopes

Conventional telescopes cannot focus X-rays, because if X-rays fall vertically onto a mirror, they will simply penetrate it by passing through the spaces between the atoms. Early instruments used a collimator (a grid placed in front of the detector), to limit the field of view, and thus give at least some directional information.

More recent satellites such as the Einstein Observatory (1978) and Exosat (1983) carry "grazing incidence" telescopes. These utilize the fact that X-rays striking a surface at a very shallow angle do not "see" the gaps between the atoms, and therefore bounce off the surface. Concentric arrays of curved tubes – rather like deep mirrors with their centers missing – are used as grazing incidence mirrors to "steer" X-rays to a focus. To date, such instruments have provided resolutions as good as 2 arcsecs.

The Basic Properties of Stars

Brightness...Distance measurements...Apparent magnitude, absolute magnitude and luminosity...Color, temperature and the spectrum...Dividing stars into spectral classes...Weighing and sizing stars...The "Hertzsprung-Russell" diagram – its significance... PERSPECTIVE...The mystery of the "wobbling" dog star...Stellar record breakers

Even in the largest Earth-based telescopes stars appear as mere points of light. Nevertheless, by applying their knowledge of physics and chemistry to the analysis of such facts as observation does reveal, astronomers have been able to achieve a detailed understanding of the nature of the stars.

Brightness

The brightness of a star as an Earth-based observer sees it is given by a quaint system of stellar "magnitudes", according to which the faintest stars have the highest values and the brightest stars the lowest. A conspicuous star such as Spica has a magnitude of 1 as seen from Earth, and is described as a "first magnitude" star. Polaris, the Pole Star, is of magnitude 2, while the faintest star visible to the naked eye under good conditions is of magnitude 6. There are 15 stars brighter than magnitude 1. Vega, for example, is of magnitude 0. Stars brighter than this have negative magnitudes. Sirius, the brightest star in the sky, is of magnitude −1·45, while the Full Moon and the Sun have magnitudes of −12·6 and −26·7 respectively.

Measuring stellar distances

It is possible to measure the distances of relatively nearby stars by the parallax method. This consists of making observations of a star's position relative to the more distant background stars at intervals of six months – that is, when the Earth is at opposite sides of its orbit. The two observing sites are thus 300 million kilometers apart, and for a nearby star this baseline is long enough to reveal a small shift in its position. The maximum shift of the star from its mean position in the sky is known as the annual parallax, and it is possible to calculate the star's distance by applying simple trigonometry to the triangle made up of Earth, Sun and star. The angles are always very small, and the angle of parallax grows even smaller with increasing distance.

The annual parallax of a star at a distance of 3·26 light years would be precisely one arcsec (one second of angular measurement). This distance is known as the "parsec" (for a *par*allax of one arc*sec*). In fact, no known star apart from the Sun is as near as that.

Parallax provides the first step in measuring distances in the universe, but because it is difficult to measure such small angles, errors begin to become significant at a range of only 20 parsecs, and the method is not effective at distances greater than 100 parsecs (326 light years). Less direct methods are available for measuring greater distances, but these depend on the accuracy with which the distances of nearby stars have been measured by parallax. Astronomers expect the Hipparchus satellite, due to be launched in the late 1980s, to produce parallaxes at least ten times more accurate than ground-based telescopes can achieve.

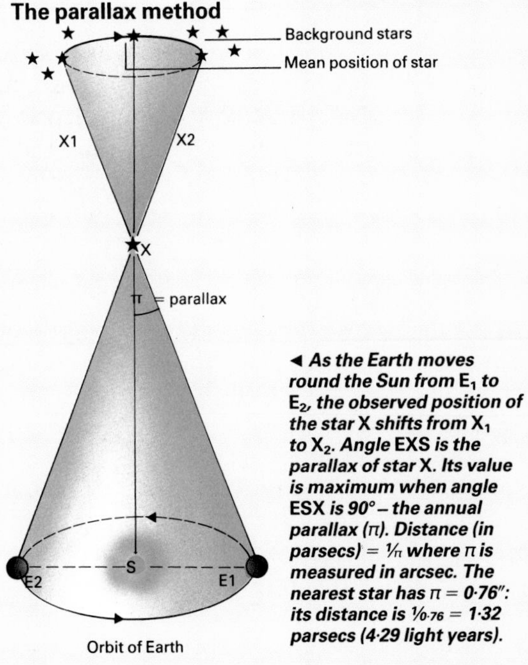

◄ As the Earth moves round the Sun from E_1 to E_2, the observed position of the star X shifts from X_1 to X_2. Angle EXS is the parallax of star X. Its value is maximum when angle ESX is 90° – the annual parallax (π). Distance (in parsecs) = $1/\pi$ where π is measured in arcsec. The nearest star has π = 0·76": its distance is $1/0{\cdot}76$ = 1·32 parsecs (4·29 light years).

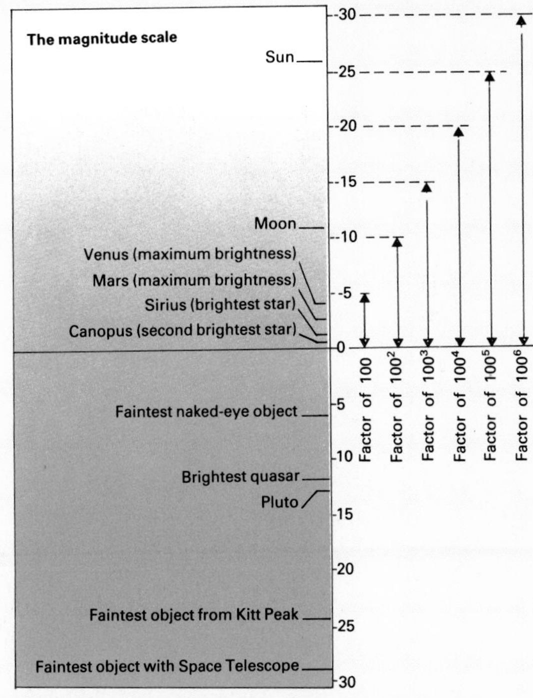

Explaining the scale of stellar magnitudes
The magnitude scale (above) is logarithmic: it is based on the fact that a difference of 5 magnitudes corresponds to a factor of 100 in brightness. A difference of 1 magnitude corresponds to a factor of 2·512. Thus a first magnitude star is 2·512 times brighter than a second magnitude star, and 2·512 × 2·512 = 6·310 times brighter than a third magnitude star, and $(2{\cdot}512)^5$ = 100 times brighter than a sixth magnitude star. Although this scale appears strange, it does in fact reflect the way in which the human eye responds to different levels of brightness.

Astronomers use a star's color and brightness to unlock the secrets of its age, its chemical and physical composition, and its life story

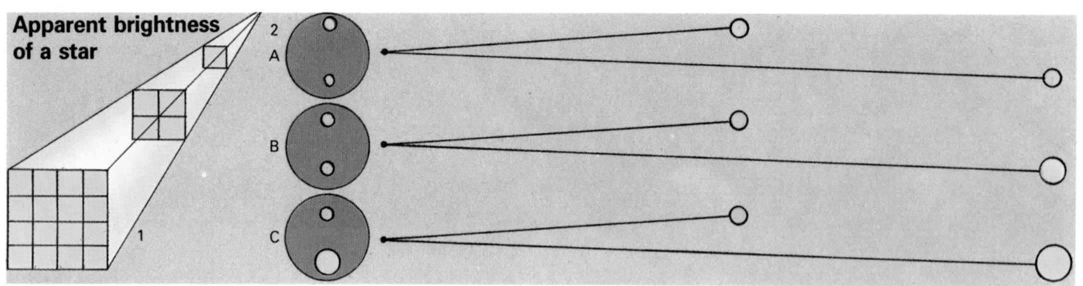

◄ *The apparent brightness of a light source decreases with the square of its distance as light is spread over progressively larger areas (1). If two stars have the same luminosity (2) the more distant looks fainter (A). Two stars seem equally bright if the farther is the more luminous (B). If the farther star is very luminous it appears the brighter (C).*

Apparent magnitude, absolute magnitude and luminosity

The apparent magnitude of a star is a measure of the amount of its light that reaches the Earth. This depends on a number of factors such as the star's distance, whether or not any of its light is absorbed in space, and its luminosity. The luminosity of a star is the amount of energy which its entire surface radiates in one second. This output is measured in watts, just like the power output of an electric light or heater. The Sun radiates a little under 4×10^{26} watts, and is a fairly average star in this respect.

If all the stars were at the same distance, their apparent magnitudes would be a true guide to their real luminosities; it is thus helpful to consider the apparent magnitudes that they would have if they all lay at a distance of ten parsecs. The *apparent* magnitude that any given star would have at this distance is known as its ***absolute*** magnitude. For example, the Sun has an absolute magnitude of 4·8, so that if it were moved to a distance of ten parsecs, its apparent magnitude would be 4·8 – and it would barely be visible on an average night. Comparing a star's absolute magnitude (the quantity of light it emits) with the apparent magnitude (the quantity which reaches Earth) makes it possible to calculate its distance.

Color, temperature and the spectrum

The color of a star is a guide to its temperature. There is a relationship, known as the Wein law, according to which the higher the temperature of a star, the shorter the wavelength at which it shines most brightly. A star such as the Sun, with a temperature of just under 6,000K, emits most strongly at the middle of the visible spectrum and so appears yellowish. Cooler stars appear red, and hotter stars appear white or blue. The coolest stars of all emit mainly at infrared wavelengths, while the hottest stars peak in the ultraviolet. Astronomers define the color of a star by reference to its color index – the difference between its magnitudes measured at two different wavelengths.

Spreading starlight out into its constituent wavelengths produces a spectrum consisting of a continuous rainbow band of color, together with a pattern of dark absorption lines appearing at a number of particular wavelengths. Sometimes, as with very hot stars, bright (emission) lines are superimposed on the spectrum.

Atoms in the outer layers of a star absorb light coming from the hotter, denser interior at wavelengths which depend upon the chemical elements present. Each element produces its own characteristic "fingerprint" pattern of lines at a number of known wavelengths, so that identification of the lines gives a clue to the chemical composition of the outer layers of a star. The spectrum requires careful interpretation, but can yield a great deal of information about such factors as temperature, chemical composition, density, rotation rate, and the presence of magnetic fields.

Stellar luminosity – the technical details

Astrophysicists define the luminosity of a star as the amount of energy emitted by one square meter of surface, multiplied by the total surface area (in square meters) of the star. The value given for the luminosity of a star depends on two factors, its radius and its temperature, and is expressed as watts.

The energy output per square meter depends on the fourth power of a star's surface temperature, and the surface area depends on the square of its radius. For example, a star with a temperature of 6,000K emits 16 times more energy from each square meter of its surface than a star with a temperature of 3,000K. If two stars have the same radius, the hotter one will be the more luminous of the two. On the other hand, if two stars of different sizes have the same temperature, the larger of the two will be the brighter.

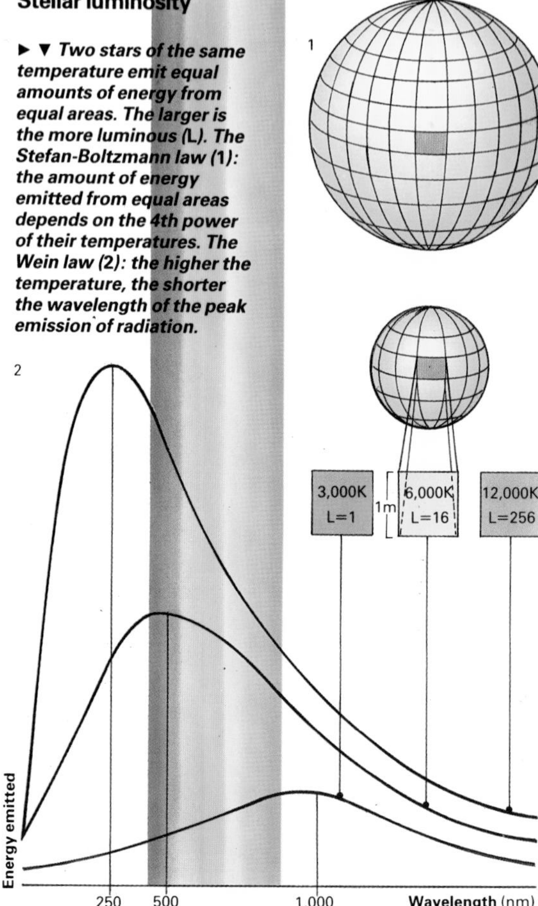

► ▼ *Two stars of the same temperature emit equal amounts of energy from equal areas. The larger is the more luminous (L). The Stefan-Boltzmann law (1): the amount of energy emitted from equal areas depends on the 4th power of their temperatures. The Wein law (2): the higher the temperature, the shorter the wavelength of the peak emission of radiation.*

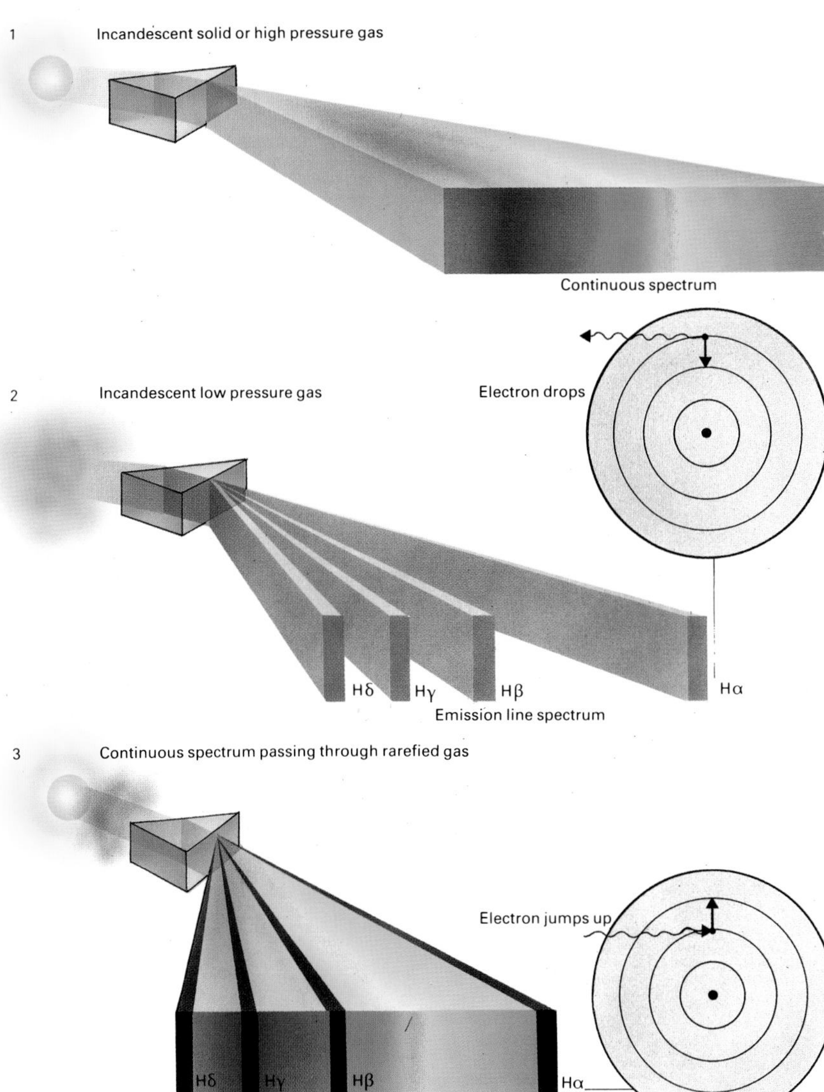

◄ ***The English astronomer William Huggins (1824-1910) was one of the pioneers of stellar spectroscopy. He analyzed the spectra of 50 bright stars, and concluded that some of the same elements were present in the Sun. His work led directly to the first measurements of a star's radial velocity.***

Spectral class

*Stars are classified according to the appearance of their spectra. In the Harvard Classification they are graded in order of decreasing temperature into the following principal classes: O, B, A, F, G, K, M, with subsidiary classes R, N, S for cool stars and the class W for very high-temperature "Wolf-Rayet" stars. This curious sequence of letters can be recalled with the aid of a dreadful mnemonic, which runs: (Wow), O*h *B*e *A F*ine *G*irl, *K*iss *M*e *R*ight *N*ow, *S*weetie. *The sequence is alphabetically chaotic because there were several major revisions during the research period.*

Each class is divided into ten subsections denoted by one of the numbers between 0 and 9, again in order of decreasing temperature. The Sun, for example, is a type G2 star.

O- and B-types are bluish in color, the O-types having temperatures in the range 35,000K to 40,000K. A-types are white, F-types creamy, G-types yellow, K-types orange and M-types red, with temperatures around 3,000K.

Additional letters are used to denote special peculiarities – for example, "e" denotes the presence of emission lines in the spectrum. Within a given class, a highly luminous star will be larger than a less luminous one, and its outer layers will be more rarefied. The lower pressures in the more tenuous atmospheres of the larger stars produce narrower, sharper lines than those produced in more compact stars.

▲ ***A hot dense body emits a mixture of all wavelengths of light which is spread out by a prism into a band of colors – a continuous spectrum (1). In a hot, low density gas, many atoms have electrons in high energy levels. When an electron drops to a lower level it emits light of a characteristic wavelength – an emission line spectrum – (2). When a continuous spectrum passes through a low density gas, electrons in low energy levels absorb light at certain wavelengths and superimpose dark lines on the spectrum – a solar spectrum (3).***

► ***Examples of the principal spectral classes arranged in order of decreasing temperature – a major factor in determining which lines are prominent.***

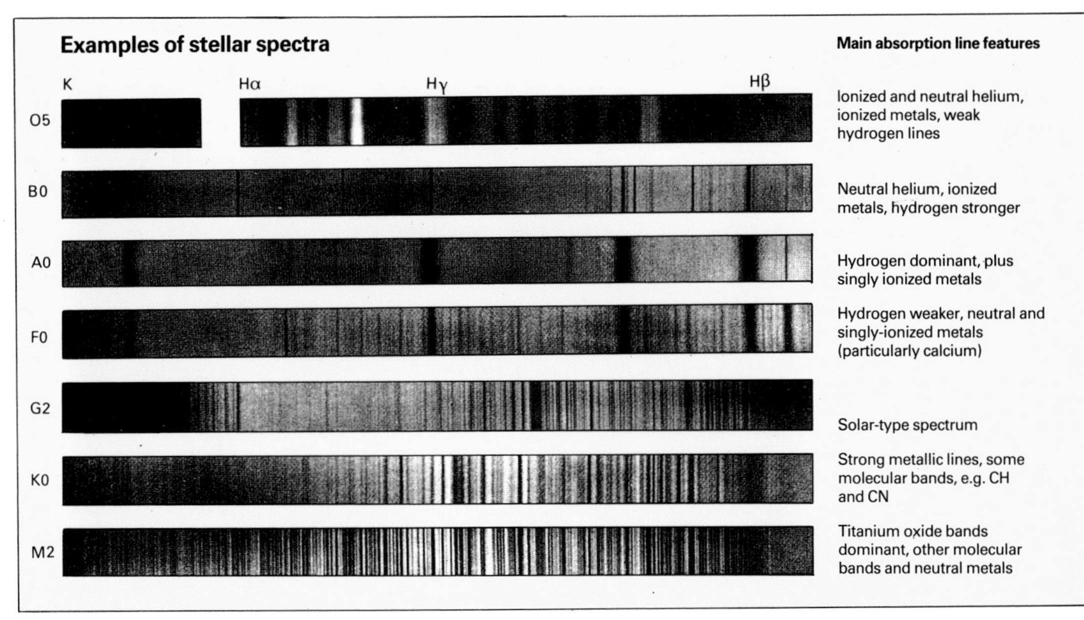

The vital statistics of stellar bodies show that many of them have remarkably little in common – except that they are all called stars

Stellar masses

It is possible in principle to "weigh" a star by observing the effect of its gravity on neighboring bodies. Thus, working out the force necessary to keep the planets in their orbits at their known distances and speeds makes it possible to calculate the mass of the Sun. Planets of other stars, even the closest, are too faint to see, and it is only possible to measure the mass of a star if it is a member of a binary system (a binary is a pair of stars which revolve round each other). Analyzing measurements of the orbital period of a binary, and of the separations between the stars, can give the masses of the stars (➧ page 161).

The majority of stars are less massive than the Sun, but astronomers think that the lowest possible mass for a fully fledged star is about 0·08 solar masses. Few stars have more than ten times the Sun's mass, but the most massive stars may be more than 100 solar masses.

Stellar diameters

Although scientists can deduce the diameter of a star from measurements of its luminosity and temperature, direct measurements of stellar diameters are very hard to achieve. In theory the largest telescopes should be able to resolve some of the nearest large stars into visible disks, but in practice atmospheric turbulence rules this out.

Stellar record breakers

The most massive star known at present is probably Plaskett's Star, HD 47129, in the constellation of Monoceros (the Unicorn). It is a binary system; each component is about 55 times as massive as the Sun, while the primary has a radius of 25 times that of the Sun and at least 50,000 times the Sun's luminosity.

The most powerful star known is the unique Eta Carinae, which is 6,400 light years away and perhaps 4,000,000 times as luminous as the Sun; even this may be something of an underestimate. In the Large Magellanic Cloud, immersed in the nebula 30 Doradus, is the extraordinary HD 38268. If it is a single star, it could be ten times as powerful as Eta Carinae and far more massive, but it is quite likely that it is made up of a compact group of highly luminous objects.

Of the largest stars, the system of Alpha Herculis, made up of a red supergiant with a double companion is enveloped in a huge cloud of gas of perhaps 250,000 million kilometers while other red supergiants have immense diameters – at least 400,000,000km in the case of Betelgeux in Orion.

Among stellar glow-worms, pride of place must go to the red dwarf RG 0050-2722 in Sculptor, with an absolute magnitude of 19. Some "brown dwarfs", such as the recently-discovered companion of Van Biesbroeck 8 (21 light years away) may be even feebler; their cores have never become hot enough for nuclear reactions to be triggered off.

▼ Edward Pickering (1846-1919) and his "harem" at Harvard College Observatory developed the system of spectral classification at the beginning of this century.

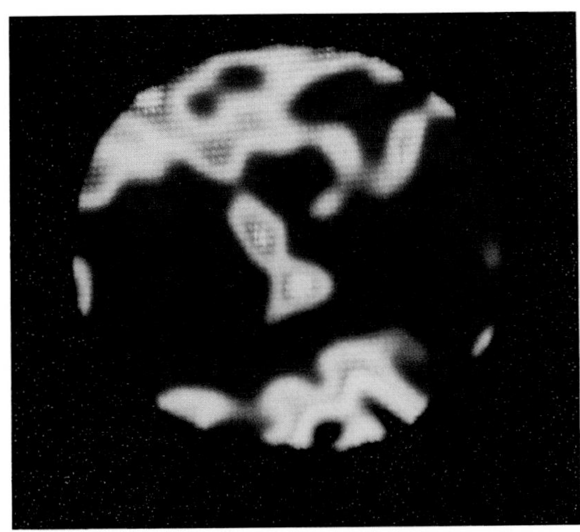

▲ The disk of the star Betelgeux, a red supergiant in Orion, was the first (other than the Sun) to be resolved. Astronomers at Kitt Peak National Observatory produced this picture in 1974 using the newly developed speckle interferometry technique. Some surface structure is clearly discernible; earlier photographs had shown only a featureless blur.

The first attempts to measure stellar diameters were made with the Michelson stellar interferometer, which used two mirrors attached to a long beam mounted at the front of a telescope. The idea was to examine a star and to produce at the focus an interference pattern (◀ page 143), which, when analyzed, would give the diameter of the star. A more advanced instrument, the intensity interferometer set up at Narrabri, Australia, has used two 6·5m movable segmented mirrors to focus light onto photomultipliers. Correlating the intensities of the signals produced by the two mirrors has enabled observers to obtain stellar diameters as small as 0·0004 arcsec.

Very short photographic exposures of star images show that the blurred image produced by atmospheric turbulence consists of a large number of little blobs, or "speckles", formed by individual cells of air in the atmosphere. Careful analysis of a large number of short exposures can yield information on the diameter and surface structure of a star down to the theoretical resolving power of the telescope. This technique is known as "speckle interferometry."

The way in which a star's light fades away if the Moon passes in front of it can also provide information on stellar diameters.

Stellar diameters range from thousands of millions of kilometers down to about ten kilometers – a truly enormous spread.

▲ The stellar intensity interferometer at Narrabri Observatory, Australia, has two 6·7m reflectors mounted on circular tracks. It is sensitive enough to measure the diameters of many main-sequence stars.

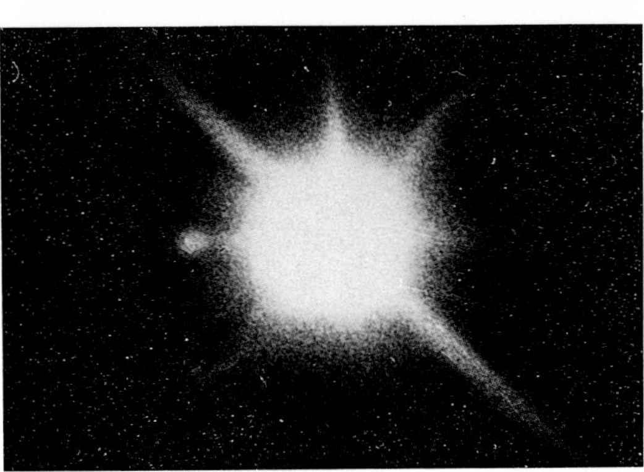

◀ In this photograph of Sirius and its white dwarf companion, the bright star is highly over-exposed – the spikes are photographic effects. Its dim companion is barely visible, yet the mass ratio is only 2·5 to 1.

The mystery of the wobbling Dog Star...

The brightest star in the sky is Sirius, in Canis Major (the Great Dog). It is not highly luminous – only 26 times more powerful than the Sun – but it is a mere 8·6 light years away, which is closer than any other bright star except Alpha Centauri. Its spectrum is of type A, and it is pure white.

Friedrich Wilhelm Bessel (1784-1846), a German astronomer, had found that Sirius was not moving steadily against the background of more remote stars. He observed a very slight, very slow "wobbling", and attributed this to the gravitational effect of a companion star, too faint to be seen in the telescopes of the time. In 1862 Alvan Clark (1832-1897), a famous American telescope-maker, was testing a large new refractor when he saw a faint speck of light close to Sirius. This proved to be the companion, almost exactly where Bessel had predicted. Since Sirius is often known as the Dog Star, it was natural that the companion should be nicknamed the Pup.

...and the dense white Pup

The Pup has only one ten-thousandth of the luminosity of Sirius itself, and astronomers assumed that it would be dim and red. It came as a major shock when in 1915 Walter Sidney Adams (1876-1956), an American astronomer at Mount Wilson, obtained a spectrum which showed that the Pup was in fact white, and had a surface temperature much higher than that of the Sun.

If the Pup were both hot and faint, it would have to be very small, and in this case its known mass could be explained only if it was also extremely dense. In fact Sirius B, the Pup, is smaller than Uranus or Neptune but has a density some 60,000 times that of water: it is a white dwarf, an aging star which has used up its main sources of energy and now shines feebly as it loses heat to its outer layers. Today many white dwarfs are known, some of them even smaller and denser than the Pup.

► *The Hertzsprung-Russell diagram compares the surface temperatures or spectral classes of stars with their luminosities or absolute magnitudes. It revolutionized the study of stellar evolution.*

The Hertzsprung-Russell diagram

This diagram, which is of crucial importance to an understanding of stellar evolution, shows the relationship between luminosity and temperature for the stars. The vertical scale shows luminosity (or an equivalent quantity such as absolute magnitude), and the horizontal scale gives temperature (or an equivalent quantity such as spectral class or color index). Scientists use the luminosity of the Sun as a unit of luminosity, so that the Sun belongs on the diagram as a point of luminosity 1 and temperature 5,800K. If a large number of stars are plotted according to their luminosities and temperatures, it becomes clear that most stars lie within a band which slopes from upper left (high temperature, high luminosity) to lower right (low temperature, low luminosity). This band is known as the "main sequence".

Stars off the main sequence – stellar outsiders

Not all stars lie on the main sequence. For example, some stars of low temperature but high luminosity lie above it and to the right. These stars are known as red giants because they are cool red stars of enormous diameter (stars of the same temperature emit the same amount of energy per square meter of surface, so if one outshines another it must be bigger). A typical red giant has a radius about one hundred times larger than the Sun's, a luminosity one thousand times greater and a surface temperature of about 3,000K.

White dwarfs are stars which lie below and to the left of the main sequence. Despite their high temperatures – 10,000K or more – they have less than one thousandth of the Sun's luminosity. A typical white dwarf has about one hundredth of the Sun's diameter and is comparable in size with the Earth. White dwarfs contain as much material as the Sun within about one millionth of the volume, and therefore are about a million times denser than the Sun.

Neutron stars are a great deal smaller than white dwarfs. The diameter of a typical neutron star is only 10-20km, and stars of this type are so dense that a teaspoonful of their material would weigh between one hundred million and one billion tonnes. By contrast, the mean density of matter in the outer layers of a red supergiant is about one ten-thousandth of the density of air at sea-level.

Stellar properties

Type of star	Diameter (Sun = 1)	Luminosity (Sun = 1)	Surface temperature (K)
Red Supergiant (MOI)	500	30,000	3,000
Red giant (K5III)	25	200	3,000
Main sequence stars:			
05	18	500,000	40,000
B0	7	20,000	28,000
A0	2.5	80	9,000
F0	1.35	6.3	7,400
G0	1.05	1.5	6,000
G2 (Sun)	1	1	5,800
K0	0.85	0.4	4,900
M0	0.63	0.06	3,500
M5	0.32	0.008	2,800
White dwarf	0.01	0.001	10,000
Neutron star	10^{-5}	–	10^{6}

▲ *E. Hertzsprung 1873-1967*

► *Relative sizes of stars of various types are compared with the size which a black hole would have if it had a mass equal to that of the Sun (▶ page 177).*

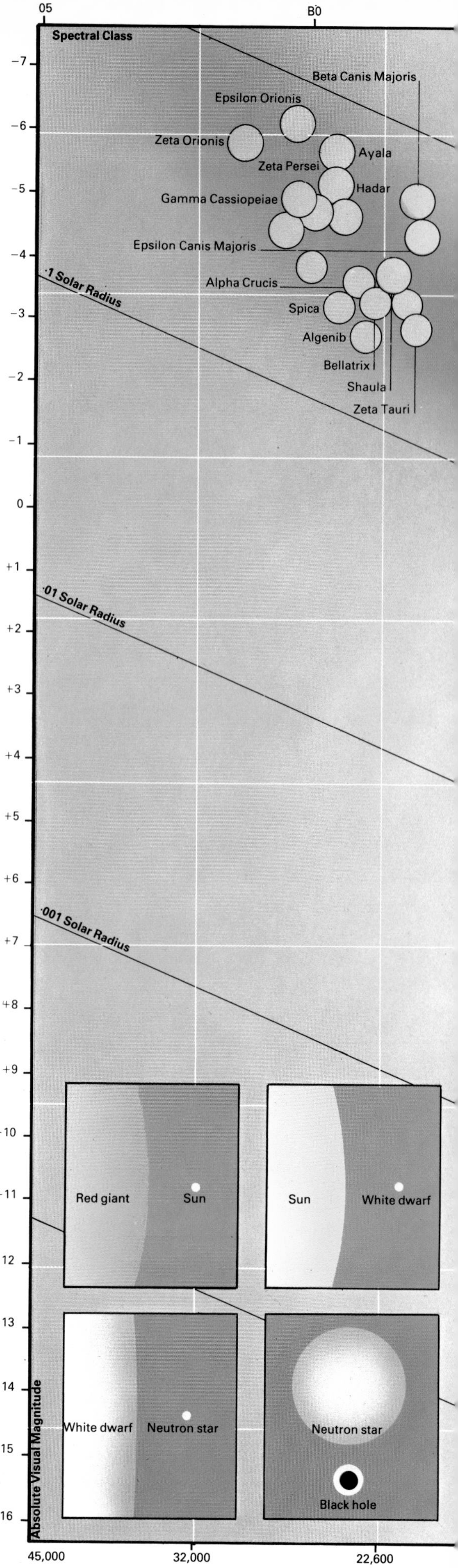

05 B0
Spectral Class
Beta Canis Majoris
Epsilon Orionis
Zeta Orionis
Ayala
Zeta Persei
Hadar
Gamma Cassiopeiae
Epsilon Canis Majoris
Alpha Crucis
Spica
Algenib
Bellatrix
Shaula
Zeta Tauri
·1 Solar Radius
·01 Solar Radius
·001 Solar Radius
−7 −6 −5 −4 −3 −2 −1 0 +1 +2 +3 +4 +5 +6 +7 +8 +9 +10 +11 +12 +13 +14 +15 +16
Absolute Visual Magnitude
Red giant
Sun
Sun
White dwarf
White dwarf
Neutron star
Neutron star
Black hole
45,000 32,000 22,600

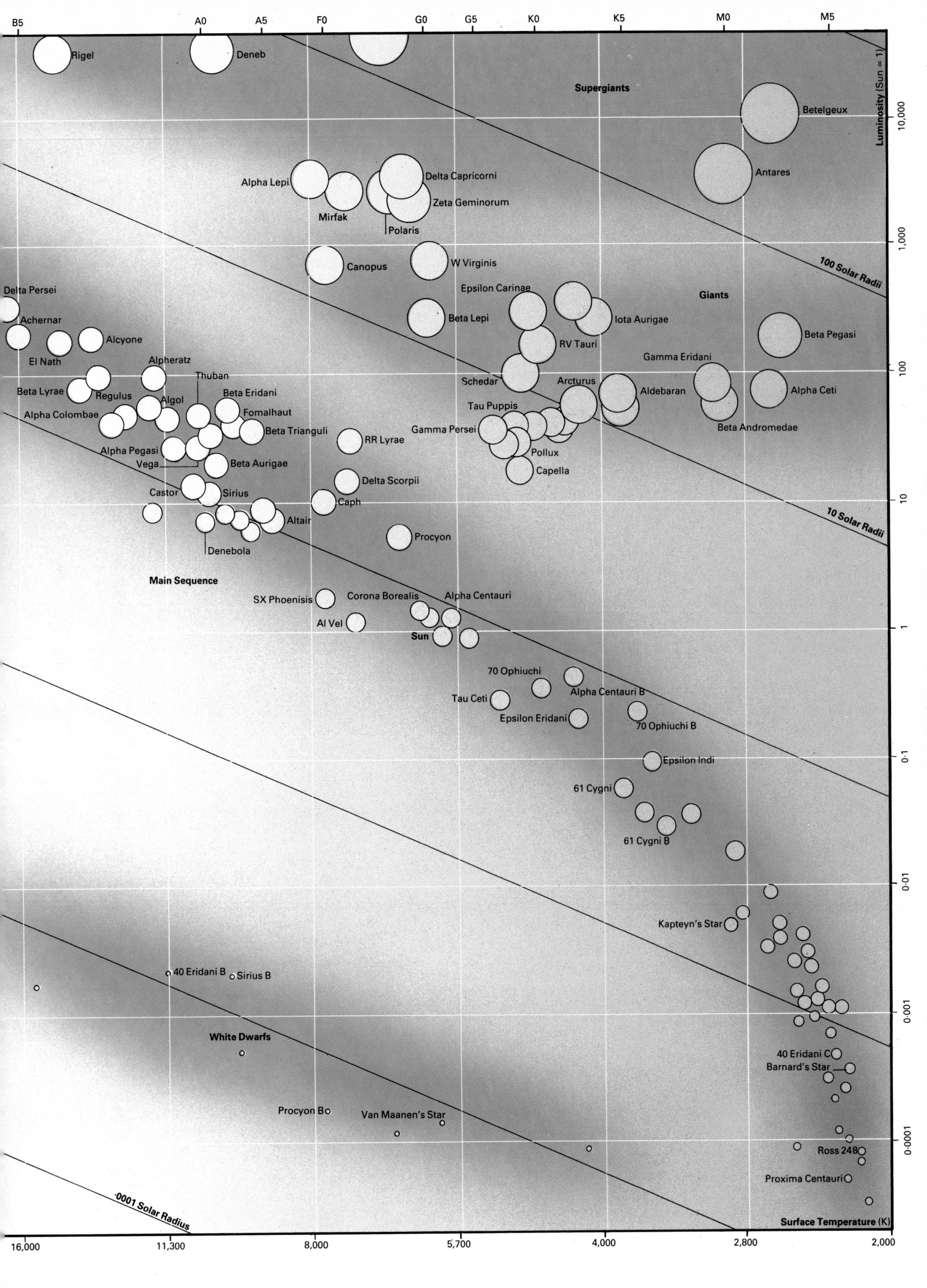
B5
A0
A5
F0
G0
G5
K0
K5
M0
M5
Rigel
Deneb
Supergiants
Betelgeux
Luminosity (Sun = 1)
10,000
Alpha Lepi
Delta Capricorni
Zeta Geminorum
Mirfak
Polaris
Antares
1,000
Canopus
W Virginis
100 Solar Radii
Delta Persei
Epsilon Carinae
Giants
Achernar
Beta Lepi
Iota Aurigae
Alcyone
RV Tauri
Beta Pegasi
El Nath
Gamma Eridani
Alpheratz
100
Thuban
Schedar
Arcturus
Beta Lyrae
Beta Eridani
Aldebaran
Alpha Ceti
Regulus
Algol
Fomalhaut
Tau Puppis
Alpha Colombae
Beta Andromedae
Beta Trianguli
Gamma Persei
RR Lyrae
Alpha Pegasi
Pollux
Vega
Beta Aurigae
Capella
Delta Scorpii
Castor
Sirius
10
Caph
10 Solar Radii
Altair
Procyon
Denebola
Main Sequence
SX Phoenisis
Corona Borealis
Alpha Centauri
Al Vel
1
Sun
70 Ophiuchi
Alpha Centauri B
Tau Ceti
Epsilon Eridani
70 Ophiuchi B
Epsilon Indi
0·1
61 Cygni
61 Cygni B
0·01
Kapteyn's Star
40 Eridani B
Sirius B
0·001
White Dwarfs
40 Eridani C
Barnard's Star
Procyon B
Van Maanen's Star
0·0001
Ross 248
Proxima Centauri
·0001 Solar Radius
Surface Temperature (K)
16,000
11,300
8,000
5,700
4,000
2,800
2,000

See also
Inside the Sun 97-104
The Sun's Outer Layers 105-12
Birth, Life and Death of Stars 153-60
Binary and Variable Stars 161-8
Collapsing and Exploding Stars 169-76

The motion of stars in space

The stars are all moving through space, although their distances are so great that their motions are imperceptible to the human eye even over centuries and millenia. Relative to the Earth, their velocity consists of two components – "proper motion" across the line of sight, and "radial motion", directly towards or away from the Solar System.

In referring to proper motion, astronomers give the angle through which a star travels across the sky in a year (after allowing for effects such as parallax). The angle is always very small. Barnard's Star (➧ page 246), six light years away, has the largest known proper motion (10·31 arcsec per year) and crosses an angle equal to the apparent diameter of the Moon in 180 years. Most proper motions are far smaller than this. If they know the distance of a star, astronomers can convert proper motion into speed across the line of sight (transverse velocity). In the case of Barnard's star this is 88km per second.

Radial motion and the Doppler effect

The Doppler effect makes it possible to measure a star's radial velocity. If a source of light is moving away from an observer, the light waves are "stretched out" and the wavelength reaching the observer is longer than the wavelength emitted by the source. This happens because each successive wavecrest is emitted from a slightly greater distance away, and so takes longer than its predecessor to reach the observer. Fewer wavecrests per second arrive at the observer's position than the source is emitting. Conversely, if a source is approaching the observer, the waves are "squeezed up", and the observer perceives shorter wavelengths. A similar effect occurs in the pitch of sound when a source is approaching or receding.

The spectral lines that occur due to the different chemical elements in a star have precisely known wavelengths. If a star is receding from the Earth, all the lines in its spectrum will occur at longer than normal wavelengths (they will be red-shifted), while if it is approaching, the wavelengths of its spectral lines will be shorter than normal (blue-shifted). Astronomers can measure the velocity of approach or recession by analyzing the blue shift or red shift in a star's spectrum.

The radial velocity of Barnard's Star is –108km per second, which implies that it is approaching the Solar System at this speed (a "+" sign denotes recession). Combining radial and transverse velocities shows that Barnard's Star is moving relative to the Solar System at about 140km per second and will make its closest approach, at a distance of 3·85 light years, in just under 10,000 years' time.

Stellar motions

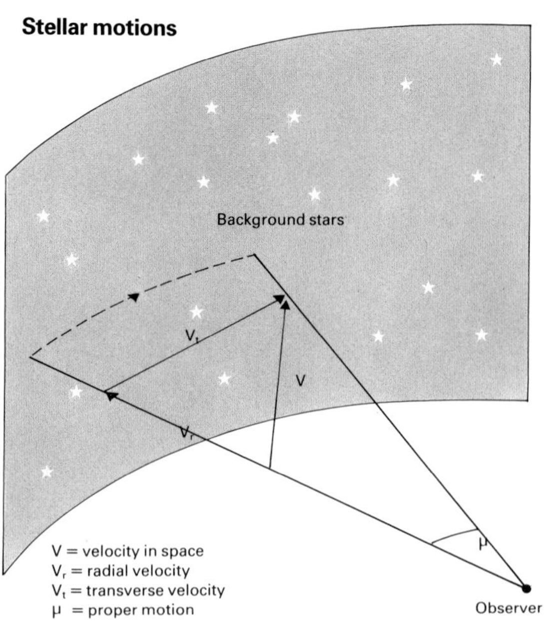

Spotting stellar movement

The stars were once called "fixed stars", because they seemed to remain stationary with respect to each other; to the naked-eye observer the constellation patterns remain unaltered for many lifetimes. Proper or individual motions of the stars were only first appreciated in 1710, by Edmond Halley. He was able to show that the stars Sirius, Procyon and Arcturus had moved perceptibly against their background since the time of Ptolemy.

All proper motions are small; however, modern techniques provide very accurate measures. Photographic atlases, such as that compiled at Palomar Observatory, provide the basic data here; comparison with atlases compiled years later show up the proper motions clearly. The artificial satellite Hipparchus, devoted entirely to this problem, will provide data in 2½ years which would take 50 years by Earth-based equipment. It is, of course, true that it is only relatively near stars which show detectable proper motions. On the other hand radial or toward-or-away measures can be made spectroscopically even for remote galaxies, where the velocities are tremendous – in some cases appreciable fractions of the velocity of light.

The Doppler effect

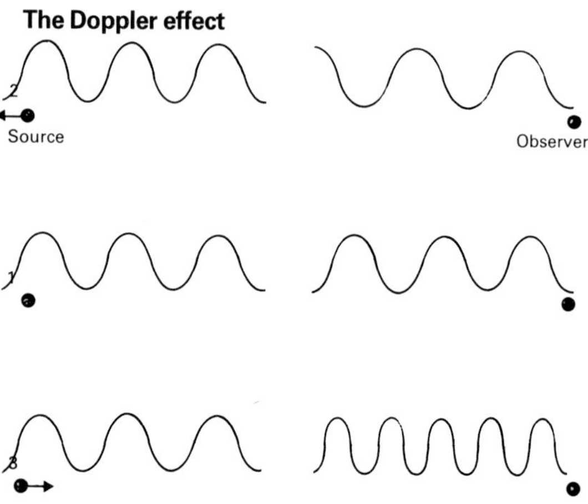

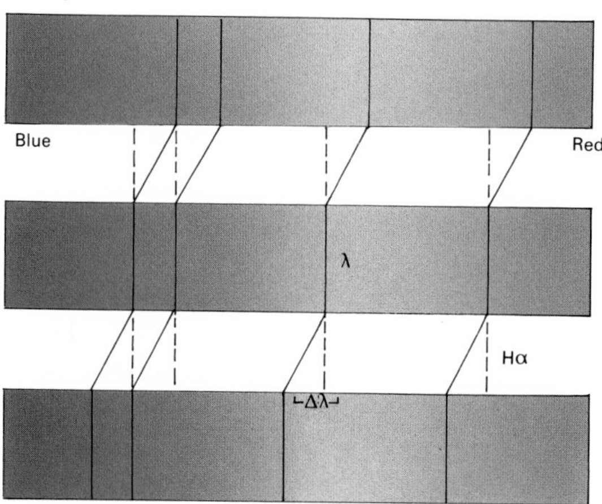

◄ Light reaching an observer from a stationary source has the same wavelength (1) as the light emitted from that source, and the lines in its spectrum have normal "rest wavelengths". Light arriving from a receding source is stretched in wavelength and the spectral lines appear at longer wavelengths – nearer to the red end of the spectrum by an amount proportional to the radial velocity (2). If the source is approaching, its light waves are compressed and the spectral lines appear at shorter wavelengths (3).

Birth, Life and Death of Stars 16

Stellar evolution – piecing together the whole story... What powers the nuclear furnace?...Ending with a bang...Why stars may occur in clusters...A special class of stars...PERSPECTIVE...Early attempts to discover stellar fuels...Investigating the new stars in Orion

Astronomers have never been able to witness the life of a single star from beginning to end, because the time scale is too great. Instead they have analyzed a broad cross-section of stars, some young, some middle-aged and some very old, and have combined their data to compile a reasonably complete evolutionary sequence for most stars.

A star begins life as a cloud of dust and gas which, triggered by some mechanism which scientists have not fully understood, begins to collapse under its own gravitational attraction. Once started, the process continues very rapidly, exerting pressure on the material at the center of the cloud, where the temperature rises. The collapsing cloud, or "protostar", becomes extremely brilliant for a while, but then begins to fade as collapse continues still farther – but more slowly – and the central region becomes hotter and hotter. At a temperature of about 10 million K, nuclear fusion begins to take place within the core, providing the energy which will keep the star shining for many millions of years – although not to the end of its life.

Everyday life on the stellar main sequence

When nuclear reactions are well under way in its interior, a star ceases to collapse because the explosion outwards of hot gases counteracts the inward force of gravity. A star in this state of balance is said to be on the "main sequence", because this is how the majority of stars spend most of their visible lives. A one solar-mass star (one which has the same mass as the Sun) takes about 50 million years to reach the main sequence, and remains there until the hydrogen in its core is spent. More massive stars arrive on the main sequence much sooner and less massive stars much later. Those of less than 0·08 solar masses never reach this stage at all, but merely go on shrinking to become first brown and then black dwarfs.

The declining years of a typical star

When all the hydrogen in a star's core has been used, nuclear fusion reactions cease to counteract the force of gravity and the star begins to contract under its own weight, increasing in temperature as it does so. This process transfers enough heat to the shell of material around the core – which does still contain hydrogen – to start a new series of fusion reactions. This hydrogen "burning" zone spreads outwards from the original core, depositing the helium created by fusion as a kind of "ash" in the center of the star.

The energy output from the expanding hydrogen zone increases, so that the star becomes brighter and perceptibly larger, even though its core continues to shrink. Eventually the core temperature will reach about 100 million K, and at this stage a new fusion process begins, in which the nuclei of helium atoms fuse to form carbon nuclei. This process, the "triple-alpha" reaction, sustains the star as a red giant.

The cornerstone of modern astrophysics

In 1914 a Danish astronomer, Ejnar Hertzsprung (1873-1967), and an American, Henry Russell (1877-1957), published the first of a series of diagrams that was to revolutionize the study of stellar evolution. Working independently, the two astronomers had discovered that there was a correlation between the brightness (absolute luminosity) and the color (spectral class) – and hence surface temperature – of stars. Plotting these two values on a graph gives the diagram known as the Hertzsprung-Russell (or H-R) diagram. Drawing such a diagram for a large number of stars shows that most lie within a broad band of decreasing brightness and surface temperature. This band is now known as the "main sequence". Most stars move across the main sequence, or on to it and back again the same way; once on, they tend to stay there for a long time.

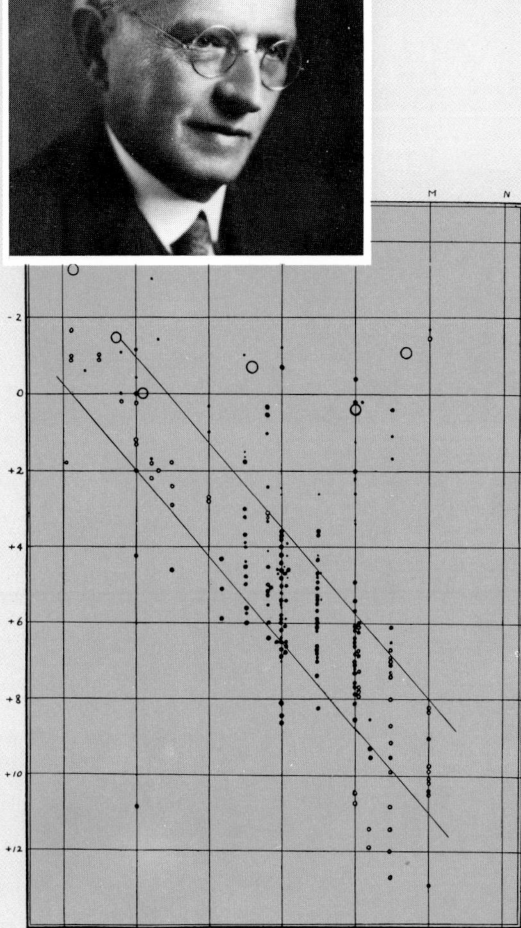

▲ Henry Russell, one of the co-originators of the graph known as the Hertzsprung-Russell diagram, is seen here with the first one. On it he has plotted absolute luminosity on the vertical axis and color on the horizontal axis, and has highlighted with diagonal lines the tight grouping now known as the main sequence. Similar diagrams can be drawn for specific groups or types of stars, and offer astronomers valuable insights into their nature.

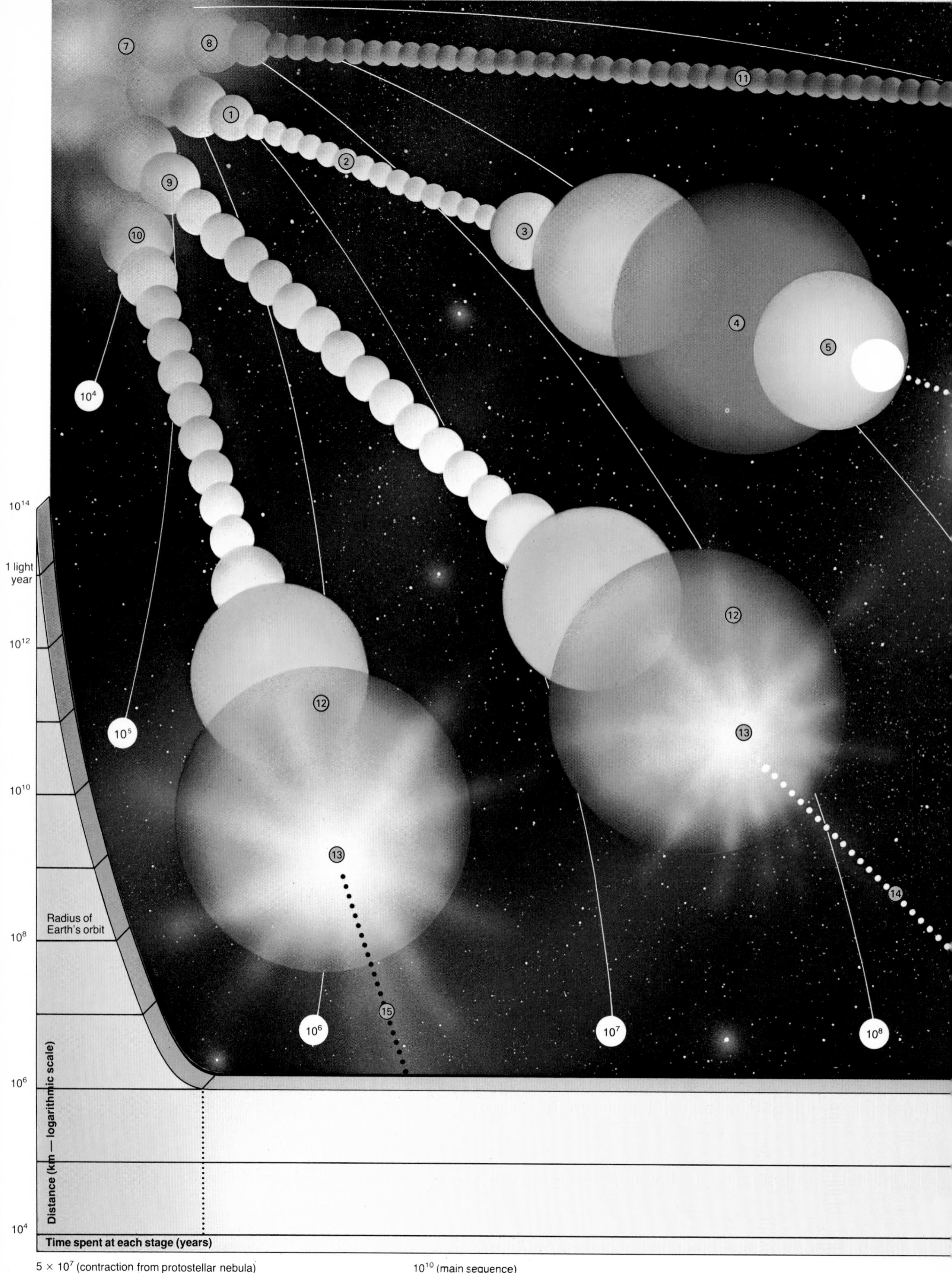

5×10^7 (contraction from protostellar nebula)

10^{10} (main sequence)

◄ How four stars of different mass evolve. The evolutionary paths are plotted against an approximate timescale in years and on the H-R diagram below. The lifespan of our Sun (one solar mass) lies between the long-lived low mass star (0·05 solar mass) and the massive stars (10 and 30 solar masses).

A star the size of the Sun contains enough hydrogen "fuel" to keep it going on the main sequence for about 10 billion years, whereas more massive stars have shorter main-sequence lifetimes because they use up their fuel very much faster. For example, a star ten times more massive than the Sun consumes fuel not ten, but 5,000 times faster, and burns itself out within about 20 million years (naturally it is also 5,000 times more luminous than the Sun). The most massive stars of all evolve to red giants within a million years; at the other end of the scale the low-mass stars, if they evolve to the main sequence at all, will outlive the Sun many times over.

The end-product of the triple-alpha reaction (➧ page 165), which sustains red giant stars, is carbon. Eventually this carbon begins to clog the core of the star, so that it ceases to generate energy. The core shrinks once again and this time may trigger helium-"burning" in the surrounding shell. In the most massive stars further reactions can take place leading, eventually, to the formation of iron in the core. This is the last in the series of possible energy-releasing fusion reactions, and is very uncommon. In stars having the same mass as the Sun or less, fusion probably ceases with the production of carbon.

When all its fuel is spent, gravity crushes a solar-mass star until it is a "white dwarf" – a hot body comparable in size with the Earth, and with a mean density as high as one billion kilograms per cubic meter. Astronomers expect most stars, including the Sun, to become white dwarfs which in time will fade and cool to become black dwarfs.

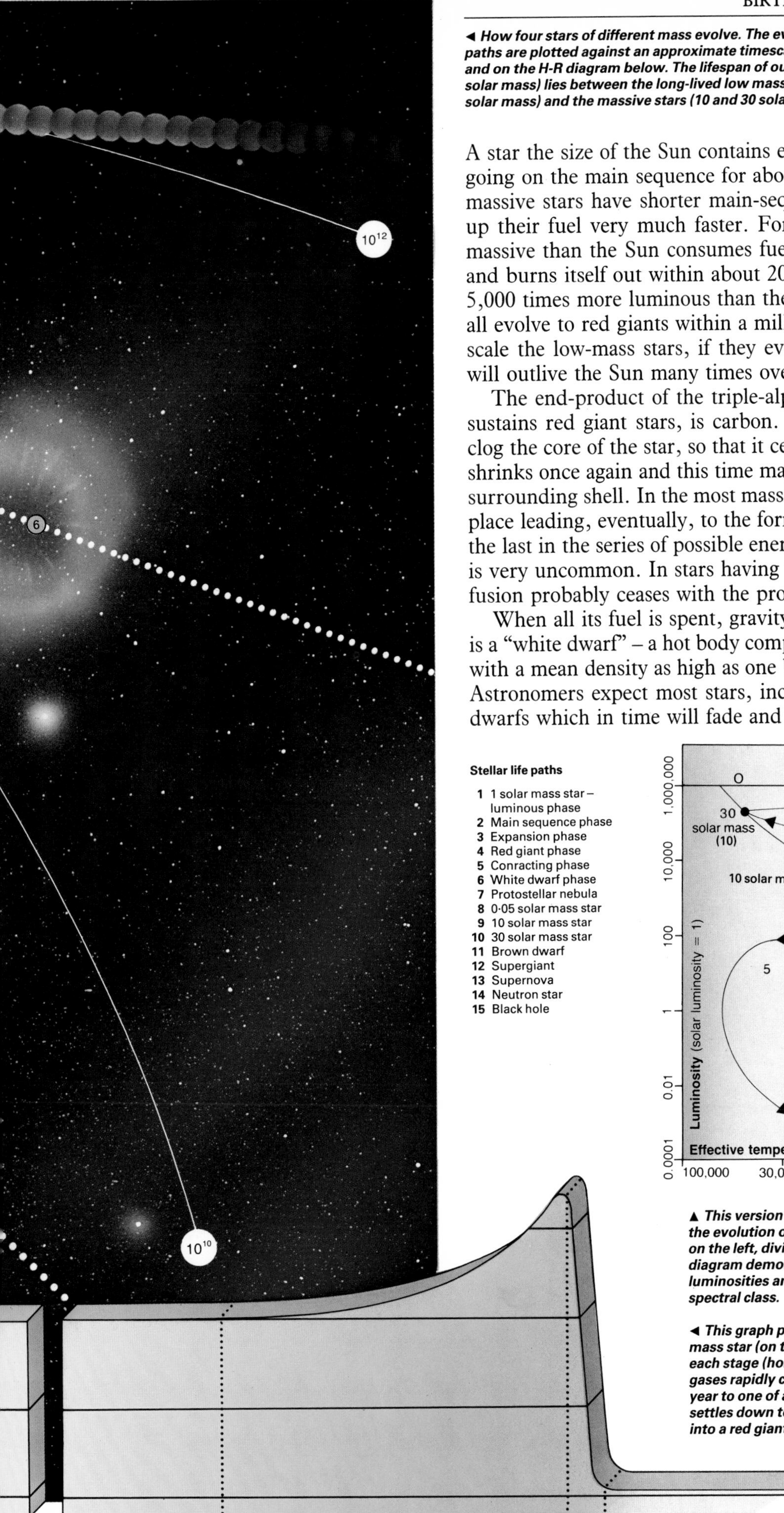

▲ This version of the Hertzsprung-Russell diagram shows the evolution of the same four star types as the illustration on the left, divided into the same numbered stages. The diagram demonstrates the relationship between the luminosities and temperatures of the stars and defines their spectral class.

◄ This graph plots in relief the changing radius of a solar-mass star (on the vertical axis), against the period in years of each stage (horizontal axis). The original cloud of dust and gases rapidly collapses from a radius of more than a light year to one of a few astronomical units, then more slowly settles down to life as a main-sequence star. Later it expands into a red giant, then shrinks to become a white dwarf.

The eventual fate of stars

A star of up to about 1·4 solar masses is likely to end its days as a white dwarf, and there is evidence that stars considerably more massive than this can shed enough matter late in life (by blowing off planetary nebulae, for example) to come within the limit. If a star is above this mass limit when it runs out of fuel, the most probable outcome is that it will suffer a violent supernova explosion which blasts its outer layers of material into space, and compresses its core to form a body even denser than a white dwarf – a "neutron star".

There is also a mass limit for a neutron star, although physicists are not sure exactly where this limit lies – probably somewhere between 2 and 5 solar masses. Once a more massive stellar remnant begins to collapse, nothing can halt the process: it continues to collapse in on itself until it becomes a point of infinite density. Before it reaches this stage, the force of gravity at the surface of the collapsing star becomes so great that it prevents light itself from escaping. Thus the star becomes literally invisible and forms – if current widely accepted views are correct – a black hole.

A closer look at the birth of stars

Scientists do not know why interstellar gas and dust clouds should start to collapse and form stars in the first place. These clouds are cool and of low density – typically one atom per cubic centimeter – but still have enough internal pressure to resist collapsing under their own gravity. Rotation of clouds and the presence of magnetic fields within them would also tend to forestall collapse. Nevertheless, collapse does occur, and there must be an explanation. One suggestion is that such clouds become compressed during an encounter with the spiral arm of a galaxy, or by collision with another cloud, and that an event of this kind may be all it takes to start the process of collapse.

Inside an evolving star

Red giant
4b
4a
6a
White dwarf
3 Post main-sequence
2 Main-sequence
1 Protonebula
Hydrogen burning core
High-mass star, late giant phase
Expanding envelope
Hydrogen burning shell
Contracting inert helium core
6b
Neutron star

▲ This diagrammatic sequence shows the nuclear reactions inside typical stars, numbered as in the "evolutionary tracks" of pages 154-5. In the protonebula (1) energy is released by contraction. On the main sequence (2) a star generates energy in its core by nuclear fusion – "burning" hydrogen as fuel; once this is exhausted the core shrinks (3). In its red giant stage, helium is converted to carbon in the star's core (4a), which is surrounded by a hydrogen-burning shell (4b). Thereafter collapse produces a white dwarf (6a) or a neutron star (6b).

▲ Sir Arthur Eddington, seen here talking to Albert Einstein, was one of the greatest of pioneer astrophysicists. He was the first to suggest that the transmutation of elements might be the process by which the stars kept shining for thousands of millions of years. Einstein once said it was worth learning English simply in order to talk to Eddington.

Stellar fuel...hydrogen and oxygen?

Around the turn of the century there were many theories concerning the fuel that kept the stars shining. In 1908 Hermann Helmholtz calculated that if the Sun were burning and were made of oxygen and hydrogen, combustion would keep it shining for 3021 years. Even if this seems rather over-precise, Helmholtz was right in saying that "known chemical forces are so completely inadequate... that we must quite drop this hypothesis."

...or cold meteorite swarms?

An altogether different theory had been summed up by Sir Norman Lockyer in 1888: namely, that the stars were connected with nebulae consisting of swarms of cold meteorites. The energy output of the stars would be maintained by constant meteoritic bombardment. The great British physicist Lord Kelvin even calculated how long the infalling of the planets could keep the Sun burning (if the planets are being steadily "braked" by friction against the tenuous interplanetary medium, then they will eventually spiral into the Sun). Kelvin found that the impact of all the planets combined would keep the Sun shining for no more than 46,000 years, which meant that again the timescale disproved the theory.

...or the mutual annihilation of particles?

Astronomers had identified two definite types of red and orange stars – giants and dwarfs – and the earliest Hertzsprung-Russell diagrams seemed to indicate that red giants were very young stars that had not yet joined the main sequence, while red dwarfs were very old. Russell himself believed that the source of stellar energy was the mutual annihilation of protons with electrons – an idea that was consistent with this view of stellar evolution. But calculation showed that the mutual annihilation process would sustain the stars for millions of millions of years, which was far too long – and yet another theory failed to survive the timescale test.

...or the transmutation of elements?

The British physicist Sir Arthur Eddington realized that the mutual annihilation of electrons and protons cannot occur in the way envisaged by Henry Russell. In 1927 he pointed out that there was another possible process – the transmutation of elements. He calculated that one helium nucleus could be formed from four hydrogen nuclei, that energy would be released in the process, and that this could provide enough energy to keep the Sun burning for about 10,000 million years.

...and, at last, the complete explanation

The problem was finally solved in 1939 by Hans Bethe, a German astronomer working in America, who realized that a straightforward transformation process was inadequate. Bethe was returning home to Washington by train after a conference at Cornell University, when he started to jot down some thoughts. Before he reached Washington he had solved the main problem. In the more massive stars, hydrogen is indeed converted into helium, but by the use of one other element – carbon – as a catalyst in a carbon cycle.

This new theory made it possible to take a fresh look at the H-R diagram. Scientists could now see that a star condenses out of nebular material, and is at first unstable, but eventually settles down to join the main sequence at a point determined by its initial mass. After it has used up all its available hydrogen "fuel", it develops into a giant located at the upper right of the H-R diagram before embarking on its final decline. Thus Betelgeux and other red giants, instead of being the stellar infants that they were once considered to be, are now regarded as cosmic geriatrics.

The carbon cycle

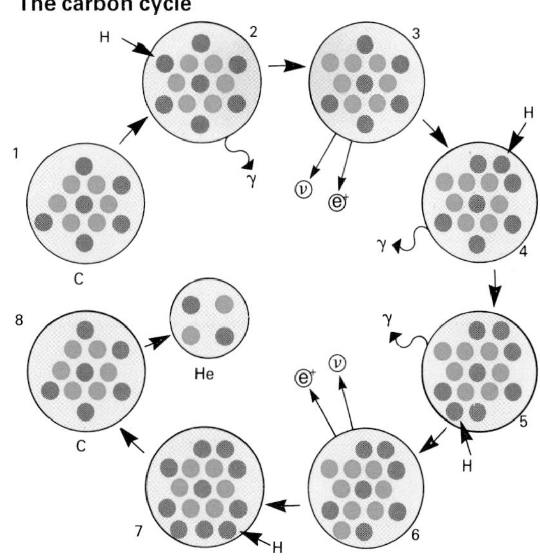

◄ Energy in the more massive stars is generated by the "carbon cycle", in which carbon is a catalyst. The carbon nucleus (1) changes its identity before emerging at the end. An extra proton is captured producing a lightweight nucleus of nitrogen (atomic mass 13) and a photon (a "particle" of radiation) is expelled (2). Next a positron and a neutrino are expelled, making carbon-13 (3). The capture of another proton produces a normal nitrogen-14 nucleus and more radiation is released (4). Next oxygen-15 (5) and nitrogen-15 (6) are produced. Finally a proton is added to the nitrogen-15 (7) which splits the nucleus to form a helium nucleus and a carbon nucleus that is recycled (8).

Neutron
Proton
Positron e^+
Electron e^-
Neutrino ν
Photon γ

The triple-alpha reaction

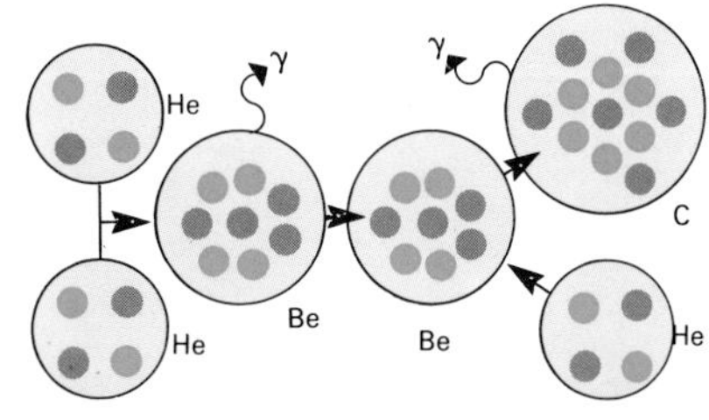

◄ Energy in red giants is created by the "triple-alpha" reaction. In the first step 2 helium nuclei combine to form a nucleus of beryllium, and radiation is released. Next another helium nucleus is added and a carbon nucleus is produced.

► The Lagoon Nebula in Sagittarius is a complex cloud lit up by highly luminous young stars, part of a cluster which is probably less than two million years old. The numerous dark blobs, some only a fraction of a light year in diameter, are known as Bok globules, after Bart J. Bok. They are dense dusty clouds which will probably collapse to form stars.

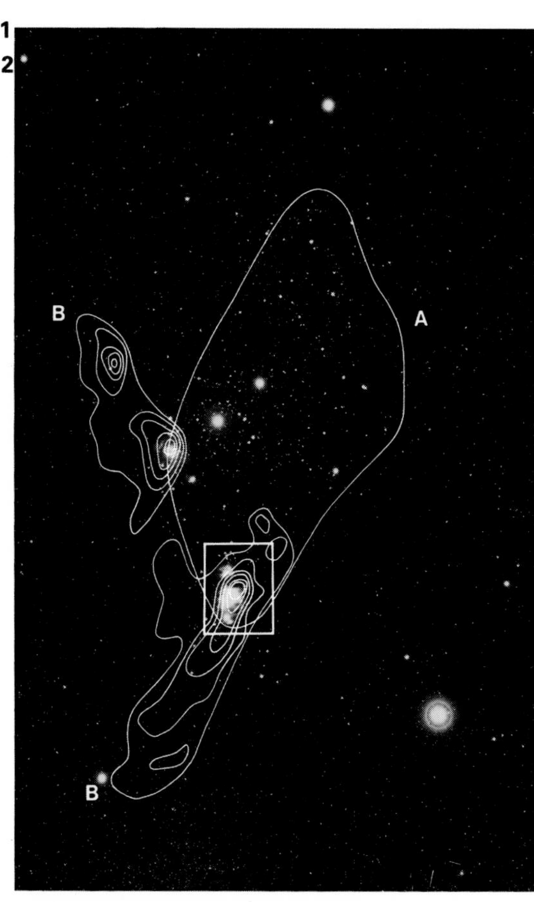

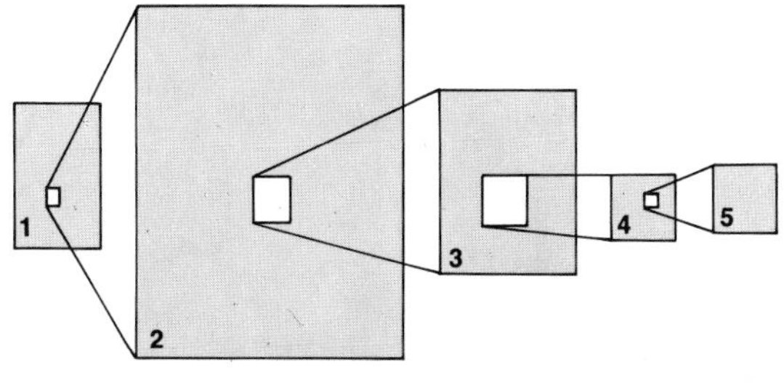

► *This scale drawing shows how the white rectangle on each of the numbered images relates to the next in the sequence. It starts with Orion as viewed with the naked-eye and ends with an infrared detail.*

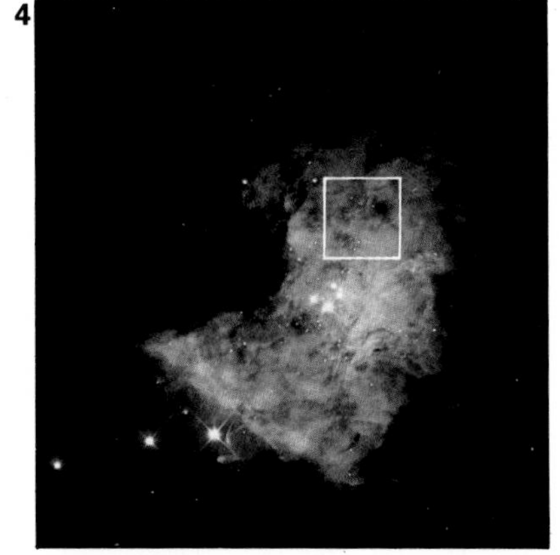

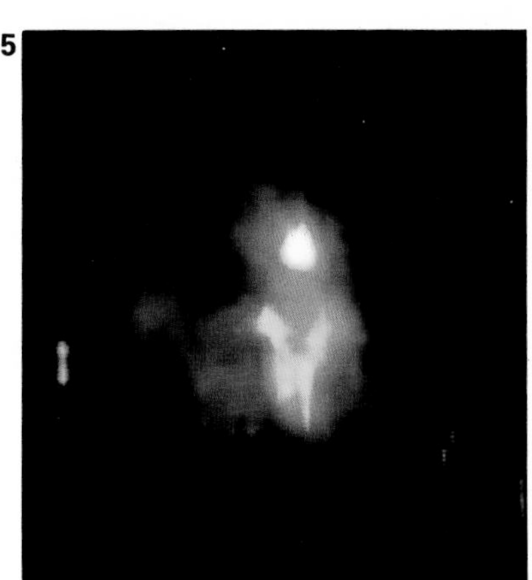

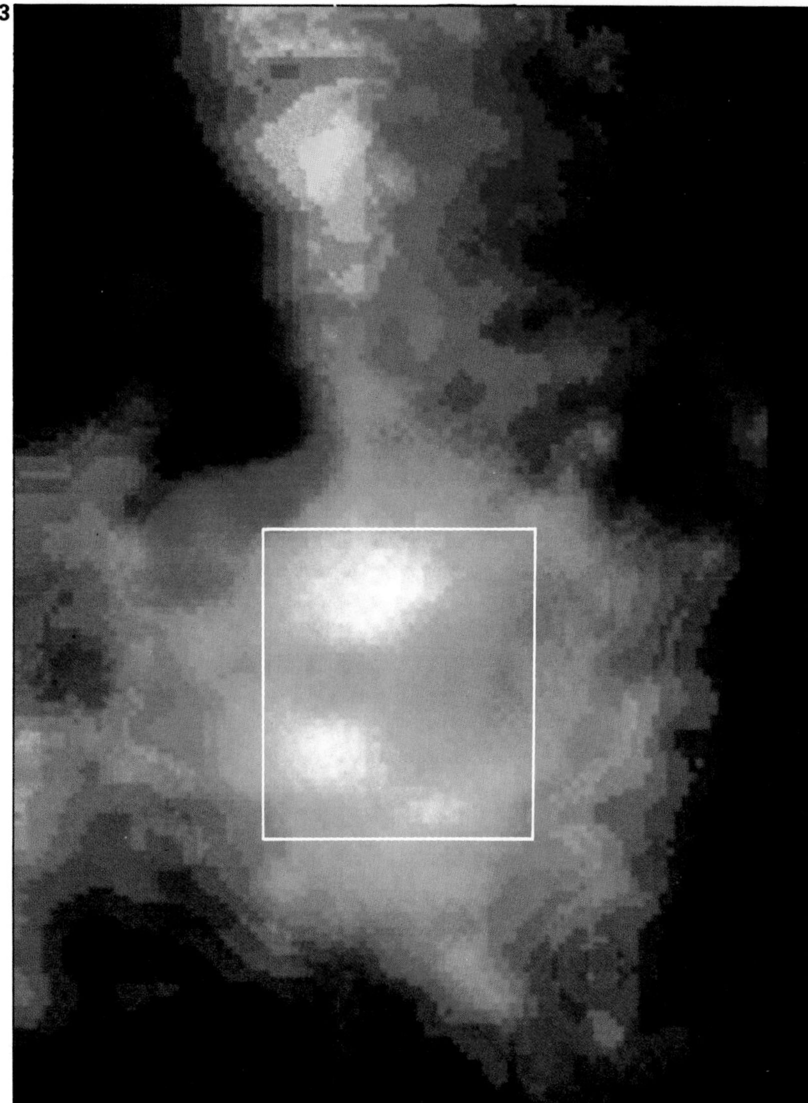

Orion – birthplace of new stars

Within the constellation of Orion (1) there is an expanding batch of O and B-type stars less than 10 million years old and known as the Orion 1 Association (1A). The complex includes two giant molecular clouds (B) each of which measures over 150 light years in diameter and contains more than 100,000 solar masses of material. Most of this is molecular hydrogen which is too cool to emit detectable radiation (the cloud contours have been drawn following the microwave emission from carbon monoxide mixed in with the hydrogen).

The Orion Nebula or M42 (2), the best known of the many emission nebulae, lies at a distance of about 1,600 light years. The concentration of gas which coincides with the nebula in the line of sight, but which actually lies behind it, is the Orion Molecular Cloud 1 (OMC1) of around 1,000 solar masses. The false-color radio map (3) shows the most intense area of carbon monoxide emission from OMC1; blue represents the most intense emission and green the least.

The visually brightest region of OMC1 (4) also coincides with the most intense infrared emission. It is illuminated by a group of four O-type stars, the Trapezium, probably formed within the last 100,000 years. Within this area the strongest emissions arise from a few compact sources. The false-color infra-red view (5) shows one of these sources known as the Becklin-Neugebauer object (colored white), believed to be a single star in the process of formation. The more complex source is the Kleinmann-Low nebula, a region of protostars.

See also
The Origin of the Solar System 121-8
The Basic Properties of Stars 145-52
Collapsing and Exploding Stars 169-76
Black Holes 177-80
Between the Stars 181-8

The formation of star clusters

Once a cloud of thousands of solar masses has started to collapse, it will break up into fragments, the subsequent collapse of the fragments yielding individual stars or clusters of stars. The more massive fragments evolve rapidly into brilliant O and B-type stars, many of which will later become supernovae. These stars heat the surrounding gases, causing them to expand. This expansion, together with the strong stellar winds which blow from these stars in their youth and with blasts from supernovae, send shock waves through surrounding clouds and can cause new bursts of star formation.

As a protostar collapses, the inner part shrinks more rapidly than the outer part and a cocoon of dust particles forms round the emerging star, hiding it from view. (However, the young star heats the dust cocoon sufficiently for it to emit infrared radiation, and this may be detectable.) Most stars emerge from their cocoons in the pre-main-sequence phase, when the dust is either vaporized, blown away, or coagulates into larger lumps – and perhaps planets.

A class of young stars

Astrophysicists have identified a particular type of star that is cool, varies erratically, and is found in the neighborhood of dense, dusty molecular clouds. Stars of this class (named "T Tauri" stars after a famous one in Taurus ➧ page 168) produce strong stellar winds and rotate rapidly – both characteristics of very young stars – and most have ages between a few tens of thousands and a few million years. Their infrared variation indicates the presence of surrounding dust, and supports the view that these are stars emerging from their cocoons but which have not yet reached their stable main-sequence phases.

Stars appear to form in batches within molecular clouds, perhaps in a repeating series of events which continues until the cloud is dissipated. The remnants of batches formed only tens of millions of years ago are seen today as youthful star clusters. Such clusters will disperse with time in the same way as the Sun's original cluster has dispersed – so that it is no longer possible to identify any cluster within which the Sun may have formed when it first came into existence nearly five billion years ago.

The Infrared Astronomical Satellite

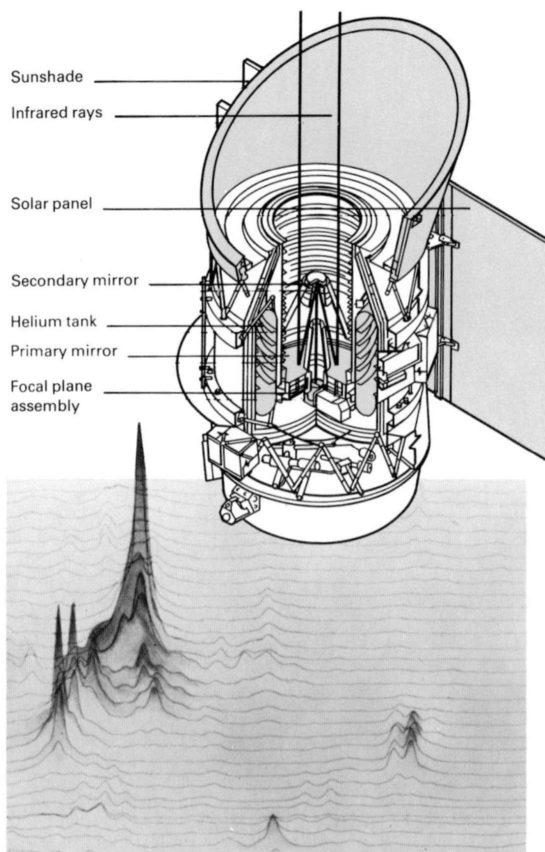

▲ This contour map from IRAS shows the strength of infrared emission at a wavelength of 100 micrometers in the Large Magellanic Cloud, our nearest neighbor galaxy. The high peak corresponds to the Tarantula nebula – a massive cloud that contains many large stars in the process of formation. Other discoveries by IRAS included the planetary system round Vega.

Infrared radiation – the key to stellar births

Newly born stars, cocooned in dense clouds of dust, reveal their presence by the emission of infrared radiation from these clouds. This radiation is invisible to the human eye, and to make matters worse, most of what there is becomes absorbed by water vapor in the atmosphere. One of the best ground-based telescopes, the United Kingdom Infrared Telescope (UKIRT), is sited 4,300m up on the top of Mauna Kea in Hawaii, where it is above 90 percent of the atmospheric water vapor.

However, the Infrared Astronomical Satellite (IRAS), launched on 26 January 1983, has attained sensitivities almost 1,000 times better than UKIRT. To minimize the infrared radiation emitted by the telescope itself, the system was cooled to 2K by liquid helium; the usable lifetime of the satellite was limited by the time taken for the helium to boil off. The satellite weighed 1,076kg and surveyed the sky at wavelengths from 8 to 119 micrometers.

◄ The Pleiades are a cluster of young stars. Although with the naked eye only six can be seen easily, there are several hundred stars in the cluster. Less than 50 million years old, they lie within a volume of space some 30 light years in diameter, about 400 light years distant. The brighter stars are surrounded by hazy nebulosity due to their light reflecting off dust particles.

Binary and Variable Stars

17

Describing double stars – astrometric, spectroscopic and eclipsing...Origins and behavior of double stars... Describing stars which vary in brightness – regularly or erratically...Stars that suddenly flare up or completely explode...PERSPECTIVE...Where mythology and astronomy coincide...Variable stars – the "standard candles" that determine intergalactic distances...The next candidate for a cataclysmic stellar outburst?

A binary consists of two stars which revolve round each other under their mutual gravitational attraction. More than half of all the stars exist as binaries or in multiple systems of up to six members. The orbital periods of binaries range from less than a day to hundreds of years, depending on the masses and separations of the stars.

In a "visual binary" the two stars are far enough apart to be seen telescopically as separate stars. Astronomers have identified about 70,000 visual binaries to date, most of them having separations of 10-100 astronomical units and orbital periods of decades or centuries. (An "optical double" consists of two stars which appear close together in the sky, but which actually lie at very different distances and are not physically connected.)

Astrometric binaries

The center of mass of a binary moves through space in a straight line, while the stars themselves revolve round the center of mass. Even if one member of a binary is too faint to be seen, the wobbling motion of the brighter companion gives its presence away: as it weaves to and fro, its position relative to the center of mass varies by an amount which depends on the relative masses of the two stars. A star which reveals the existence of an invisible companion in this way is known as an "astrometric binary". The faint white dwarf companion of Sirius was discovered in this way (◀ page 149), and several nearby stars, notably Barnard's star, wobble by such small amounts that their companions must be planets rather than stars.

Astrometric binaries

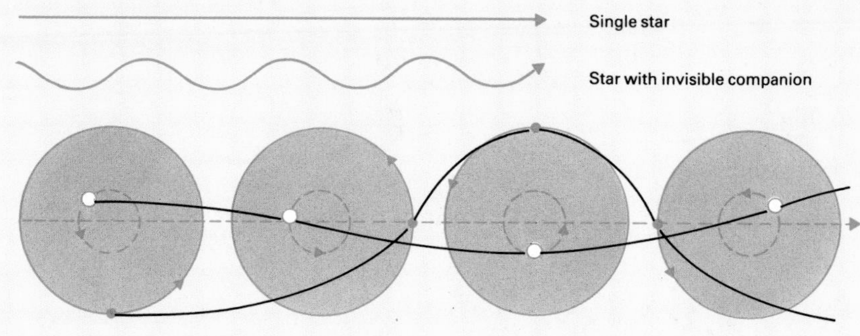

▲ *Of the two stars making up an astrometric binary, one is too faint to be seen. Both revolve around the center of mass of the system, the brighter and more massive lying closer to the center of mass. The combination of this motion with the straight-line motion of the center of mass, results in the visible star following a wobbly track across the sky.*

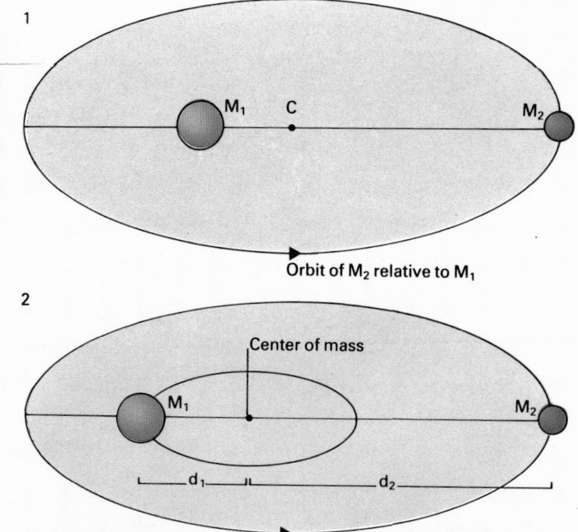

▲ *In a typical binary (1) the fainter companion (having mass M_2) seems to orbit round the more massive primary (M_1). In fact (2) both stars follow elliptical paths round the center of mass (C). The ratio of the two masses $M_1/M_2 = d_2/d_1$.*

Binary stars: scientific fact...

If the mean separation (a) of two members of a binary is given in astronomical units, and their orbital period (P) in years, it is possible to find the combined mass of the pair by applying Newton's law of gravitation to their motions. The combined mass is given by the simple relationship

$$M1+M2 = a^3/P^2$$

where M1 and M2 are the masses of the stars in solar masses. For example, if a binary has a period of 20 years and a mean separation of 10 astronomical units, the combined mass is $10^3/20^2 = 1000/400 = 2{\cdot}5$ solar masses.

In fact, each star moves round the center of mass of the binary – a point which lies between the stars. If the two stars are of equal mass, the center of mass will be midway between them. However, if the masses are unequal, then just as a weightlifter's bar carrying unequal weights must be balanced at a point closer to the heavier weight, so the center of mass of the binary will lie closer to the more massive star.

...and ancient mythology

Some stars which appear to alter in brightness are not true variables, as was first thought: they are eclipsing binaries, of which the most famous example is Algol, in Perseus (▶ page 163). In mythology, Perseus was the hero who rescued the Princess Andromeda from the sea-monster Cetus. The monster was about to gobble her up when Perseus appeared, riding on his winged sandals, and turned it to stone by showing it the head of the Gorgon, Medusa, whom he had killed. Algol's position in the sky lies in Medusa's head, and the star "winks" in a sinister fashion every 2·87 days when the darker component of the binary pair blots out most of the light from the brighter one. Algol is known, naturally enough, as the Demon Star. Yet the "winking" of Algol was unknown before 1669, so that its position in Medusa's head is purely fortuitous.

Binary stars are particularly rewarding subjects for the watcher with a small telescope

Spectroscopic binaries

In most binaries the stars are so close together that an Earth-based observer cannot see them as individuals. A "spectroscopic binary" is one which looks like a single star, but its spectrum consists of the combined spectra of two stars. As the stars revolve round each other, star A may be approaching the Earth while star B is receding. Due to the Doppler effect, the spectral lines of star A are blue-shifted, while those of star B are red-shifted. As the stars continue in their orbits they cross the line of sight, after which star A begins to recede while star B begins to approach. The spectral lines of A then become red-shifted and those of B become blue-shifted. The spectral lines from each star will thus oscillate to and fro in wavelength, so that the combined spectrum betrays the presence of two stars.

If one star is too faint for its light to register in the combined spectrum, the orbital motion of the visible star will nevertheless cause periodic variations in the wavelength of its spectral lines, and these variations reveal the presence of a binary companion. Stars of this type are known as "single-line binaries".

A

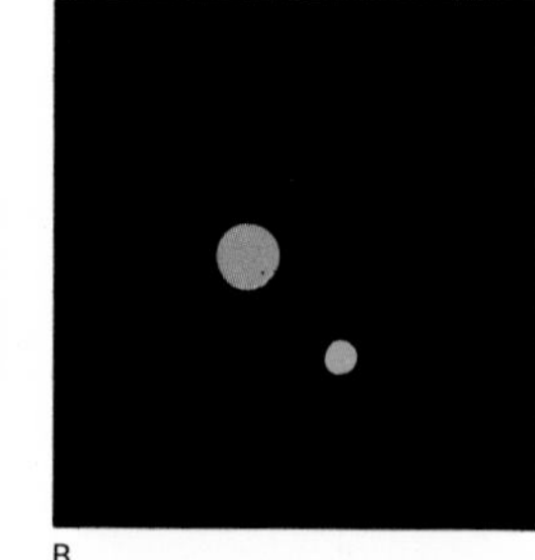

B

C

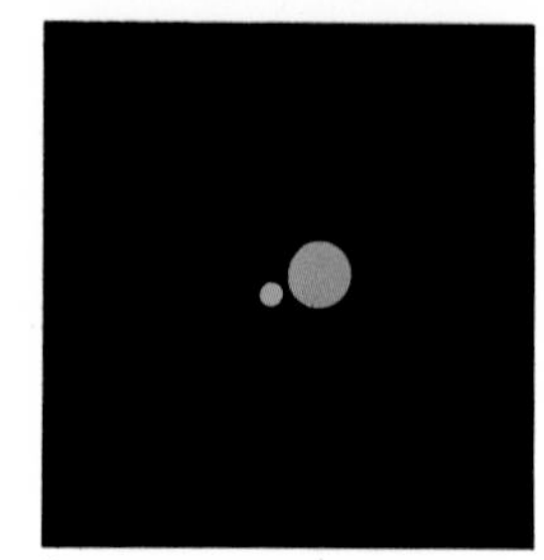

D

▲ *These visual binaries can be resolved by small telescopes (south is at top as seen in an astronomical telescope).*
(A) β Cygni – stars of magnitudes 3·2 and 5·5 separated by 34 arcsec (″).
(B) γ Andromedae – magnitudes 2·2 and 5·5; separation 10″.
(C) β Orionis (Rigel) – magnitudes 0·1 and 7·0; separation 9″.
(D) α Scorpii (Antares) – magnitudes 1·0 and 6·5; separation 2·9″.

Spectroscopic binaries

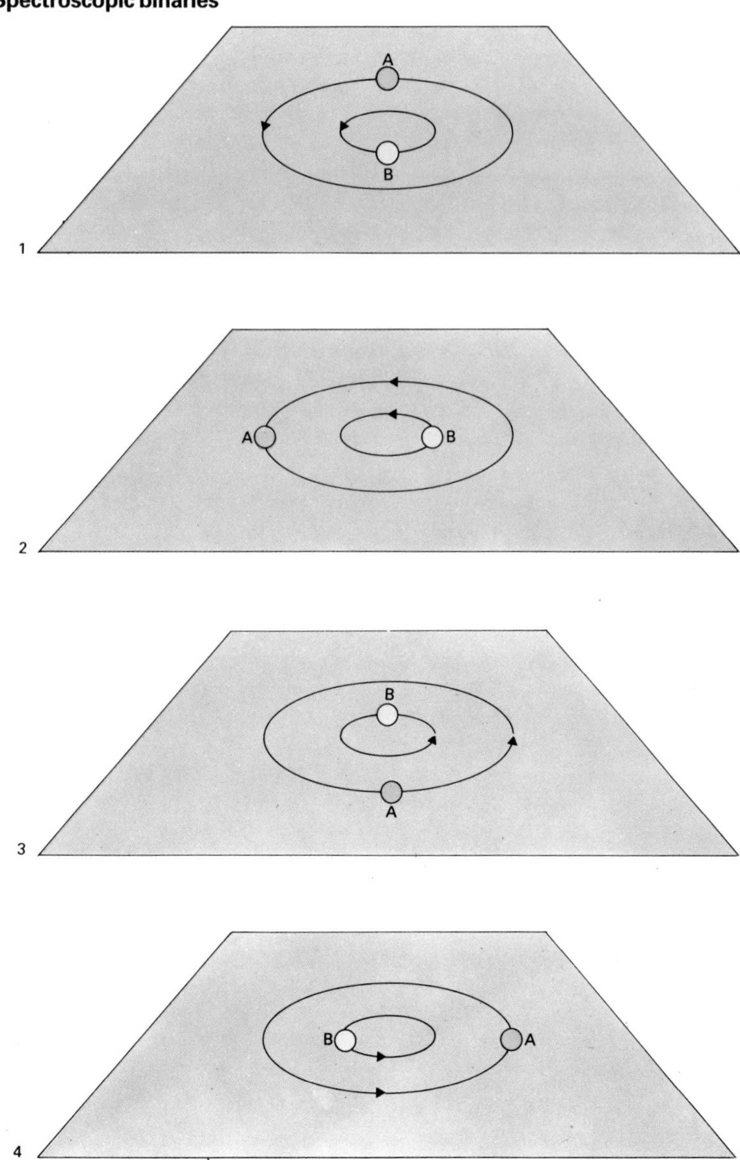

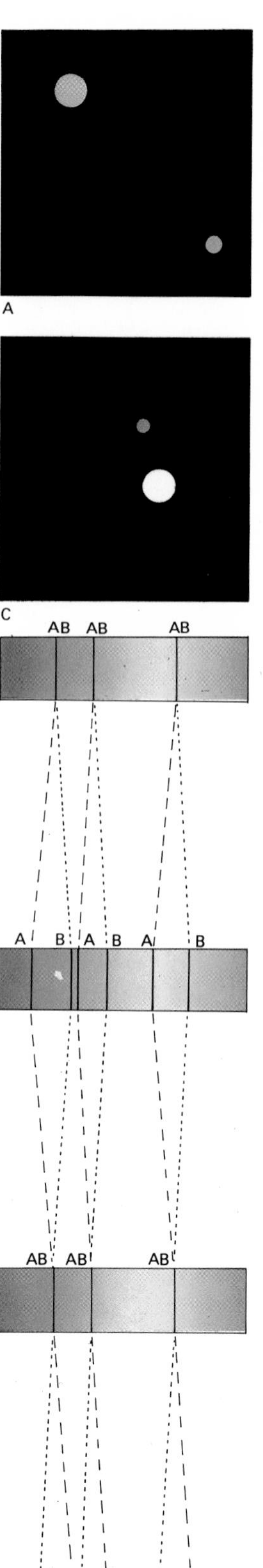

◄ *Spectroscopic binaries. At stages (1) and (3) the stars are moving across the line of sight. There is no Doppler shift of their lines and the two sets merge at the same wavelengths. At (2) and (4) one star is approaching while the other is receding. One set of lines is blue-shifted and the other is red-shifted, so that the two sets are seen in the combined spectrum.*

► *Eclipsing binaries. (1) If two identical stars totally eclipse each other the minima in the light-curve are equally deep, but (2) if the orbital plane is tilted to the line of sight, the eclipses are partial and the minima shallower. (3) If the stars are unequal in size or luminosity the depths of the successive minima are different. The flat bottom of the minima shows the duration of the total (or annular) phase. (4) Unequal stars with partial eclipses: A is hotter and brighter than B. The primary minimum occurs when the dull star (B) eclipses the bright star (A).*

Eclipsing binaries

If the orbital plane of a binary is edge-on to the line of sight (or very nearly so), each star will pass in front of the other alternately, causing eclipses, so that what appears to be a single star shows regular changes in brightness. Such a star is called an "eclipsing binary". Usually an eclipsing binary is also a spectroscopic binary, and astronomers can obtain a great deal of information about the components. For example, plotting brightness against time gives a curve known as the "light-curve", and the shape of this reveals whether the eclipses are partial or total. Also, since measurements of the Doppler effect in the stars' spectra yield their orbital speeds, it is possible to calculate the size of each star from the duration of the eclipses and the time taken to fade from maximum to minimum brightness. If one star is more luminous than the other, the drop in brightness will usually be greater when the dimmer star eclipses the brighter. The dips in the light-curve will be unequal, the deeper of the two being known as the primary eclipse.

The best-known eclipsing binary is Algol, in Perseus. The primary eclipse is easy to see, as the brightness drops by about one magnitude.

Three types of close binary

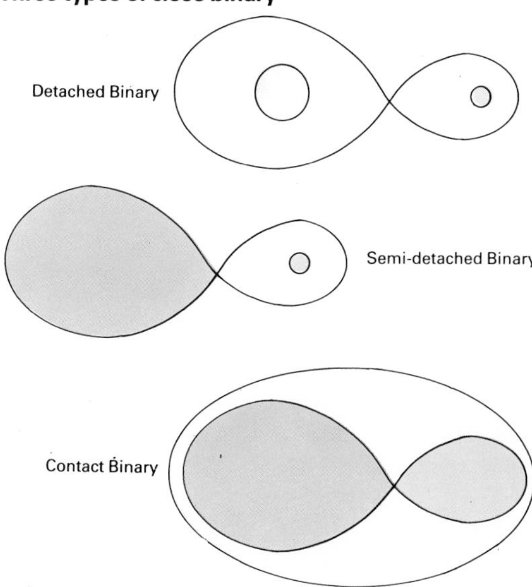

Eclipsing binaries

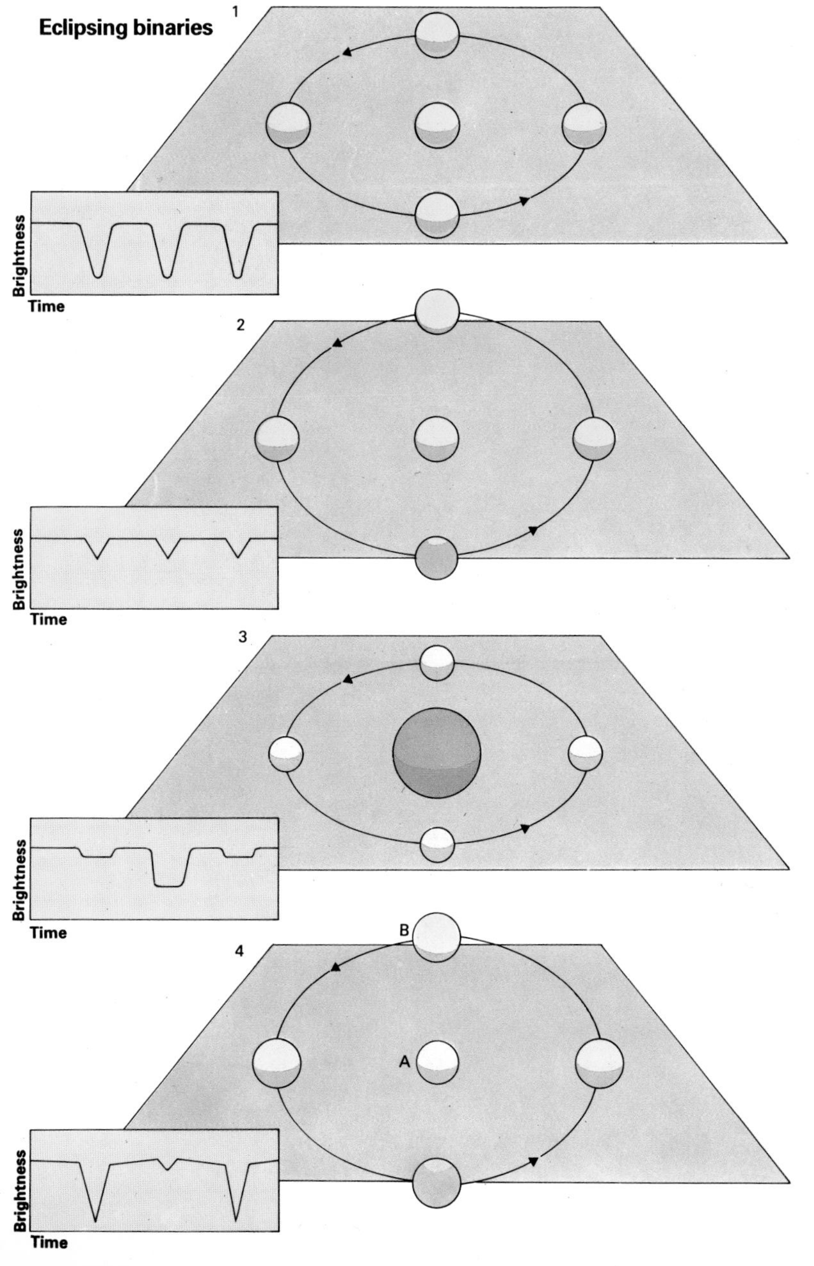

Life histories of binary stars

If the stars making up a binary are far apart, they were probably born as two separate protostars, but were sufficiently close together for gravity to keep them in orbit round each other. But many of the short-period, close binaries may have resulted from the splitting apart of a single protostar which, as it collapsed, spun too fast to hold itself together.

There is a three-dimensional figure-of-eight shape, known as an equipotential surface, around the two stars in a binary, and the force of gravity is equal at all points on this imaginary surface. The two parts of the curve, each containing one star, are called "Roche lobes" after Eduard Roche (1820-1883). Matter which lies inside a lobe remains under the influence of the star in that lobe. The nearer a star comes to filling its Roche lobe, the more distorted its shape becomes due to the gravitational attraction of its companion.

Some possible variations

When both components of a binary are on the main sequence they lie well inside their lobes, but if one star is more massive than the other it will evolve more rapidly towards the red giant stage. As it expands it may grow to fill, or even to exceed, its Roche lobe, and matter will flow from its surface through the crossover point of the figure-of-eight (the "inner Lagrangian point") towards the secondary. Although some material from the red giant may escape altogether from the system, substantial amounts of matter can be dumped on the secondary – enough to make a significant difference to its mass and to speed up its evolution. A system in which one star fills its lobe is called a "semi-detached binary." If both stars fill their lobes they form a "contact binary"; in addition to the eclipses, both of the stars show continuous brightness variations due to their changing apparent shapes as they revolve round each other.

Evolution may continue to a stage where the original primary becomes a collapsed star and the secondary becomes a red giant, filling its Roche lobe and dumping material onto its companion.

The identification of one type of variable star led, eventually, to the realization that our Galaxy is just one among many millions

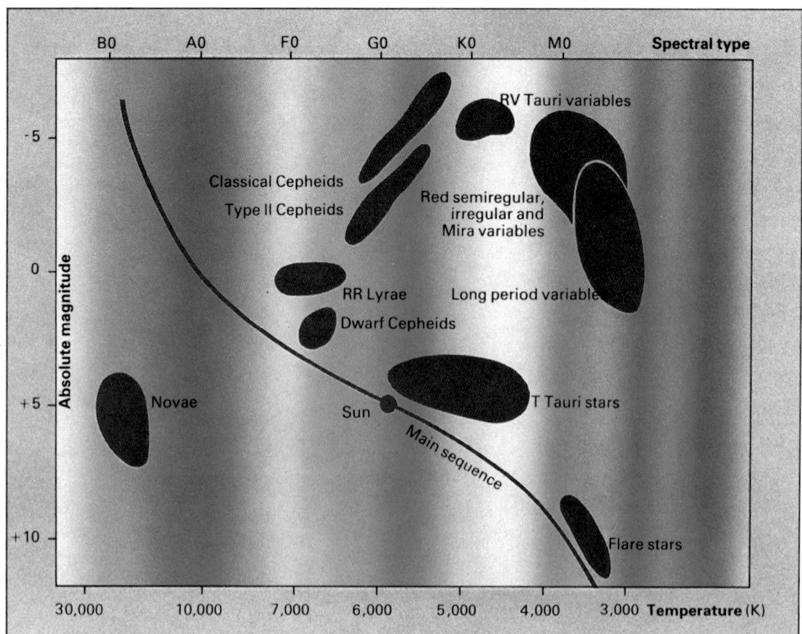

◀ **Variables have specific locations on the H-R diagram. Thus T Tauri stars are young pre-main sequence, and pulsating variables are old, becoming unstable late in their lives.**

▼ **The graphs show how the luminosity, temperature and radius of a Cepheid fluctuate throughout its cycle. The doppler effect in its spectrum reveals the expansion and contraction.**

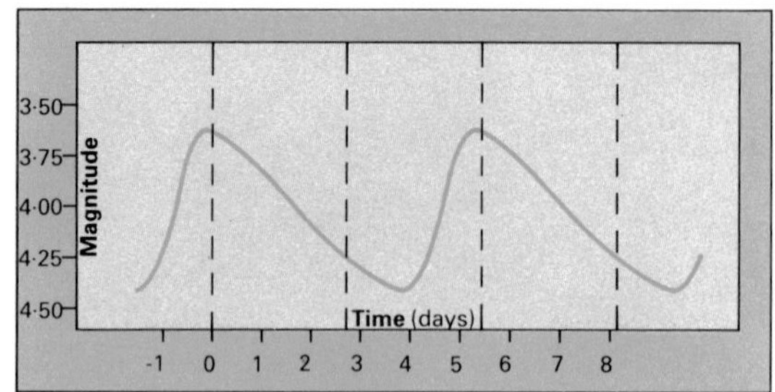

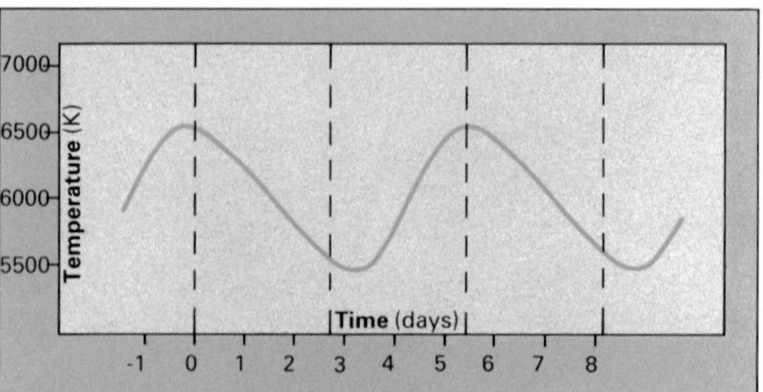

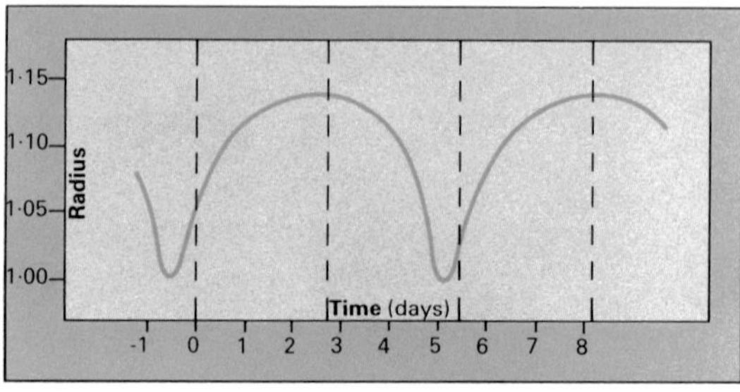

Variable stars

Variable stars are stars which vary in brightness. If such variations are due to some external influence – eclipsing binaries being the most obvious example – they are called "extrinsic variables"; if they vary because of genuine fluctuations in light output they are known as "intrinsic variables".

Astronomers classify variable stars according to how they vary. One group consists of stars that vary in a regular periodic manner, and these are subdivided into short-period variables and long-period variables according to whether their periods are shorter or longer than about 80 days. Semi-regular variables show variations of both period and amplitude, while irregular variables fluctuate in an erratic and unpredictable way. Most variables are pulsating or eruptive in nature.

Pulsating variables

These are stars which expand and contract as their brightness varies. The best-known examples are the Cepheid variables, named after δ Cephei, the first star of this type to be recognized. They vary in a regular and characteristic way, increasing in brightness rather more rapidly than they decrease, and having periods of between 1 and 60 days. The radius of a typical Cepheid varies by 10-20 percent, its light output fluctuates by about one magnitude, and there are also variations in its temperature and spectral class.

Pulsating variables "vibrate" at their natural frequency rather as a bell rings when struck. It is the periodic storage and release of energy in a layer some way below the star's surface that sustains these vibrations. The layer consists of partly ionized helium (some of the helium atoms have lost electrons) and electrons. As the star contracts the ionization increases (more electrons are stripped off) and the layer becomes more effective at trapping outgoing radiation. Eventually the stored radiation builds up enough pressure to push out the outer layers of the star, which then cool down and become less opaque, so allowing the stored radiation to escape. The star then shrinks and initiates another cycle of expansion and contraction. The pulsation cycle continues for as long as the right conditions exist inside the star.

Two types of Cepheid variables

Type I Cepheids are very bright F and G-type giants and supergiants with absolute magnitudes of −2 to −6. They are relatively young massive stars which have evolved away from the main sequence. Their period of variation is related to their brilliance by a "period-luminosity law": the longer the period the greater the luminosity. Cepheids are so bright that they are visible at very great distances, and observers can pick them out in other galaxies by their variability. Measuring their periods enables scientists to work out their luminosities or absolute magnitudes using the period-luminosity law, and comparing their apparent and absolute magnitudes can then reveal their distances.

Type II Cepheids, also known as W Virginis stars, show a similar pattern of behavior but are about two magnitudes fainter. They are older low-mass stars which have evolved away from the main sequence and have reached the stage where they are "burning" helium in their cores. RR Lyrae stars are a closely related type, commonly occurring in globular star clusters. Their mean absolute magnitudes are about 0·5 and their periods range from 0·3 to 1 day.

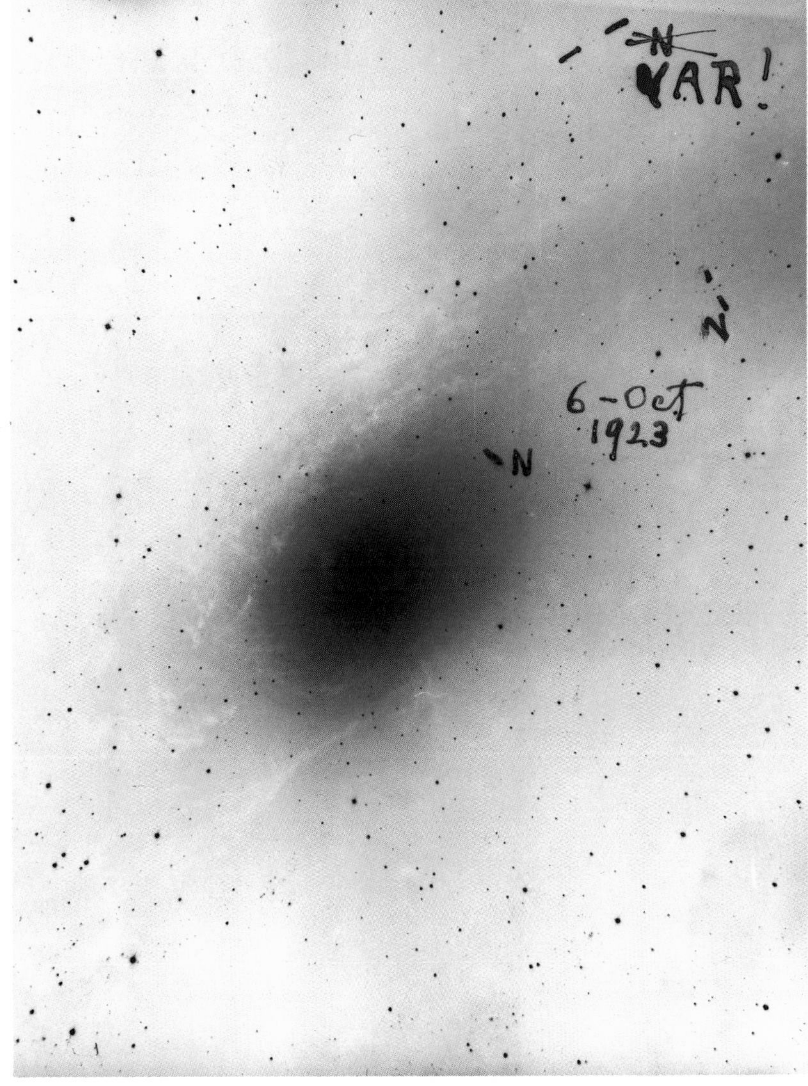

Cepheid variables and intergalactic distances

The parallax method of distance determination (◀ page 145) works well for distances up to about 300 light years, but breaks down after that because the shifts become too small to be measurable. It is then necessary to use less direct methods, most of which involve using spectroscopes to work out the real light output of a star: from this, it is possible to work out its distance.

However, there are some short-period variable stars, known as Cepheids after δ Cephei, the brightest member of the class, which reveal their luminosity by the manner in which they brighten and fade. The longer the period, the more powerful the star. Cepheids thus act as "standard candles", and because they are so luminous they are visible over great distances.

A crucial discovery

In 1912 an American astronomer at Harvard, Henrietta Leavitt (1868-1921), was studying some photographs of the Small Magellanic Cloud. This looks rather like a detached part of the Milky Way, and although it was known to be a distinct system, astronomers still believed that it was contained within our Galaxy. Miss Leavitt found Cepheid variables in the Cloud, and realized that the brighter Cepheids were those with the longest periods. For practical purposes it was reasonable to regard all the stars in the Cloud as lying at the same distance from Earth, and therefore the longer-period stars really were the more powerful.

The discovery was even more important than Miss Leavitt appreciated, because it led on to final proof that the so-called spiral nebulae are external systems, millions of light years away. In 1923 the American astronomer Edwin Hubble (1889-1953) detected Cepheids in some of the spirals. He determined their distances and found that they were much too remote to belong to our Galaxy. This could only mean that the systems in which they lay were also external. Hubble estimated the distance of the Andromeda spiral as 900,000 light years, which he later amended to 750,000 light years. Herschel, long before, had suggested that the spirals might be "island universes", but had been unable to obtain any proof.

▲ *Edwin Hubble found the first Cepheid variable in a spiral galaxy while studying this historic plate. His triumphant "VAR!" marks the event.*

◀ *Miss Henrietta Swan Leavitt, who discovered Cepheid variables in the Small Magellanic Cloud.*

▶ *The best-known example of a long-period variable is Mira Ceti, in the constellation of Cetus, which varies in brightness between magnitudes 3 and 9 over a period of about 332 days as shown by its light-curve. On each cycle it disappears from naked-eye view for more than 200 days; its temperature varies between about 2,600K and 1,700K. Long-period variables such as Mira Ceti are red M-type giants, so cool that they emit most strongly in the infrared.*

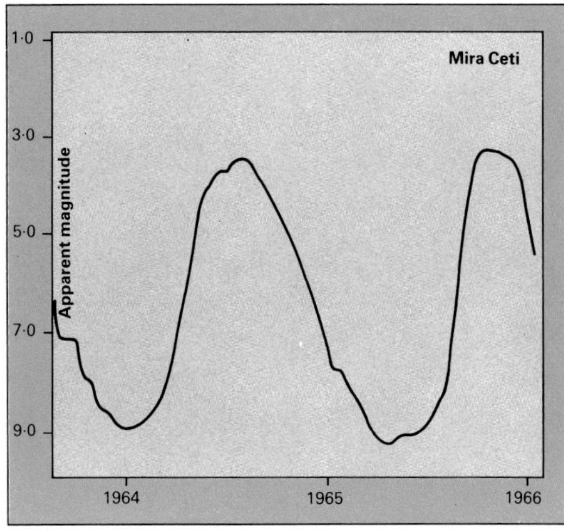

The next cataclysmic outburst?

Eta Carinae is a very peculiar star which lies at a distance of some 6,800 light years from Earth, within the spectacular η Carinae nebula. Despite its great distance, during the 1840s η Carinae became for a time the second brightest star in the sky, reaching an apparent magnitude of about −1. Thereafter it faded to about magnitude 7. The star itself is no longer visible, but lies within a compact cloud of gas and dust, known as the Homunculus nebula, which is expanding at a rate of about 500km per second. Physicists believe that the star is at least a hundred times more massive than the Sun, and several million times more luminous. Dust grains heated by the star make η Carinae the brightest star in the sky at infrared wavelengths of around 10 micrometers.

Astronomers are not certain whether η Carinae is a massive young star which has not yet reached the main sequence, or a highly evolved star approaching the end of its life. However, recent observations of nitrogen-rich blobs of matter which may have been expelled from the star support the view that it has evolved sufficiently far to produce nitrogen by nuclear reactions and that it cannot, therefore, be a protostar. If η Carinae is well advanced in its evolution, then, because it is so massive, it is likely to end in a cataclysmic outburst – a supernova. When this happens, it will probably outshine Venus in our skies.

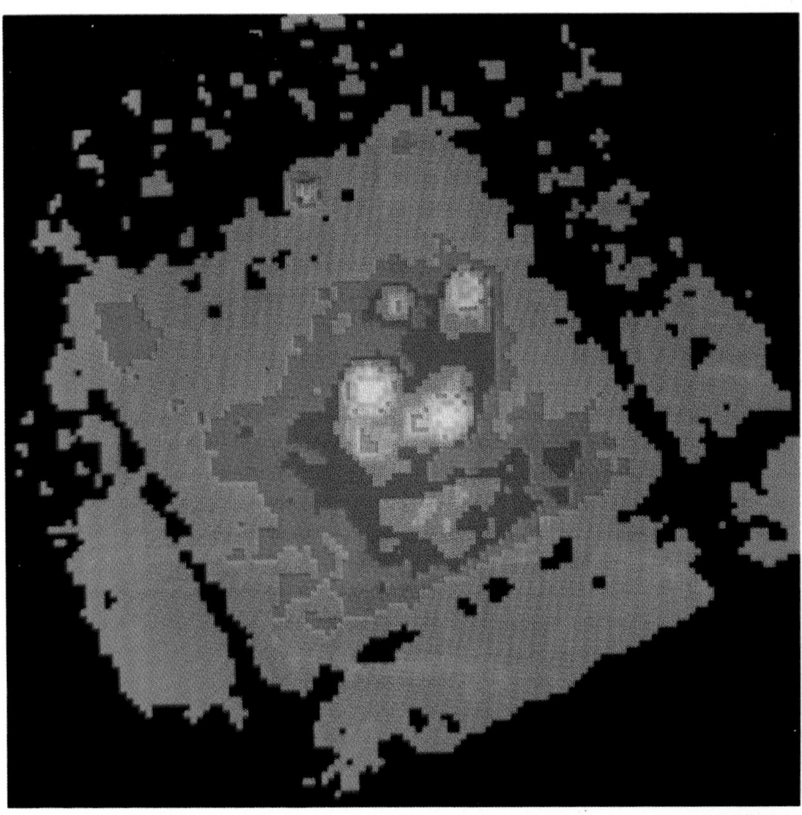

Four views of η Carinae. In true color (left) the Carina nebula is a staggering 300 light years across. The two X-ray views (above and below) are approximately in the same scale as the optical view and show η Carinae in the center. In the color-coded version blue is the least intense, white the most. The nebula around η Carinae, called the "Homunculus" – the "little man" (right) – contains as much matter as ten Suns.

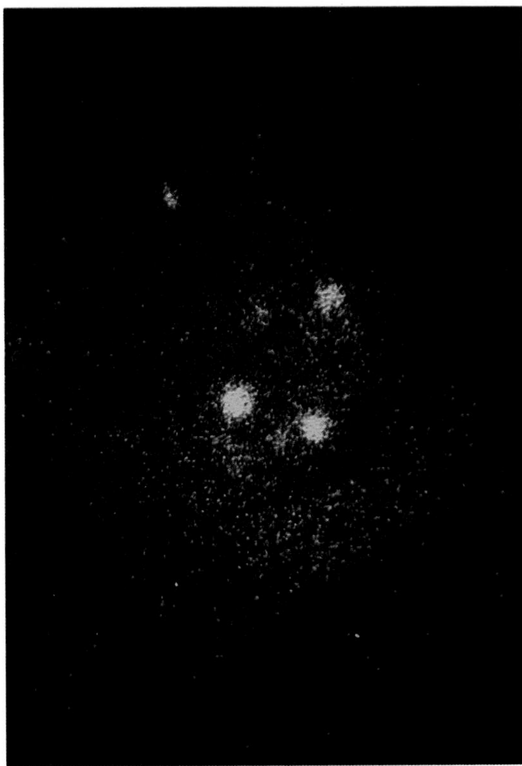

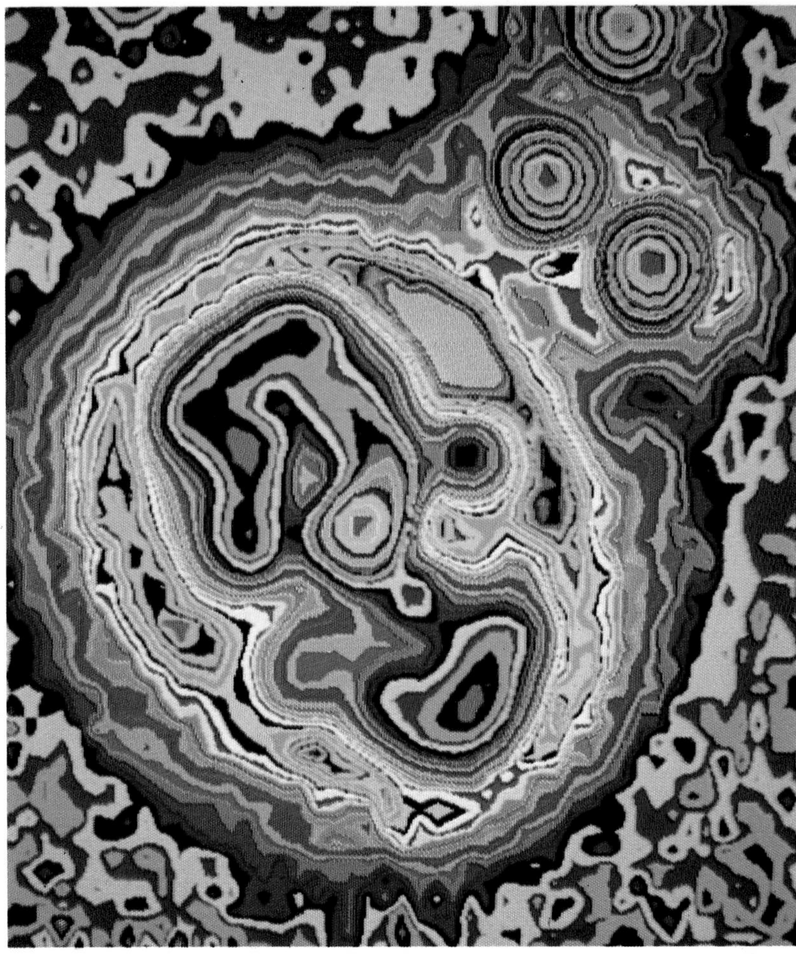

See also
Inside the Sun 97-104
The Basic Properties of Stars 145-52
Birth, Life and Death of Stars 153-60
Collapsing and Exploding Stars 169-76
Black Holes 177-80

Irregular and eruptive variables

Among the different kinds of irregular variables the group known as T Tauri stars is particularly interesting. Their spectra show that these are young, rapidly rotating stars which are blowing into space as much as 10^{-7} solar masses of material per year. They are enveloped in clouds of gas and dust, and their variability may be due to a combination of intrinsic and extrinsic factors, such as flares in their atmospheres and fluctuations in the density of the swirling clouds of enveloping dust.

Flare stars, otherwise known as UV Ceti stars, are cool dim M-type stars on the lower end of the main sequence which flare up suddenly once or twice a day. Typically they increase in brightness by one or two magnitudes in seconds, then fade back to normal in a few minutes. Astronomers think that the flares are similar to, but more spectacular than, flares which occur on the Sun due to the sudden release of energy stored by localized powerful magnetic fields.

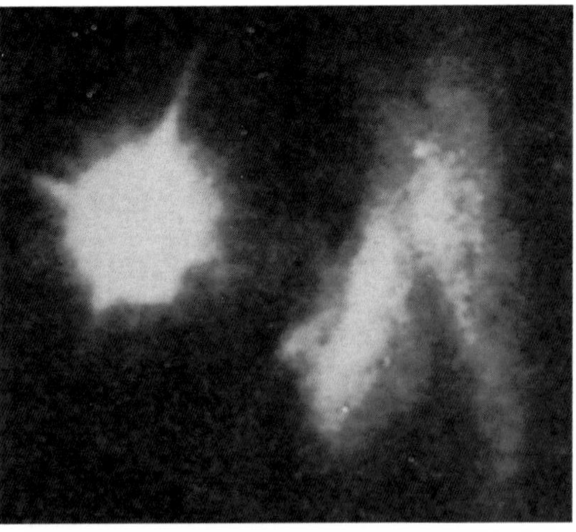

▲ T Tauri seen with neighboring nebulosity (right).

Novae

A nova ("novae" is the plural) is a star which flares up within a matter of hours, or at most a few days, increasing in brightness by a factor of between ten thousand and one million. The nova then declines back to its previous luminosity, over a couple of months, in the case of Nova Cygni, or a number of years. A typical nova expels a shell of matter containing up to one hundred-thousandth of the star's mass.

Novae occur in close binary systems in which one member is a hot compact body such as a white dwarf. The larger component dumps material onto the compact star, or into a disk of matter around it, and this material undergoes violent explosive burning to result in the flare up (▶ page 172). Recurrent novae repeat this pattern at intervals of a few decades, while so-called dwarf novae show repeated nova-like outbursts, brightening by two to five magnitudes, at intervals of tens or hundreds of days.

Supernovae

Supernovae are exploding stars which can become ten billion times more luminous than the Sun and, for a short time, can outshine entire galaxies. The explosion completely disrupts the star, throwing most of its material into space and leaving at most a tiny collapsed remnant (▶ page 174).

Type I supernovae occur among older stars and Type II supernovae among younger stars. The light-curves of Type I supernovae show a steep rise in brightness to an absolute magnitude of about −19 followed by a fairly slow decline of about 0·5 magnitudes per month. Type II supernovae show a similarly sharp rise, but are about two magnitudes fainter at maximum, and go through a period of faster, more irregular fading, before settling to a long, slow decline.

R Coronae Borealis stars

Among the most peculiar of variables, stars like R Coronae Borealis have been described as being rather like "novae in reverse" because they suffer sudden and irregular drops in brightness, declining by as much as ten magnitudes (a factor of 10,000) and then returning to normal between the events. They are supergiants with carbon-rich atmospheres and it is possible that the sudden minima result from the accumulation of clouds of carbon dust which are eventually blown away, allowing the star to revert to its normal brightness.

▼ Typical light curves and periods are shown for a T Tauri star (1), a nova (2), a Type 1 supernova (3) and R Coronae Borealis (4).

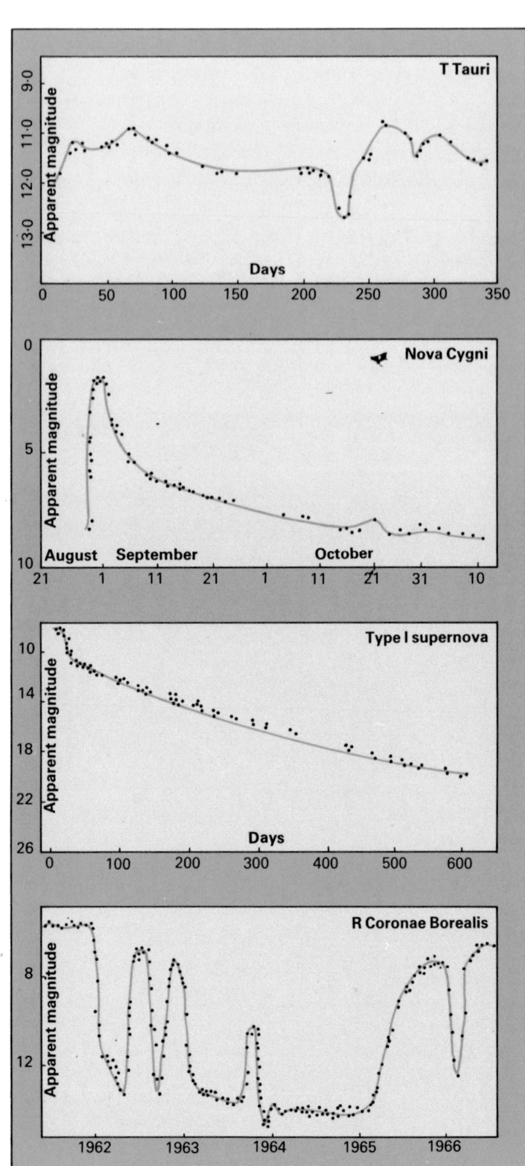

Collapsing and Exploding Stars

What happens when stars run out of fuel?...From planetary nebula to white dwarf...The most famous nebulae...Neutron stars and pulsars...What makes a pulsar tick?...The fastest pulsars of all...PERSPECTIVE... A puzzling celestial object...Discovering the first pulsar

Scientists believe that when most stars run out of nuclear fuel, they will become white dwarfs. These are stars comparable in size with the Earth, but with densities of between 10^8 and 10^9kg per cubic meter, which cool down over many billions of years, eventually to become dark black dwarfs.

Inside the core of a star, matter is completely ionized, and electrons move around separately from the atomic nuclei. There is a physical principle, known as the Pauli Exclusion Principle, which states that there is a limit to the number of electrons which can be contained within a small volume of space. Accordingly, as a dying star shrinks under its own weight and its electrons are squeezed closer and closer together, the electrons move faster and faster, building up a powerful pressure which, at a high enough density, becomes sufficiently great to prevent further contraction. Material in this state is said to be "degenerate"; the pressure which supports white dwarfs against gravity is known as "electron degeneracy pressure." A cubic centimeter of white dwarf material, transported to Earth, would weigh about one tonne.

The formation of neutron stars

If the mass of a collapsing star or stellar remnant exceeds 1·4 solar masses – the Chandrasekhar limit – electron degeneracy pressure is unable to support the weight of the star, and it collapses into a much denser state. As density builds up towards 10^{17}kg per cubic meter, positively charged protons and negatively charged electrons combine to form electrically neutral neutrons. Atomic nuclei dissolve into neutrons and protons which combine with electrons to add yet more neutrons to the "neutron sea."

Eventually the collapse is halted by the pressure exerted by degenerate neutrons, which behave at these higher densities in a similar way to electrons in white dwarfs. The resultant body is a neutron star, typically about 10km in radius and with a density of between 10^{17} and 10^{18}kg per cubic meter. A cubic centimeter of neutron star material, if brought to the Earth, would weigh between a hundred million and a billion tonnes.

A neutron star has a bizarre structure with an atmosphere squeezed by gravity to a thickness of about 1cm. This probably lies on top of a crystalline layer about 1km thick which, in turn, lies on top of a neutron fluid interior. The more massive neutron stars may also have a small solid central core.

Gravity overwhelms neutron degeneracy pressure if the mass of a collapsing star or stellar remnant exceeds a limit, which, although open to some doubt, almost certainly lies nearer to two than to five solar masses. Beyond this limit, continuing collapse leads to the formation of a black hole (➧ page 177).

White dwarf

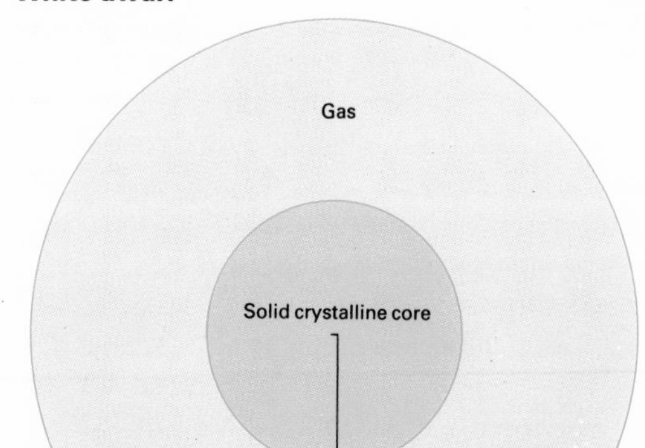

Neutron star

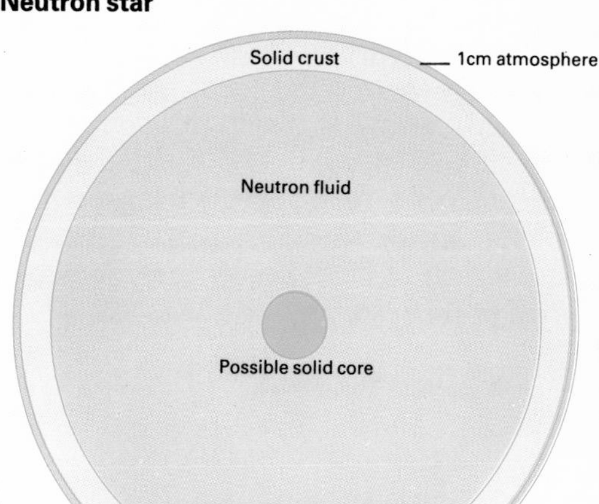

The peculiar properties of white dwarfs

In an ordinary main sequence star the "perfect gas law" applies: pressure depends on density and temperature. If the internal temperature increases, the pressure increases and the star expands. If the interior cools down, the star contracts. The material of a white dwarf is called degenerate because it behaves differently: its pressure depends only on density. For this reason a white dwarf does not shrink as it cools down. Even when it has become a cold black dwarf it will be little different in size from when it first became a white dwarf. In addition, like a neutron star, the more massive it is, the smaller it is.

White dwarfs formed from stars of similar mass to the Sun are likely to be composed mainly of carbon – the end product of the nuclear reactions which took place when they were red giants. Higher-mass stars, which have shed enough mass to allow their cores to become white dwarfs, will have gone through further nuclear reactions and will, therefore, produce white dwarfs made of heavier elements such as silicon or even iron. Whether or not a white dwarf has a solid crystalline core depends on its composition and temperature.

Despite their name, planetary nebulae are neither planets nor nebulae, but old stars which have thrown off their outer atmospheres

Planetary nebulae

Throughout their lives, stars lose matter from their atmospheres in the form of "stellar winds". During the main sequence phase the mass loss is very low, but when a star swells up to become a highly luminous red giant or supergiant, its surface gravity drops and mass loss may become as great as 10^{-5} solar masses per year. Many astronomers believe that at the end of this giant phase, a final burst of nuclear energy blows a shell of gas into space, and that this shell contains up to a few tenths of a solar mass of material. This wispy, expanding cloud is known as a planetary nebula. The nebula absorbs ultraviolet radiation from the central star, which may have a temperature of about 100,000K, and reemits it as visible light: this process is known as fluorescence.

Stellar winds, together with the ejection of a planetary nebula, may carry away 80 percent of a star's mass. This can bring a star of up to 6-8 solar masses down below the "weight limit", so that as the nebula expands, thins out and becomes invisible after about 10,000 years, the central star fades to become a white dwarf, instead of ending its days explosively as a supernova.

Some astronomers have proposed that a planetary nebula is itself produced by stellar winds. The suggestion is that the stellar wind in a giant blows gently at about 10 kilometers per second until the hot stellar core is exposed. Wind speed then increases to about 1,000 kilometers per second and the high-speed wind catches up with and plows into the slower-moving matter which was emitted earlier. This causes a shell of high-density gases to pile up, emitting light by the process of fluorescence.

▲ **This planetary nebula, NGC 6302, is believed to be the result of an unusually violent eruption from the central star. Most of the material has been expelled into two oppositely-directed lobes. This "bipolar" appearance is common among planetary nebulae (although usually less obvious) and may be due to the preferential funneling of matter along the axis of rotation or the magnetic axis of the star. Bipolar outflows occur in a wide variety of astronomical phenomena.**

The "shell ejection" model

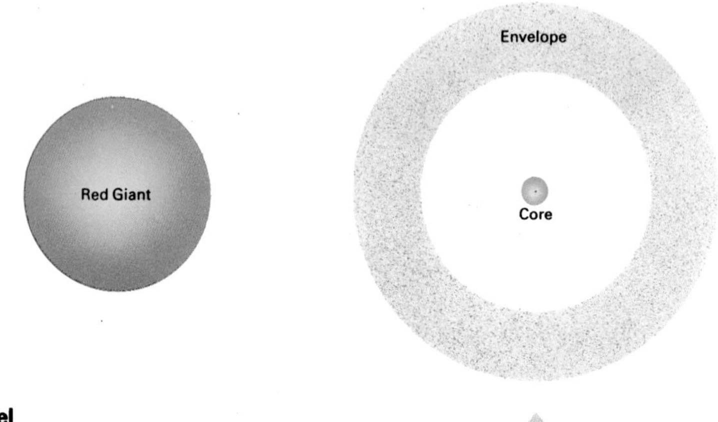

The "snowplow" model

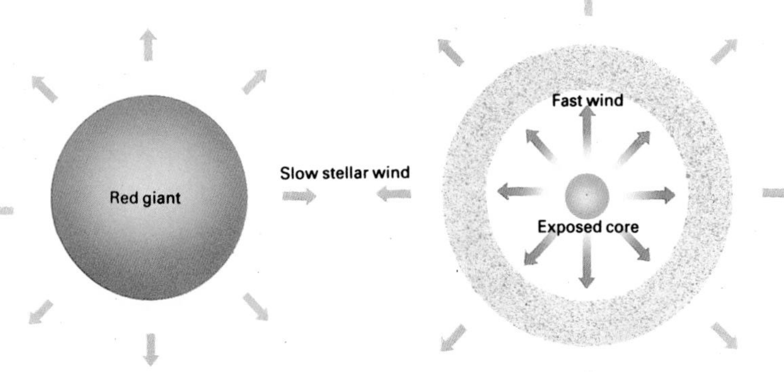

◄ **According to the "shell ejection" model a planetary nebula is an envelope of unburnt hydrogen ejected from an old red giant as the final climax to a series of pulsations. In the "snowplow model" substantial amounts of mass are lost from a red giant in a relatively slow-moving stellar wind. When the star's hotter interior is exposed a faster wind "blows" and plows into the slower wind to produce a compressed luminous shell.**

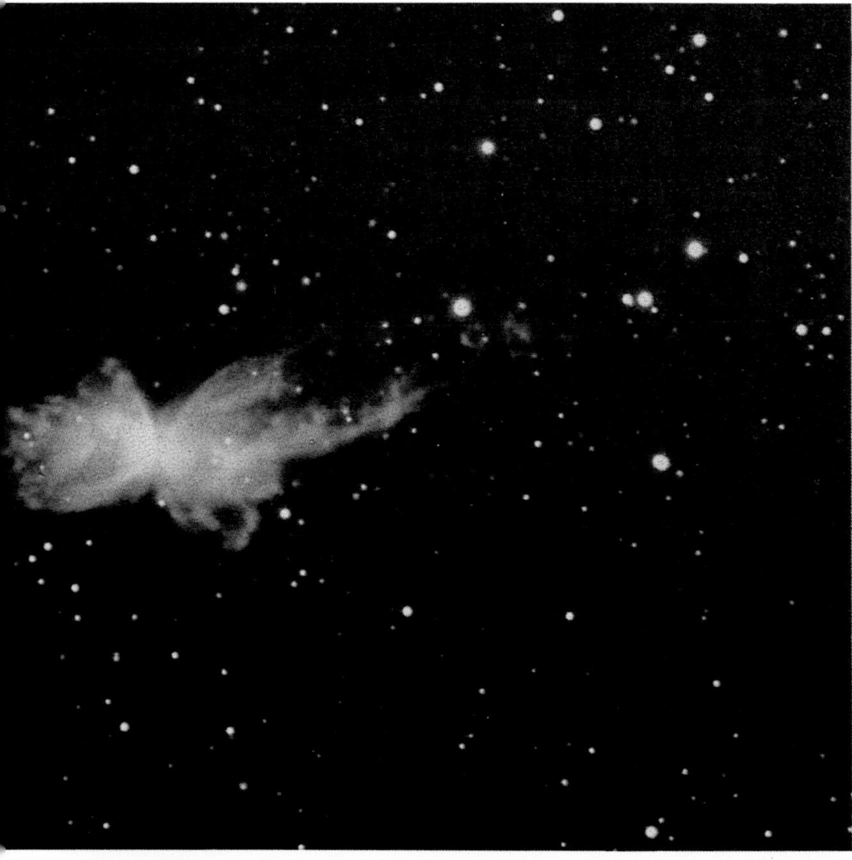

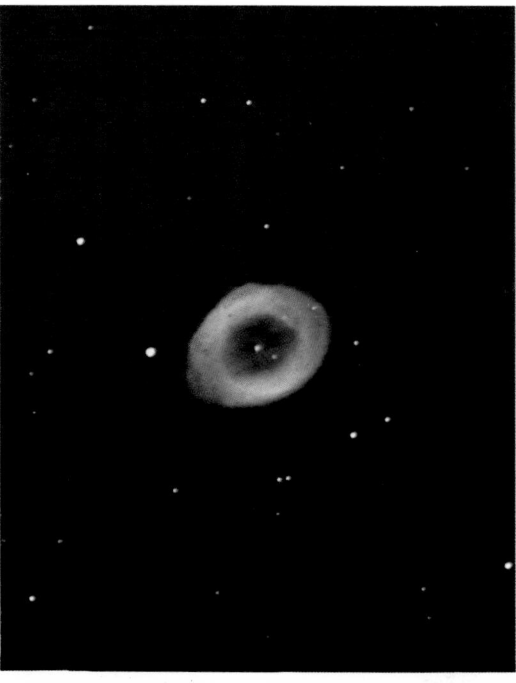

▼ ***The Helix nebula is invisible to the naked eye despite being the nearest planetary nebula to the Earth. This photograph is from the Anglo-Australian telescope at Siding Spring.***

▲ ***The most famous of planetary nebulae is the Ring nebula in the constellation of Lyra. In this picture, the colors represent areas within the same temperature range.***

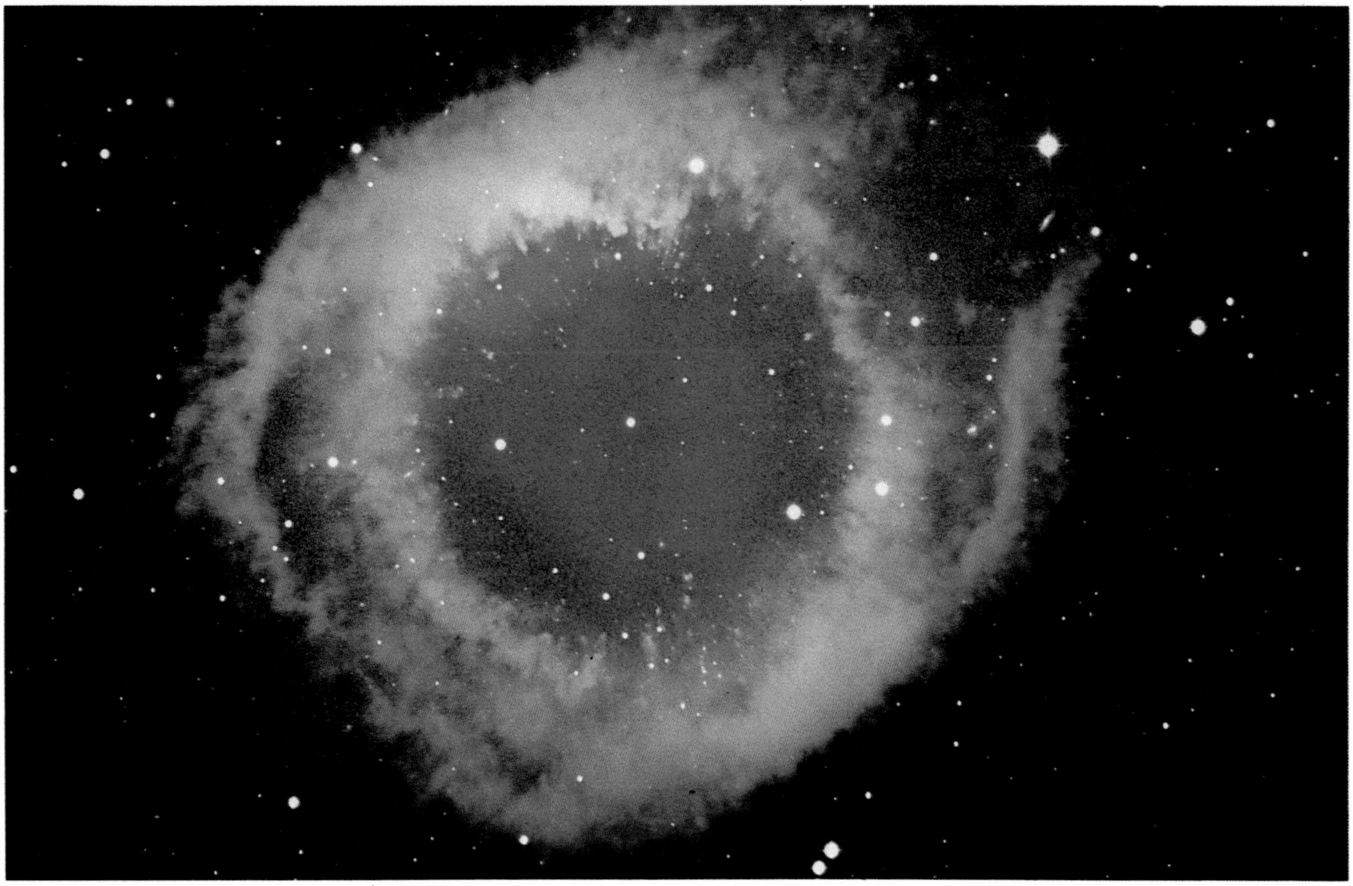

Novae and Supernovae

Stellar explosions

1 Star A is more massive than B
2 A expands: transfers mass
3 A becomes white dwarf
4 B becomes red giant
5 White dwarf accretes from B
6 Fuel builds up on white dwarf
7 *Nova* explosion on surface
8 Shell blows out from white dwarf
9 If enough mass is accreted carbon-burning starts in the white dwarf's core
10 *Type I supernova* explosion completely destroys white dwarf
11 Empty supernova remnant
12 If A is a high-mass star it can become *Type II supernova:*
13 Start of core collapse
14 Whole core collapses
15 Shockwave moves out from core
16 Envelope blown away, core becomes neutron star
17 Binary: neutron star and star B
18 B becomes red giant, material flows to neutron star
19 H and He build up on neutron star
20 Explosive helium fusion causes X-ray burst
21 Accretion of fuel resumes

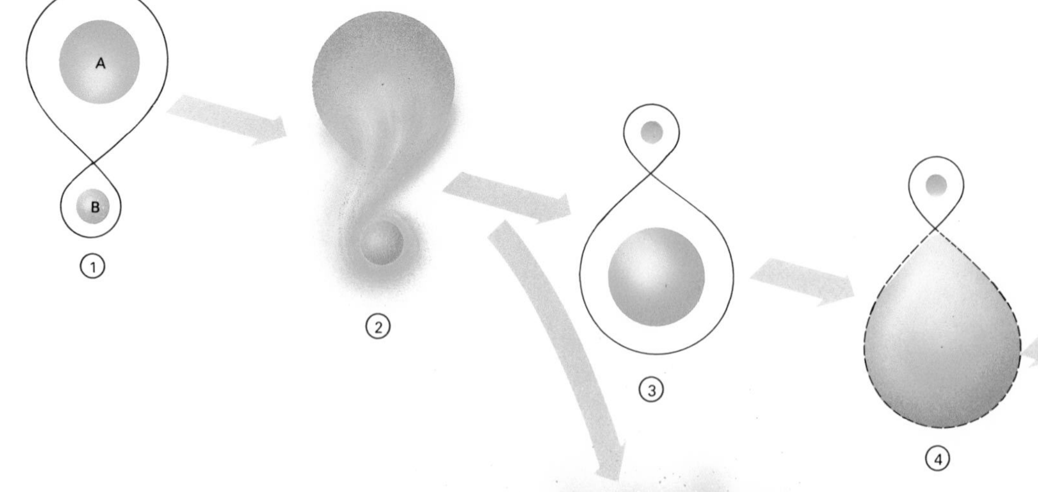

Novae

Astronomers believe that a nova occurs in a close binary system in which one star is a white dwarf and its companion is evolving from the main sequence to the giant phase. As the star expands it becomes distorted, and matter flows from it into a circulating disk, known as an "accretion disk", around the white dwarf. Gas spirals down from the accretion disk onto the surface of the white dwarf, and eventually builds up a temperature of some 20 million K, at which point a run-away nuclear reaction causes an explosive outburst on the surface. The shell of matter blasted away usually fades from view before it is large enough to be resolved in telescopes, but in a few cases, such as Nova Persei, 1901, the shell could be seen shining where it collided with the interstellar gas. The most spectacular nova of recent times was Nova Cygni, 1975, which flared up by about 19 magnitudes.

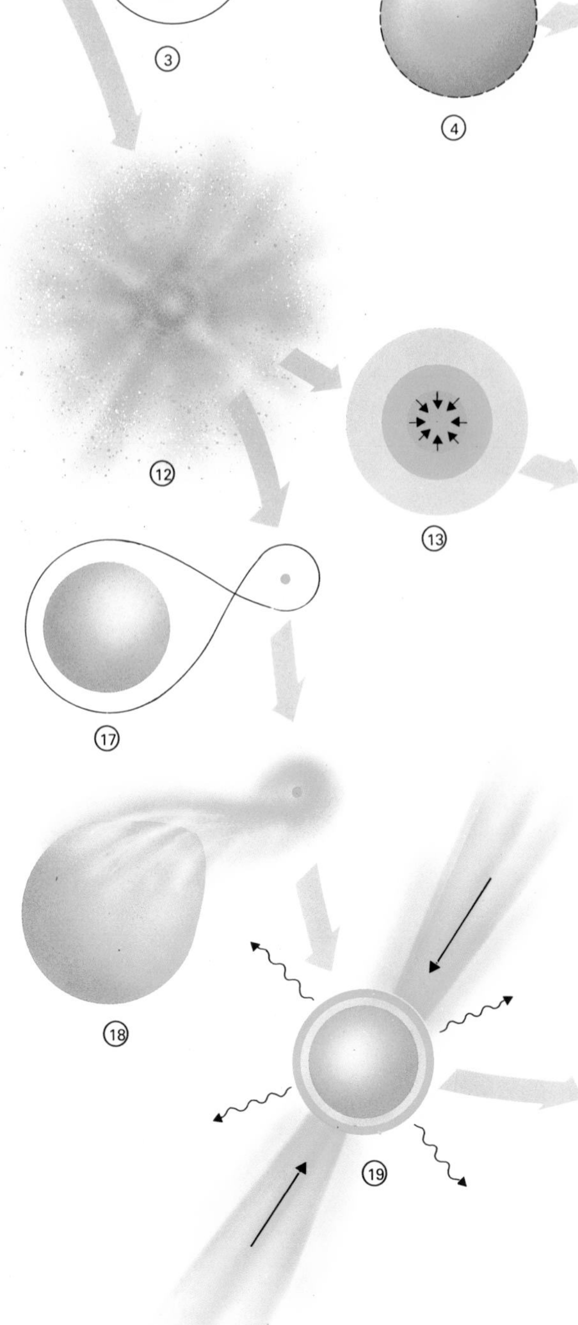

Supernovae

The most spectacular of all stellar catastrophes, supernovae are exploding stars which at their peak can shine as brilliantly as 10 billion Suns. They are rare events: whereas in an average galaxy about 25 novae occur every year, only two or three supernovae per century are expected throughout the observable universe. No supernova has been seen in the Milky Way system for more than three centuries, the most recent being those seen by Kepler in 1604 and Tycho Brahe in 1572.

There are two principal classes of supernova, known as Type I and Type II. Their behavior is different and they involve quite dissimilar types of star and detonation mechanisms. Type I supernovae, which are several times more luminous than Type II events, occur among old stars, while Type II supernovae involve much younger stars.

Much remains to be discovered about the mechanisms which detonate these cosmic conflagrations, but a Type I event probably involves the complete destruction of a white dwarf in a close binary system, while a Type II event probably occurs when a high-mass star runs out of fuel and its core collapses to form a neutron star. A Type II supernova occurs whether or not the star is in a binary system. Both types of event produce a turbulent expanding cloud of debris. The best known supernova remnant is the Crab nebula, the remains of a supernova which was seen to occur in the year 1054, and a source of all kinds of radiation from gamma rays to radio waves.

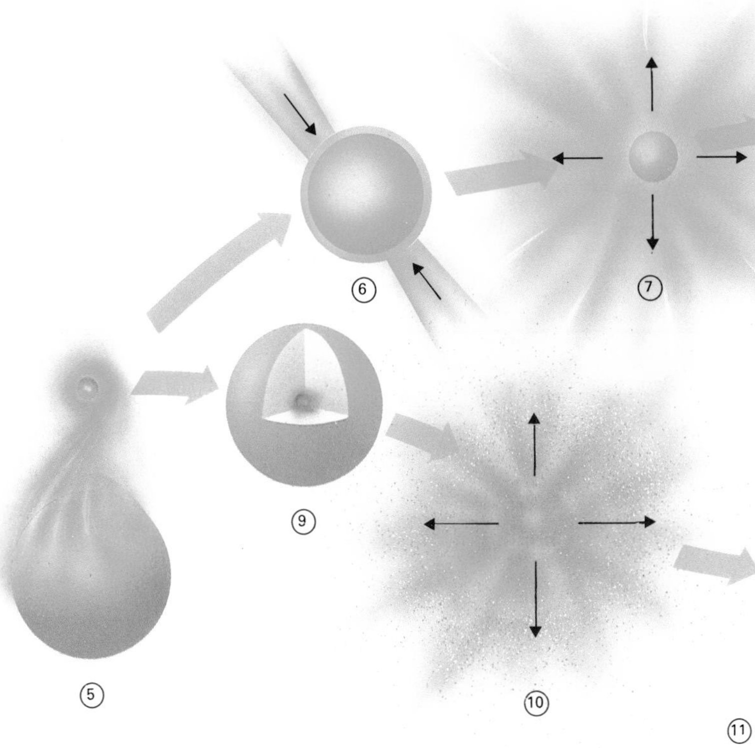

Type I supernovae
Type I supernovae probably involve the detonation of a white dwarf in a close binary system. If sufficient material is dumped on a white dwarf by its companion to bring it very close to the Chandrasekhar limit, the central region will be compressed and heated to the temperature at which a carbon-burning nuclear reaction can begin. Degenerate matter, unlike an ordinary gas, does not expand when heated; instead the internal temperature of the white dwarf surges upwards, the reaction rate increases explosively and the entire star is blown apart. The cloud of debris expands at about 10,000km per second, reaching peak brightness in about two weeks. According to one theory, the explosion converts up to one solar mass of material into radioactive nickel. As this decays it supplies energy to the expanding cloud at a rate which fits in with the observed rise and slow decline of the supernova's luminosity

Type II supernovae
These events probably involve short-lived stars more massive than 6 to 8 solar masses, since less massive stars may well succeed in losing enough material to come below the weight limit for white dwarfs. Such stars burn a succession of nuclear fuels in their cores, but if the core becomes converted into iron, no more energy can be released by fusion reactions (♦ page 155).

The core collapses, heavy nuclei disintegrate, and protons combine with electrons to form neutrons. According to one theory, the inner part of the core "overshoots" its neutron star density and rebounds, sending a powerful shock wave out through the star. If the shock wave reaches the envelope, the outer part of the star is blown away in a supernova explosion and the collapsed core remains as a neutron star; but if the blast wave is dispersed before it gets that far, the rest of the star will fall in on top of the core to form a black hole.

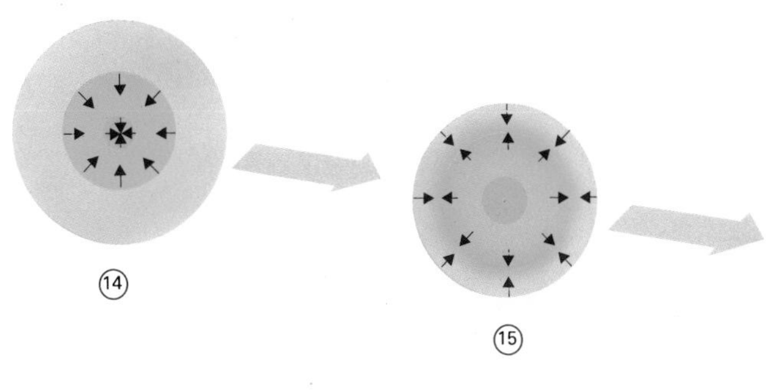

Other products of stellar collapse
If a close binary contains a neutron star or black hole, gas falling from its companion and spiralling inwards is accelerated to very high speeds by gravity, and will attain temperatures of about 100 million K. Matter at these temperatures emits X-rays, and many of the X-ray sources in our Galaxy turn out to be binaries of this kind.

If the neutron star is young and has a strong magnetic field, material may be channeled down to form hot spots near its magnetic poles; and if these do not coincide with the poles of rotation, this will give rise to pulsar-like behavior as the star spins.

If the neutron star is an old one, with a weak magnetic field, material can accumulate on its surface until it detonates in a nova-like event to produce an X-ray burster, which flares up in a few seconds and declines in a few minutes. The event is repeated at irregular intervals of a few hours to a few days. Most bursters are found in regions where old stars abound. Rapid bursters – which produce several thousand bursts per day – are probably due to matter falling into the accretion disk in irregular blobs. Neutron stars also seem to be responsible for at least some known gamma ray sources.

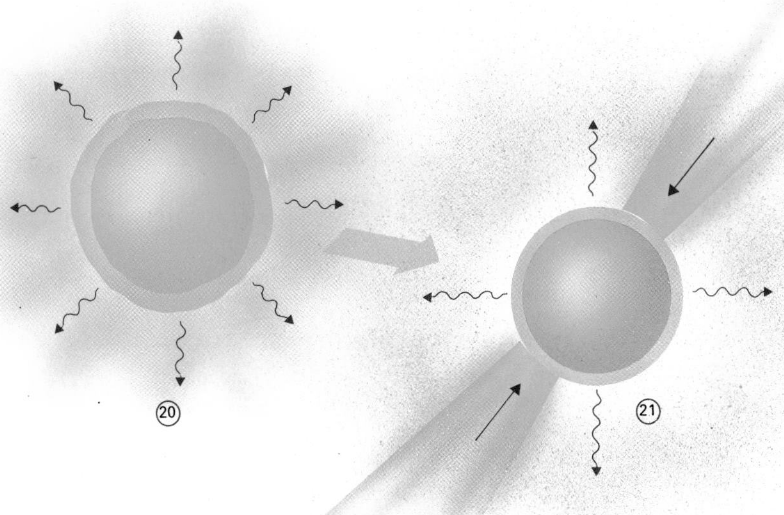

Swirling gas clouds, some of them lit by rapidly rotating neutron stars, mark the sites of the greatest explosions ever witnessed

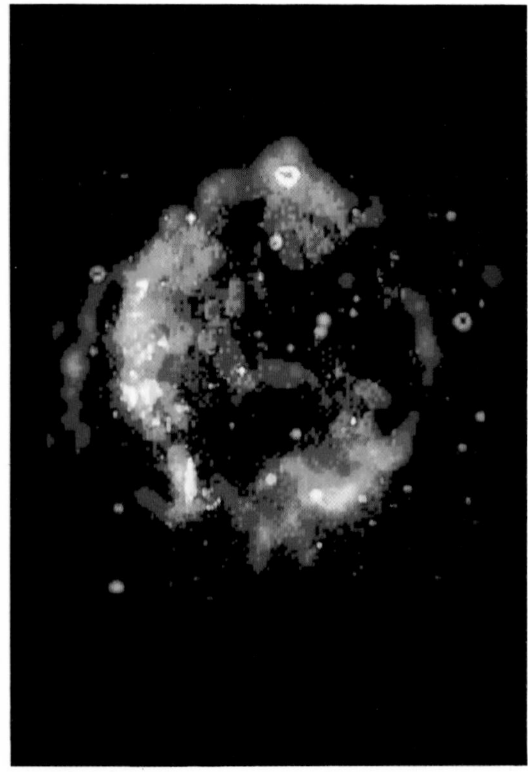

▲ ***This composite picture of a supernova remnant in Cassiopeia has been constructed from radio (blue) and X-ray (green) images superimposed on an optical (red) picture.***

► ***The Crab nebula is all that is left of a supernova explosion in Taurus, recorded by Chinese observers in 1054. The Crab pulsar now illuminates the swirling clouds of gas.***

Pulsars

Pulsars are sources of radio waves, and in some cases also of visible light and X-rays, which flash on and off at intervals ranging from a few seconds to a tiny fraction of a second. Each pulse lasts for only a few percent of the period (the interval between successive pulses) so that from the observer's point of view, the pulse is off for most of the time. Although the strength of successive pulses is variable, the period is constant to within one part in ten million, apart from a slow steady increase of one part in a few thousand per year, and occasional sudden short-lived reductions known as "glitches".

Astronomers believe that a pulsar is a rapidly rotating neutron star which emits radiation in a narrow beam that sweeps round like the beam of a lighthouse as the star rotates. Each time a beam points towards the Earth, a pulse can be detected.

Both the Crab nebula and the Vela supernova remnant contain pulsars, and this fits well with the idea that neutron stars are created in supernova explosions. Although other supernova remnants do not seem to contain pulsars, this may be because the beam does not point in the direction of the Solar System, or because the explosion was asymmetrical and hurled the neutron star out of the expanding cloud. Furthermore, it seems likely that some supernovae (particularly of Type I) will not produce neutron stars at all. Supernova remnants like the Crab nebula shine because they are powered by a flow of energy from the neutron star, whereas older, more distended remnants like the Cygnus loop shine where they collide with the interstellar gases.

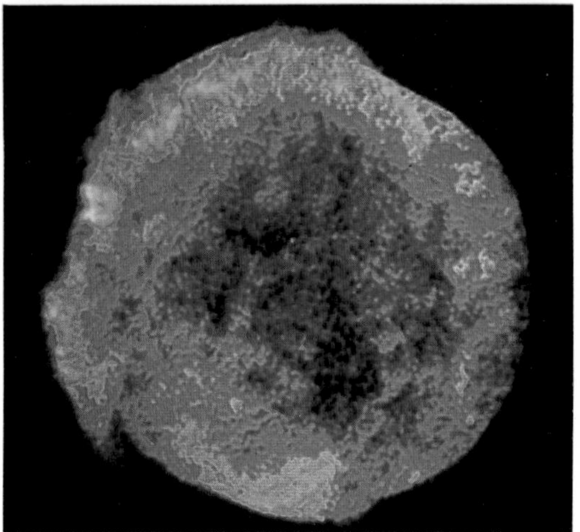

▲ ***The remnants of Tycho's star, which exploded in November 1572, differ from the Crab nebula in that they form an expanding hollow sphere of gases with no neutron star or pulsar at the center illuminating the clouds. Tycho's star (or Tycho's supernova) has long since faded from view, but is still bright at radio and X-ray wavelengths.***

► ***The Vela supernova remnant, only a part of which is shown here, has been expanding into space since the explosion of a star about 10,000 years ago. Like the Crab nebula it contains a pulsar. The visible light comes from the interstellar gas as it is heated by the passage of the shock wave from the original explosion.***

See also
The Basic Properties of Stars 145-52
Birth, Life and Death of Stars 153-60
Black Holes 177-80
Between the Stars 181-8
The Evolving Universe 229-44

The "lighthouse" model for explaining pulsars fits the facts rather well. All stars rotate, and if a rotating object contracts, it spins faster. A neutron star has collapsed so far that it spins very rapidly indeed. Furthermore, collapse will compress the magnetic field at the surface of the star to a strength of some 100 million tesla – about a million million times stronger than the Earth's magnetic field. Electrons moving at high speed in this powerful field emit radiation. Scientists do not know exactly what produces pulsar beams, but one possibility is that they are produced in "hot spots" above the magnetic poles of the neutron star. Another possibility is that particles caught up in the magnetic field emit the radiation as the field sweeps them round above the star's equator at nearly the speed of light.

The interaction between a neutron star's magnetic field and the surrounding matter acts as a brake to slow down its rotation, and this accounts for the slow increase in period. It is possible to estimate the age of a neutron star by measuring its period and the rate at which this is increasing – in general, the shorter the period, the younger the pulsar. Although most pulsars seem to be no more than a few million years old, the surrounding remnants fade much faster than this, and consequently most pulsars occur in isolation. The Crab pulsar, with a period of 0·033 seconds, is the fastest conventional pulsar.

The fastest pulsars of all

Two "millisecond" pulsars were discovered in 1982 and 1983, one with a period of 1·557 milliseconds (corresponding to 642 pulses per second) and the other with a period of 6·1 milliseconds (170 pulses per second). Despite their very short periods they both seem to be old pulsars – at least 100,000 years old – and show no sign of supernova remnants. Astronomers were at first puzzled by these objects, but it seems likely that both were formed in binary systems and, because of their orbital motions, material flowing from the companions onto the neutron stars has added rotational motion, increasing their rates of spin and reducing their periods. Observational confirmation that one of them has a binary companion supports this view.

A puzzling celestial structure

The object known to radio astronomers as W50 is an elongated supernova remnant measuring some 200 by 600 light years. At its center lies a starlike object, named SS433, which is a source of visible light, radio waves and X-rays. It is a binary in which an ordinary star is losing matter to a collapsed companion. Its spectrum shows staggering red and blue shifts which astronomers believe are due to two jets of matter being expelled at speeds of about 80,000km per second. The jets seem to be perpendicular to the accretion disk around the collapsed body, and may account for the elongated appearance of the radio-emitting cloud. Far larger jets and beams occur in active galaxies and quasars (▸ page 205), and SS433 may be a small-scale version of the phenomenon. Study of this relatively nearby object may provide clues to events occurring in some violently disturbed galaxies.

▼ Radiation from a pulsar may beam out along the magnetic axis of a rapidly rotating neutron star, or it may originate above its equator from particles swept by its magnetic field.

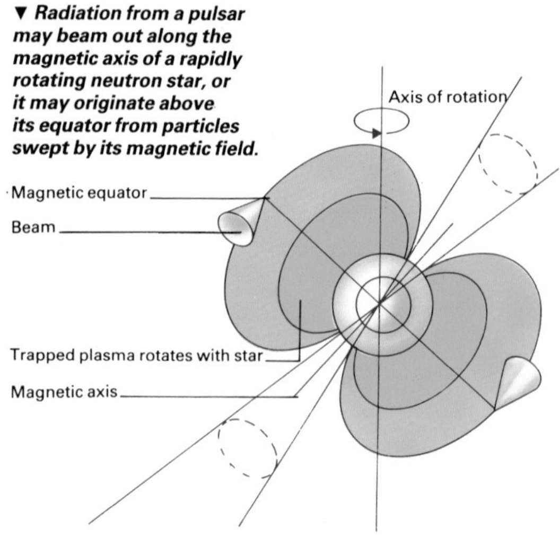

The discovery of pulsars

The first pulsar was discovered quite by chance when Miss Jocelyn Bell, a member of Professor A. Hewish's team at Cambridge, announced that she had found a faint source "flashing" rapidly and regularly. The team had set up the radio equipment for a completely different experiment, and their initial reaction was one of disbelief. Thinking (briefly) that the signals might be from a distant civilization, they delayed announcing the discovery of the pulsar, designated CP 1919, until 29 February 1968.

The discovery of a pulsar in the Crab nebula followed soon after in January 1969. Astronomers at the Steward Observatory in Arizona decided to make a visual search for it. They had a week's telescope time, but failed to find it – in fact, their equipment had been wrongly connected. By chance the observer due to use the telescope next was delayed, and the pulsar-hunters gained an extra night. This time they found the Crab pulsar in exactly the right position and flashing at the right rate. It later emerged that Rudolph Minkowski (1895-1976) had recorded it in 1942, but had not realized its significance.

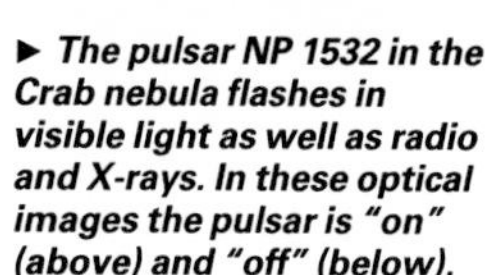

▲ Jocelyn Bell-Burnell discovered the first pulsar in November 1967. It emitted a series of pulses of 0·05 seconds' duration every 1·33730 seconds.

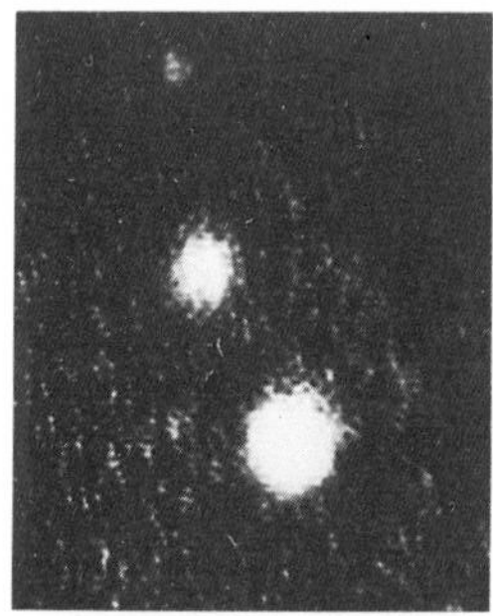

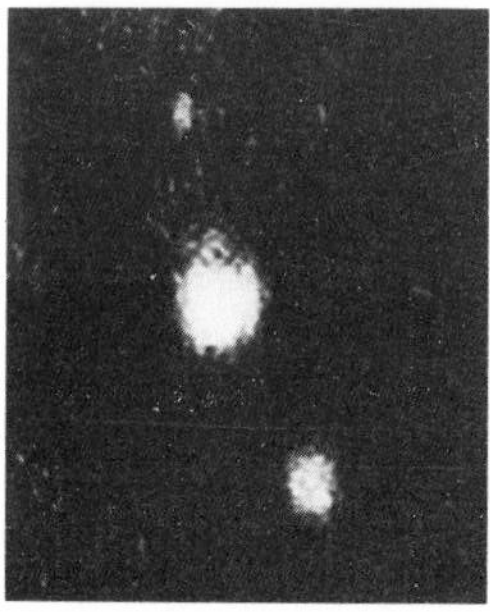

► The pulsar NP 1532 in the Crab nebula flashes in visible light as well as radio and X-rays. In these optical images the pulsar is "on" (above) and "off" (below).

Black Holes

Describing a black hole...Black holes as cosmic energy generators...Black holes and X-ray emissions – the evidence so far...Supermassive or mini black holes?... Can black holes evaporate?...PERSPECTIVE...The "effect" of a black hole on a passing astronaut... Black holes and fringe ideas

A black hole is a region of space into which matter has fallen and from which nothing, not even light itself, can escape. If a lump of matter is compressed, the force of gravity and the value of escape velocity at its surface increase. If it is squeezed within a radius known as the Schwarzschild radius, escape velocity exceeds the speed of light and gravity prevents light from escaping into space. Thereafter gravity overwhelms all other forces and the body collapses to a point of infinite density known as a "singularity". A black hole is the region of space around the singularity, with a radius equal to the Schwarzschild radius, within which gravity is so powerful that nothing can move outwards. Matter or radiation can fall in and be sucked down to the central singularity, but nothing can get out. The term "black hole" is entirely appropriate, as it is a "hole" in the sense that matter can fall in, and "black" in the sense that no light, radiation or matter can get out. The boundary of a black hole is called the "event horizon", because no means exists by which information about events occurring within it could be communicated to the universe outside.

The Schwarzschild radius R_s (in kilometers) for a mass M can be calculated from the simple formula $R_s = 3M/M_\odot$, where $M_\odot$ denotes the mass of the Sun. The Schwarzschild radius for the Sun is about 3km, while that of the Earth is about 1cm, and there is no natural process in the present-day universe which can compress either of these bodies sufficiently to form a black hole. However, one way in which a black hole can be formed is by the collapse of a massive star which has run out of nuclear fuel, and which is too massive to become either a white dwarf or a neutron star. Such a star will collapse without limit, for there is no known force capable of halting the process, but before collapsing to infinite density it will pass inside its Schwarzschild radius and disappear from view. A black hole formed from the collapse of a 10 solar-mass star would have a radius of some 30km.

A black hole

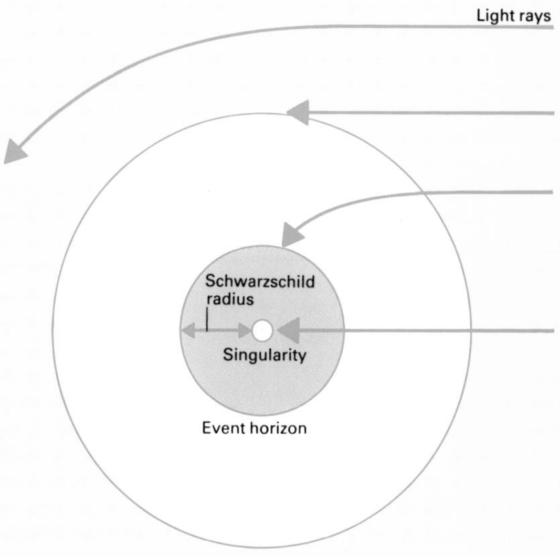

▲ ***The boundary, or event horizon, of a non-rotating black hole has a radius equal to the Schwarzschild radius (R_s). Not even light can escape from within this region. A passing ray of light will be deflected by the gravitational field of the black hole. If it approaches to 1·5 R_s it may be bent into a circular orbit (orbiting light rays form the "photon sphere") while if it comes still closer it will fall into the event horizon.***

Object	Mass	Schwarzschild Radius (R_s)	Density when object reaches R_s (kg/m²)
Small mountain	10^{12}kg	$1{\cdot}5\times10^{-15}$m	10^{56}
The Earth	6×10^{24}kg	9mm	10^{30}
The Sun	2×10^{30}kg ($=1M_\odot$)	3km	10^{19}
100 million Suns	$10^8M_\odot$	3×10^8km (=2 astronomical units)	10^{-3} (= water density)
Entire Galaxy	$10^{11}M_\odot$	0·03 light years	10^{-3}

NB density figures are rounded to the nearest power of 10

The making of a black hole

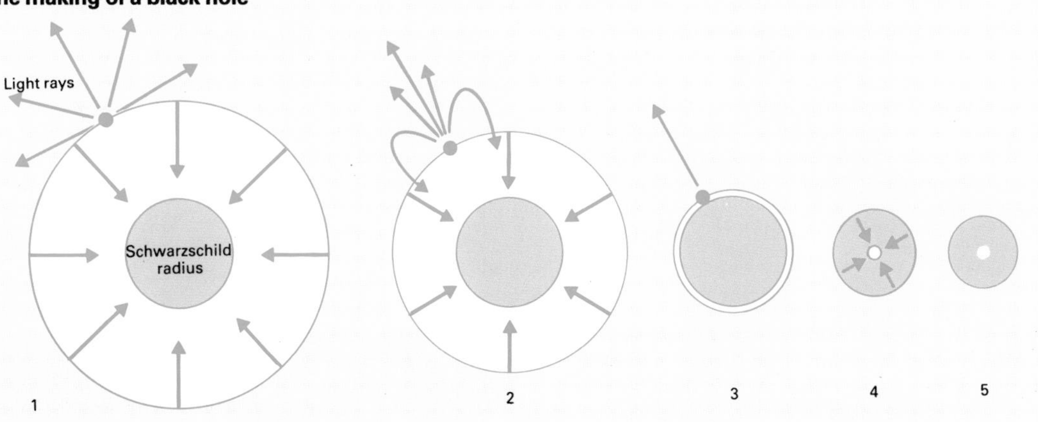

◀ ***As a star begins to collapse (1) light can escape from its surface at any angle, but as the collapse accelerates (2) the surface gravity increases rapidly and rays of light emitted at shallow angles are bent back. Just before it reaches its Schwarzschild radius (3) only vertical rays can escape. As soon as the star collapses inside that radius gravity prevents any light from escaping and the star immediately vanishes from view (4). An event horizon is formed and the entire mass of the star then falls into a singularity (5).***

Black holes as cosmic energy generators

Since stars rotate, it is reasonable to expect black holes formed from collapsing stars to be rotating rapidly. Outside the event horizon of a spinning black hole there is a region, called the "ergosphere", within which nothing can avoid being dragged round in the direction of the black hole's rotation – it is as if space itself were spinning round with the black hole. In principle it is possible for particles to enter the ergosphere and, if they are moving fast enough, escape again with more energy than they had originally. It has been speculated that the ergosphere of a black hole could be used as an energy source.

Black holes and X-ray emissions

Matter falling in towards the event horizon of a black hole is accelerated until it approaches the speed of light, releasing copious quantities of energy before vanishing across the event horizon. If a black hole is a member of a close binary system, matter dragged in from the companion star falls into a circulating accretion disk. Friction in the disk causes matter to spiral in towards the event horizon, and the high temperatures generated by the infalling matter result in the emission of X-rays from the inner part of the disk.

Most astronomers agree that some cosmic X-ray sources are of this type. The best evidence for this view comes from Cygnus X-1, a powerful, rapidly fluctuating X-ray source in the constellation of Cygnus. Cygnus X-1 coincides in position with a hot blue supergiant of 20-30 solar masses, which has an invisible companion of some 9-11 solar masses – far in excess of the maximum possible mass for a neutron star. The X-ray emission seems to come from an accretion disk of material swirling round the invisible companion drawn from the visible star. Although there is no conclusive proof, the evidence is strong that Cygnus X-1, at a distance of some 8,000 light years, contains a black hole.

Similar cases include the binary X-ray source Circinus X-1 and the highly variable LMC X-3, in the Large Magellanic Cloud. LMC X-3 seems to be a member of a binary system comprising a B-type main sequence star and a dark object of between 6 and 14 solar masses.

A rotating black hole

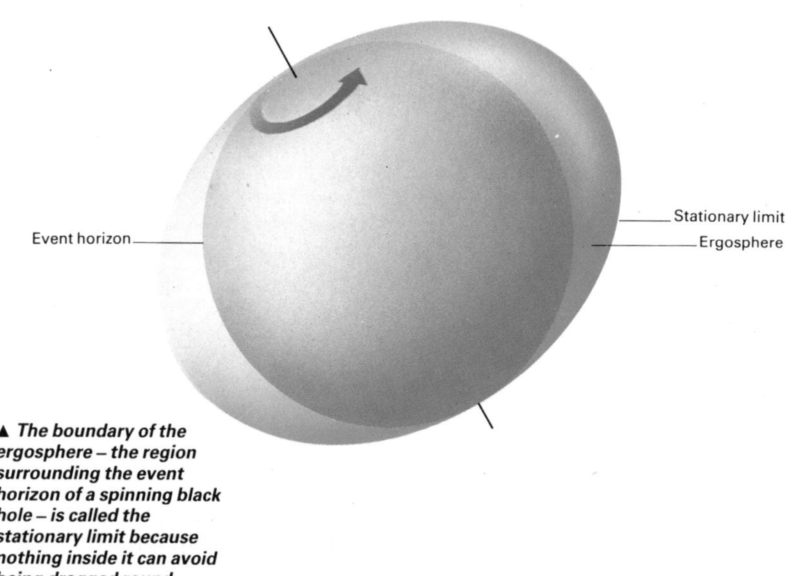

▲ The boundary of the ergosphere – the region surrounding the event horizon of a spinning black hole – is called the stationary limit because nothing inside it can avoid being dragged round.

An astronaut stretched on the cosmic rack

Although the star itself has vanished from view, the resultant black hole still exerts a powerful gravitational influence on its surroundings, and one way in which this would make itself felt is in strong tidal effects on neighboring matter.

For example, if an astronaut were falling feet first towards the hole, his feet, being closer to the black hole than his head, would be subject to stronger gravitational force, and this difference in attraction would stretch him apart on a cosmic "rack" of ever-increasing severity. Close to the event horizon of a 10 solar-mass black hole the tidal force would be equivalent to that which a person would experience swinging from a bridge with the entire population of London or New York dangling from his ankles. Any material body approaching the event horizon of such a black hole would be torn to shreds by "tidal forces".

In the illustration (below) the arrows indicate schematically the gravitational forces acting on the astronaut's head and feet. The larger the black hole the weaker are the tidal forces near its event horizon. Strong tidal forces would not occur until well inside a supermassive black hole.

The cosmic space rack

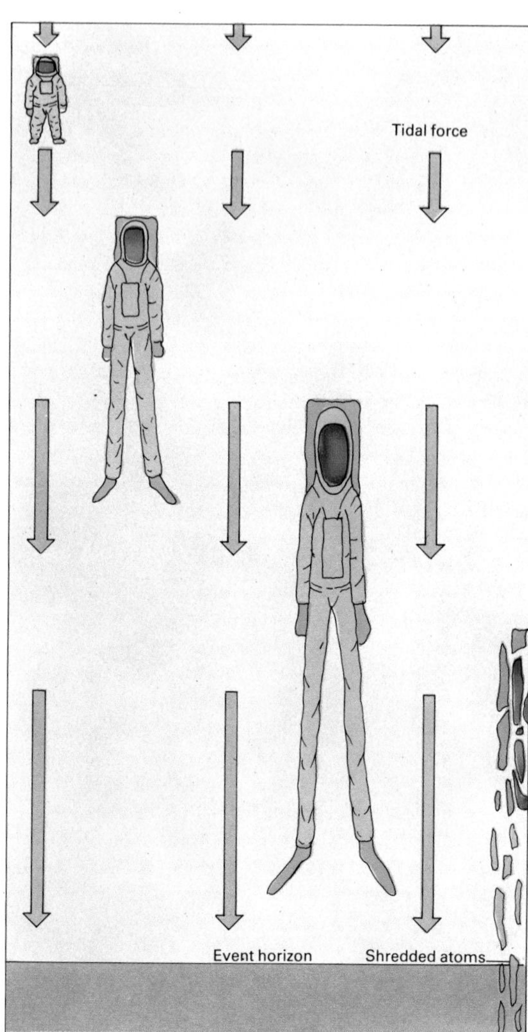

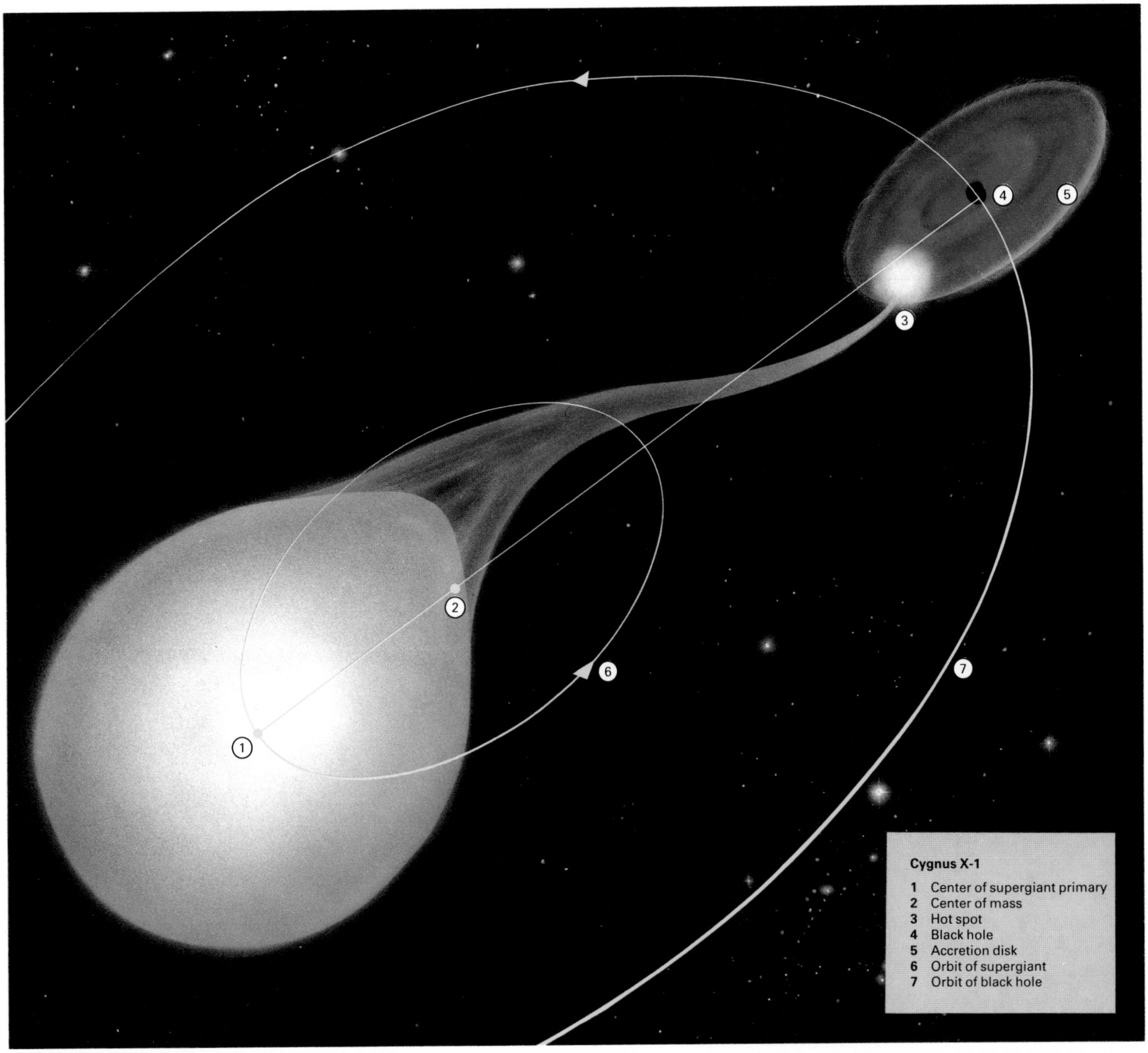

Cygnus X-1

1 Center of supergiant primary
2 Center of mass
3 Hot spot
4 Black hole
5 Accretion disk
6 Orbit of supergiant
7 Orbit of black hole

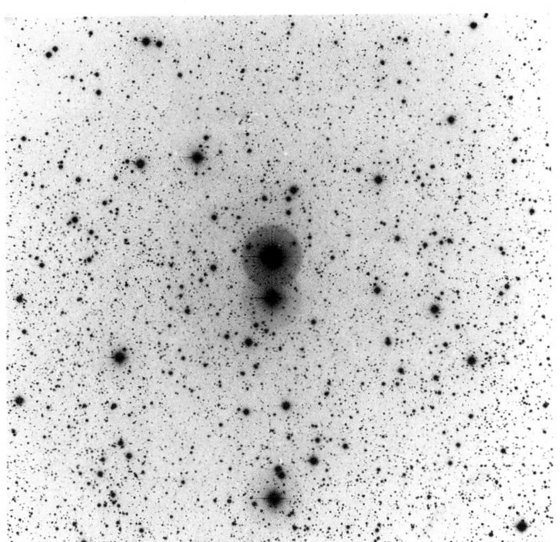

▲ Cygnus X-1 is thought to be a binary – of 5·6 day period – comprising an O or B supergiant and an invisible companion of 9-15 solar masses – believed to be a black hole. Because of the rotation of the binary, matter torn from the star does not fall directly into the hole but instead joins a fast-spinning disk of matter around the hole. The temperature of this "accretion disk" is so high that X-rays are emitted. Matter on the inside of the disk eventually spirals in through the event horizon.

◄ ► Cygnus X-1 – an optical view (left) and an X-ray view (right).

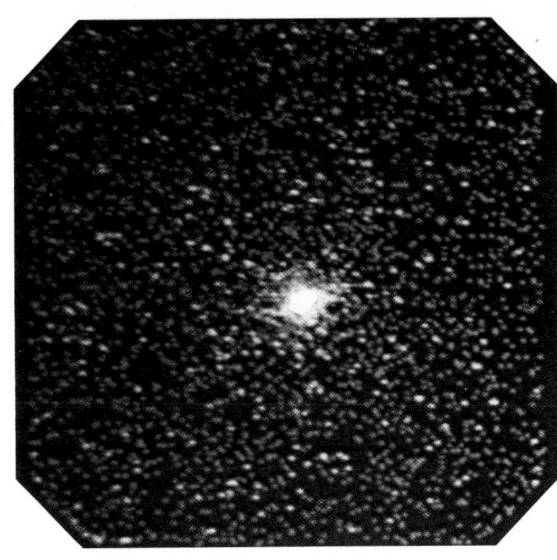

See also
Birth Life and Death of Stars 153-60
Collapsing and Exploding Stars 169-76
Between the Stars 181-8
Space, Time and Gravitation 225-8
The Evolving Universe 229-44

At the frontiers of knowledge

In principle a black hole could be made from any mass of material provided it can be compressed within its Schwarzschild radius. It is perfectly possible for black holes of millions or even billions of solar masses to exist, and many scientists believe that objects of this kind exist in the cores of violently active galaxies and quasars, and even in the core of our own Galaxy. It has also been suggested that if the universe began in a hot dense state, mini or "primordial" black holes with masses of only a few billion tonnes might have been formed in the extreme conditions which existed in those early instants.

An intriguing suggestion, made by Professor Stephen Hawking, is that black holes may not be as black as they were thought to be, but that particles may be able to leak out of them at a very slow rate. For a stellar-mass black hole the leakage would be utterly negligible, but, since the suggestion is that the leakage rate is inversely proportional to mass, a mini black hole would radiate particles at a significant rate. The more mass it lost, the greater the evaporation rate would become, until, finally, the mini black hole would explode in a blast of gamma rays and exotic nuclear particles. Nobody has ever observed an event of this kind, and mini black holes remain strictly a theoretical concept; but if such an event is ever seen to occur, it will be a most dramatic demonstration of the link between gravity, the universe and the exotic world of particle physics.

Black holes probably exist in a wide variety of forms throughout the universe; the evidence in favor of both stellar-mass and supermassive black holes, although not indisputable, is quite convincing.

◄ Professor Stephen Hawking of the University of Cambridge has made fundamental advances in our understanding of gravity and black holes. In 1971 he proposed the possible existence of mini-black holes and in 1974 suggested that particles might be able to leak out of black holes.

▼ When a particle and antiparticle pair is formed in the gravitational field of a black hole one member of the pair may fall in while the other escapes and appears to an outside observer as a particle emitted by the hole.

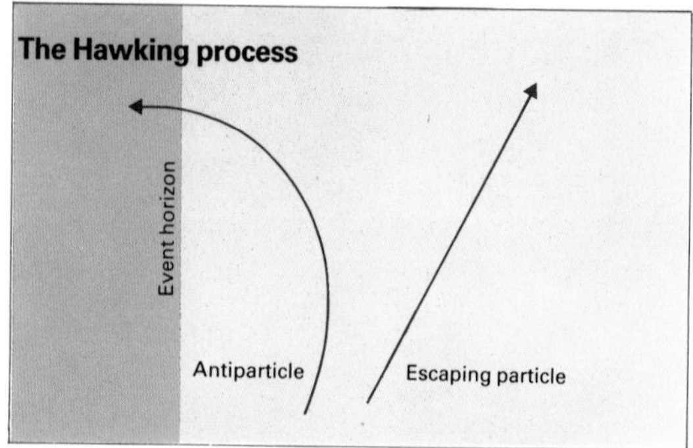

► The graph illustrates Hawking's suggestion that the rate at which particles can leak out of a black hole increases as its mass decreases.

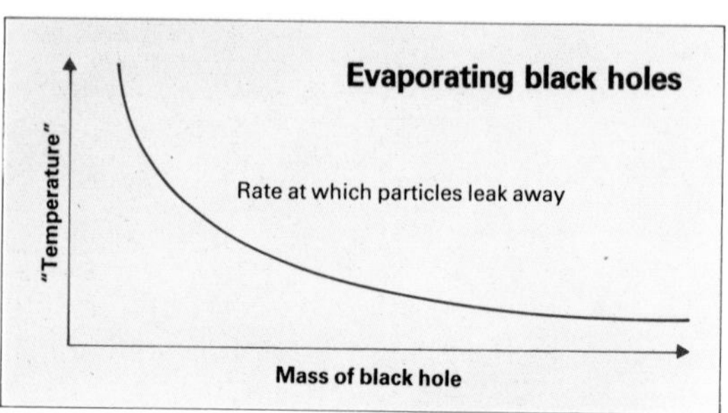

An exploding mini-black hole

▲ A mini-black hole would leak particles at a rate which rapidly escalated in the final stages until it exploded in a blast of particles and gamma rays.

Can particles leak out of black holes?

The "uncertainty principle", proposed in 1927 by Werner Heisenberg, implies that on a subatomic scale we cannot simultaneously measure both energy and time precisely. The more accurately one is known the more uncertain the other becomes. In 1974 Stephen Hawking applied this concept to black holes. Over a fleetingly short instant of time the uncertainty principle allows enough energy to appear in a tiny volume of space to create a particle and its antiparticle. These almost immediately collide and annihilate so that their brief existence cannot be observed and there is no net creation of particles in the universe. However, this process is enhanced in the powerful gravitational field of a black hole, and the strong tidal forces can sometimes cause one member of a pair to fall into the black hole and the other to escape, as if, to a distant observer, it had emerged out of the event horizon. This process takes energy from the gravitational field of the hole and so reduces the black hole's mass to a point where, given sufficient time, it could evaporate completely.

Between the Stars

Interstellar dust – the constituents...Luminous clouds – the types...How to establish their nature, extent and distance...PERSPECTIVE...Identifying celestial objects by numbers...The most spectacular examples yet seen... Molecules in interstellar space

◀ ***Regarded in his day as the foremost astronomical observer in France, Charles Messier was called the "comet ferret" by Louis XV. Today he is remembered for his catalog of nebulae.***

The popular belief is that the space between the stars is a vacuum, and by terrestrial standards it is. It does, however, contain a mixture of gases and small particles, known as interstellar dust or grains. Its average density is about 10^{-21}kg per cubic meter – equivalent to one hundred million million millionth of the density of air at sea-level. Hydrogen is the main constituent, and for every million hydrogen atoms there are about 120,000 helium atoms, a few hundred each of nitrogen, carbon and oxygen, about 100 each of neon and sulfur, and a few heavier atoms such as iron, calcium, sodium and potassium.

Interstellar matter is clumpy, and contains localized clouds where the density is thousands of times higher than average. These clouds are where new stars are born.

Luminous nebulae (***nebula*** is the Latin word for "cloud") such as the Great nebula in Orion (M42) provide evidence for the existence of gases in space. These clouds shine because they contain young stars which emit strongly at ultraviolet wavelengths. This energetic radiation knocks electrons out of atoms to produce positively charged ions (atoms with one or more electrons missing) and free electrons which roam about among the sparsely scattered atoms. When an electron is recaptured by an atom it enters an outer orbit round the nucleus, then drops down to successively lower orbits. As it does so it emits light with a wavelength which depends on the difference in energy between the two orbits. In this way a luminous nebula emits light at a number of wavelengths characteristic of the chemical elements it contains. In other words, it emits an emission line spectrum.

Cataloging the night sky

The first astronomer to compile a catalog of star clusters and nebulae was Charles Messier (1730-1817), a Frenchman from Lorraine. Messier was mainly interested in comets, of which he discovered 13 between 1760 and 1798. However, he was frequently misled by small cloudy-looking objects which were not comets at all, but clusters and nebulae – and he was not in the least interested in these. On the contrary, they wasted an immense amount of his time, and eventually he decided to make a list of them as "objects to avoid". Thus in 1781 he completed a list of 103 objects, including some of the most famous, and astronomers still use Messier's numbers to identify these objects today. The Crab nebula is Messier (M) 1, the Andromeda spiral is M31, the Orion nebula M42, and the Pleiades M45.

Four Messier objects – M46, M47, M91 and M102 – are "missing", and it is possible that they actually were comets, while M73, in Aquarius, is merely an arrangement of four faint stars. His catalog included only the brighter clusters and nebulae. William Herschel discovered and listed many more. Much later, in 1888, the Danish astronomer J.L.E. Dreyer (1852-1926) published his "New General Catalogue" (NGC) of clusters and nebulae. He included all of Messier's objects and most of Herschel's, so that, for instance, M31 is NGC 224, and M1, the Crab nebula, is NGC 1952.

Why nebulae shine

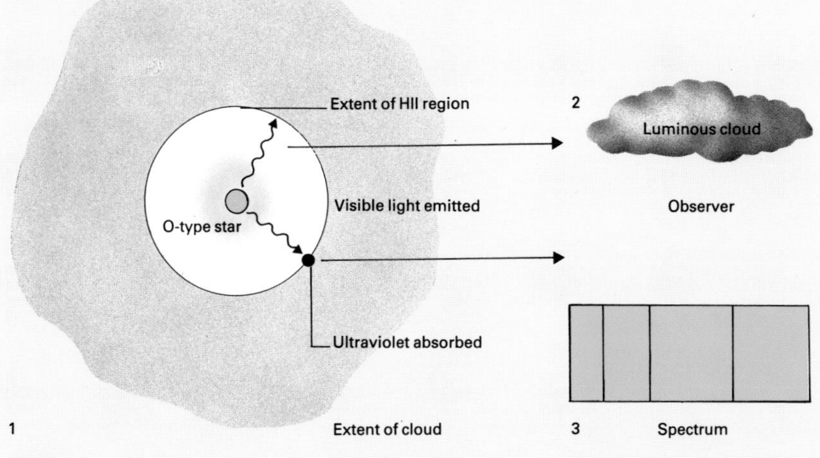

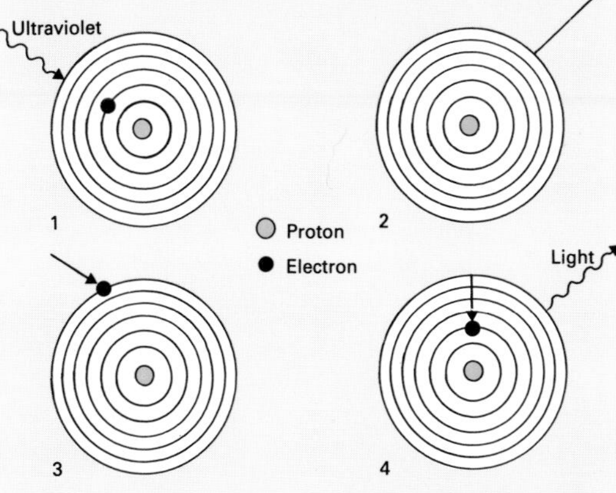

◀ ***If a gas cloud (1) contains a hot O or BO star whose UV light is absorbed by the gas and reemitted as visible light a luminous nebula (2) is seen which has an emission line spectrum (3).***

▲ ***If an atom in the cloud absorbs UV light (1) it may become ionized (2). An electron recaptured by an atom (3) drops to a lower energy level (4) and emits light at a particular wavelength.***

Huge, shapeless clouds of gas and dust, now drifting in space, may be the next generation of stars

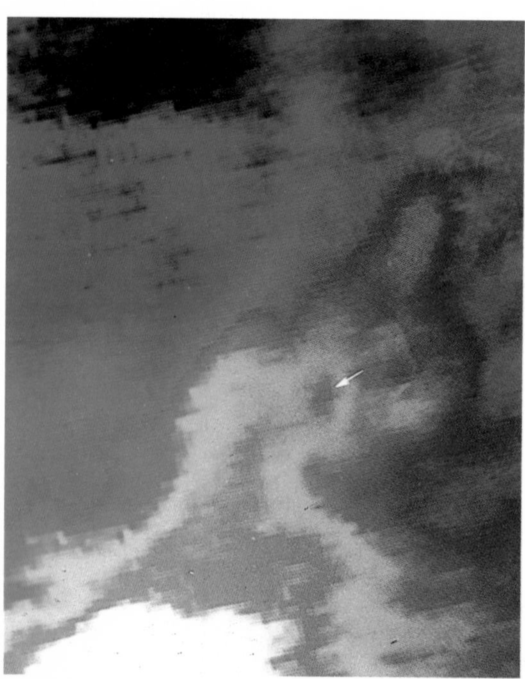

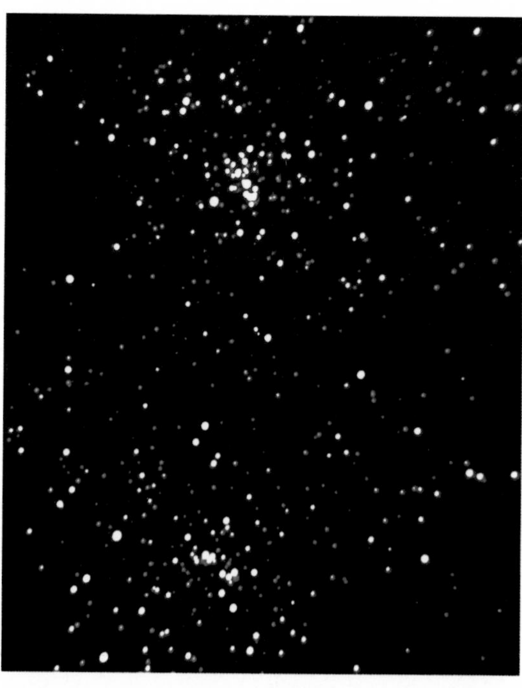

◄ **At 7,000 light years, the double star cluster h and χ Persei, also known as the Sword Handle, is visible to the naked eye. Each cluster contains about 350 stars.**

◄◄ **This false-color infrared image shows a patch of warm dust which is probably a protostar of about 1 solar mass within the dust cloud, Barnard 5.**

► **The Rho Ophiuchi dark clouds lie some 700 light years away. The mixture of gas and dust includes dark clouds, reflection nebulae and emission nebulae.**

▼ **John Herschel described this southern sky open cluster as "a casket of variously colored precious stones", hence its name Jewel-Box. The bright red star is κ Crucis, a red giant. The whole cluster lies at a distance of 7,800 light years from the Solar System.**

Even invisible clouds have now yielded up secrets about the objects which lie beyond them

Discovering interstellar lines

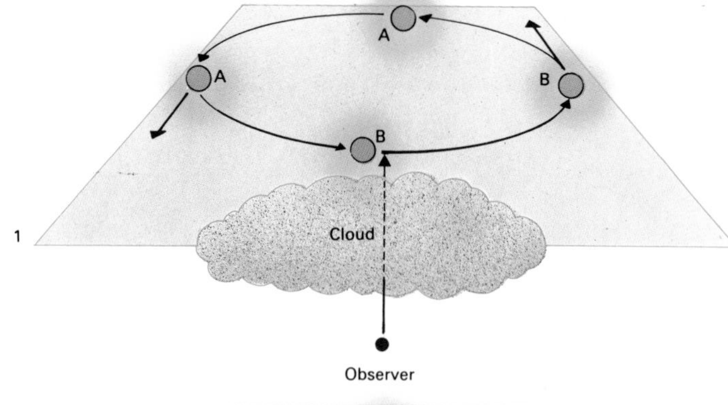

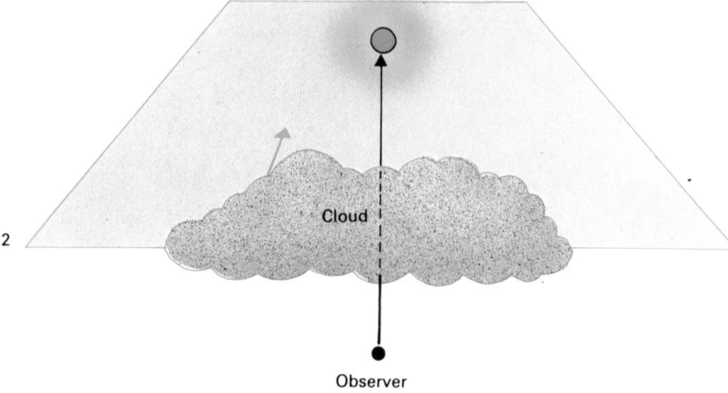

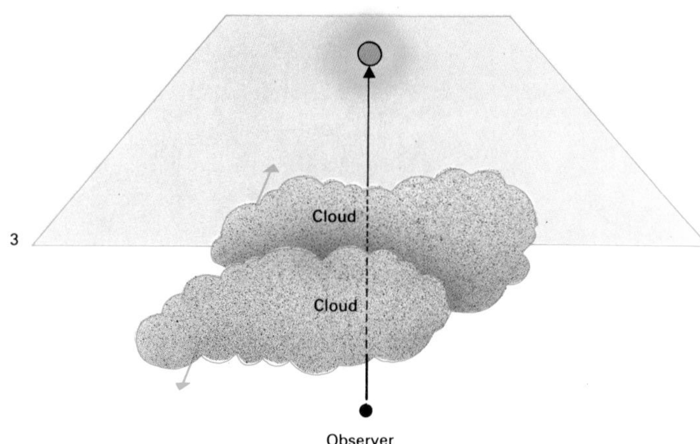

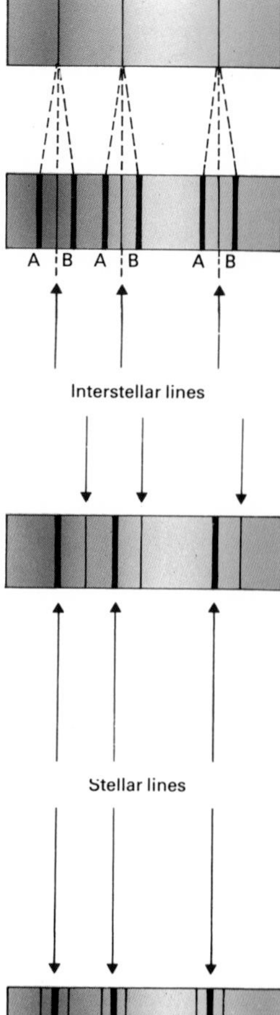

◄ Identifying interstellar lines. A stationary cloud lies between the observer and a spectroscopic binary (1). The two stars revolve round each other, each alternately approaching and receding from the observer and the resultant spectral lines shift to and fro about their mean wavelengths because of the Doppler effect. Those lines superimposed on the spectrum by the absorption of light in the cloud remain at a fixed wavelength and thus can be distinguished from the stellar lines. Interstellar lines were first identified by this means in the spectrum of δ Orionis in 1904. If the cloud is receding from the observer, and approaching the star (2), the interstellar lines produced within the cloud will be redshifted to wavelengths longer than those produced in the atmosphere of the star. If two clouds lie between the observer and the star, one approaching and the other receding (3), two sets of interstellar lines – one blue-shifted and the other red-shifted – will be superimposed on the stellar spectrum.

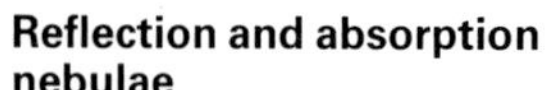

Reflection and absorption nebulae

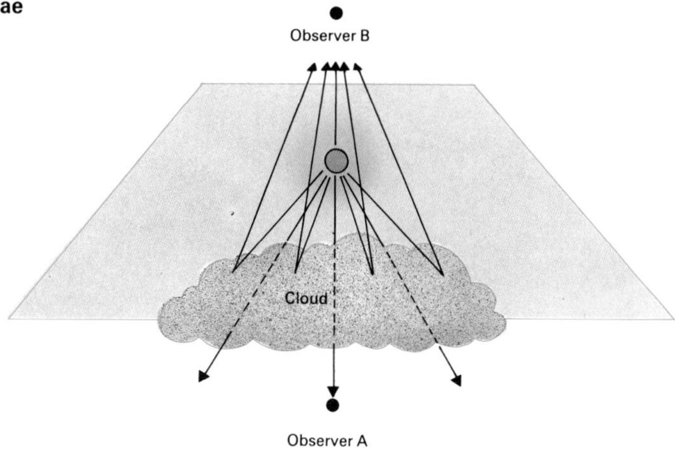

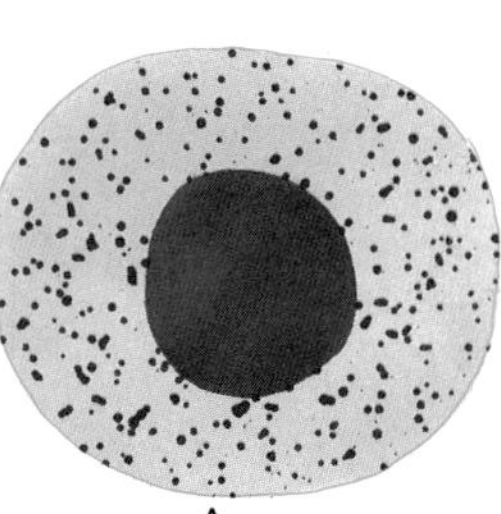

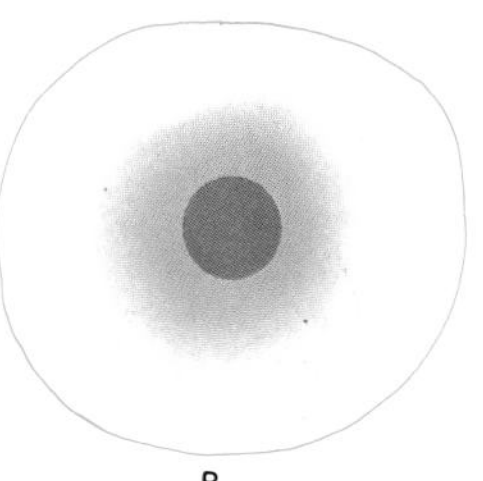

◄ ► If a dust cloud lies between the observer and the stars it will show up as a dark patch against the starry background (A). If dust lies close to a star, the star's light will be reflected towards the observer and a bright reflection nebula will be seen (B). The spectrum of a reflection nebula is similar to that of the star but is usually slightly bluer because blue light is more strongly scattered by dust.

Luminous nebulae are also known as HII regions. (This term means that a proportion of their principal constituent, hydrogen, is ionized; un-ionized, or neutral, clouds of hydrogen are known as HI regions.) The hot stars embedded in HII regions maintain them at temperatures of around 10,000K. In the Orion nebula, which is lit up by the four stars called the Trapezium, hydrogen emits mainly red light, while the stronger ultraviolet radiation nearer to the stars ionizes heavier gases, notably oxygen, which emits strongly in the green part of the spectrum. This gives the inner part of the nebula a greenish hue.

The only part of the cloud to shine is the area lying close to the hot stars, but interstellar clouds are often much larger than the visible HII regions. There are also numerous clouds which do not shine at all, because they do not contain stars of the right type. If an invisible cloud lies between an observer and a star it will absorb light at certain wavelengths and leave its "fingerprint" on the spectrum of the star. The Doppler effect provides a way of distinguishing such interstellar absorption lines from those produced in the star, for if the cloud and the star are moving at different speeds, the absorption lines produced by the cloud will be shifted in wavelength relative to those produced in the star. A special case of this occurs if there is a cloud in front of an eclipsing binary, when the interstellar lines remain at constant wavelengths while the lines produced by the stars shift to and fro in wavelength (◀ page 162). If several clouds moving at different speeds lie in the line of sight, the spectrum of a star includes several sets of interstellar lines with different Doppler shifts. Analysis of these lines gives important information about the clouds.

Interstellar dust

Dust reveals itself in a number of ways. Clouds containing enough dust to obliterate the light from background stars show up against the starry background as dark nebulae. The best known of these are the Coal Sack (a near-circular dark cloud some 25 light years in diameter, located in the Southern Cross) and the distinctive Horsehead nebula (a striking dust cloud highlighted against a luminous nebula just south of Orion's belt). Clearly defined dark clouds range in size from about 25 light years in diameter down to small, dense globules, less than one light year in diameter and containing less than one solar mass. The denser globules may represent clouds which are collapsing to become protostars, and many of the dark clouds lie in vast complexes where star-formation is known to be taking place.

Molecules between the stars

Microwave and infrared observations reveal the presence of complex molecules (groupings of atoms) in dense clouds where the chances of atoms meeting and sticking together are better than average, and where there is more dust to shield molecules from ultraviolet starlight, which otherwise would break them apart. Molecules have a physical structure: they rotate around axes of symmetry, and also vibrate, and changes in their rotational and vibrational states result in the absorption or emission of microwave and infrared radiation respectively.

Studying the interstellar hydrogen

Radio astronomy is a powerful tool in the study of interstellar gas. Neutral hydrogen emits radiation at a wavelength of 21·1 centimeters (a frequency of 1,420 megahertz) as a result of the tiny energy difference which exists between the two possible spin states of the electron in the hydrogen atom (below). An electron can be visualized as spinning like a tiny globe, and if it is spinning in the same direction as the proton in the atom's nucleus, the energy of the atom is a little higher than if the electron and proton are spinning in opposite directions. As electrons in cool clouds of hydrogen change between these two spin states, they emit or absorb at this wavelength of 21·1 centimeters, and studying this wavelength reveals the distribution of cool hydrogen throughout the Galaxy.

Detecting radio waves

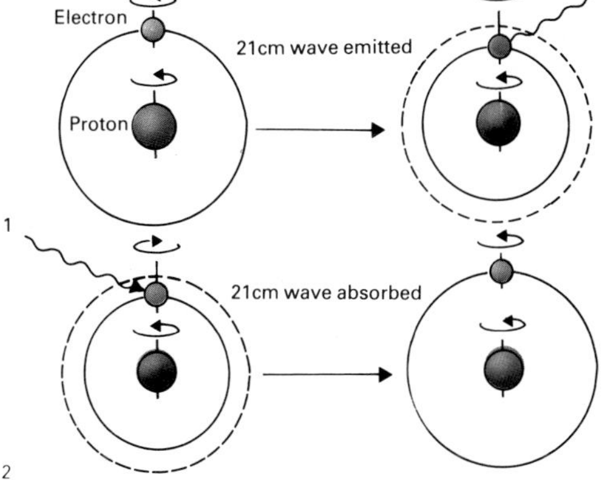

▲ Radiation of 21cm wavelength is emitted by a neutral hydrogen atom (1) when the spin of its electron changes from parallel to that of the nucleus (the higher energy state) to opposite to the nucleus (the lower energy state). The opposite process produces absoption (2)

Creating interstellar dust

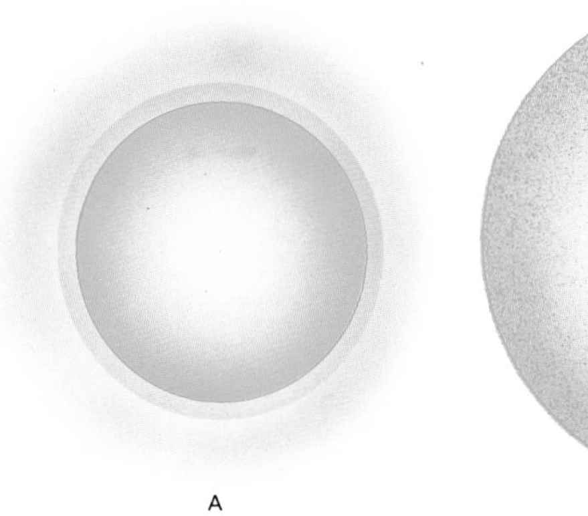

◀ A possible process for forming interstellar dust. A cool pulsating red giant (A) expands; its atmosphere cools (B) and solid particles of substances like carbon and silicates form. These are blown into space by light pressure as the star contracts and heats up (C).

The Horsehead nebula – the best-known dark nebula – is aptly named

▲ The center of the Galaxy lies in the direction of the constellation Sagittarius, but at optical wavelengths it is obscured by dust clouds. The blue patches are starlight reflected by the dust; red betrays the presence of hydrogen clouds.

◄ The Coal Sack nebula in the constellation of the Southern Cross contains solid particles (not just gases), and these absorb the light from more distant stars. The Coal Sack is one of the best examples of a dark nebula.

► One of the most striking objects in the sky is the Horsehead nebula in Orion, a dark nebula which stands out in silhouette against the brighter and more distant emission nebula. The bright star is Alnitak (d Orionis), and below it is a hot gas cloud, NGC 2024.

See also
The Origin of the Solar System 121-8
The Basic Properties of the Stars 145-42
Birth, Life and Death of Stars 153-60
Binary and Variable Stars 161-8
The Milky Way Galaxy 189-96

A view spoiled by dust

Recent infrared observations made by the IRAS satellite (◀ page 160) show that wispy clouds of dust, called "interstellar cirrus", occur throughout the Galaxy. Although they are cool by terrestrial standards, with temperatures of 10-100K, they emit strongly at infrared wavelengths of about 100 micrometers.

The interstellar medium probably consists of a very tenuous general distribution of dust in addition to the large proportion which is contained in relatively dense localized clouds. The effect of the dust is to scatter and absorb starlight, dimming the distant stars by an amount which increases with distance in proportion to the amount of dust in the line of sight. On average, starlight is dimmed by 1-2 magnitudes for every 1,000 parsecs (3,260 light years). So much dust lies between the Sun and the center of the Galaxy that only one photon of visible light in every million million succeeds in penetrating this dusty veil. Consequently the center is not visible at optical wavelengths.

The dust affects shorter wavelength light, such as blue and ultraviolet, to a greater extent than longer wavelength light, such as red. The amount of dimming is approximately inversely proportional to wavelength, so that longer wavelengths – infrared and radio waves – are hardly affected at all and can pass right through the Galaxy. Blue light being weakened more than red, starlight becomes progressively "redder" as it passes through more and more dust. This does not mean that all distant stars appear red in color, but that a progressively smaller proportion of a star's short-wave light reaches Earth.

Composition of interstellar dust

Measurements of the absorption spectrum of the dust (the amount of dimming at different wavelengths) give clues to the size and composition of the grains. Typical grain sizes seem to be between 0·1 and 1 micrometers – comparable with or smaller than the wavelengths of visible light – and this is why they are so effective at scattering visible light. The chemical composition of the grains is a matter of debate: some are probably made of graphite (a form of carbon) and others seem to be silicates. Some have icy mantles, and those which lie in dense clouds probably have coatings of complex molecules.

An alternative, highly controversial interpretation of the absorption spectra, propounded by Sir Fred Hoyle and Chandra Wickramasinghe, is that they are frozen bacteria. Although most astronomers disagree, the fact remains that interstellar molecules are mainly organic and more complex structures doubtless remain to be found.

Interstellar molecules

Name	Formula
Methylidyne	CH
Cyanogen	CN
Methylidyne ion	CH^+
Hydroxyl	OH
Ammonia	NH_3
Water	H_2O
Formaldehyde	H_2CO
Carbon monoxide	CO
Hydrogen	H_2
Hydrogen cyanide	HCN
Methanol	CH_3OH
Formic acid	HCO_23
Silicon monoxide	SiO
Acetaldehyde	CH_3CHO
Hydrogen sulfide	H_2S
Dimethyl ether	CH_3OCH
Ethanol (ethyl alcohol)	CH_3CH_2OH
Sulfur dioxide	SO_2
Ethyl cyanide	CH_3CH_2CN
Nitric oxide	NO
Glycine	$C_2H_5O_2N$
Cyano-octatetrayne	HC_9N

Molecules in interstellar space
Physicists have identified over 50 different species of molecule in space, ranging from simple ones like hydroxyl, water and ammonia to more complex organic molecules such as formaldehyde, formic acid and ethyl alcohol. The heaviest molecule discovered so far (and new ones are still being found) is an extraordinary chain of nine carbon atoms with a hydrogen atom at one end and a nitrogen atom at the other (HC_9N) – a species which does not occur on Earth.

An intriguing aspect of interstellar molecules is that the majority are organic – they are chains made up mainly of carbon, hydrogen, nitrogen and oxygen, and include some of the elementary building blocks from which much more complex living cells are constructed. The most common molecule is molecular hydrogen, but this is difficult to detect other than at ultraviolet wavelengths.

A view spoiled by dust

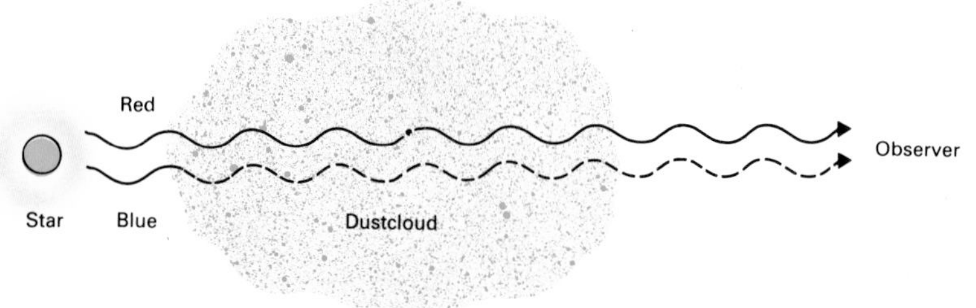

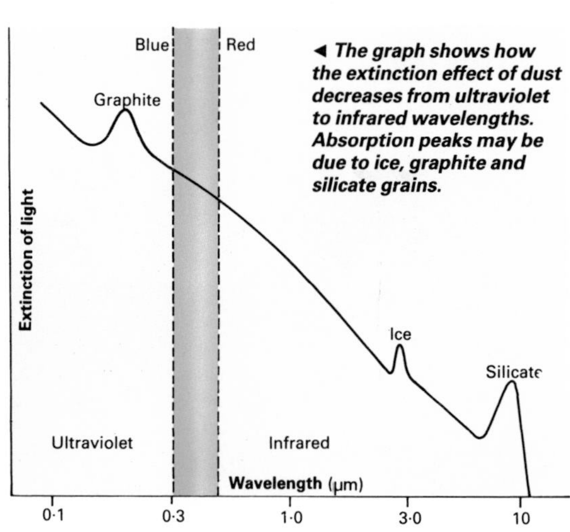

◀ *The graph shows how the extinction effect of dust decreases from ultraviolet to infrared wavelengths. Absorption peaks may be due to ice, graphite and silicate grains.*

The Milky Way Galaxy

21

The size and structure of our Galaxy...Globular star clusters...The Sun's orbit in the Galaxy...The galactic corona...The spiral arms fully revealed...The galactic center – an intriguing puzzle...The evolution of the Galaxy...PERSPECTIVE...The first observations...The Milky Way seen in five wavelengths...Estimating the age and size of the Galaxy..."The great debate" – were the spiral nebulae within our Galaxy?

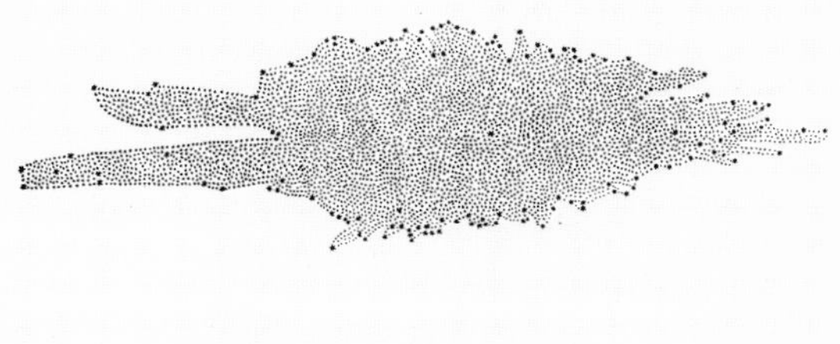

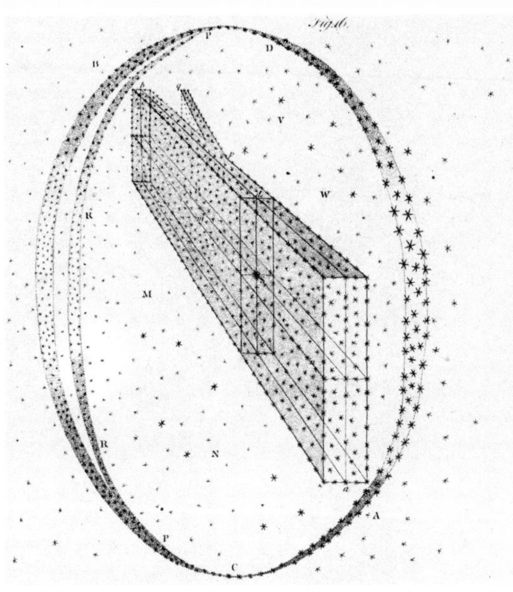

▲ ***William Herschel's diagrams of the Milky Way.***

Our Galaxy contains more than 100 billion stars, most of which are concentrated, together with clouds of gas and dust, into a flattened system 80,000 to 100,000 light years in diameter.

The greatest concentration of stars is in the nuclear bulge, a flattened sphere some 15,000 light years in radius. Surrounding this bulge is a disk 40,000 to 50,000 light years in radius but less than 3,000 light years thick, which, in addition to stars, contains most of the gas and dust in the Galaxy. Outside this is a nearly spherical distribution of widely separated stars and massive globular star clusters extending to a radius of some 75,000 light years. This is known as the "halo". Recent observations have produced evidence to suggest that the halo is itself surrounded by a tenuous but much more extensive "corona" of invisible matter which extends to a distance of between 200,000 and 300,000 light years, and which may contain up to 90 percent of the total mass of the Galaxy.

It is difficult to determine the size and structure of the Galaxy because the Solar System lies inside the galactic disk, and clouds of dust obscure the view towards the center. That the basic shape is a flattened disk is obvious from the appearance of the Milky Way – the faint band of starlight which encircles the celestial sphere and is readily seen on a clear moonless night. A telescope reveals the Milky Way to be composed of millions of individual stars spread out in a thin but extensive disk. In recent years, observers have studied the Milky Way at wavelengths ranging from gamma rays to radio waves, infrared and radio waves being particularly valuable in the study of galactic structure since they are unaffected by the dust.

The Galaxy's basic structure

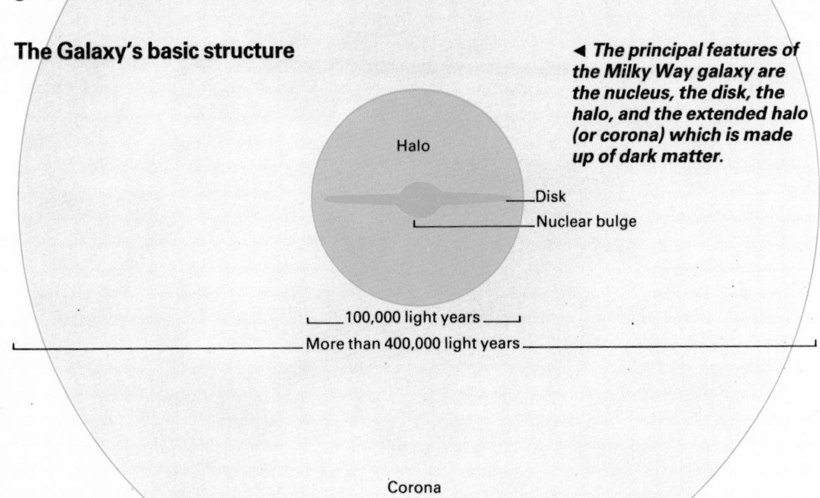

◄ ***The principal features of the Milky Way galaxy are the nucleus, the disk, the halo, and the extended halo (or corona) which is made up of dark matter.***

Exploring the stars of the Milky Way

On a clear dark night when the sky is studded with stars it is easy for a person without prior knowledge to think that they number in their millions. Yet appearances are deceptive: in all, naked-eye observers will be lucky to see as many as 2,500 stars at any one time.

Using binoculars or a telescope changes the situation. The total rises to millions, and to count them is a physical impossibility. Therefore, in order to estimate the number of stars in the Galaxy, at the end of the 18th century the German-English astronomer William Herschel (1738-1822) (◆ page 138) decided to count the stars in selected areas and average out the results. While making his survey, he confirmed that stars were concentrated into the band of the Milky Way and tended to be increasingly sparse in other directions. On this evidence he envisaged a star system shaped like a "cloven grindstone", which is a reasonably accurate description of the shape of the Galaxy.

Other important observations that Herschel made in the course of his survey concerned the nebulae. Noticing that the "starry nebulae" tended to avoid the region of the Milky Way, while the "irresolvable nebulae" (which he correctly took to be gas clouds) were common there, he suggested that the former might be external star systems. But the idea met with only a lukewarm reception and Herschel himself later expressed grave doubts about this crucially important observation.

Astronomers are now looking into what was, until quite recently, one of the obscurest regions in space – the center of our own Galaxy

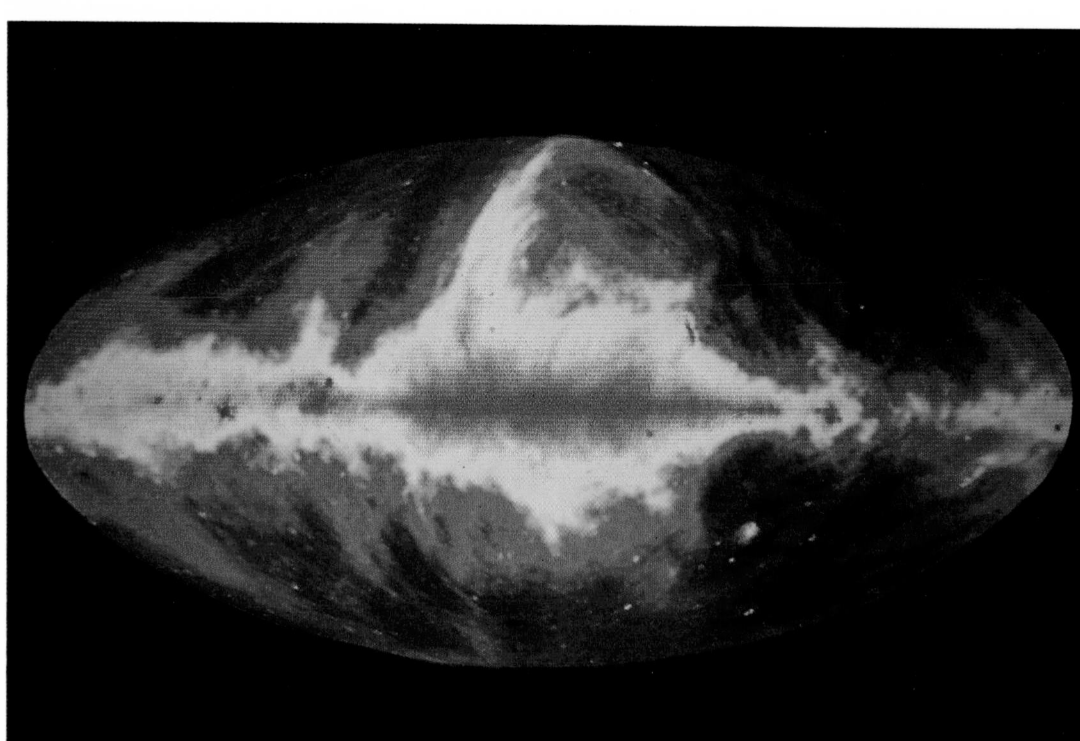

◄ *This map of the radio sky is the result of 15 years' work with three radio telescopes. Emissions are strongest in the plane of the Milky Way: the black regions are dark at radio wavelengths. Since these wavelengths can penetrate the interstellar dust, this map gives an accurate, hitherto unobtainable portrait of our Galaxy.*

► *The X-ray sky is shown here in the same projection as the Lund Observatory map. Two X-ray bursters (* ***♦ page 173*** ***) are shown, that at top left as seen by three detectors, the one at lower right (in the Small Magellanic Cloud) by a single one. The map was made by the Vela 5B satellite in 1972, and the positions of the X-ray bursters added later.***

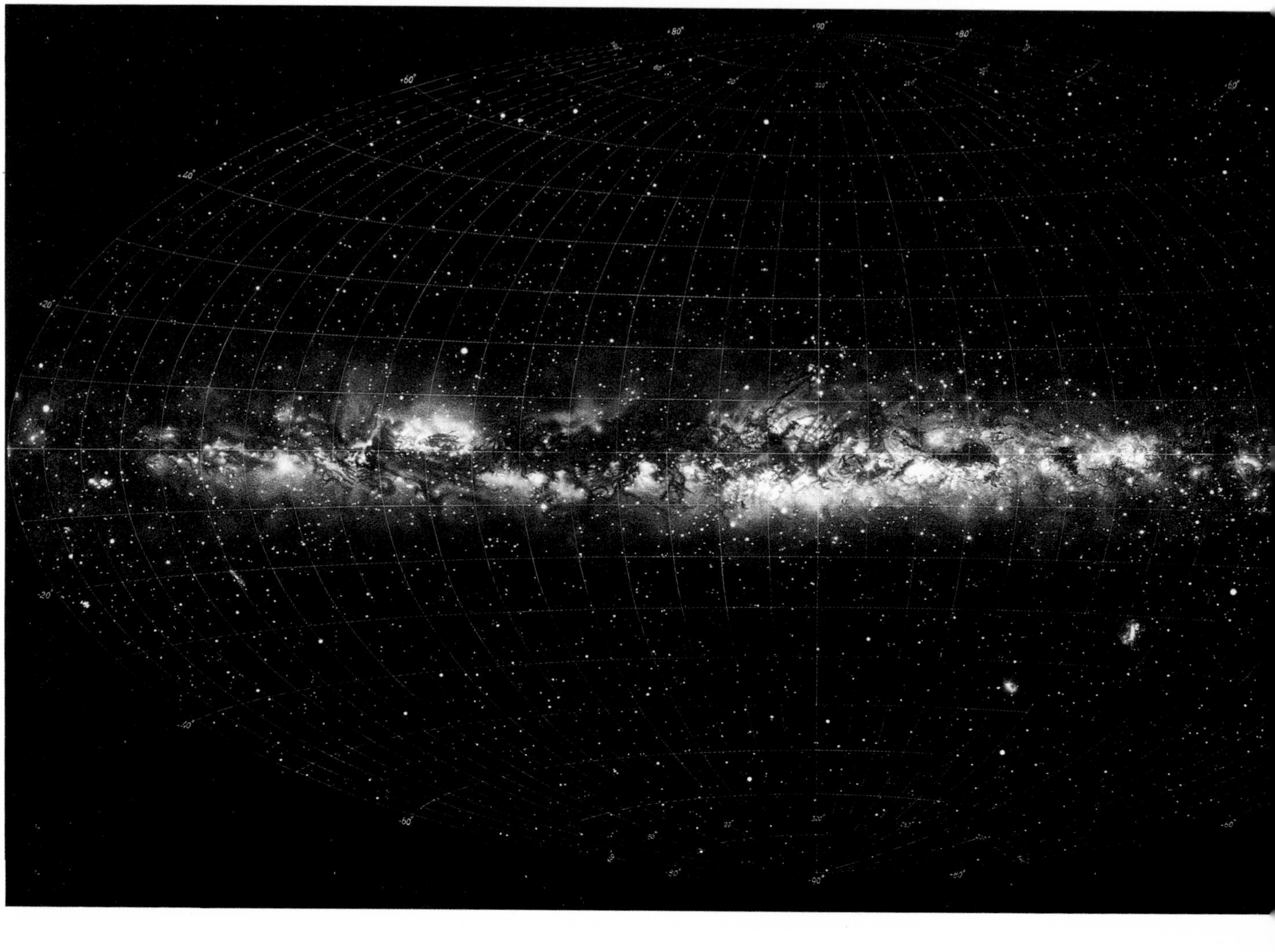

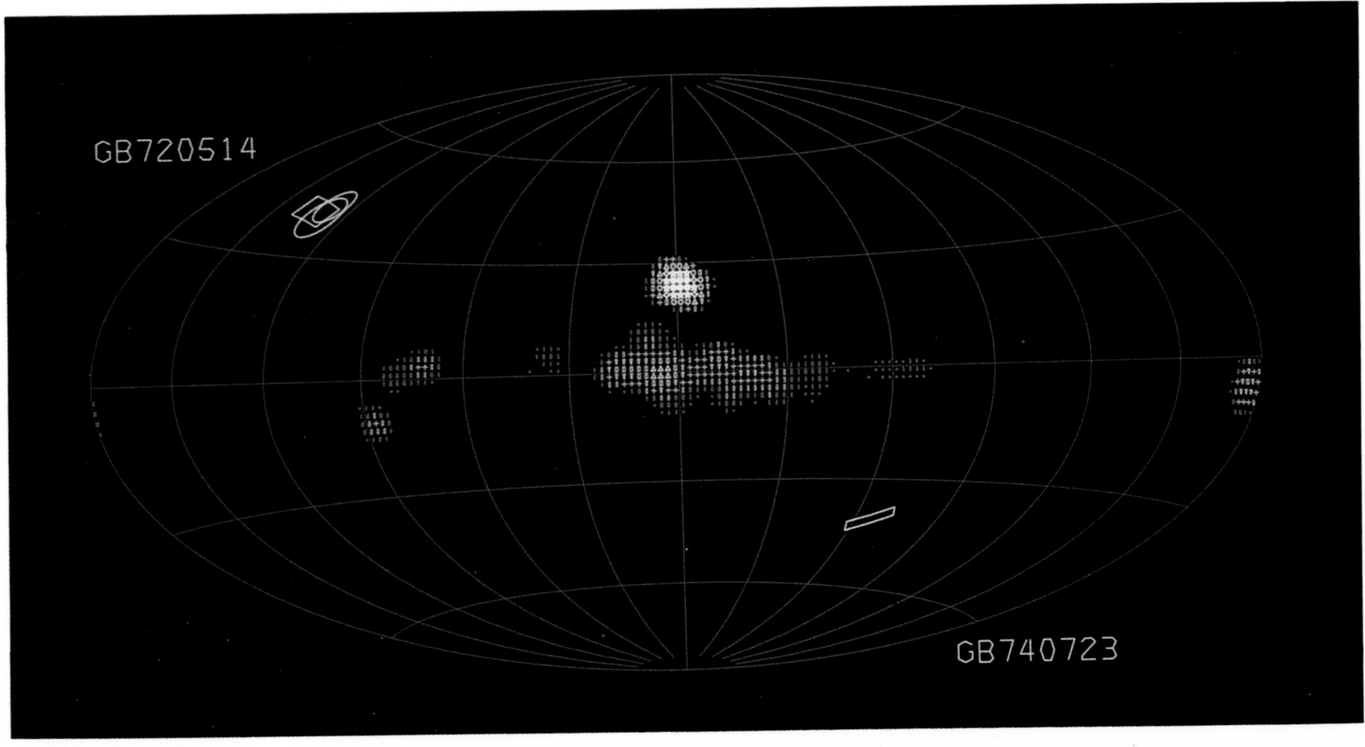

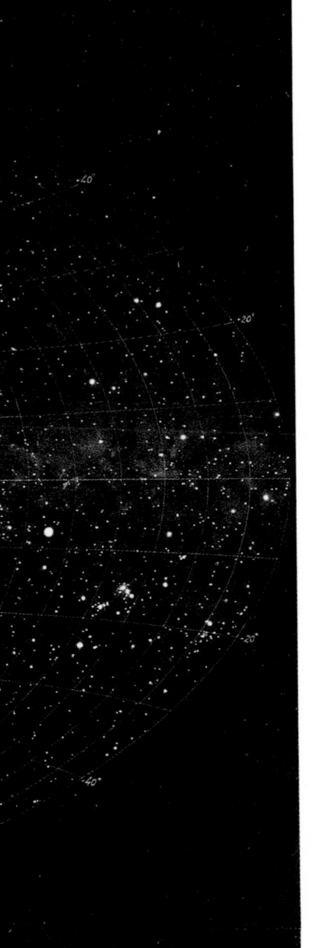

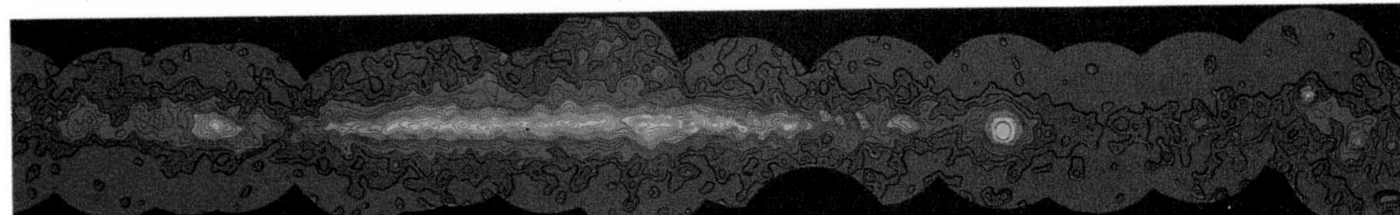

▲ In this gamma-ray map of the Milky Way, made by the European satellite COS-B, the colors show the intensity of the radiation, yellow being the areas of strongest emission.

◄ This composite map of the whole sky was compiled from photographs by a team of astronomers at the Lund Observatory in Sweden. The constellation of Auriga is at the extreme left and right, while Sagittarius is in the center. The extreme 360° "fish eye" view causes distortions, but each rectangle covers an equivalent area.

► This false-color infrared image obtained by the IRAS satellite shows the distribution of dust along the plane of the Milky Way and across the galactic center. The temperature of the dust ranges from about 30K (red) to about 250K (blue).

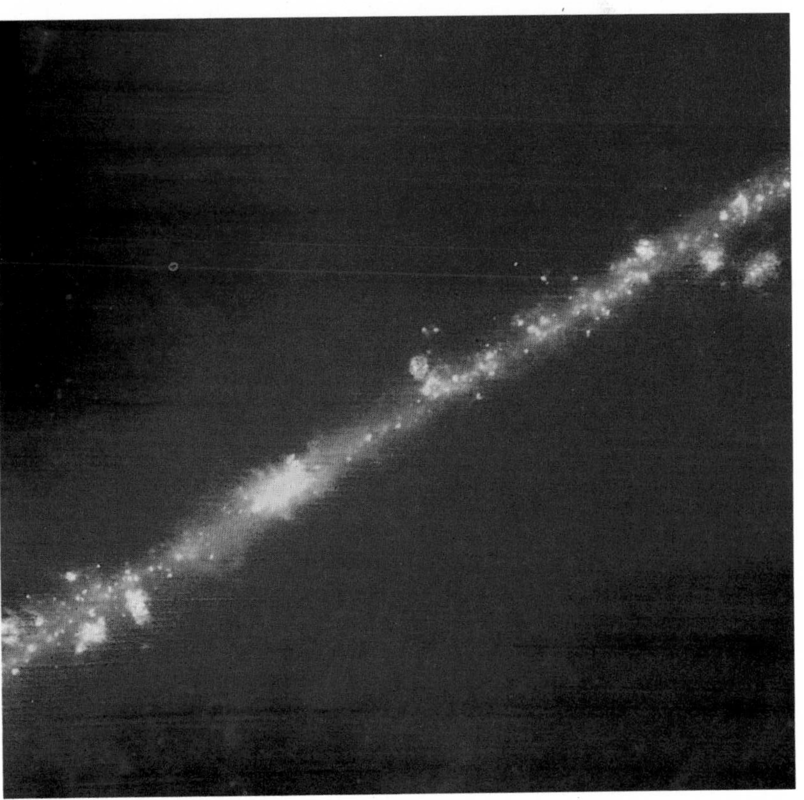

Recent studies suggest there may be ten times more material in the Milky Way than anybody has so far been able to observe

Globular star clusters and the scale of the Galaxy

The galactic halo contains about 200 globular star clusters, and these are distributed fairly uniformly round the galactic center. A typical globular cluster contains hundreds of thousands of stars and, because many of them lie above or below the plane of the Milky Way, they are not heavily obscured by dust. Some of the stars are of the RR Lyrae type – variable stars of known luminosity (◆ page 164). Comparing the apparent brightness of an RR Lyrae star with its known luminosity can reveal its distance – and hence the distance of the cluster in which it lies. Studies of the distance and distribution of globular clusters show that the Sun is about 28,000 light years away from the center of the Galaxy, which lies in the direction of Sagittarius.

Almost all the young O and B-type stars, the dense clouds in which stars are forming, the illuminated HII regions and the young open star clusters lie in the galactic disk. Rather confusingly, the younger stars, which contain higher proportions of heavy elements and which lie close to the galactic plane, are called "Population I" stars, while the older stars of the halo and bulge are known as "Population II."

The rotating Galaxy

The Galaxy is rotating, with the Sun moving at about 230km per second in a near-circular orbit round the galactic center and taking about 220 million years to complete each orbit. Most Population I stars in the disk move in the same direction as the Sun, also in near-circular orbits, but many of the Population II stars follow elliptical orbits at quite steep angles to the galactic plane.

◀ Dr Harlow Shapley, the first astronomer to work out accurately the size of the Milky Way, was Director of the Harvard College Observatory for over thirty years.

▼ M13, the brilliant globular cluster in the constellation of Hercules, was discovered by Edmond Halley in 1714. The most prominent globular cluster in the northern hemisphere, it is 22,500 light years away.

Shapley's determination of distance

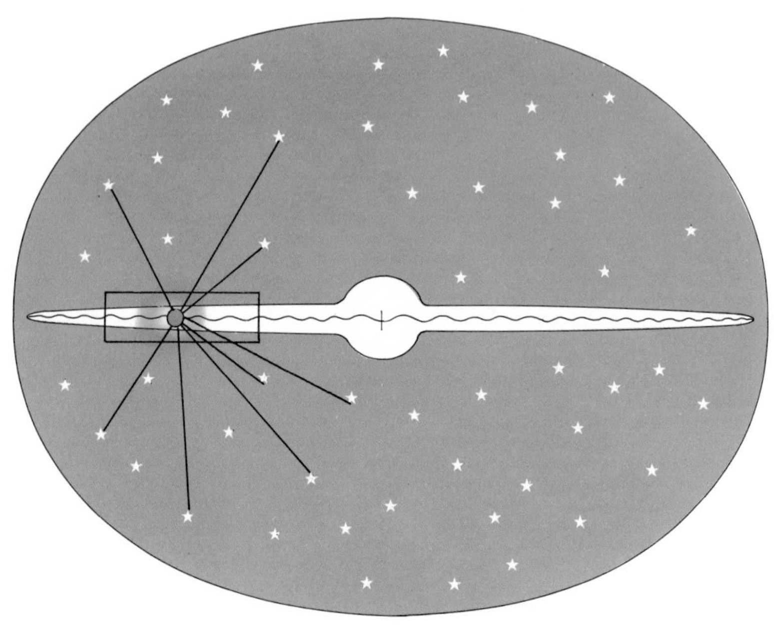

▲ Interstellar dust is concentrated in the plane of the Galaxy and restricts optical astronomers to observing a limited part (boxed). Most globular clusters lie above and below the galactic plane and thus are not heavily obscured by dust. Shapley assumed that the center of the system of globular clusters coincided with the center of the Galaxy. He then estimated the Galaxy's size by measuring their distances.

Harlow Shapley and the size of the Galaxy

Most of the credit for accurately measuring the size of the Galaxy must go to the American astronomer Harlow Shapley (1885-1972), who joined the staff of Mount Wilson Observatory in 1914. Shapley was aware of the relationship between the period of a Cepheid variable and its real luminosity: this had been established by observations of the short-period variables in the Small Magellanic Cloud made by Henrietta Leavitt (◆ page 157). Studying the Cepheid variables in the globular clusters, Shapley found that they lay round the boundary of the main Galaxy. He also deduced from their uneven distribution (most of the globular clusters lie in the southern hemisphere) that instead of being centrally placed in the Galaxy, the Sun was about 50,000 light years out towards the edge.

The Great Debate of 1920

Shapley's determination of the size of the Galaxy was of the correct order. However, at that time he still believed the spiral nebulae to be contained in the Galaxy. This view was challenged by another American astronomer, Heber Doust Curtis (1872-1942), then of the Lick Observatory. Curtis believed that the spirals were extragalactic, and that the Galaxy itself was very much smaller than Shapley had estimated. The whole problem was made the subject of a "great debate" between Shapley and Curtis, held at a meeting of the National Academy of Sciences in 1920. The debate itself was inconclusive, but history has shown that Shapley was closer in his estimate of the size of the Galaxy, and wrong in believing the spirals to be members of it, so the final verdict can be seen as a draw.

Estimating the age of the Galaxy

Globular clusters, in common with the halo and the nuclear bulge, contain old, metal-deficient stars that were born early in the history of the Galaxy, when it was composed almost entirely of hydrogen and helium, and before the heavier elements generated inside stars were blown into space in supernova explosions. The ages of the clusters are revealed in their H-R diagrams, which show that the only main-sequence stars they contain are old, low-mass stars. All others have evolved to the red giant stage and beyond. Studies of this kind show that the Galaxy must be at least 12 billion years old.

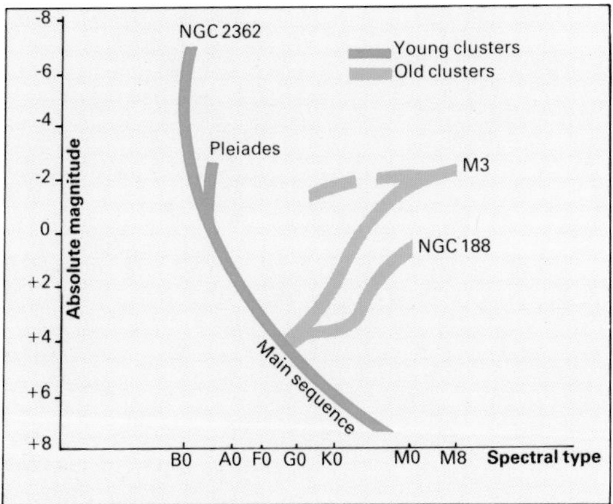

Stars closer to the galactic center have shorter orbital periods than those which are farther away, but they do not behave like planets traveling round the Sun. The mass of the Galaxy is spread out over a vast volume and the speed of a star is determined not only by its distance, but also by the mass which lies between it and the center. The greater the distance of the star, the greater the mass acting upon it.

Close to the galactic center, velocities are quite high. They decrease towards the outer fringe of the nucleus, and then farther out they increase to about 230km per second at the Sun's distance, and to some 300km per second at a radius of 60,000 light years.

The galactic corona

The high speeds of stars in the outermost parts of the galactic disk imply that a large fraction of the Galaxy's mass lies in the fringes of the system. If this were not so, the speeds of stars would begin to decrease at distances beyond that of the Sun. One high-speed RR Lyrae star, nine globular clusters, 3 dwarf elliptical galaxies and the two Magellanic Clouds are known to lie at distances between 65,000 and 200,000 light years from the galactic center. If, as appears to be the case, they are all part of the halo, then it is possible to calculate from their speeds how massive the Galaxy must be in order to prevent these objects from escaping into intergalactic space. This mass turns out to be between 1,000 billion and 2,000 billion solar masses.

This mass consists of neither luminous stars nor neutral hydrogen, or it would be visible. Possibly it consists of cool planet-sized lumps of matter, old dead stars, or very low-mass stars too faint to be seen.

▲ H-R diagram showing that faint stars in galactic clusters are on the main sequence.

► All the stars within our Galaxy are following orbits centered on the galactic nucleus. Stars closer to the center overtake the Sun because they are moving in smaller orbits. Stars farther out lag progressively farther behind. The spiral pattern is a "density wave" rotating independently at about half the Sun's rate. Stars and gas clouds pass periodically into and out of the spiral arms.

▼ The rotation curve shows how the speeds of stars and gas clouds, orbiting the galactic center, vary with distance from the center.

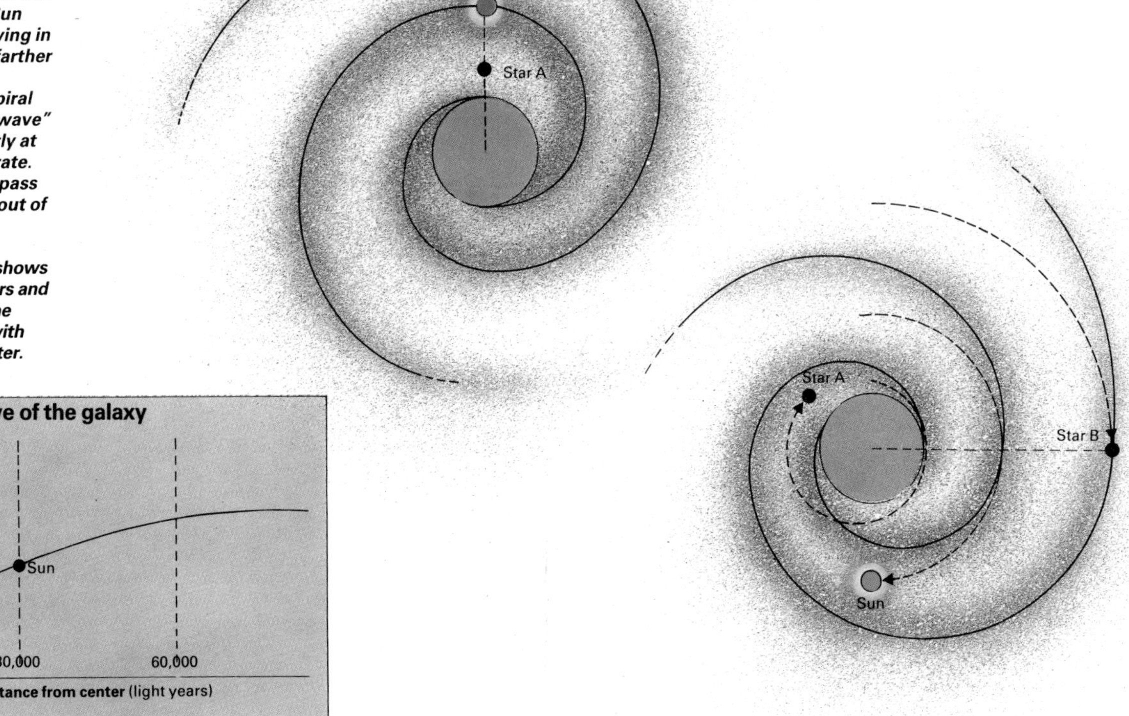

Rotation curve of the galaxy

Velocity (kms^{-1})

Sun

30,000 60,000

Distance from center (light years)

A

100,000 light years

C

B

10,000 light years

D

E

1,000 light years

100 light years

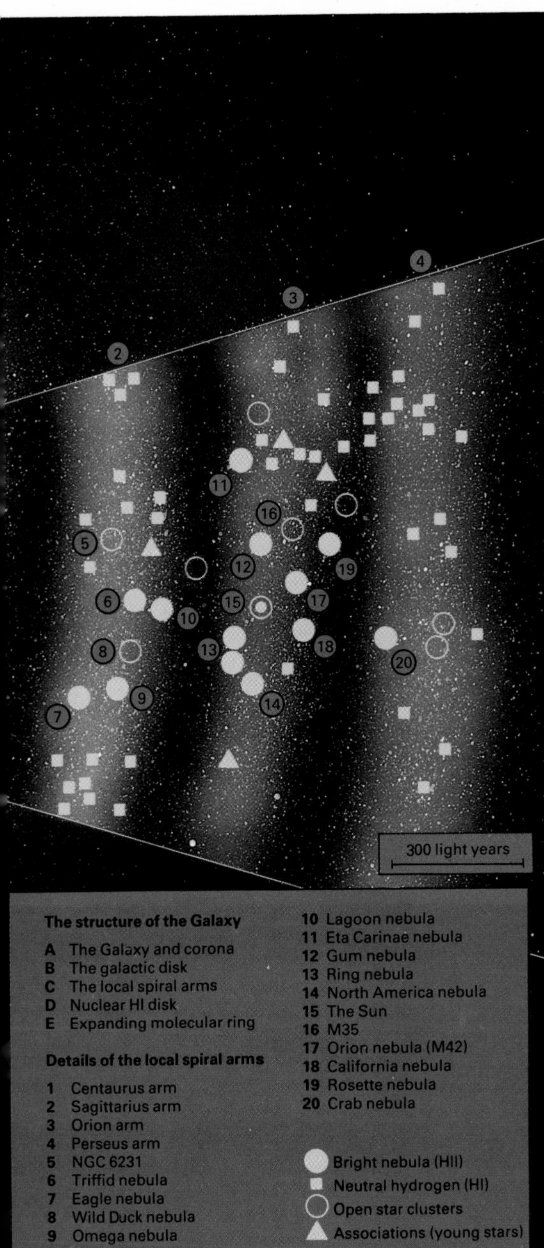

◀ *The structure of our Galaxy. The spiral structure in the Sun's locality is deduced from optical observations; the large-scale distribution of gas is revealed by radio observations. The true core of the Galaxy coincides with a compact radio source which may contain a black hole.*

The spiral structure of the galactic disk

Like many other galaxies, the disk of the Milky Way has a lumpy, coiled structure as if the stars and gas clouds were arranged along the arms of a spiral spreading out from the nuclear bulge. The basic scheme has two arms extending from opposite sides of the nucleus, and wound round quite tightly into a Catherine-wheel shape. An alternative view suggests a four-arm spiral, although in practice observations do not show a simple, clear-cut pattern.

The positions of the spiral arms in the Sun's locality can be picked out optically by studies of the distribution of bright O and B-type stars, HII regions (◆ page 185) and galactic clusters. The Sun is near the inside edge of a spiral arm, known as the Orion arm. The Perseus arm lies about 6,000 light years farther out from the center and the Sagittarius arm some 6,000 light years farther in. The Carina arm meets the Sagittarius arm at the Carina nebula and there is evidence for another – the Centaurus arm – closer again to the galactic center.

Neutral hydrogen extends in a thin disk to about twice the Sun's distance from the galactic center, but most of the molecular hydrogen and the dense molecular clouds (◆ page 188) are concentrated into a broad ring with its inner and outer edges at 12,000 and 25,000 light years respectively from the center. This is where most of the current bout of star-formation is taking place.

The nuclear bulge

Very little hydrogen is present within a radius of about 12,000 light years of the galactic center, but there are many old red stars, increasing in concentration towards the center. There is an expanding ring of gas – known as the "3 kiloparsec arm" – lying at a distance of just under 10,000 light years from the center, and containing some 30 million solar masses of gas. This may have been expelled from the nucleus some 30 million years ago, and was possibly responsible for sweeping much of the gas and dust out of the central regions. Closer in, there is an expanding disk of atomic and molecular hydrogen apparently tilted at an angle of nearly 20° to the galactic plane, and at a radius of about 100 light years there is a ring of massive molecular clouds and ionized clouds heated to some 10,000K by hot young stars. There is another, slightly cooler ring about 30 light years from the center, while inside a radius of 10 light years there are vast numbers of stars and fast-moving ionized clouds.

▼ *The center of the Milky Way: this radio photograph shows a region measuring ten light years across.*

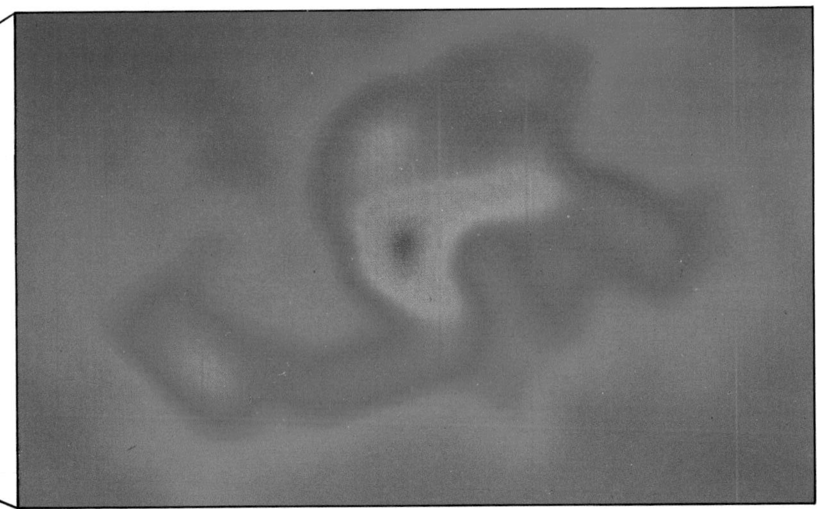

▶ *The optical photograph shows lanes of dust hiding the Galaxy's center at visible wavelengths.*

See also
Overview of the Universe 129-32
Galaxies 197-204
Active Galaxies and Quasars 205-16
Clusters of Galaxies 217-24
The Evolving Universe 229-44

The galactic center

The nature of the galactic center presents an intriguing puzzle. It is marked by a powerful radio source, known as Sagittarius A. Much of the radio emission comes from a region of space less than 10 astronomical units across – that is, a region considerably smaller than the Solar System. This central region also includes an X-ray source and a collection of infrared sources, many of which are cool giant stars.

In order to explain the very high speeds of orbiting gas clouds close to it, the center must contain about 5 million solar masses of material within a radius of one light year. This suggests that the center of the Galaxy may contain a black hole of several million solar masses into which gas and stars may fall. If such a black hole does exist, it could be the underlying energy source responsible for some of the emission from the galactic core.

The evolution of the Galaxy

Since the older Population II stars have a near-spherical distribution around the galactic center, the conventional view is that the Galaxy originated as a near-spherical, slowly rotating cloud of hydrogen and helium, and that the first objects to form were the massive globular clusters. As time went by the rest of the gas settled into a spinning disk within which newer generations of stars have been born, and continue to be born, while the old stars still trace out orbits through the halo in which they were formed.

However, astronomers do not really understand the evolution of galaxies, and there are many alternative theories. One suggestion is that supermassive stars formed first in the nucleus of the Galaxy, then exploded, scattering forth the material to make the disk and halo. The various clouds which seem to be moving out through the nuclear bulge do seem to hint at a succession of (modestly) violent events in the Galaxy's history.

Star formation

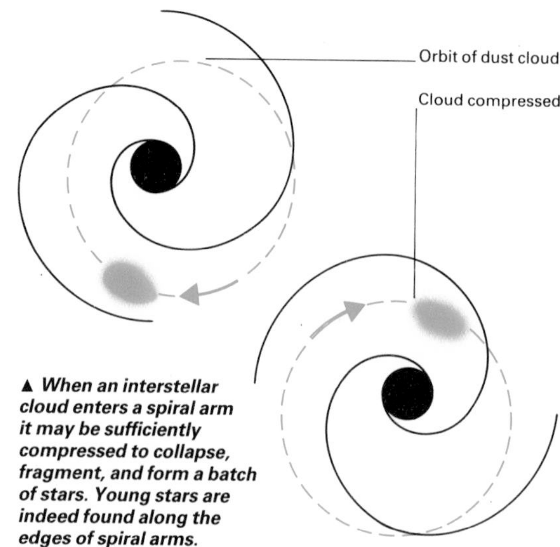

▲ ***When an interstellar cloud enters a spiral arm it may be sufficiently compressed to collapse, fragment, and form a batch of stars. Young stars are indeed found along the edges of spiral arms.***

Locating the gas clouds

Unlike optical astronomers, whose view is blocked by dust, radio and millimeter-wave astronomers can plot the distribution of gas clouds throughout the Galaxy. Clouds closer to the galactic center have shorter orbital periods than the Sun, while those farther out have longer periods. The radial velocity of a cloud relative to the Sun depends on its distance from the galactic center, and on the angle between Sun, cloud and galactic center. This velocity can be measured from the Doppler shift (◂ page 152) in the cloud's radiation, and it is then possible to determine the location of the cloud.

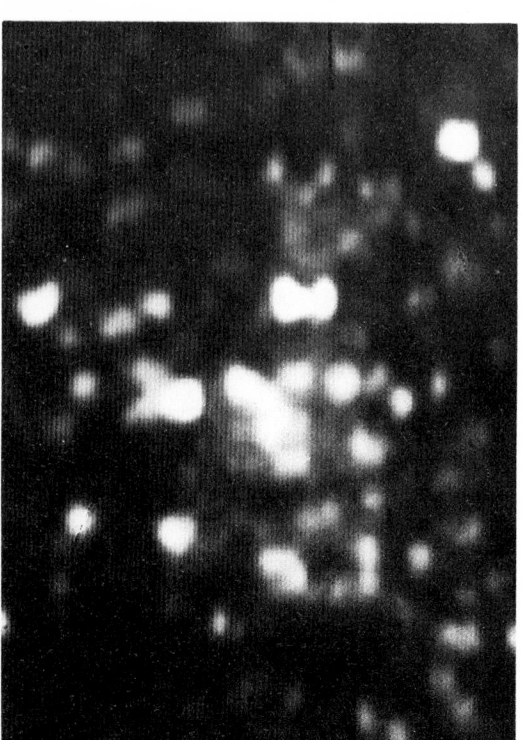

▲ ***Computer-processed view of the Galactic core.***

Analyzing the spiral structure

► ▼ ***Radio emissions from three hydrogen clouds: the cloud in ring B is stationary relative to the Sun and the received radiation has a wavelength of 21cm. The cloud in ring A is pulling ahead and receding; its radiation is red-shifted. The cloud in ring C is approaching; its radiation is blue-shifted. The relative speeds of clouds show the distribution of hydrogen in the Galaxy.***

Sun
C
B
A
Line of sight

Signal strength
A
B
C
21-Δλ 21cm 21+Δλ

Galaxies

22

Measuring the distances of the galaxies...Our nearest galactic neighbors...Classifying the types of galaxy... Calculating their masses...The evolution of galaxies... PERSPECTIVE..."The great debate" continued...The pioneers of galaxy classification

Within range of present-day telescopes there are billions of galaxies. Some are swirling spiral systems like the Milky Way, but others are very different, and there is a wide variety of types. Our nearest neighbors are a motley collection of some thirty galaxies comprising the "Local Group." Its senior members apart from our Galaxy are the massive Andromeda spiral, the smaller spiral in the constellation Triangulum and the two Magellanic Clouds, which lie within 200,000 light years and may be regarded as companions of our Galaxy.

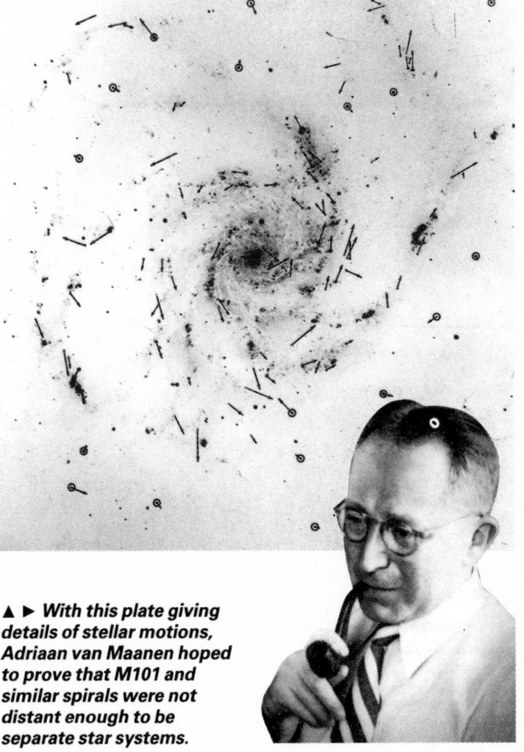

▲ ▶ With this plate giving details of stellar motions, Adriaan van Maanen hoped to prove that M101 and similar spirals were not distant enough to be separate star systems.

Measuring galactic distances

It is possible to determine the distance of galaxies by identifying within them conspicuous objects, such as highly luminous stars, star clusters or HII regions, which are similar to well-known objects in the Milky Way system, and by then computing either a "luminosity distance" or a "diameter distance."

To obtain a luminosity distance, astronomers must estimate the absolute magnitude or brightness of an identifiable object, and compare this with its apparent brightness. Cepheid variables (◀ page 164) provide a basic yardstick out to about 12-15 million light years. Luminous supergiants and novae extend the range to some 30 million light years, while observations of rare supernovae extend the range of distance measurement to billions of light years.

Diameter distance relies on the measurement of the apparent diameters of objects such as globular clusters and HII regions, and comparison of these with the known size of similar objects in the Milky Way. Needless to say, accuracy diminishes with increasing distance.

There is a relationship between the distance of a galaxy and the redshift in its spectrum. This is known as the Hubble Law (▶ page 230), and astronomers can use it to assign distances to the remotest galaxies; but the uncertainty in this method may be as much as 100 percent.

Arguments about the spirals

One of the quarrels that punctuate the history of astronomy arose between the Dutch astronomer Adriaan van Maanen (1884-1946) and his great American colleague Edwin Hubble (1889-1953), concerning the nature and distance of the spiral nebulae. Hubble believed that they were "island universes" as originally suggested by William Herschel, but van Maanen, who had taken photographs of some of the spirals, claimed to have detected slight movements in some of the stars they contained. If this were correct, the spirals could not possibly be outside our Galaxy; however, van Maanen later conceded that his measurements had been incorrect.

Astronomers have now probed much farther: modern equipment has revealed objects well over 10,000 million light years away – perhaps not far from the very edge of the observable universe.

Selected bright galaxies in order of distance

Name or Catalog Number	Distance (thousands of light years)	Diameter (thousands of light years)	Mass* (Solar Masses)	Apparent Magnitude	Absolute Magnitude	Type
Milky Way	—	100	2×10^{11}	—	−21	Sb
Large Magellanic Cloud	170	30	1×10^{10}	0·1	−18·7	Ir I/SB?
Small Magellanic Cloud	200	16	2×10^{9}	2·4	−16·7	Ir I
Sculptor System	280	4	3×10^{6}	7	−12	E
Fornax System	560	7	2×10^{7}	7	−13	E
NGC 6822	1,500	7	3×10^{8}	8·6	−15·6	Ir
NGC 205	2,100	7	8×10^{9}	8·2	−16·3	E5
M32	2,200	4	3×10^{9}	8·2	−16·3	E2
M31, Andromeda	2,200	130	3×10^{11}	3·5	−21·1	Sb
M33, Triangulum	2,400	50	1×10^{10}	5·7	−18·8	Sc
Maffei I	3,300	?	2×10^{11}	11	−20	S0
M82	10,000	23	3×10^{10}	8·2	−19·6	Ir II
M81	10,500	100	2×10^{11}	6·9	−20·9	Sb
M51 (Whirlpool)	13,000	65	8×10^{10}	8·2	−19·7	Sc
Centaurus A	16,000	30	1×10^{12}	7	−20	E0p
M101 (Pinwheel)	20,000	200	3×10^{11}	7·5	−20·3	Sc
M83	27,000	100	$?10^{12}$	7·2	−20·6	SBc
M104 (Sombrero)	40,000	30	5×10^{11}	8·1	−22	Sa
M87 (Virgo A)	50,000	40	3×10^{12}	8·7	−22	El

* Mass figure excludes dark extended haloes

Ferdinand Magellan's famous clouds

Our nearest neighbors in the Local Group of galaxies – the Large Magellanic Cloud (LMC) and the Small Magellanic Cloud (SMC) – are visible to the naked eye in the southern hemisphere. The LMC is about 170,000 light years distant, 30,000 light years in extent and contains about ten billion stars, while the SMC, at a distance of 200,000 light years, contains about a billion stars and is 16,000 light years in diameter. Both are irregular galaxies with little obvious structure.

The LMC contains a massive HII region – the Tarantula nebula – which is lit up by a cluster of more than a hundred O and B-type stars spread over a volume of space some 200 light years in diameter. The overall extent of this spectacular nebula is about 900 light years and it contains some 500,000 solar masses of gas. The IRAS satellite has shown that a vast region of star-formation lies behind the visible nebula, together with a huge cloud of neutral hydrogen. A great many old Population II stars also exist in the neighborhood of the nebula, which suggests that star-formation has been going on in this region throughout the galaxy's history. Although the Tarantula nebula lies off to one side of the main bar-like structure which contains most of the stars, some astronomers believe that it is the nucleus of this ill-defined galaxy, because it is such a massive and long-lived object.

A belt of fast-moving gas, known as the Magellanic stream, envelopes the two Magellanic clouds and stretches over the south pole of the Milky Way. This stream may be composed of gas which was dragged from the clouds when, in the relatively recent past, they made a close approach to the Milky Way system, or it may be a cloud of intergalactic gas caught in the gravitational fields of the Galaxy and the Magellanic clouds. If current ideas are correct, the Magellanic clouds and stream are all part of the massive galactic halo, but there has been a suggestion that the LMC is a separate system which has made a close approach and will eventually recede into the distance.

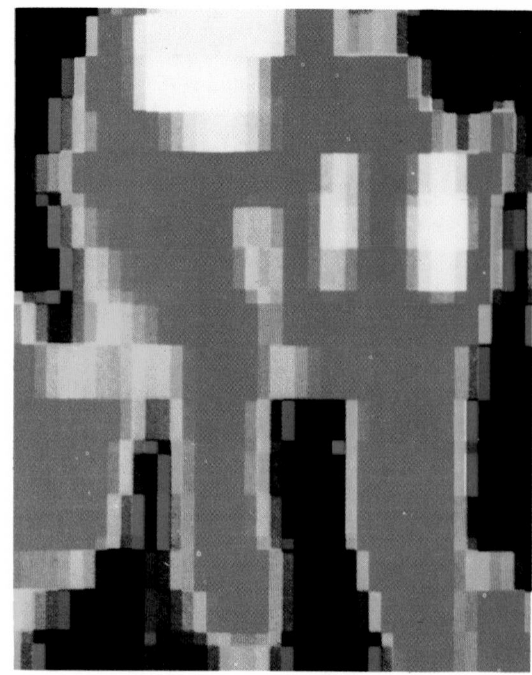

▲ This infrared scan shows the region in the Large Magellanic Cloud which is known variously as 30 Doradus or the Tarantula nebula. Within this cloud of gas and heated dust lurks possibly the most massive and most luminous star known. Its mass has been estimated at 2,500 solar masses and its luminosity at 100 million Sun power.

► The Large Magellanic Cloud is the Milky Way's nearest neighbor galaxy; this photograph shows it in true color.

▼ The Small Magellanic Cloud, like its large namesake, is a splendid galaxy to view in a small telescope; but it is visible only to observers in the southern hemisphere.

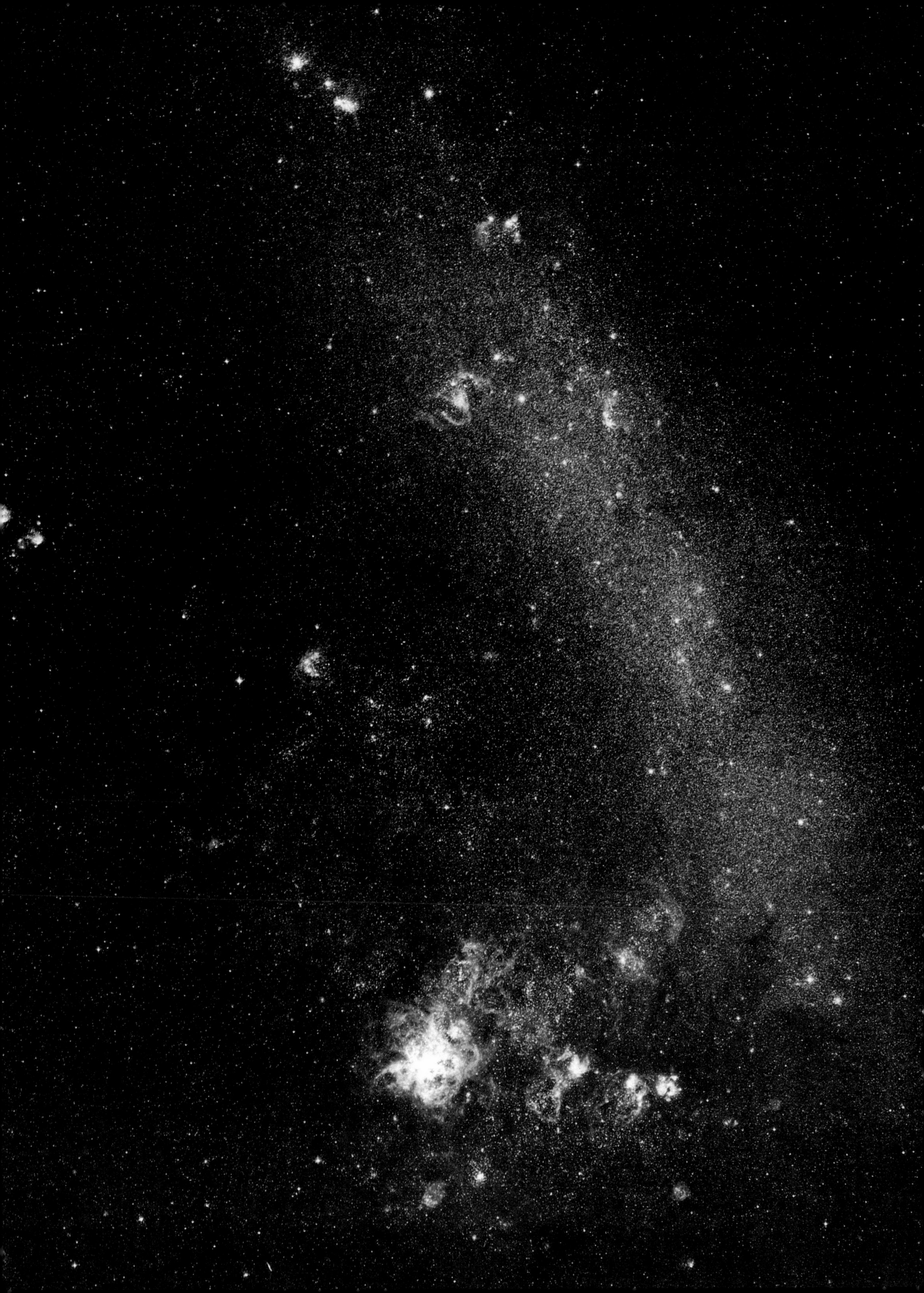

Astronomers classify galaxies by their shape, but do not really know what the different shapes indicate, nor how they evolve

The classification of galaxies

Galaxies are classified according to their shapes. Several classification schemes exist, but the simplest, due originally to Edwin Hubble, recognizes three fundamental forms – elliptical, spiral and irregular – and arranges them in a sequence known as the "tuning fork" diagram.

The letter "E" denotes elliptical galaxies, and is followed by a number between 0 and 7 according to the degree of flattening of the apparent ellipse. If "a" denotes the major axis and "b" the minor axis, the number is given by $10(a-b)/a$. For example, if the major axis is twice as long as the minor axis (that is, a = 2b), the class would be given by $10(2-1)/2 = 10/2 = 5$, and the galaxy would be of type E5. Spherical galaxies are denoted by E0, and no true ellipticals are flatter than E7.

Galaxies with spiral arms occur in two broad types – spirals (S) and barred spirals (SB). The former type have spiral arms which emerge from the nuclear bulge, while in the latter the arms emerge from the ends of what looks like a bar of stars and interstellar matter straddling the nucleus. Each type is then subdivided according to the size of the nucleus and the tightness of the spiral pattern. Of the spiral galaxies, Sa have the most tightly wound arms and the largest nuclear regions (in proportion to their overall size), Sb galaxies have a less tight spiral pattern and a relatively smaller nuclear bulge, while Sc galaxies have loose, open arms and small nuclei. Barred spirals are similarly subdivided into SBa, SBb and SBc types. Still more information may be added; for example an additional letter "s" signifies that the arms start in the nucleus, while "r" denotes that they emerge from a ring round the nucleus.

Intermediate between the ellipticals and the spirals are the SO, lens-shaped (lenticular) galaxies, which are elliptical with a narrow disk round their "equator" but no spiral arms. Irregular galaxies, which have no well-defined nucleus or structure, are designated Irr. Galaxies of type Irr I follow the tuning fork sequence in that they give the impression of spiral arms broken up into a confused jumble. Galaxies of type Irr II are completely chaotic and often highly active.

Gas, dust and hot young stars

The average content of gas, dust and hot young O and B-type stars increases along the tuning fork from the ellipticals to the Sc-types and irregulars. Typical proportions of gas and dust are less than 0·1 percent for ellipticals, up to 20 percent for spirals and 20-30 percent for irregulars, although there are exceptions with up to about 50 percent. The gas-depleted ellipticals contain a high proportion of old red Population II stars, while spirals have Population II stars in their nuclear and halo regions together with Population I disks which become progressively more dominant from Sa to Sc. Some, but not all, irregulars contain the highest proportion of blue O and B-type stars. Star formation has almost ceased in ellipticals, but not in spirals and irregulars.

Galactic masses

The masses of galaxies range from less than a million to over ten million million solar masses for ellipticals and one billion to a thousand billion solar masses for spirals, while irregulars are all less than a few tens of billions of solar masses. Apparent diameters range from less than 3,000 light years to more than 150,000 light years. These figures refer only to the visible parts of galaxies – if many of them have massive haloes and coronas of invisible matter, the figures given for size and mass are certainly underestimates.

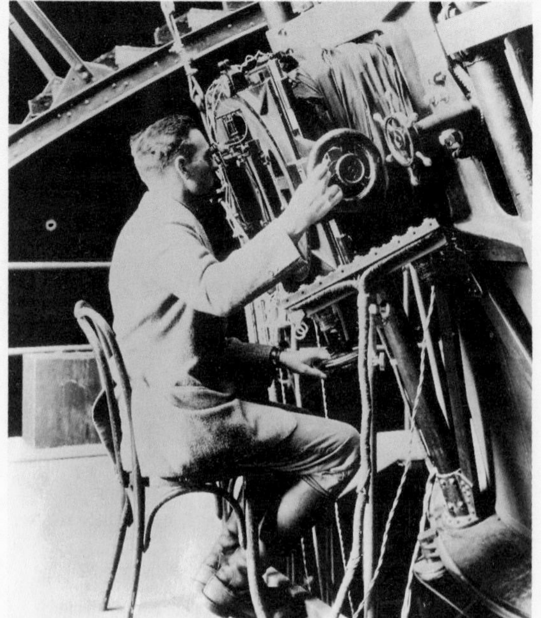

▲ ***Edwin Hubble at the Mount Wilson's 2·54m reflector.***

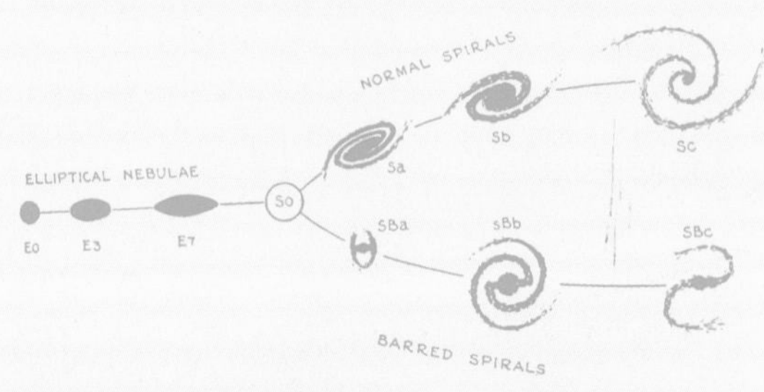

▲ ***The Hubble "tuning fork" classification of galaxies.***

The pioneers of galaxy classification

Following his discovery that the "starry nebulae" really were external systems, in 1923 Hubble set to work with his colleague Humason to draw up a scale of classification. The two American astronomers were an interesting pair. Edwin Hubble (1889-1953) served in the American army during the first world war, and also became an amateur boxing champion. Milton Humason (1891-1972) had no formal training at all; his first engagement at Mount Wilson Observatory was as a mule-driver, but he joined the scientific staff in 1920 and from then on worked closely with Hubble.

The two produced a plan which included the various types of galaxy, and were tempted to suggest that these represented different stages of evolution – that an elliptical galaxy evolved into a spiral, or vice versa. However, there were serious objections. In particular, the most massive ellipticals, such as M87 in Virgo, were much more massive than the spirals, and the idea of an evolutionary sequence had to be abandoned. As with the studies relating to stellar evolution, the situation proved to be much less straightforward than astronomers had originally thought.

The classification of galaxies

NGC 1201 – Type SO

NGC 4636 – Type EO

NGC 175 – Type SBa

NGC 2811 – Type Sa

NGC 4406 – Type E3

NGC 1530 – Type SBb

NGC 3031 – Type Sb

NGC 3115 – Type E7

NGC 1073 – Type SBc

NGC 5364 – Type SC

The Local Group comprises an assortment of irregular, elliptical and spiral galaxies of various sizes

Other members of the Local Group

Most of the galaxies in the local group are rather small elliptical or irregularly-shaped systems, but there are a few larger systems, including two spirals, M33 in the constellation of Triangulum and M31 in Andromeda. M33 is a well-defined spiral with a nucleus and spiral arms, but with a more open structure than the Milky Way system. Some 2·7 million light years distant, it is smaller than our own Galaxy, and contains about 30 billion stars.

M31, the great galaxy in Andromeda, is a spiral but is nearly edge-on to the Milky Way so that its spiral structure is not easy to see. At a distance of 2·3 million light years, it is the most distant object detectable – under good conditions – to the naked eye. Only the extensive nuclear bulge is visible to the naked eye, or through small telescopes and binoculars. The whole system, comprising nuclear bulge and disk, is larger and more massive than the Milky Way system and probably contains about 300 billion solar masses. However, when the massive corona is taken into account, the Milky Way is quite possibly the most massive member of the Local Group. M31 has two smaller elliptical companion galaxies, M32 (NGC 221) and NGC 205.

The other massive galaxy in the locality is Maffei 1, one of two galaxies discovered in 1968 by the Italian astronomer Paolo Maffei. Both are in Cassiopeia, near the plane of the Milky Way, and are therefore heavily obscured by dust. Maffei 1 appears to be about 3 million light years away, making it a member of the Local Group, but being a giant elliptical system it is quite different from any other member of the group. Only about one percent of its light penetrates to the Solar System, however, making it very difficult to analyze.

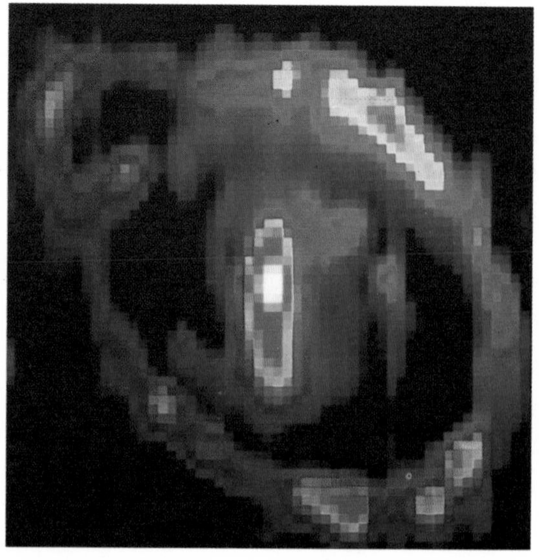

▼ *The galaxy known as M33, one of the Local Group, lies in the constellation of Triangulum. This optical photograph has enhanced colors which reveal the areas of continuing star formation (pink), regions of young stars (blue) and dust (orange) lit by older yellow stars in the center of the galaxy.*

► ▲ *M31, the great galaxy in Andromeda, is a massive spiral similar to the Milky Way. The true-color view (right) shows M31 and its two elliptical companions, M32 and NGC 205. Viewed at a shallow angle its spiral structure is not particularly obvious. The false-color infrared view (above) shows lanes of heated dust where stars are forming.*

See also
Overview of the Universe 129-32
Between the Stars 181-8
Active Galaxies and Quasars 205-16
Clusters of Galaxies 217-24
The Evolving Universe 229-44

Calculating the masses of the galaxies

The spectrum of a galaxy consists of the combined spectra of vast numbers of stars, and takes the form of a continuous spectrum with dark absorption lines. With a spiral galaxy, Doppler shifts in these lines and in radio emission lines (such as the 21cm radiation emitted by HI regions) can reveal the speed of rotation at different distances from its center. From this it is possible to calculate the mass of the visible disk and nuclear bulge. However, in many spirals, stars and gas clouds continue to move at high speeds at the outer perimeter of their visible disks, and this implies that they have massive haloes or coronas, so that such masses must be regarded as minimum values.

Spherical and near-spherical galaxies have very little net rotation and their stars move in a more or less random fashion, so that in any part of such a galaxy some stars are approaching and others are receding from an outside observer. Each line in the galaxy's spectrum contains contributions from receding (red-shifted) stars and approaching (blue-shifted) stars, and the effect of this is to make the spectral lines broad rather than narrow. The degree of line-broadening reveals the range of velocities, and, since the speed of stars depends on the gravitational pull to which they are subjected, this provides a clue to the total mass of that galaxy.

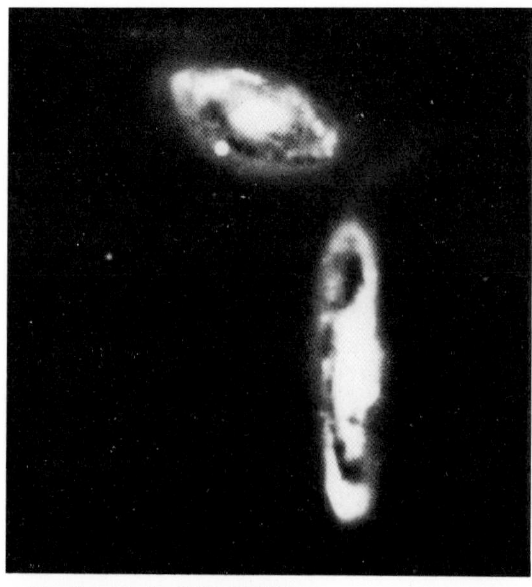

▲ *Peculiar galaxies NGC 3986/88.*

The evolution of galaxies

Astronomers do not know for certain how galaxies were born or how they have evolved. They think that each galaxy originated as a massive cloud of hydrogen and helium contracting under the action of gravity; and the type of galaxy into which it evolved probably depended on the amount of rotation in the cloud and the rate of star formation.

In a slowly collapsing cloud with little rotation, star formation would proceed quite rapidly, converting most of the gas into stars at an early stage and producing a spherical or elliptical galaxy. Galaxies of this type today are dominated by old red first-generation stars, since very little gas was left to form later generations of stars after the first bout of star formation. With a rapidly collapsing and faster spinning cloud, stars would begin to form while the cloud was nearly spherical, but the gas would quickly settle into a spinning disk. This kind of behavior would produce a tenuous population of old stars in the halo and a greater concentration in the ellipsoidal nucleus; within the disk, star formation would still proceed at a relatively slow rate.

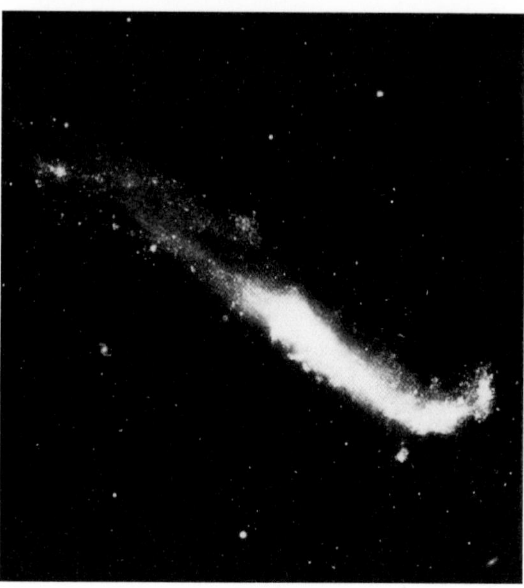

▲ *Peculiar galaxies NGC 6621/22.*

Rapid star formation

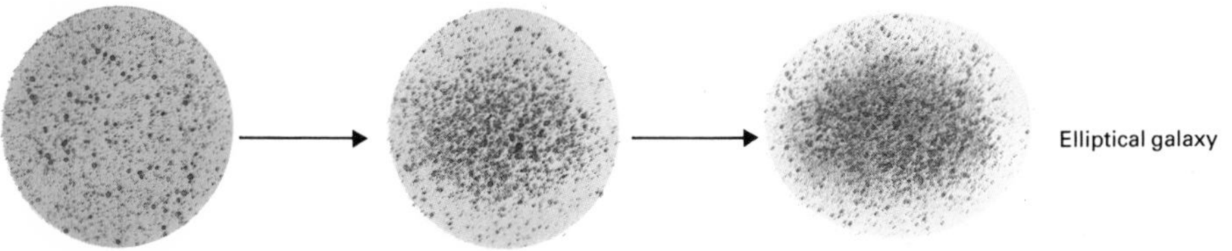

Slow star formation

◀ ***In a slowly collapsing cloud with little rotation, stars form rapidly to leave a spherical or elliptical galaxy with little gas. A faster spinning and collapsing cloud may form stars more slowly, producing a halo of old stars and a disk of gas and younger stars.***

Active Galaxies and Quasars

The massive energy emitters – the principal types... Quasars – the most powerful sources in the universe... Their energy machine – supermassive black holes... Measuring a quasar...PERSPECTIVE...The largest objects in the universe...The nearest supermassive black hole...Quasars – the most distant visible objects?... Weighing the energy machine

Astronomers have coined the term "active galaxy" to describe a wide variety of disturbed sources of light, radio waves and other radiations which emit up to a million times more energy than an ordinary galaxy, and which all seem to require the presence of a compact and immensely powerful "energy machine" at the center. Among the various types of active galaxy are Seyfert galaxies, BL Lacertae objects, radio galaxies and quasars, all of which are generally considered to be galaxies containing hyperactive nuclei.

Seyfert galaxies

The Seyfert galaxies are members of a class first recognized in 1943 by Carl Seyfert. They are S and SB galaxies with extremely bright nuclei. In short-exposure photographs they look rather like stars because so much of their light comes from the compact nucleus. Most Seyferts are strong infrared sources (this radiation probably coming from heated dust); some are X-ray sources, but in general they are not particularly active at radio wavelengths. Their central radio sources are so small that existing techniques cannot resolve them, and they show short-term variability. Analysis of the emission lines of ionized gas in their nuclei indicates that gas clouds are circulating at speeds of several thousand kilometers per second, and this suggests that the Seyferts may contain massive compact central objects.

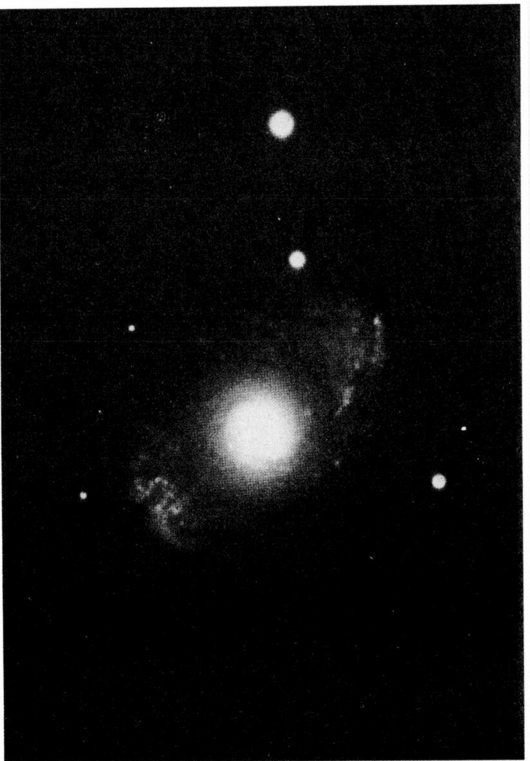

▲ *The exposure of the photographic plate was long enough to allow the spiral structure of this Seyfert galaxy, NGC 4151, to register. At shorter exposures the compact, bright central nucleus may be the only part of such a galaxy to appear, and because of this it can be mistaken for a star.*

▼ *This loose, moderately rich cluster of galaxies, about 250 million light years away, is dominated by the giant "Lenticular" galaxy IC4329 (right of center, top). The edge-on spiral just to the left of IC4329 is the Seyfert galaxy IC4329A. The nucleus of this galaxy is a strong emitter of X-rays.*

The bizarre active galaxy Centaurus A is so complex in structure that scientists used to think it consisted of two colliding galaxies

A "nearby" supermassive black hole

The nearest radio galaxy, Centaurus A, lies about 16 million light years away; optically it appears as an elliptical galaxy, designated NGC 5128. An elongated radio-emitting envelope some 3 million light years in extent surrounds the visible galaxy. There are two outer lobes, a single radio-emitting cloud 100,000 light years from the center and two inner lobes 30,000 light years from the core.

Observers have detected infalling clouds of hydrogen about 500 light years from the center, and these could provide enough energy to power the radio galaxy. Variability on a timescale of a few hours points to the existence of a compact source which could contain a black hole of up to a billion solar masses. Although Centaurus A is not highly active today, the extended radio lobes indicate a higher level of activity in the past. It may well harbor the nearest actively (if rather feebly) accreting supermassive black hole to the Earth.

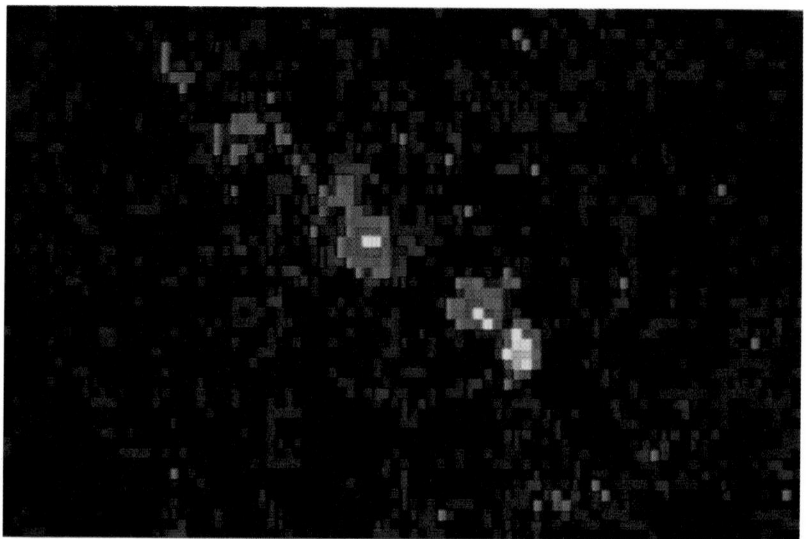

◀ *This X-ray image of Centaurus A shows a jet of X-ray emitting material extending for some 15,000 light years from the core (white spot near center) towards one of the radio lobes. The jet contains several bright knots and its X-ray emission is probably caused by the synchrotron process. The X-radiation from the core probably comes from an accretion disk of hot gas swirling round a massive black hole.*

▼ *The optical photograph of Centaurus A, also known as NGC 5128, shows older (yellow) stars at the center and younger (blue) stars on the fringes of the reddish dust cloud.*

▶ *Radio emissions at a wavelength of 21cm (the hydrogen line) show Centaurus A to be a huge radio galaxy, elongated and lobed, spanning about 2·5 million light years of space. This makes it one of the largest objects known to astronomy. The color coding on the contour map runs from purple for the faintest regions to pink for the most intense. The visible galaxy covers only the innermost pink contour.*

▼ *This radio picture from the Very Large Array in New Mexico shows in more detail the structure of the center of Centaurus A. Bright radio regions are red, dim regions blue.*

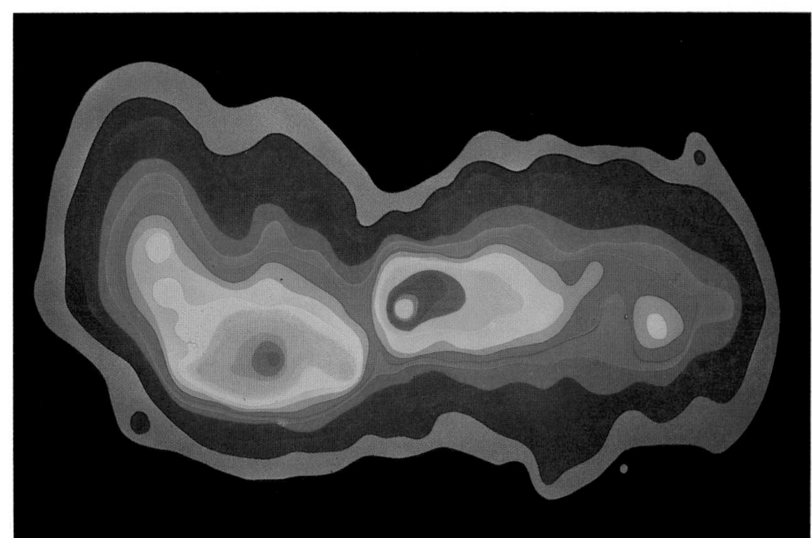

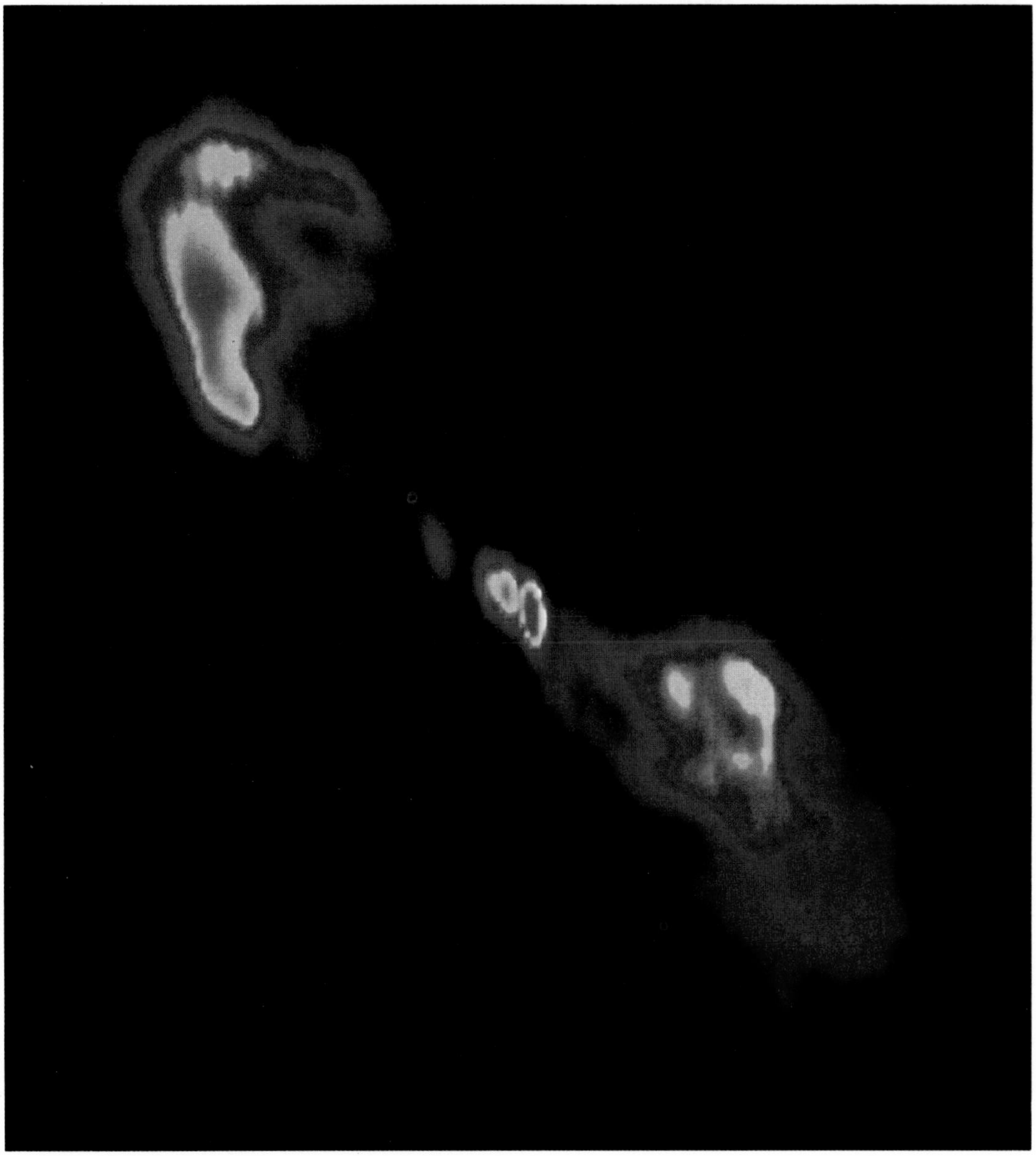

The largest single object known to science is a radio galaxy with a maximum diameter of about 20 million light years: its name is 3C 236

Radio galaxies

A radio galaxy emits more strongly at radio wavelengths than in the visible part of the spectrum. The power outputs of strong radio galaxies are typically 10^{38} to 10^{40} watts – ten to a thousand times greater than the entire luminosity of the Milky Way galaxy. Many of them have a characteristic double-lobed appearance with two radio-emitting regions located on either side of the central optical galaxy – usually an elliptical galaxy which may also contain a compact central radio source. The overall extent of the largest radio galaxies is enormous. For example, the giant radio source 3C 236 spans a maximum diameter of about 20 million light years.

Radio-emitting lobes and jets

Some radio galaxies have several roughly symmetrical lobes located at different distances from their centers along a line, and in some cases high-resolution radio maps reveal that the cores of these galaxies contain radio-emitting jets which are lined up with the outer lobes. The giant radio galaxy NGC 6251 is particularly interesting. It has a central jet some six light years long, lined up with a much more extensive jet 720,000 light years long. This, in turn, is aligned with the outermost lobes, which have a span of no less than ten million light years.

Scientists believe that the radio-emitting lobes are clouds of material ejected from the nuclei of the central galaxies in a series of violent episodes. The jets of the core represent new material which is being beamed out towards the outer lobes. Energetic electrons in these clouds, moving through magnetic fields at a large fraction of the speed of light, emit radiation because they are compelled to gyrate around the magnetic lines of force. When radiation is produced in this way, it is known as "synchrotron radiation" (➧ page 209).

The first radio galaxy to be identified was Cygnus A. Optically it appears as a faint, peculiar galaxy located in the constellation of Cygnus at a distance of about 750 million light years. Despite this enormous distance, Cygnus A is the second brightest radio source in the sky.

▲ *The object known as 3C 236 had been identified by astronomers as a radio source associated with an elliptical galaxy long before its full extent was revealed. Its radio-emitting lobes span a region of space nearly 200 hundred times larger than the diameter of the Milky Way. As with other radio galaxies, it is centered on a nucleus that is optically detectable.*

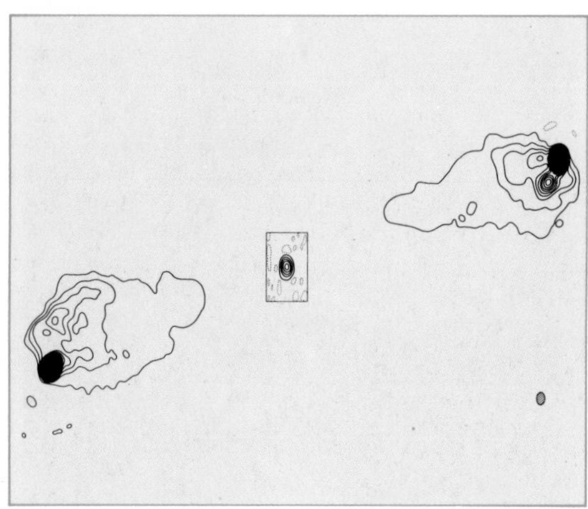

▲ *Cygnus A, a radio galaxy in Cygnus, is one of the strongest sources of radio emission in the sky. This contour map reveals "hot spots" at the extremities of the lobes.*

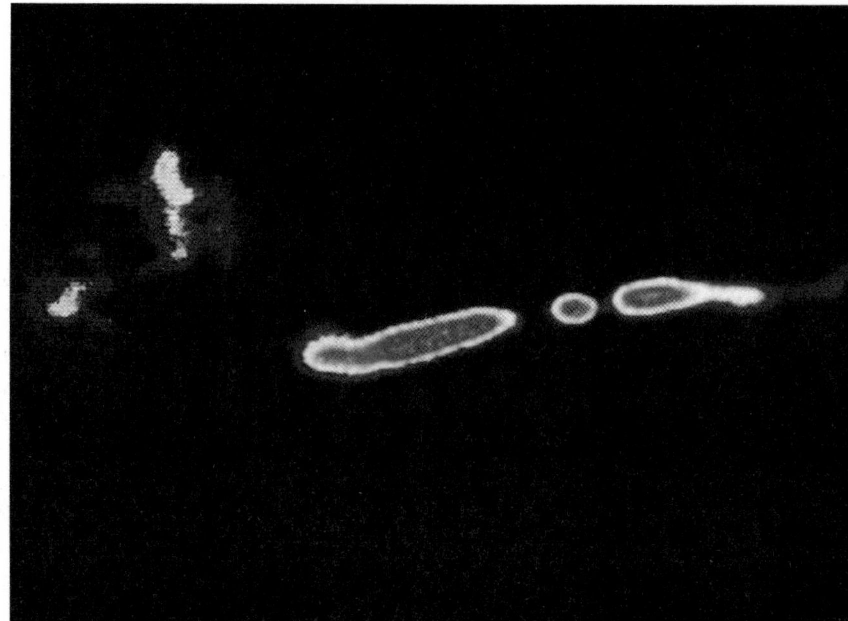

► *The object known as 3C 449 is an enormous radio galaxy. Only the central component is faintly visible to optical telescopes, but radio telescopes detect extended jets and lobes.*

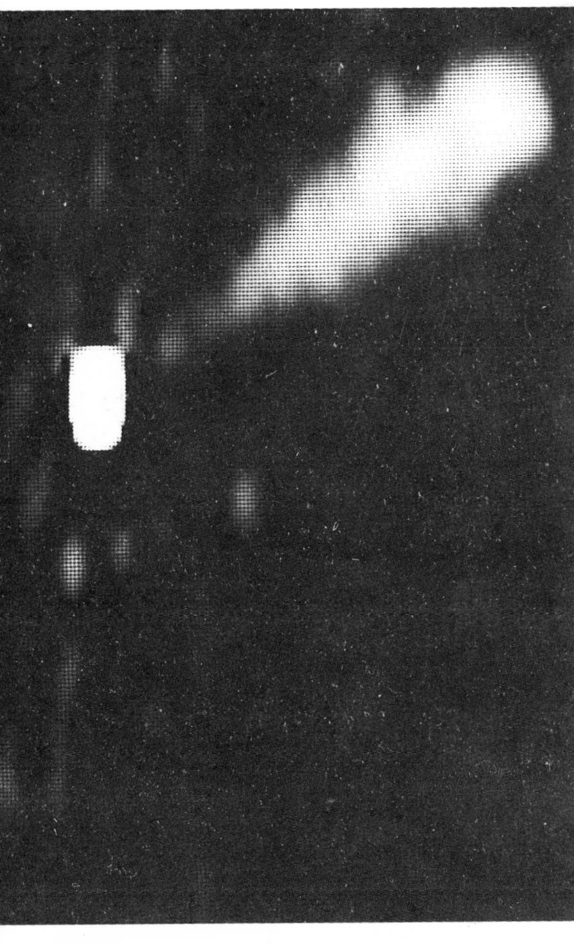

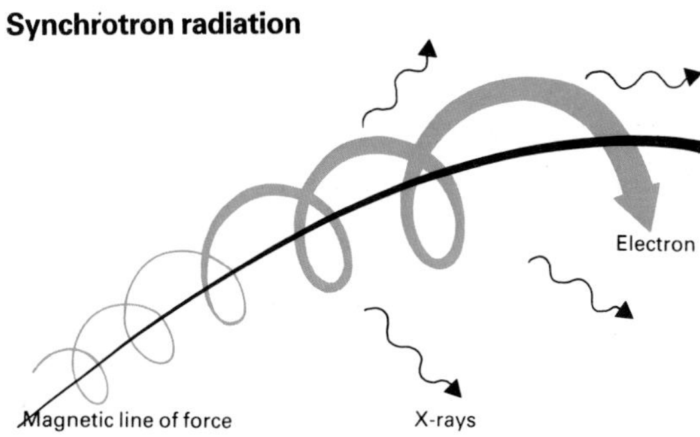

◄ *Synchrotron radiation: an electron moving in a magnetic field at a large fraction of the speed of light is forced to follow a helical path round a magnetic line of force and, because of its continual changes of motion, it emits radiation as it does so. The faster the electron the shorter the emitted wavelength.*

▼ *Our galaxy compared in size to four radio galaxies.*

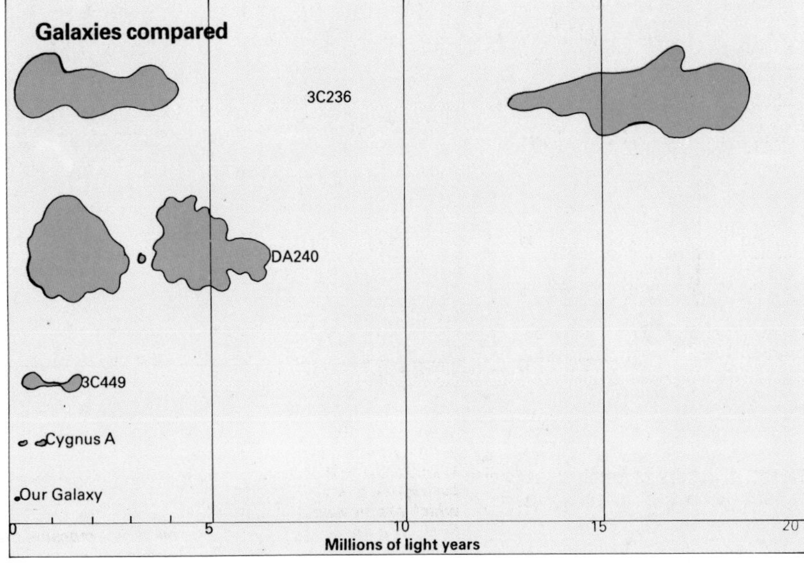

► *Despite its distance, the radio galaxy DA 240 would look wider than the Full Moon if its radio emissions were visible. Computer processing has given this simulated view.*

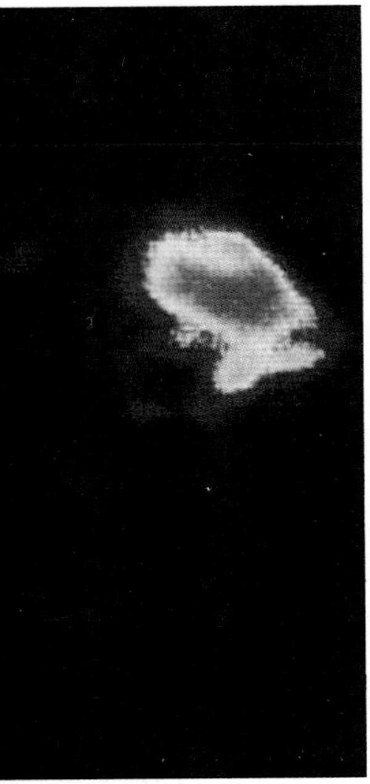

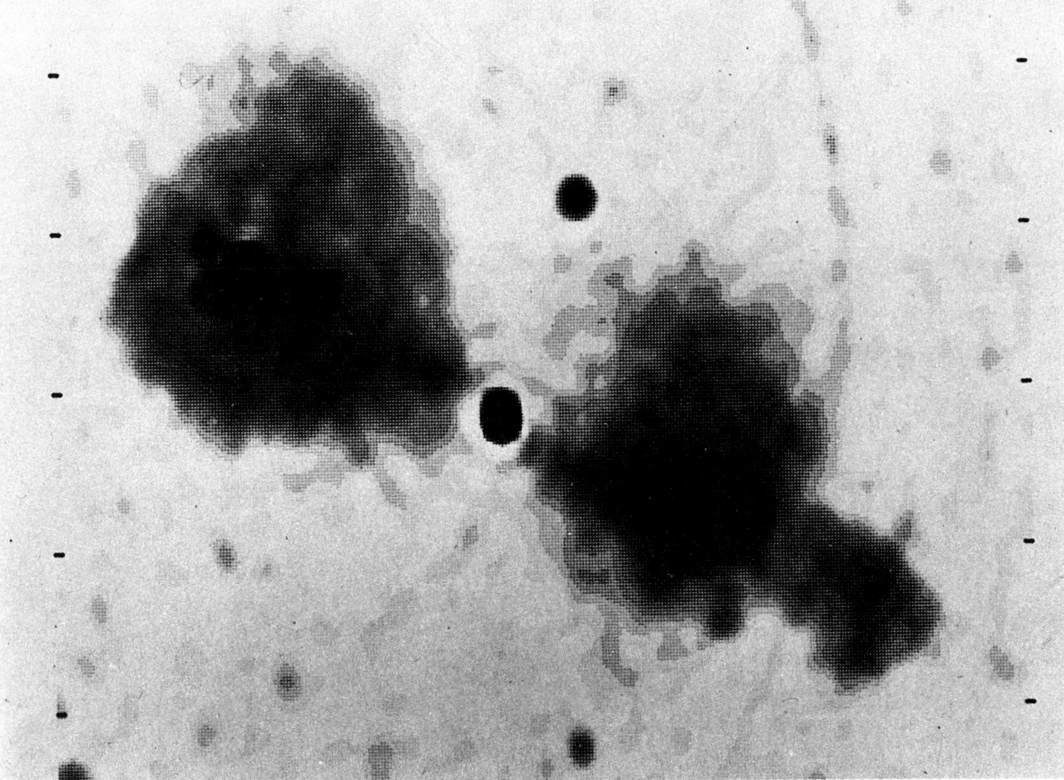

The more astronomers learn about the active galaxy M87, the more enigmatic it becomes

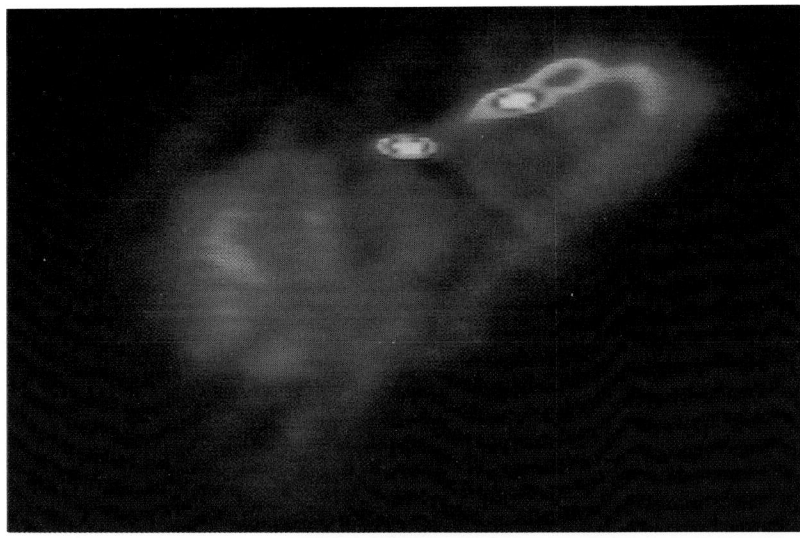

A nearby active galaxy

One of the nearest active galaxies is M87, a giant elliptical galaxy which lies at a distance of 50 million light years. Also known as Virgo A, it is the brightest radio source in that constellation, but is not particularly radio-powerful compared to some more distant radio galaxies.

M87 contains a conspicuous nuclear jet which emits all kinds of radiation from X-ray to radio, but there is no obvious sign of an accretion disk. It is not a brilliant source like a quasar (▸ page 212), and it is almost devoid of gas. However, it may contain a dead quasar. Stars in the innermost part of the galaxy are moving so fast that a central mass of about 5 billion solar masses is required to account for their motions. This central body may be a supermassive black hole, and if so, when the galaxy was much younger and there was plenty of fuel to feed the energy machine, M87 could have matched the most brilliant quasar.

▲ The radio-emitting regions of the galaxy M87 are considerably less extensive than the visible galaxy. Radio emissions are strongest in the jet, which, in the absence of image enhancement techniques, is hardly descernible within the galactic nucleus.

◄ This CCD optical photograph of M87 has been processed to reveal the structure of the remarkable jet. This is caused by synchrotron radiation.

► The spiral galaxy NGC 4319 is shown here in false color. The small appendage on the left-hand side, approximately level with its nucleus, is the quasar Markarian 205. The quasar has a redshift more than ten times greater than that of the galaxy, which implies that the quasar is over ten times further away. However, H.C. Arp argued that Markarian 205 was linked to the galaxy by a bridge of luminous matter and could not, therefore, be a much more remote background object. If this were correct the widely held view that quasars are extremely remote objects might have to be reconsidered. However, recent observations seem to have proved that Markarian 205 is after all more than ten times farther away than NGC 4319. It seems to be surrounded by a faint luminous "fuzz" with the same redshift as the quasar – probably a galaxy with Markarian 205 in its core, supporting the theory that quasars are the hyperactive nuclei of very remote galaxies.

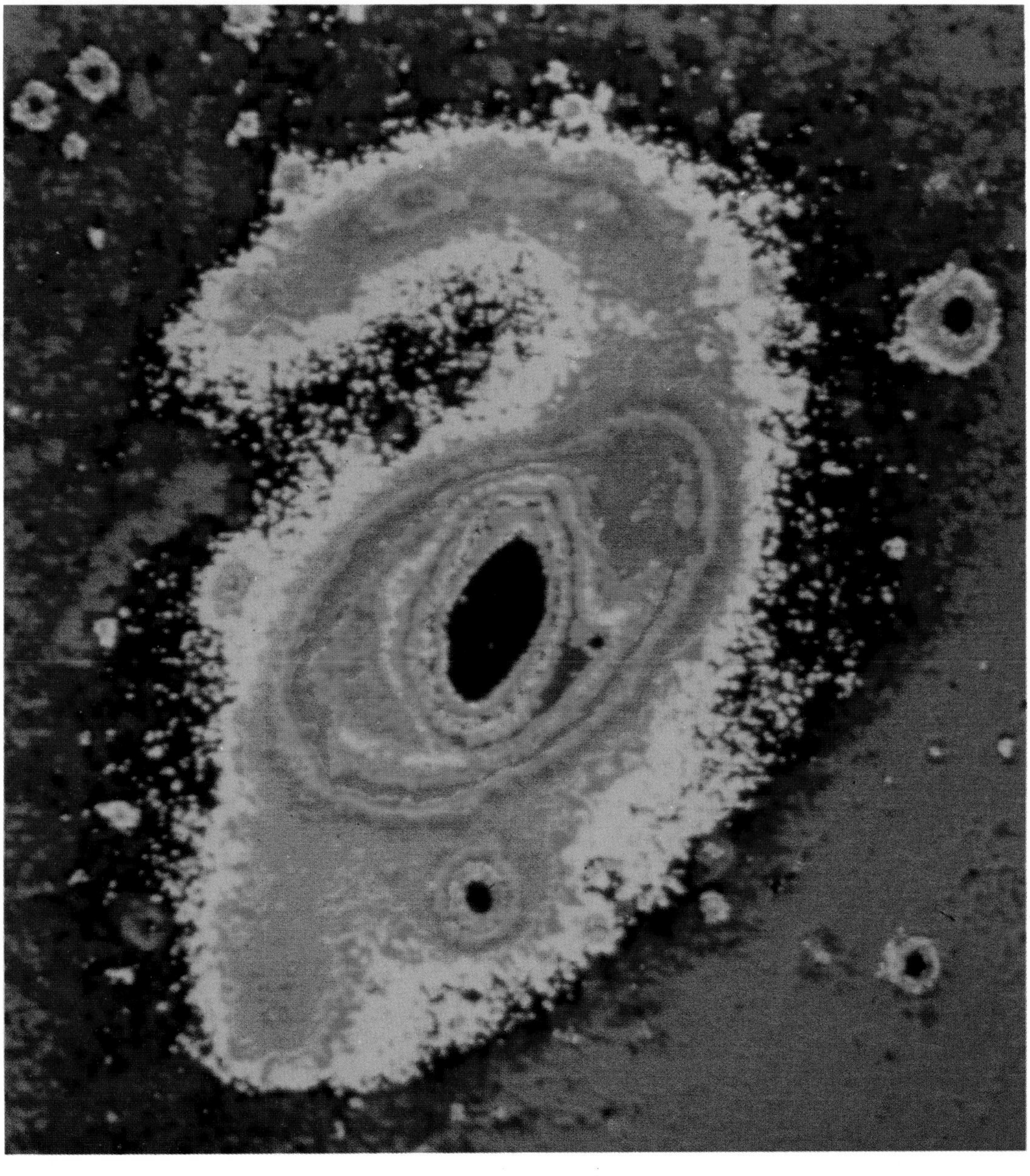

Some scientists doubt whether any object could deliver the power ratings claimed for quasars

Redshifted quasars

The redshift controversy

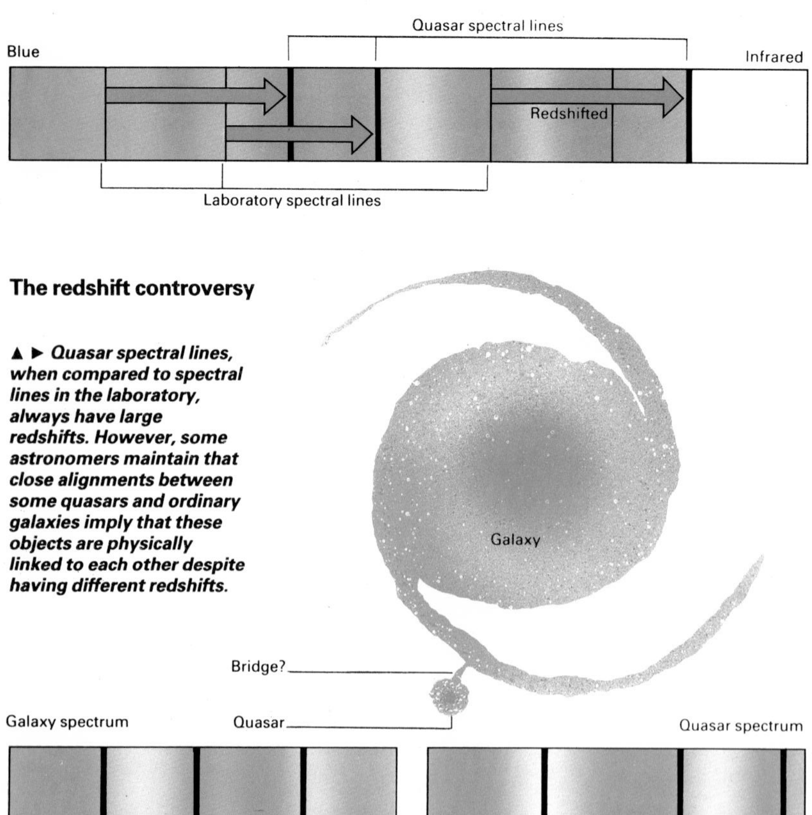

▲ ► Quasar spectral lines, when compared to spectral lines in the laboratory, always have large redshifts. However, some astronomers maintain that close alignments between some quasars and ordinary galaxies imply that these objects are physically linked to each other despite having different redshifts.

Questions about quasars

The prodigious power output of quasars is so difficult to account for that some astronomers have questioned whether they really can be at such vast distances: if they were relatively local, the power requirement would not be nearly so great. Several quasars lie very close in the sky to ordinary galaxies, and in a few cases seem to be linked to these galaxies by "bridges" of material. However, the redshifts of the quasars are much greater than those of the apparently associated galaxies. The argument runs that such close chance alignments between foreground galaxies and distant quasars would be highly improbable, and that the galaxies and quasars must be physically connected. If this were so, quasars would not be at vast distances.

Most astronomers doubt the significance of the small number of alignments. Indeed, if quasars have been ejected from galaxies, surely some of them should be heading towards the Milky Way system and would show blue shifts in their spectra? No blue-shifted quasars have been observed. Recent observations of quasars in clusters of galaxies show that they have closely similar redshifts to the cluster galaxies, and this supports the view that they are at large distances, with their redshifts being due to their velocities of recession.

Some quasars have fuzzy outlines. In a number of cases the surrounding "fuzz" has a galaxy-like spectrum and has the same redshift as the quasar. This supports the idea that a quasar is a hyperactive galactic nucleus.

Quasars

The most intriguing and controversial of all extragalactic objects, quasars derive their name from the term *quas*i stell*ar* radio source, because the first ones to be discovered were radio sources which coincided in position with starlike objects. In fact, only about one percent of these objects are strong radio sources and the terms QSO (quasi stellar object) or QSS (quasi stellar source) would be preferable. Nevertheless, the term "quasar" is in general use.

Quasars are very compact and emit radiation over a wide range of wavelengths. Most of them emit strongly at radio wavelengths, and several hundred are X-ray emitters. They vary in brightness over a timescale ranging from over a year to less than a day. Their spectra contain emission lines and sometimes absorption lines, and all of them show a large redshift. The quasar PKS 2000-330 has a redshift of nearly four – the largest so far measured – and this means that its spectral lines appear at wavelengths nearly five times longer than normal. Lines usually found in the ultraviolet are seen in the visible part of the spectrum, while optical lines are shifted into the infrared.

Most astronomers believe that quasars are remote objects sharing in the expansion of the universe so that, as with galaxies, their redshift is a Doppler shift due to the speed at which they are receding from us (➧ page 229). If this is so, they must lie at colossal distances, and some of them are among the most distant visible objects. The redshift of PKS 2000-330 suggests that it is receding at more than 90 percent of the speed of light and must be at a distance of about 13 billion light years. To be conspicuous at such vast distances, quasars must emit from 100 to 10,000 times more energy than an ordinary galaxy.

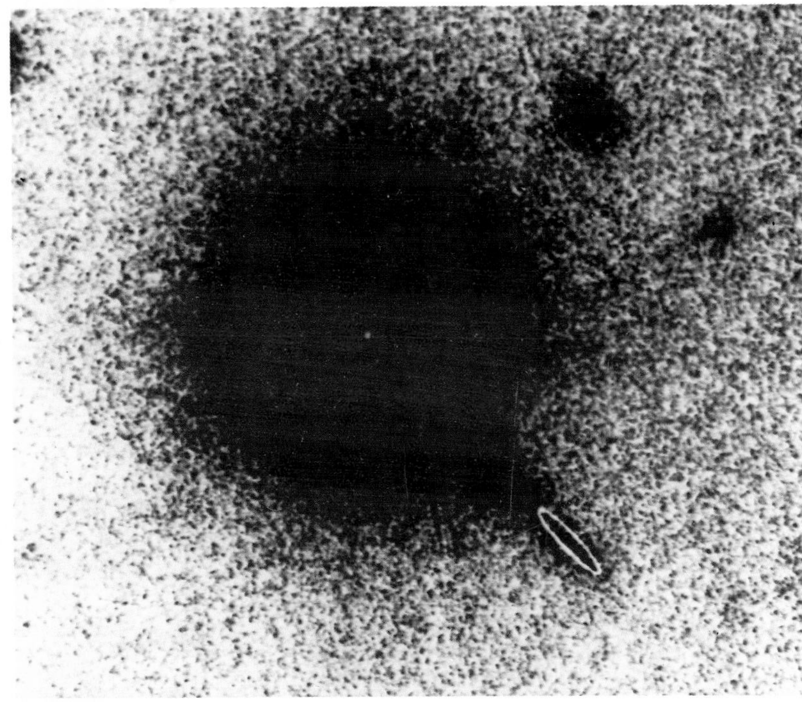

▲ *Quasar 3C 273: the dot and the oval represent discrete components of this source.*

◄ *Einstein Observatory X-ray picture of 3C 273.*

▼ *BL Lacertae objects are point-like optical and radio sources which often show variability of greater degree and on a shorter timescale than quasars themselves. Their redshifts are rather hard to measure, but in those cases where spectral lines have been seen, large redshifts have been found. Some of these quasar-like objects have been shown to lie inside galaxies.*

Estimating the size of a quasar

The compact appearance of quasars suggests that they must be much smaller than ordinary galaxies, and their rapid variability confirms this, for a source of light (or any other kind of radiation) cannot vary in brightness in a period of time much less than the time which light takes to cross the source. For example, consider the surface of a luminous sphere with a radius of 300,000km (the distance that light travels in one second) which increases in brightness and then fades back to normal in a fraction of a second. If a distant observer could resolve the sphere, he would see a disk which would brighten first at its center and then progressively towards the edge. Because of the extra distance from the edge of the disk to the observer, the edge itself would finally brighten up one second after the center, by which time the center of the disk would already have faded back to normal. If the source cannot be resolved at the observer's distance, and appears as a star-like point, the flash will be spread out over a second. Thus the timescale of variability sets an upper limit to the size of the energy source, so that if a quasar varies in brightness in a year, its energy source cannot be more than a light year across; if it varies in a day, it must be about a light day in size. One of the X-ray quasars, OX169, varies in X-ray brightness by a factor of two or three in less than 100 minutes, and this implies that the energy source is smaller than the orbit of Saturn.

On the face of it quasars, which typically emit about 10^{40} watts, seem to radiate the power of a thousand galaxies from a region of space less than one billionth of the diameter of a normal galaxy. How can so much energy be generated in such a tiny volume?

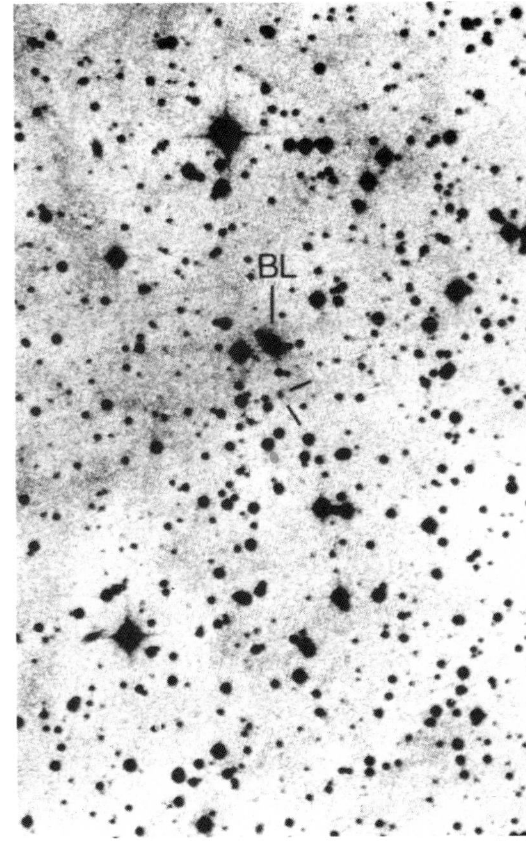

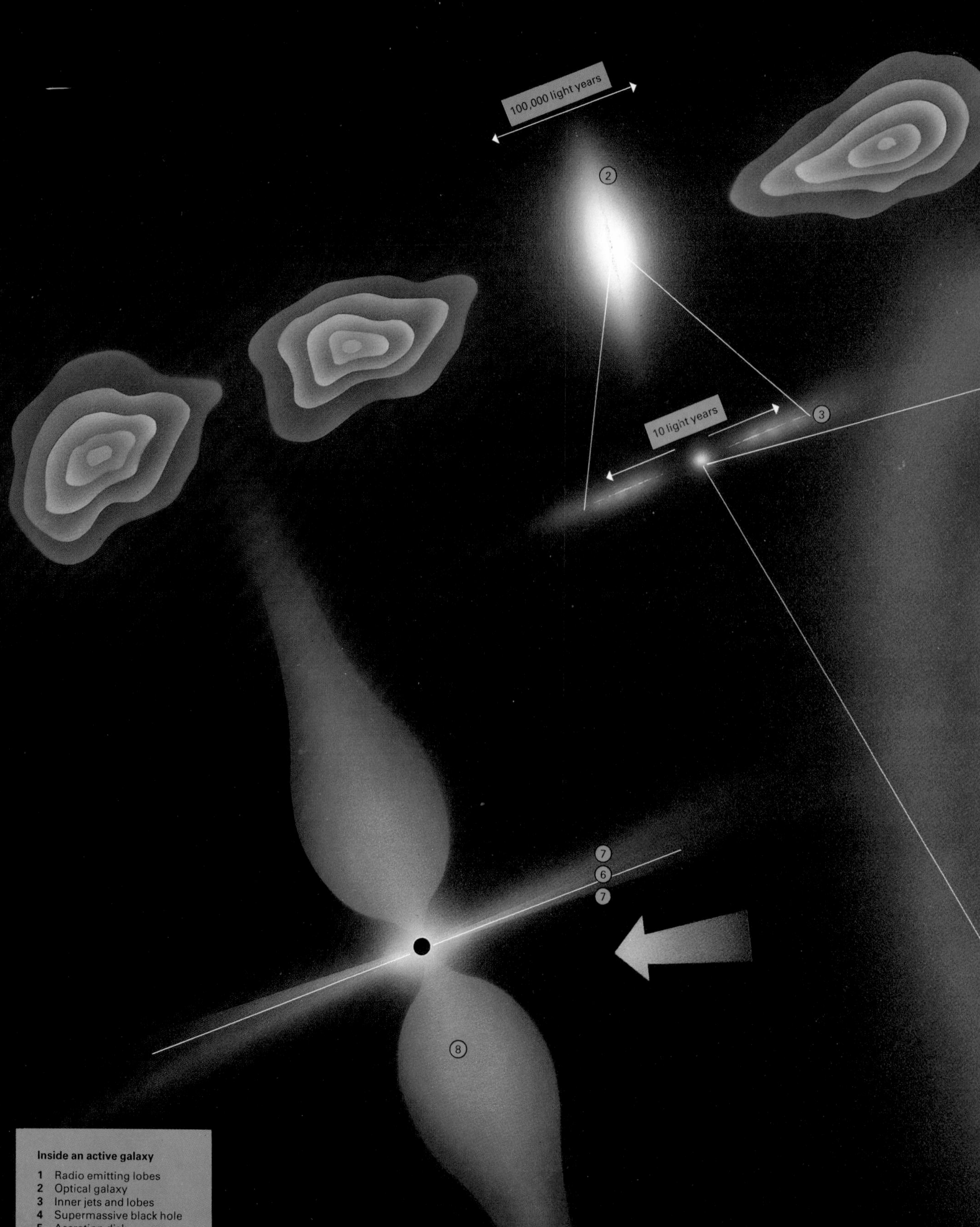

Inside an active galaxy

1 Radio emitting lobes
2 Optical galaxy
3 Inner jets and lobes
4 Supermassive black hole
5 Accretion disk
6 Axis of black hole
7 Expelled jets
8 Inner accretion disk

◀ *The energy machine inside an active galaxy. All active galaxies require immensely powerful compact central energy sources. Many astronomers believe that these must be supermassive black holes surrounded by swirling disks of gas, into which gas and possibly shredded stars are falling.*

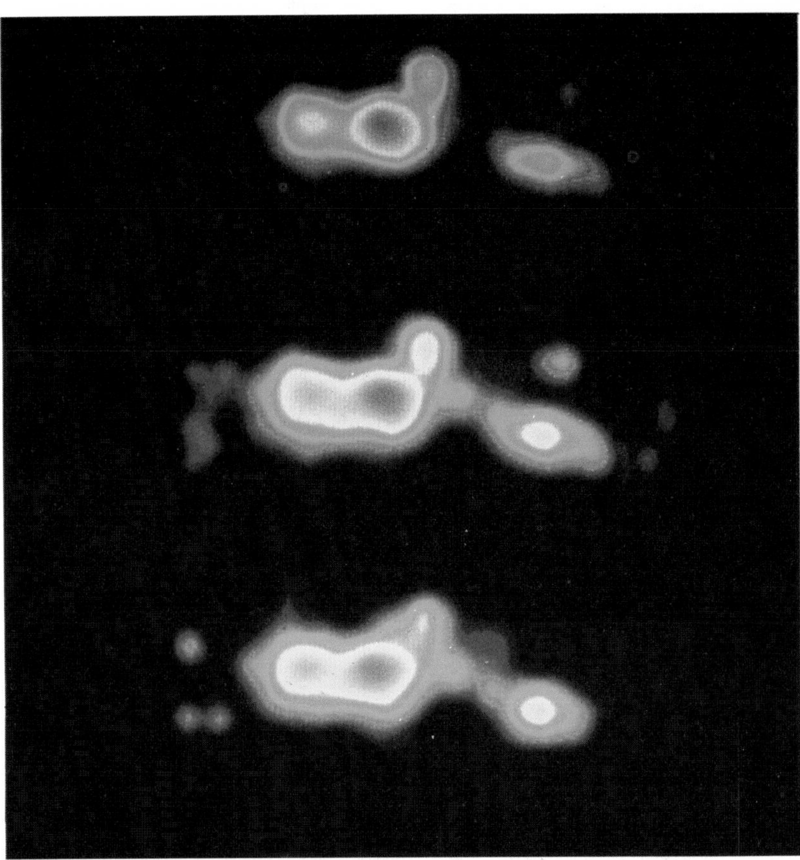

▲ *Radio maps of the Seyfert galaxy NGC 1275.*

Active galaxies – the energy machine

The evidence strongly suggests that all forms of active galaxy, from Seyferts to BL Lacertae objects, quasars and the nuclei of radio galaxies, are different manifestations of the same basic phenomenon, differing in degree rather than in fundamental nature. Each can be described as a galaxy with a hyperactive compact nucleus.

They all require the presence of a variable, compact and immensely powerful energy source. The most convenient source would be a supermassive black hole accreting matter from its surroundings in the form of both gas and the shredded remains of stars, torn apart by tidal forces near its event horizon. Matter falling under the gravitational attraction of a black hole acquires a prodigious amount of energy, which is released as radiation as it plows into the accretion disk of hot gas swirling round the hole. The energy released in this way could easily be equivalent to that which would be released if ten percent or more of the incoming mass were converted into energy in accordance with Einstein's's formula $E=mc^2$.

Operating at this sort of efficiency, a supermassive black hole could power a typical quasar by digesting no more than a few solar masses of material per year. An accreting supermassive black hole would be the most luminous type of object in the universe.

An alternative view is that the energy machine is a massive object which is prevented from becoming a black hole by its rapid rotation, or is supported by a powerful magnetic field. Such objects have been termed "spinars" and "magnetoids." However, objects such as these would in time lose enough energy to collapse into a black hole, so the black hole model remains the most useful, at least for the present.

See also
Overview of the Universe 129-32
Between the Stars 181-8
Galaxies 197-204
Clusters of Galaxies 217-24
The Evolving Universe 229-44

Mechanics of a supermassive black hole

The amount of energy released from the region around a black hole would depend on the mass of the black hole and the amount of material available to feed it, while variability in its output could be due to a number of factors such as fluctuations in the flow of fuel, or the presence of hot spots in the circulating accretion disk. It is possible to estimate the maximum mass of the black hole from the variability in energy output (◀ page 178). Since there is a limit to the amount of material that can be funneled into a black hole, it must exceed a minimum mass value in order to sustain a high luminosity. If the hole is too small, radiation pressure due to the concentrated output of energy will blow away the infalling material. For an output of 10^{39} watts a black hole of about a hundred million solar masses is required, and for 10^{40} watts, a billion solar masses. Excessively massive black holes would swallow stars whole, without tidally disrupting them, and this process would not efficiently produce X-ray emitting matter.

Can our Galaxy become a quasar?

Taking account of all these factors, the optimum mass for a black hole capable of powering a quasar seems to be between 100 million and one billion solar masses. Less spectacular objects, such as Seyfert galaxies, require less massive black holes.

An active galactic nucleus which ran short of gas or stars would no longer be able to power its energy machine and would switch off or become faint. If the Milky Way galaxy has a central black hole, it is not in a very active phase at present, but if more "fuel" were available in the past it could perhaps have behaved like the nucleus of a Seyfert galaxy and, indeed, could "switch on" again in the future if enough material were to accumulate in the central region. The Galaxy does not contain a black hole sufficiently massive to power a quasar.

Black holes and the jets and lobes

The quasar phase is probably a transient one in the life histories of a small proportion of galaxies – those which contain sufficiently massive black holes. The quasar phase probably lasts, on average, from 10 million to 100 million years, and comes to an end when the black hole has mopped up all the available mass or when it has become so massive that it digests whole stars without tearing them apart.

The black hole theory provides a plausible explanation of the jets and beams seen in radio galaxies and some quasars. If a spinning black hole resides at the center of a large thick accretion disk, the vast amount of energy released at the inner part of the disk will drive some matter away from the vicinity of the hole. It is much easier for matter to escape along the lines of least resistance – perpendicular to the accretion disk – than to fight its way out through the extensive disk itself. In these circumstances material would be squirted out in two narrow jets departing in opposite directions along the rotation axis of the black hole, and as the electrons in the beams are accelerated through magnetic fields they emit synchrotron radiation at wavelengths ranging from X-ray to radio.

A supermassive spinning black hole would act like a huge cosmic gyroscope, its axis maintaining a constant direction in space over very long periods of time. Consequently the jets would continue to point in the same direction for very long periods, and this would account for why the small-scale jets and the large radio lobes remain lined up over distances, in some cases, of many millions of light years.

Weighing the energy machine

The nearest bright Seyfert galaxy, NGC 4151, is "only" 50 million light years distant. Ultraviolet observations of its compact core have revealed three different high-speed gas clouds at different distances from its center and moving at speeds between 14,000km per second for the innermost and 4,000km per second for the outermost. Flare-ups in the core successively illuminate the three clouds, and the time delay between the flare-ups of the clouds allows their distances from the center to be established. It is then possible to calculate the mass of the central body from the distances and velocities of the orbiting clouds. A figure of 100 million solar masses was announced in 1983, but subsequent analysis suggested that the data fitted better to a mass of one billion solar masses. Either figure is consistent with the supermassive black hole hypothesis.

The central region of NGC 4151

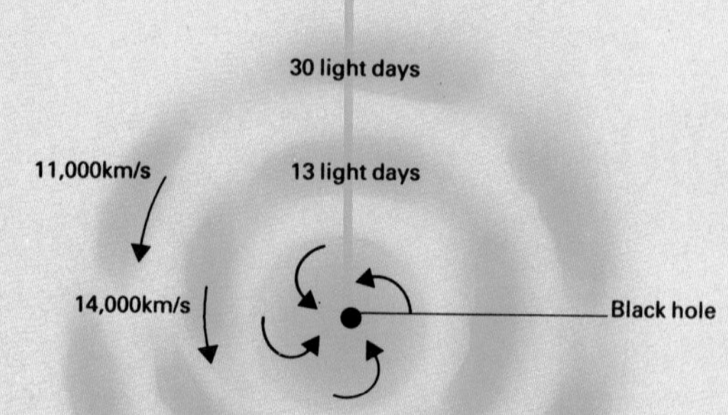

▲ *The photograph shows the bright nucleus and relatively faint spiral arms of the Seyfert galaxy NGC 4151 inset against a view of the possible structure of that galaxy's core. The rings of gas clouds, orbiting at the speeds and distances shown, have been inferred from ultraviolet observations.*

Clusters of Galaxies

The Local Group as a typical cluster...The superclusters...Close encounters – where galaxies can collide...What holds the cluster together? – the missing mass problem...The distorting effect of intergalactic gas...PERSPECTIVE...Clusters within clusters?

Galaxies seem to be gregarious, for they normally occur in groups or clusters, and the clusters themselves seem to be loosely gathered together into still larger units called superclusters.

Groups contain anything from a few up to about a hundred member galaxies. The Milky Way system is one of the dominant members of the Local Group which contains about 30 galaxies, including 3 spirals, 4 irregulars, 4 ellipticals of moderate size (including the 2 companions of the Andromeda galaxy, M31) and an assortment of dwarf ellipticals and intergalactic globular clusters. Maffei 1, a substantial elliptical or SO galaxy lying some 3 million years from the Milky Way, may be a permanent member of the group but could well be just a temporary member or a member of another nearby group.

The most massive clusters contain up to several thousand members. As a rule the less populated clusters and groups are straggly and irregular in shape and contain galaxies of all types, with spirals usually being the most conspicuous. Massive, richly populated clusters are more structured, and have a spherical or ellipsoidal shape with a greater concentration of galaxies towards the core, where there is usually at least one very massive galaxy. They consist mainly of elliptical and SO galaxies which contain very little gas.

The ultimate cluster

Our ideas about the status of the Earth, the Sun and the Galaxy have changed dramatically over the years. All three have been shown to be very unimportant in the universe as a whole. Now the same has been found to be true of our Local Group of galaxies, despite the presence in it of three spirals (the Andromeda and Triangulum systems, and the Milky Way galaxy) and possibly a large elliptical, Maffei 1.

Is it possible that our Local Group is itself a unit in a larger system? This was the suggestion of the French astronomer Gérard de Vaucouleurs. The Virgo Cluster, at a distance of around 50,000,000 light years, is certainly much larger than ours, but it is not easy to measure its full diameter unless the redshifts in the spectra of individual galaxies can be determined very exactly. However, it may well be that our Local Group is associated with the Virgo Cluster and others, in which case it makes up part of what may be termed a supercluster.

There is definite evidence that superclusters do exist, though it cannot yet be said that the evidence is conclusive. This leads on to a larger concept still: are the superclusters themselves parts of a still greater system?

Here we are handicapped by our ignorance of the full size of the universe. The observable universe may be limited to 20,000 million light years or rather less, because at such distances a galaxy would be receding from us at the full velocity of light; but this is not to say that there are not other universes, or other parts of our own universe, which are permanently out of range. The possibility of reaching any definite conclusions in the near future hinges on our extending that range with such instruments as the Space Telescope.

◄ Abell 1060 is a typical cluster of galaxies which includes spirals and ellipticals – the galaxies appear as fuzzy elongated images. Clusters contain from a few tens up to thousands of members loosely held together by gravity. They stand out against the general background of galaxies as clumps of higher than average numbers and are distinguished by the similar redshifts of their members.

► *Galaxies are grouped together into clusters which, in turn, seem to be aggregated into even larger structures known as superclusters. The Local Group lies on the fringe of the Virgo supercluster – a collection of several thousand galaxies spanning 100 million light years.*

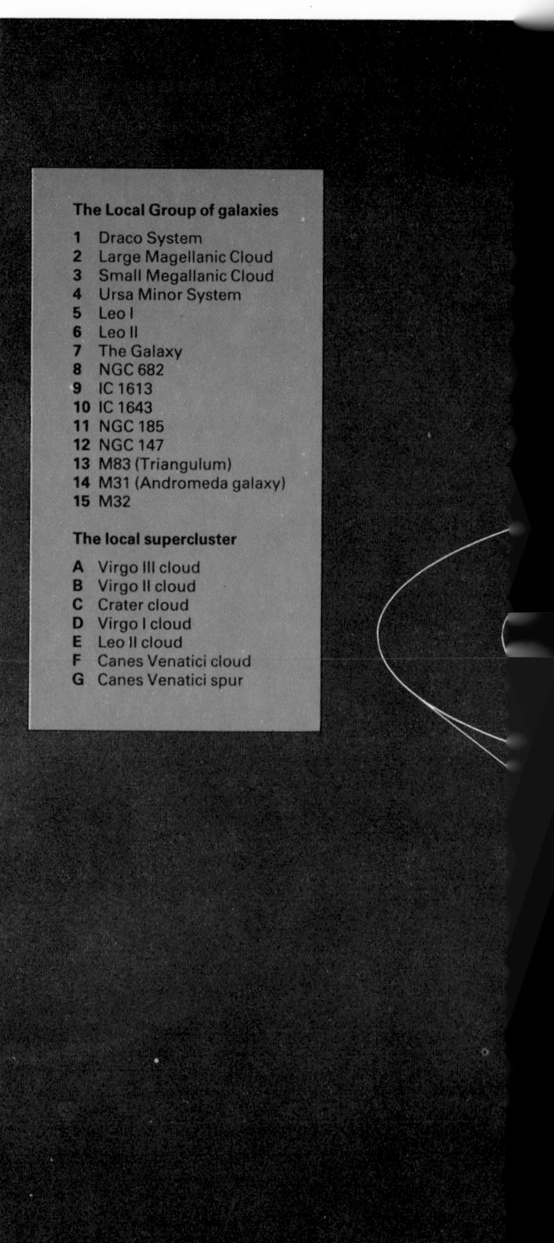

Clusters and superclusters

The nearest major cluster, the Virgo cluster, is centered on the massive elliptical galaxy M87, a galaxy which contains about 30 million million solar masses. Located at a distance of some 50 million light years, it is more than 10 million light years in diameter and contains over a thousand members. Other important major clusters within a few hundred million light years include the Coma, Centaurus, Perseus and Hercules clusters.

Superclusters contain typically about 100 clusters and groups within a radius of a hundred million light years or so. The Local Group is on the fringe of the Virgo supercluster, a vast assembly centered on the Virgo cluster. About 20 percent of the galaxies in the supercluster are contained in Virgo itself, another 40 percent lie within two huge flattened clouds on either side of the Virgo cluster, and the rest are contained in smaller clouds distributed in a roughly spherical halo around it. The Hercules cluster, some 600 million light years distant, is the center of a more remote supercluster which, nevertheless, is so large that it spans about 60° of arc in the sky.

► *The false-color computer graphics image by Brent Tully (University of Hawaii) represents the Virgo supercluster as seen by an observer located in its equatorial plane. The colored contours, from blue through yellow to red, denote progressively denser concentrations of galaxies. The white patch near the center is the dense core of the Virgo cluster.*

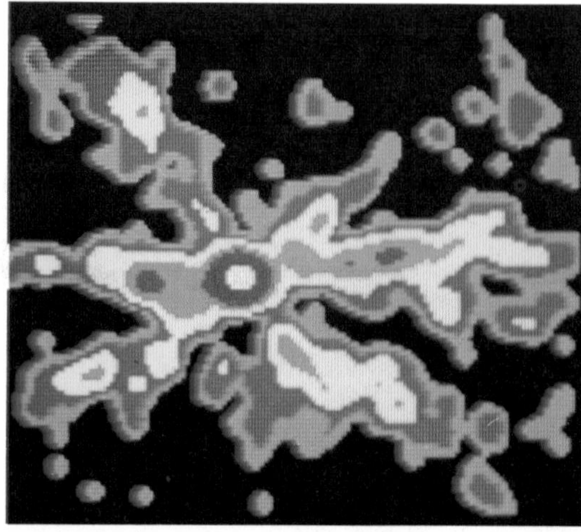

▼ *The Virgo cluster contains over 1,000 member galaxies. This image shows the central part of the cluster and includes the giant elliptical galaxy M87 (bottom left).*

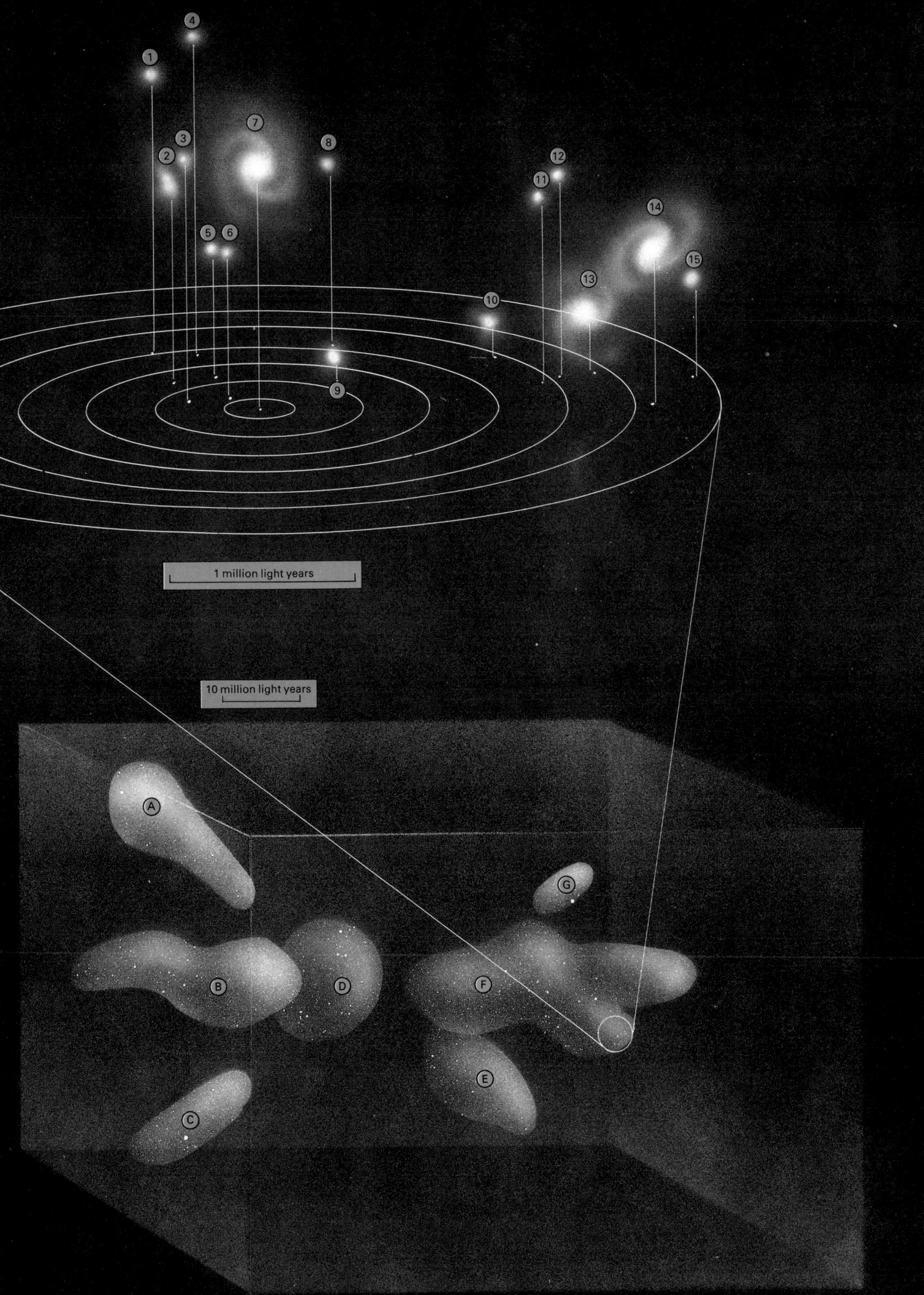

1
2
3
4
5
6
7
8
9
10
11
12
13
14
15
1 million light years
10 million light years
A
B
C
D
E
F
G

Close encounters

The average distance between galaxies in clusters is about ten galaxy diameters, and as they tumble about under their mutual gravitational attractions, close encounters and collisions can occur. Many galaxies are tidally distorted by near neighbors, and close encounters sometimes cause luminous bridges of stars and HII regions to form.

The galaxies NGC 4038 and NGC 4039 have had a close encounter. Computer simulations suggest that both were ordinary disk-type galaxies before the encounter, but that tidal interactions over hundreds of millions of years have expelled the two long, curving streams of matter from which the pair derives its name "the Antennae."

Another extraordinary interaction has produced the so-called Cartwheel galaxy, NGC 1510. It appears that a few hundred million years ago, a smaller galaxy passed straight through the center of a larger one. This caused a "splash" which sent a ripple surging out through the larger galaxy, piling up gas and starting a bout of star formation which gives this galaxy its striking rim of hot blue stars and luminous HII regions. The "rim", which measures about 170,000 light years in diameter, has been pulled off-center relative to the "hub" by the gravitational attraction of the hit-and-run galaxy, which is now some 250,000 light years away from the scene of this vast cosmic accident.

▲ *Stephan's quintet consists of five galaxies closely adjacent in the sky. Four of them represent a genuine and unusually close grouping of large galaxies, two of which (center) are interacting strongly. The fifth is a separate foreground object, not a member of the group (which is therefore more properly referred to as "Stefan's quartet").*

Gas-stripping and the evolution of galaxies

The Milky Way seems to be linked to the Magellanic Clouds by a stream of gas (the Magellanic Stream), while in a nearby group, the spiral galaxy M81 seems to be stripping gas from the neighboring irregular galaxy NGC 3077. The stripping of gas from one galaxy by another during collisions or close encounters may play an important role in the evolution of galaxies and clusters of galaxies, and it is interesting to note that gas and dust-rich spirals and irregulars tend to be more isolated or occur in smaller groups, whereas gas-depleted E and SO galaxies dominate the more densely populated major clusters in which close encounters are more common and which contain very massive central galaxies that may have grown by cannibalizing others.

◄ *The "wing" of the Fly's Wing galaxy consists of stars and gas clouds torn from the galaxy by the gravitational interaction which occurred during a close encounter with another galaxy.*

▼ *Astronomers believe that the Cartwheel galaxy was once a spiral, but that a head-on collision with a smaller galaxy knocked out the center 300 million years ago, giving the ring shape observed today.*

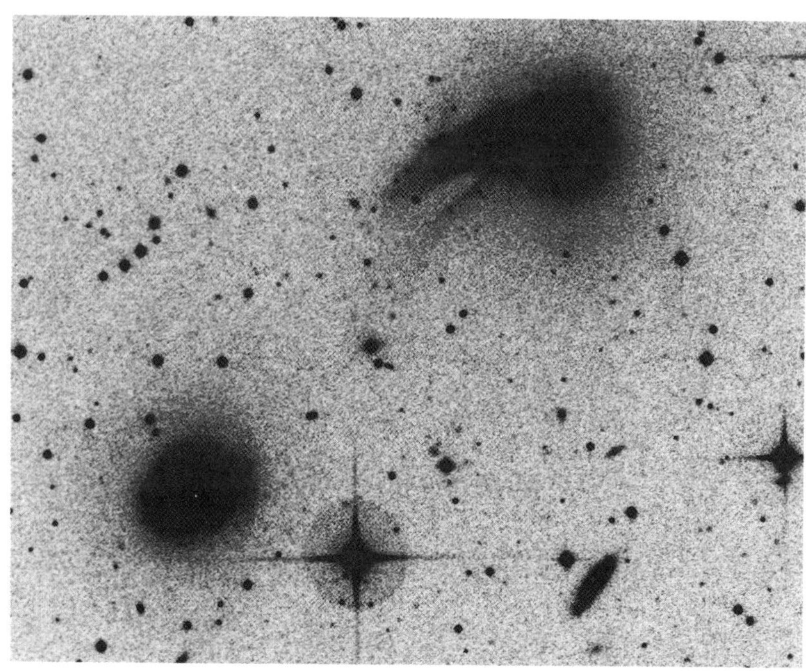

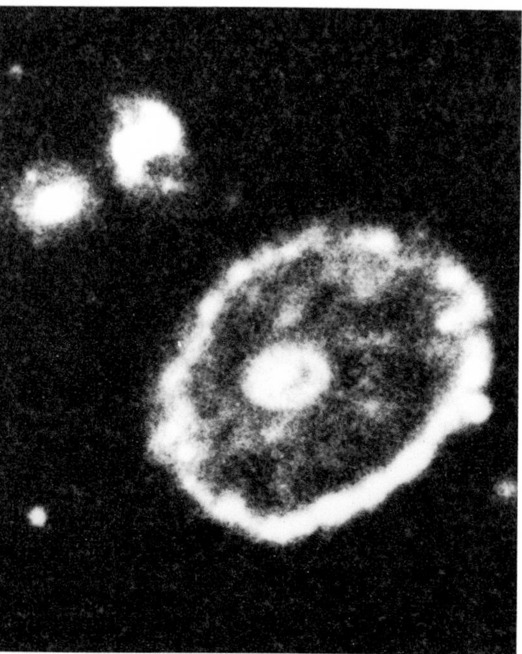

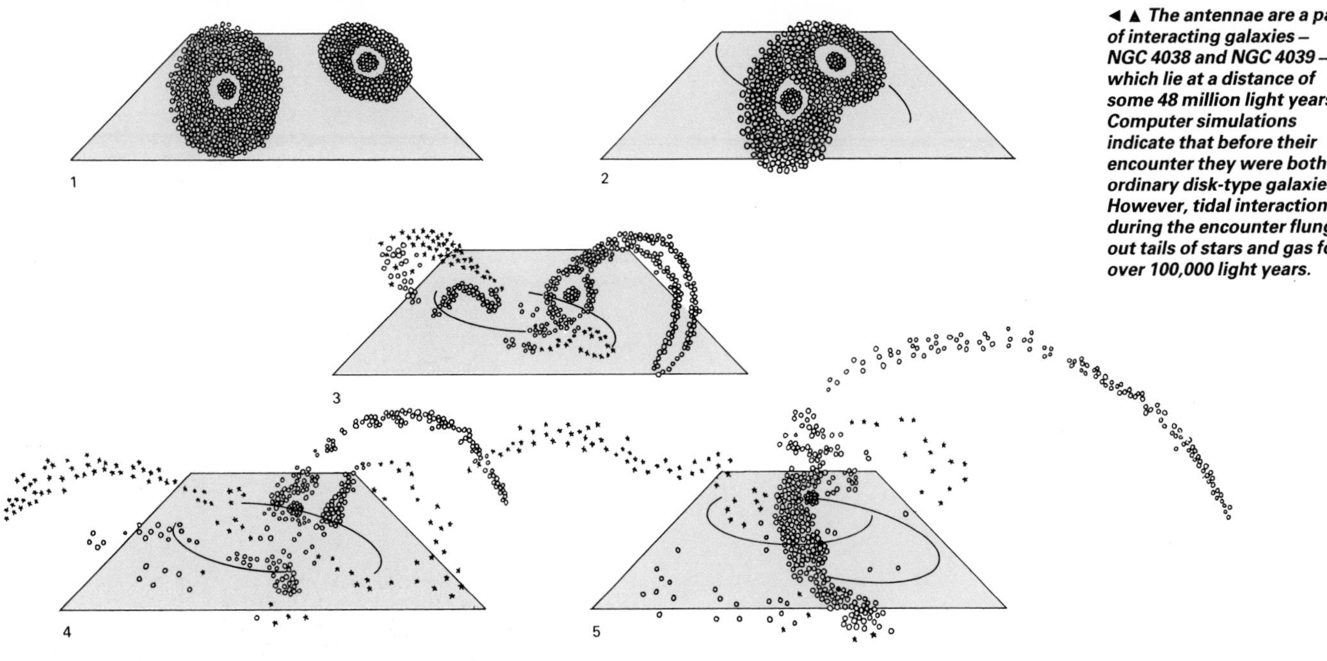

◀ ▲ ***The antennae are a pair of interacting galaxies – NGC 4038 and NGC 4039 – which lie at a distance of some 48 million light years. Computer simulations indicate that before their encounter they were both ordinary disk-type galaxies. However, tidal interactions during the encounter flung out tails of stars and gas for over 100,000 light years.***

▲ *The group of galaxies centered on M81 in the Big Dipper or Great Bear is similar to the Milky Way but more compact. The peculiar galaxy above M81 (center) is M82; the small irregular galaxy (bottom left) is the apparently unrelated NGC 3077.*

► *This 21cm radio map of two of the galaxies in the M81 group – M81 itself (right) and NGC 3077 – reveals how the two are in fact interacting. Yellow and red areas represent regions where the concentration of hydrogen is at its strongest.*

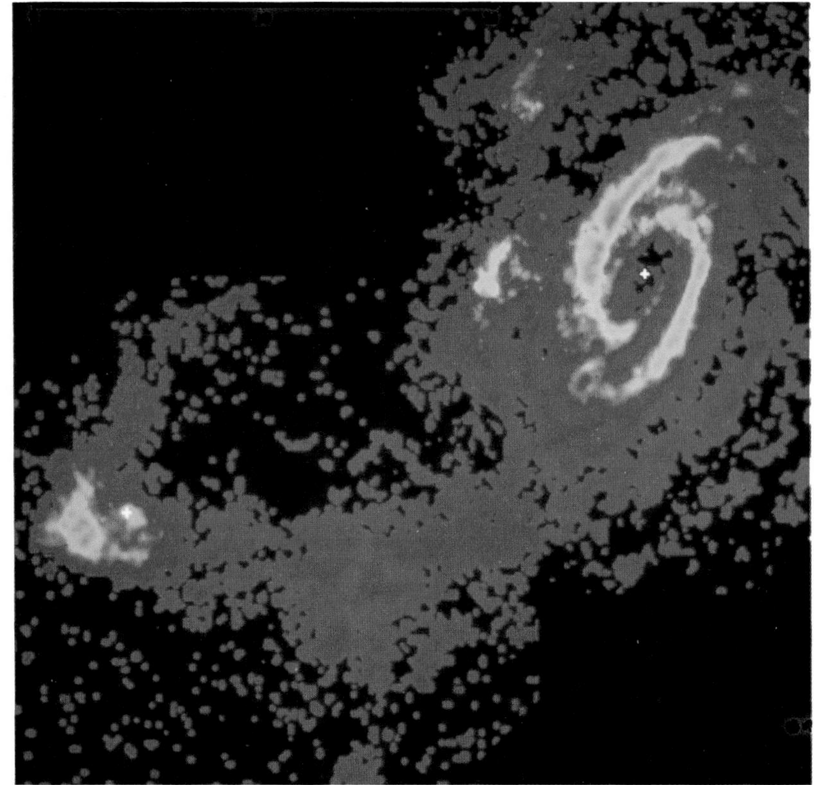

◄ *The Whirlpool galaxy, M51, appears to be interacting with its smaller companion, the barred spiral NGC 5195, via a "bridge" – an extension to one of the spiral arms.*

▼ *The Seyfert's sextet, NGC 6027, is a compact group of interacting galaxies in the constellation of Serpens. The small spiral near the center is a remote background galaxy.*

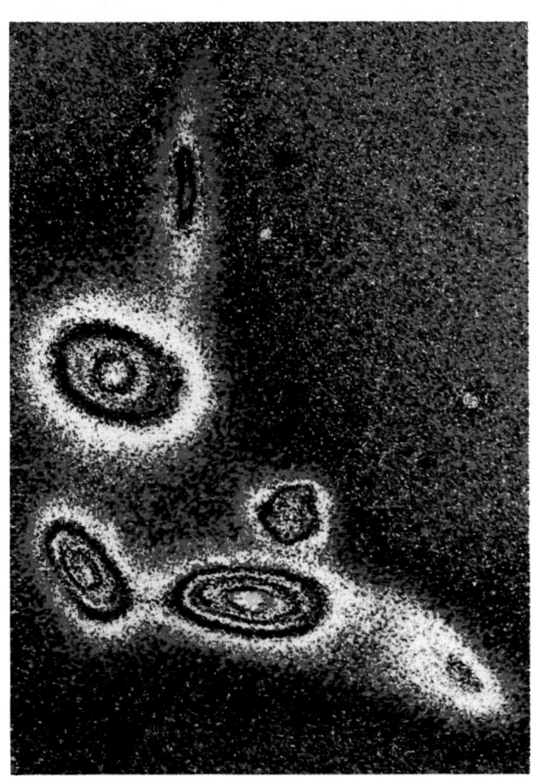

Missing mass in clusters

In order to hold itself together, a cluster of galaxies must exert sufficient gravitational attraction on each of its members to prevent them from escaping into intergalactic space. By applying a relationship known as the virial theorem to the speeds at which the galaxies are moving, astronomers can calculate the minimum mass needed to prevent the cluster from dispersing. This is known as the virial mass.

Calculating the mass of a cluster by adding together the estimated individual masses of its members always gives a value that is far too low to hold the cluster together – it usually falls short of the virial mass by a factor of ten or more. For example, the combined mass of the visible galaxies in the Coma cluster is a few hundred million million solar masses, but the mass needed to bind the cluster together is about 5,000 million million solar masses. If the mass of the cluster were really as low as the combined individual masses of the visible galaxies, the cluster would disperse within a billion years or so – a short time compared to the ages of the galaxies. However, there are so many clusters of galaxies around that they must be very long-term or permanent structures which do not disperse. Therefore there must be a great deal more mass – known as the "missing mass" – holding them together.

Although some clouds of intergalactic nuclear hydrogen have been detected, and other forms of intergalactic gas are known to exist (➧ page 224), there does not seem to be nearly enough matter in these forms to bind the clusters together. The discovery that many galaxies seem to possess massive haloes of dead stars, brown dwarfs (faint low-mass stars) or smaller lumps of matter goes some way towards accounting for the missing mass; but some astronomers believe that clusters can be held together only if they also contain vast numbers of exotic nuclear particles for which, as yet, there is no direct evidence.

See also
Overview of the Universe 129-32
Between the Stars 181-8
Galaxies 197-204
Active Galaxies and Quasars 204-16
The Evolving Universe 229-44

Intergalactic matter

A number of radio galaxies have a "head-tail" appearance, rather like a tadpole, as if their radio-emitting clouds were plowing through a resisting medium or a wind were blowing past them, compressing them on one side and drawing them out into a tail on the other. In a substantial number of clusters which are X-ray sources, the X-rays come from the space between the galaxies rather than from the galaxies themselves. This suggests that there is a tenuous gas at temperatures of 10 million to 100 million K permeating these clusters, and emitting X-rays because of these high temperatures.

Such a high-temperature gas must be highly ionized and so will interact with the electrically charged clouds of electrons which surround radio galaxies. The flow of hot gas would not slow down the massive stars in the central galaxies, but it would sweep back the electron clouds in a similar way to the solar wind blowing the tail of a comet away from the Sun. In time it could strip the clouds away from the galaxies altogether.

Galactic "exhaust" gases

The X-radiation from the Perseus cluster and the head-tail appearance of the member galaxy NGC 1265 is consistent with the presence of a high-temperature gas in that cluster with a density of about 1,000 ions per cubic meter – about one-thousandth of the density of interstellar gas in the Milky Way system. This intriguing galaxy seems at different times to have expelled several pairs of radio-emitting clouds which have been swept back into two curving arcs as the galaxy itself plowed through the intergalactic gas.

The massive galaxy NGC 5291 in Centaurus is surrounded by a disk of some hundred billion solar masses of hydrogen, and this seems to be in the process of being swept out of the galaxy as it plunges into the hot intergalactic gas and greater concentration of galaxies in the central regions of that cluster. As this happens, knots of compressed hydrogen are forming into batches of stars comparable to small galaxies.

Unanswered questions

Whether the intergalactic gas in clusters is primordial hydrogen and helium which dates from the earliest era of the universe, and which is falling into clusters, or whether it has instead been expelled from galaxies, is a matter of continuing debate. Some galaxies, such as NGC 1275, a highly luminous galaxy in the Perseus cluster, seem to be the focus of infalling streams of gas, but the presence in some clusters of highly ionized iron suggests that at least some of the gas must have been processed inside stars and expelled into intergalactic space at a later date.

One possible sequence of events is that over billions of years, after the formation of the clusters, close encounters and collisions between galaxies stripped young distended galaxies of much of their hydrogen, which had not at that stage been converted into stars. This gas was then heated by collisions and by shock waves set up as galaxies plowed through it at speeds of thousands of kilometers per second and explosive events such as supernovae took place. Supernova explosions in galaxies and in their more extensive haloes released heavier elements such as iron which, together with much of the remaining gas from the galaxies, were swept into intergalactic space by encounters with the high-temperature intergalactic gas.

▲ *The radio galaxy NGC 1265 seems to be releasing occasional gas clouds, which are slowed down by the intergalactic medium and trail behind the galaxy like smoke from an engine.*

▼ *This negative photographic print shows filaments of hydrogen, probably streaming in towards a massive black hole at the center of the galaxy NGC 1275.*

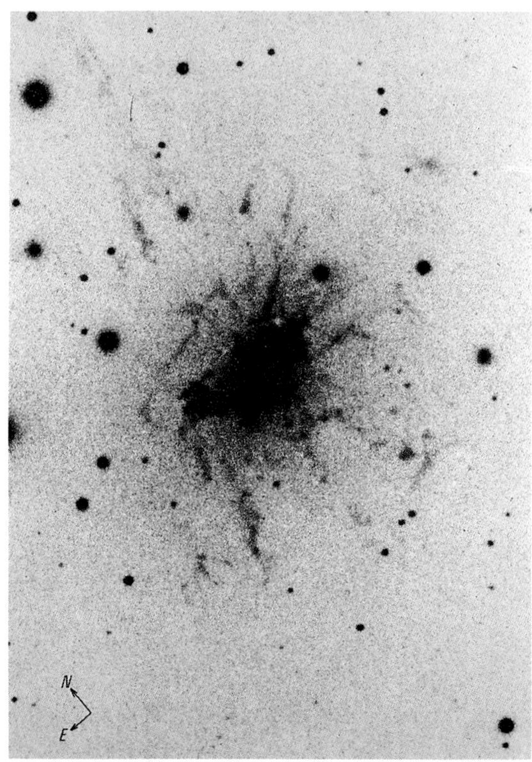

Space, Time and Gravitation

Einstein's breakthrough – the connection between mass and energy, and space and time...General relativity – the proof...Investigating gravity – does it exist as "waves"?...PERSPECTIVE...Can objects ever travel faster than the speed of light?

▲ ***Oscar Minkowski conceived the idea of four-dimensional spacetime which was central to the further development of the theory of relativity.***

▶ ***With his special and general theories of relativity, published in 1905 and 1915, Albert Einstein revolutionized the concept of space, time and gravity.***

▼ ***The jet from the quasar 3C 273 appears to be traveling faster than light.***

The traditional (Newtonian) concepts of space, time and gravity are adequate for all normal purposes. However, when considering certain exotic phenomena such as the behavior of bodies moving at almost the speed of light, the powerful gravitational fields near black holes, and the large-scale structure of the universe, scientists must adopt the ideas embodied in the special and general theories of relativity.

The special theory rests on two basic ideas (or postulates) – namely, that laboratory experiments are unaffected by the motion of the laboratory (so long as it is moving at a constant speed), and that the measured speed of light is constant regardless of the speed of the light source or the observer. This idea seems absurd: if two cars, each traveling at 100km per hour in opposite directions have a head-on collision, the speed of the impact will be 200km per hour. But if an observer is approaching a source of light at 150,000km per second, and that source is emitting light at 300,000km per second, the observer will nevertheless find, if he measures the speed of that light, that it is precisely 300,000km per second – not 450,000km per second as common sense would suggest.

If we accept these ideas, a number of consequences follow. In particular these include the phenomena of mass increase, length contraction and time dilation. As the speed of a body approaches closer and closer to the speed of light, its mass increases dramatically, its length diminishes, and time as measured by any kind of natural or artificial clock attached to the body passes more slowly than time as measured on a "stationary" observer's clock. No material body can actually attain the speed of light – as its mass would theoretically become infinitely great, there could be no means of applying sufficient energy to accelerate it to that speed.

$E=mc^2$

Another consequence of the theory is the equivalence of mass and energy. Mass (m) can be converted into energy (E), or energy into mass, in accordance with the relationship $E=mc^2$, where c denotes the speed of light. Since c is a large number, it follows that the destruction of a small amount of matter can release a very large amount of energy. This relationship provides the key to an understanding of how the stars shine, and an appreciation of the nature of events which occurred in the first seconds of the history of the universe.

In everyday experience we think of bodies as existing in three dimensions – length, breadth and height – and of time being something independent, flowing past at a steady rate. Relativity goes beyond this, embracing the concept of four-dimensional spacetime, where the three dimensions of space and the dimension of time are related so that the motion of an observer affects his perception of space (the measurement of lengths or distances) and time (the ticking of clocks).

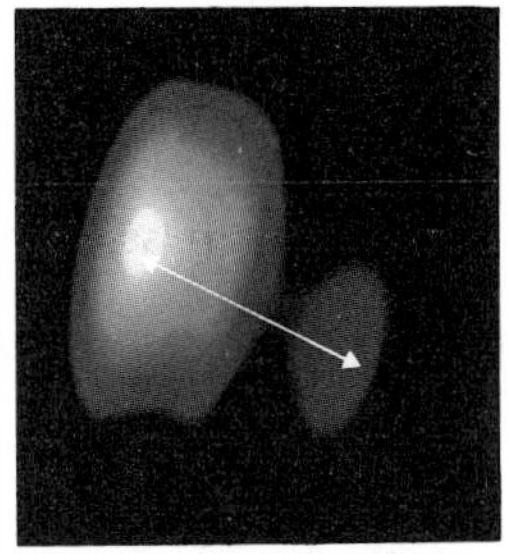

1977 62 light years

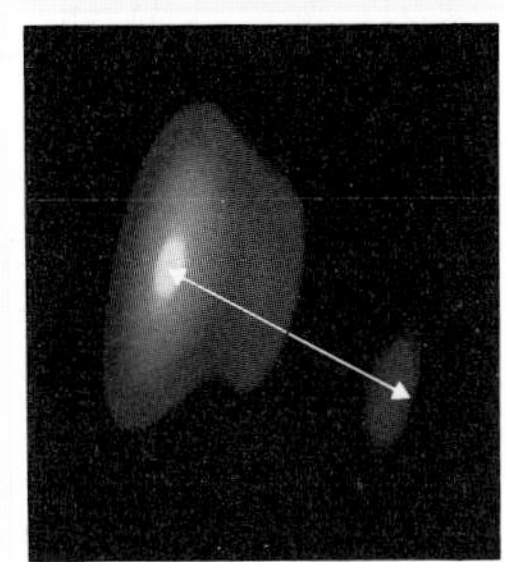

1978 68 light years

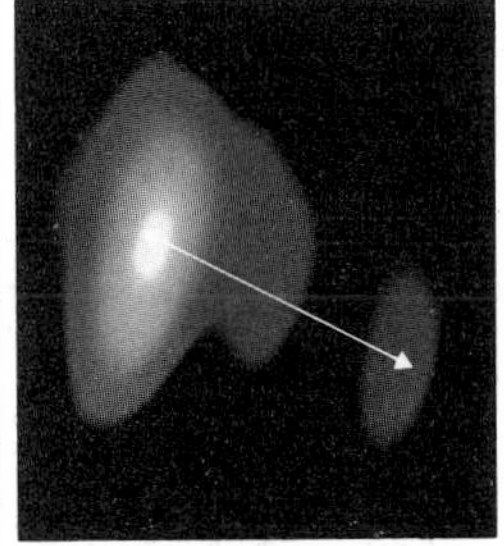

1979 77 light years

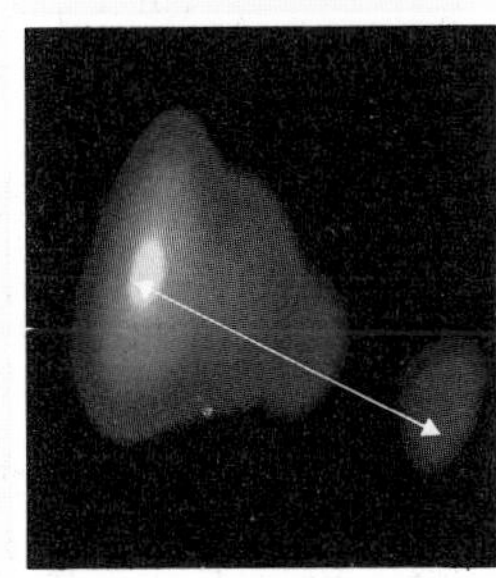

1980 87 light years

Faster than light?
High resolution radio images made in Germany and the USA show an expanding jet of material emerging from the core of the quasar 3C 273. The jet appears to have moved through 25 light years in 3 years. If this were so, matter in the jet would have traveled at about eight times the speed of light in blatant contradiction of Einstein's theory that nothing can travel faster than light. However, the apparent "faster than light" expansion is almost certainly a geometric illusion caused by the matter in the jet traveling only slightly slower than the speed of light at a small angle to the line of sight.

As one cosmologist has said, "Matter tells space how to curve, and space tells matter how to move"

General relativity

The term "general relativity" describes a theory of gravity consistent with the principles of relativity. It regards gravity not as a force which acts directly across empty space between individual bodies (the Newtonian concept), but as an apparent force which arises because space itself is curved in the presence of matter. To picture this, it is useful to think of empty space as being a flat elastic sheet. In the absence of any massive bodies, a particle, once set in motion, continues to move across the sheet in a straight line at a uniform speed. Placing a weight on the sheet to represent a large mass causes a dent, distorting the elastic sheet. A particle moving across the sheet then follows a curved path when it meets the indentation.

According to general relativity, the effect of a large mass is to distort space and cause particles – even rays of light – to follow curved paths in the vicinity of matter. Thus, in following their orbits round the Sun, the planets are following their natural paths in the curved space surrounding the Sun. As one cosmologist, Professor Archibald Wheeler, has said: "Matter tells space how to curve, and space tells matter how to move." This may seem an abstruse way of looking at gravity, and in most circumstances general relativity gives the same result as Newtonian gravitation, but there are circumstances where Einstein's theory explains phenomena which Newton's theory cannot.

Observational and experimental proof

General relativity has been checked by both observation and experiment, one of the best-known observations being the small but measurable deflection of rays of light passing close to the edge of the Sun. The bending of light by gravity allows a massive object to act rather like a lens to produce a focused image of a distant object. If the alignment is not right, rays of light from a spherical background object, passing a massive "gravitational lens", will produce two crescent-shaped images. A well-known example of this effect is the so-called double quasar 0957 +561,A,B. Astronomers believe this to be a single quasar with a foreground galaxy acting as a gravitational lens.

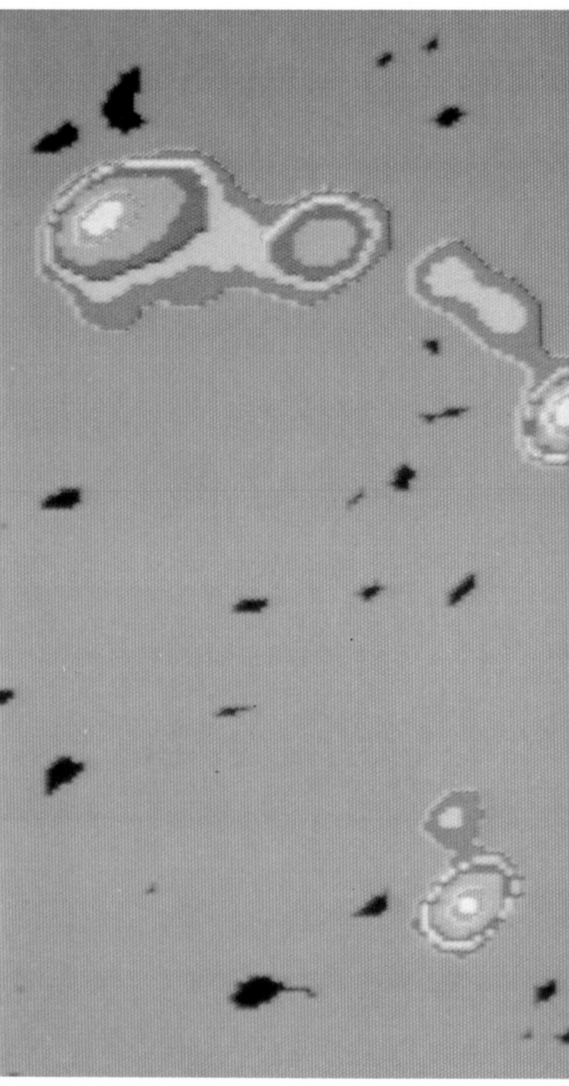

The deflection of a star

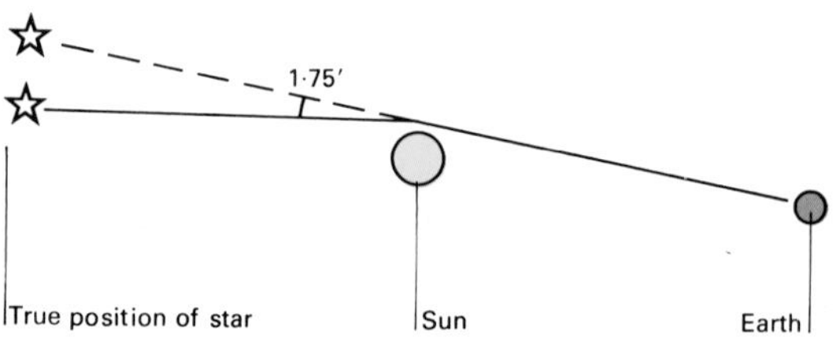

◄ ▲ According to Einstein's General Theory of Relativity published in 1915, a ray of light should be deflected when passing by a massive body. In particular, a ray grazing the edge of the Sun should be deviated by an angle of 1·75 seconds. If (above) the Sun lies closely in line with a background star, that star will be deflected from its normal position in the sky by a small but measurable amount. The bending of starlight was first detected by a British expedition headed by Sir Arthur Eddington, which observed the total solar eclipse of 29 May 1919 from Sobral, Brazil. Analysis of photographic plates taken by the expedition showed clearly that the apparent stellar positions were indeed displaced by an amount which was comfortably close to Einstein's prediction. The eclipse photograph (left) shows the positions of the stars which were used to check the predicted bending of light.

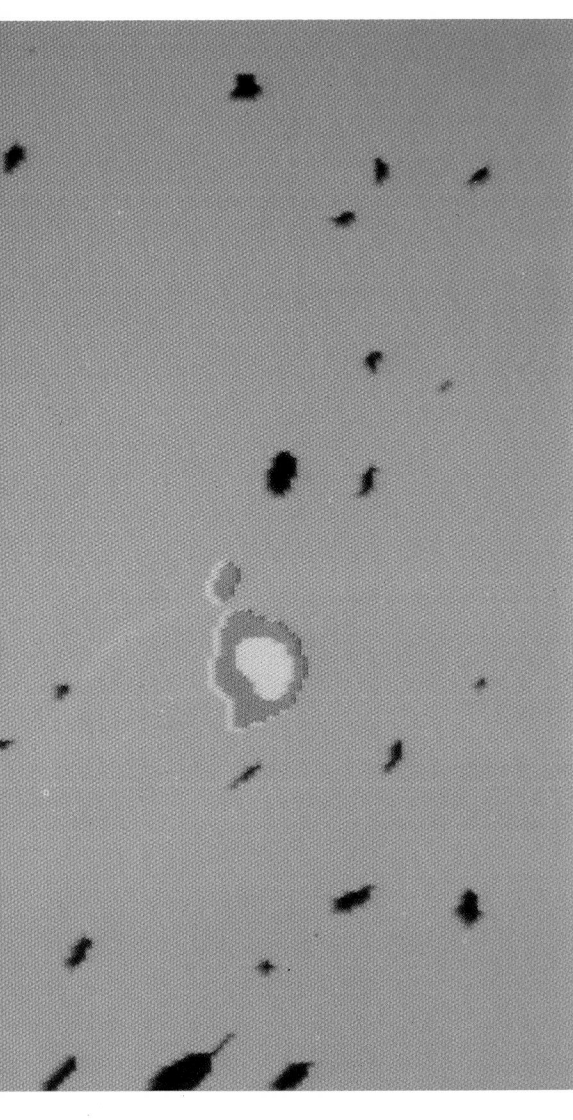

◄ ► A gravitational lens at work. Light and radio waves from a single remote quasar and its radio lobes are deflected by a massive elliptical galaxy as shown schematically (right). Seen from Earth the quasar and its image appear as the "double quasar" 0957+561.

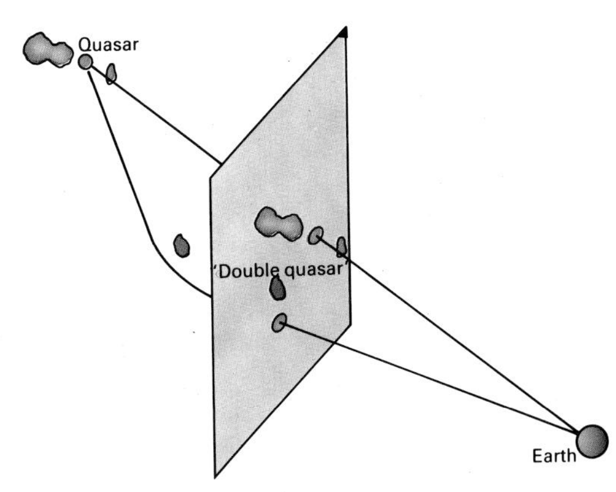

▼ This is an optical photograph, reproduced as a negative print, of the same quasar, showing its "double" appearance (taken from Mauna Kea).

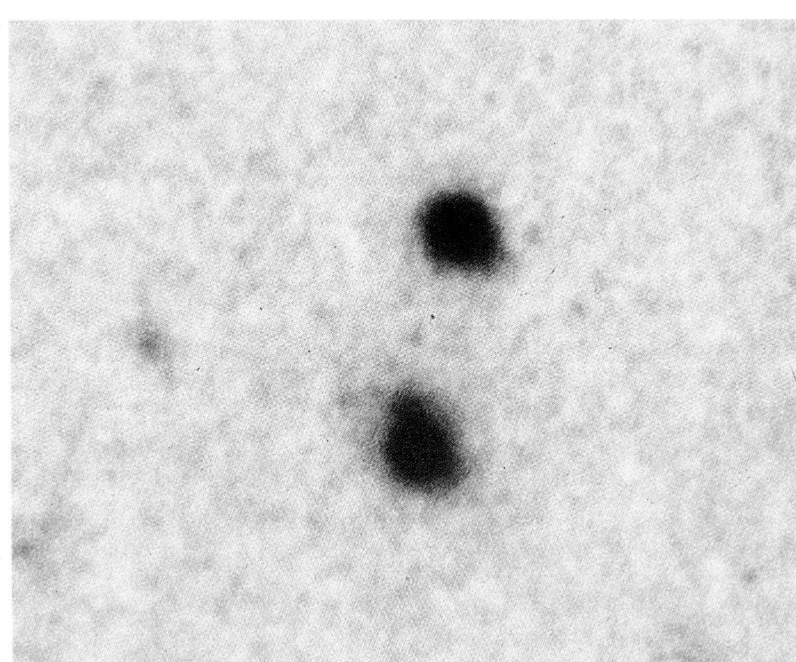

The apparent effect of gravity

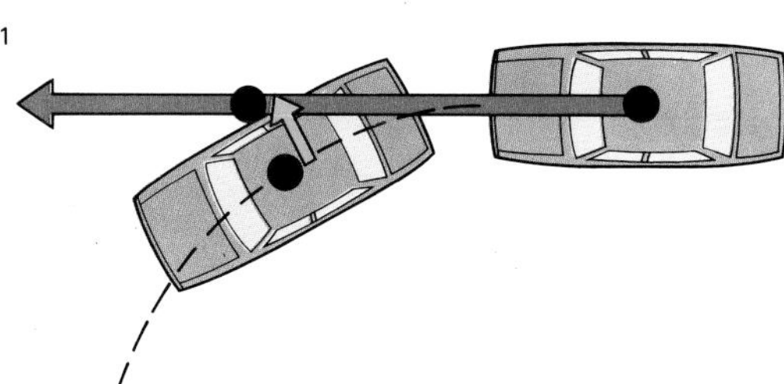

Curved space

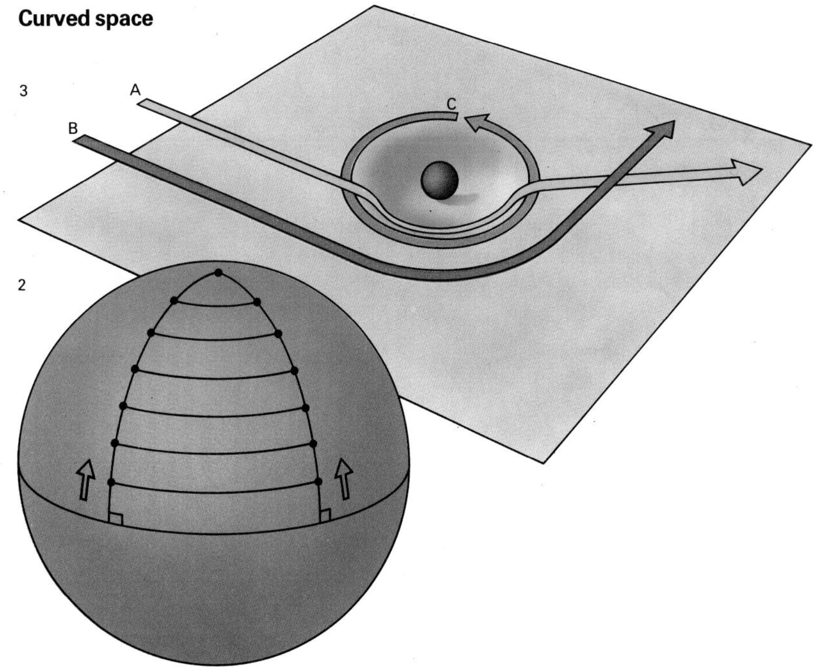

***1** Centrifugal force experienced by occupants of a cornering vehicle is an "apparent" force rather than a "real" one. Although the occupants feel they are flung to the side, in fact they have carried on in a straight line at uniform speed and the side of the vehicle has swung round to meet them.*
***2** Gravity can be regarded as an apparent force due to the curvature of space. Thus two creatures setting off in parallel directions on the surface of a sphere will eventually meet, as if they had been drawn together by a force of attraction.*
***3** The effect of curved space can be visualized by imagining space to be an elastic sheet. A heavy mass placed on the sheet produces an indentation and bodies encountering the indentation follow curved paths. A slow-moving body (C) deep in the "well" follows a closed path (like a planet round the Sun) while a faster body (B) follows an open curve. A ray of light (A) is deflected (normally) only very slightly.*

See also
Observing the Universe 137-44
Black Holes 177-80
Active Galaxies and Quasars 205-16
The Evolving Universe 229-44
Communication and Travel 245-48

Investigating gravity

Relativity predicts that light will be red-shifted and time will pass more slowly in a strong gravitational field, and observation and experiment have confirmed both of these predictions. If a body is following an elliptical orbit in the strong gravitational field of a massive object, the orbit will slowly precess round, so that the point of closest approach revolves round the massive object. The orbit of Mercury demonstrates this phenomenon, and so, to a much greater extent, does the orbit of the one known binary pulsar (a neutron star in a close binary system with another collapsed object).

Just as a vibrating electrical charge emits electromagnetic waves so, according to general relativity, a vibrating mass emits gravity waves – disturbances in the gravitational field which travel like ripples at the speed of light. Gravity waves, if they exist, will be very weak indeed, and it has so far not been possible to detect them. Likely sources of gravity waves include close binaries (like the binary pulsar), supernovae and events involving matter collapsing into black holes.

The net effect of matter in the universe is to curve space as a whole. Astronomers do not know for sure how space is curved. One possibility is that it is curved in such a way as to give a closed, finite yet unbounded universe – a universe of limited volume but with no discernible edge. In such a universe it would be possible (in principle) to set off in one direction and to arrive back at the starting point from the opposite direction without ever having come to an edge.

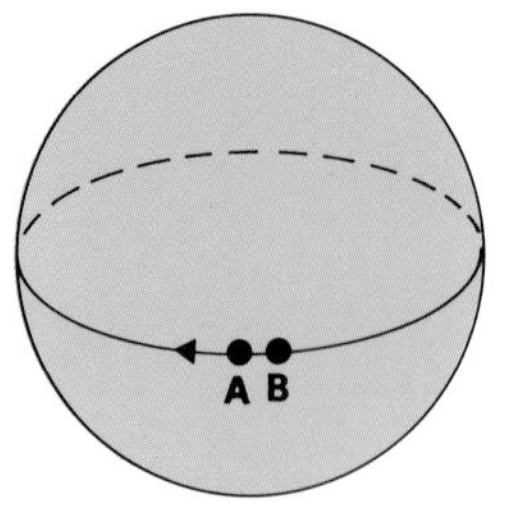

◄ The universe can be regarded as finite and unbounded. Thus, a two dimensional creature (A) whose universe is the surface of a sphere, can circumnavigate his universe and return to his colleague (B) without changing direction or encountering an edge to his universe.

Gravity waves

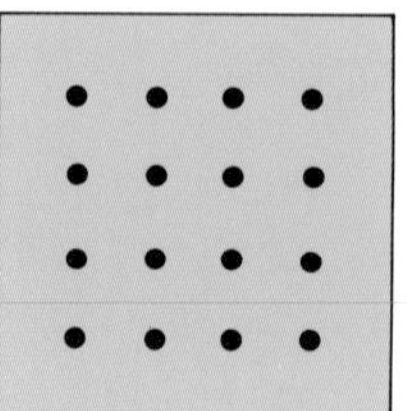

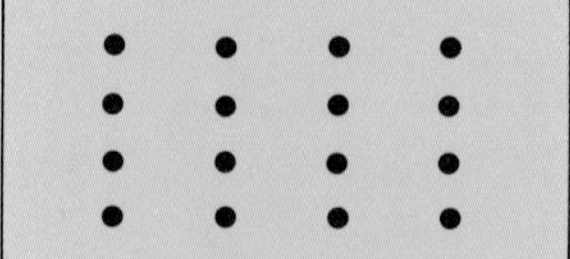

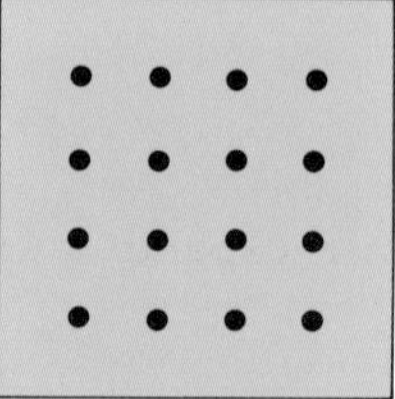

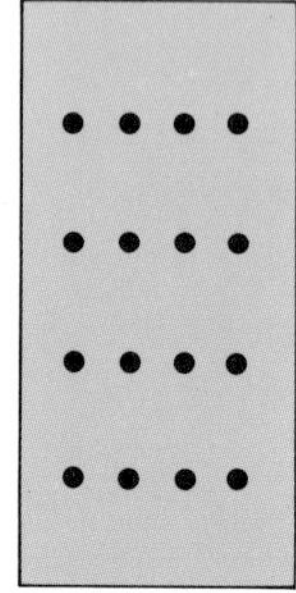

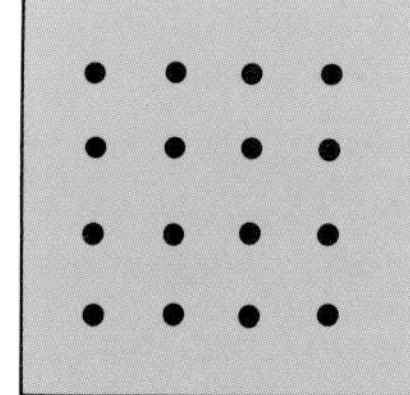

◄ ▲ As a gravitational wave passes through a square array of particles (above) the array is first squeezed vertically and stretched horizontally, then squeezed horizontally and stretched vertically. Professor J. Weber attempted to detect microscopic vibrations of this kind in a 4-tonne aluminium cylinder (left) but so far gravity waves have not been positively detected.

The Evolving Universe

When and how did the universe come into existence?... The Big Bang and its alternatives...Estimating the age of the universe...Its infancy – the hidden years when the galaxies were born...Eternal expansion or a "big crunch"...In search of dark matter...PERSPECTIVE... Hubble Time and the age of the galaxies...The four forces of nature...Is the universe open, closed or flat?

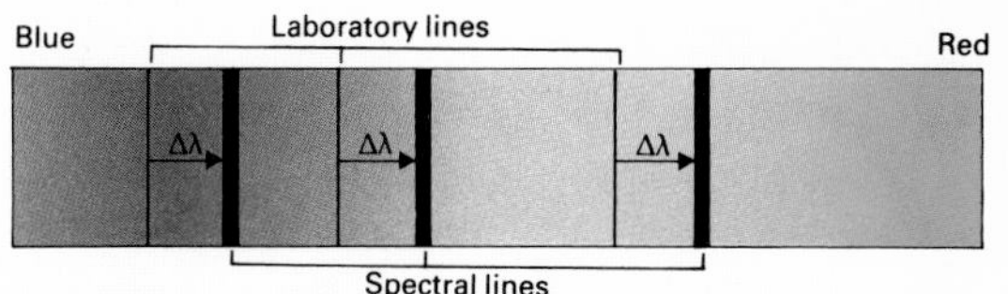

▲ *The redshift of a galaxy is obtained from the difference (Δλ) in wavelength between lines in its spectrum and lines produced by a stationary laboratory reference source.*

Cosmology is concerned with the structure and evolution of the universe as a whole. It deals with deep and fascinating questions: is space finite or infinite? Did the universe come into existence a finite time ago, or has it always existed? What forces have molded the universe to its present form and how will those forces shape its future? Will the universe exist forever, or will it eventually come to an end?

The redshifts of the galaxies

Apart from the members of the Local Group, which are held together by their mutual gravitational attractions, all of the galaxies have redshifts in their spectra. The greater the distance of a galaxy, the farther are its spectral lines shifted towards the long-wavelength end of the spectrum. Almost all astronomers believe that the redshifts arise from the Doppler effect, and that the galaxies are receding at speeds proportional to their distances. For example, if one galaxy is twice as distant as another, its speed of recession will be twice as great.

Given that the galaxies are receding, it seems natural to conclude that the universe is expanding. All the remote galaxies and clusters of galaxies appear to be receding from the Local Group in particular, as if the Milky Way were at the center of the expansion, but in fact the expansion is completely symmetrical, so that an observer on any galaxy would see all the others receding from him at speeds proportional to their distances. No galaxy can claim to be the unique "center of the universe". It may be helpful to draw an analogy with a child's balloon: think of the whole of space as the *surface* of the balloon, with many small paper spots stuck on to it to represent the galaxies. As the balloon is inflated, the pattern of spots will remain the same, but the distance between them will increase by an amount which is proportional to their separations, in accordance with a law known as the Hubble Law. Each spot "sees" the others moving away, but none can claim to be the center, for the surface of a balloon has no center as such.

Most modern cosmological theories are based on the "cosmological principle" that the universe is homogeneous and isotropic – homogeneous in the sense that on the large scale it is the same everywhere, and isotropic in the sense that it looks the same in every direction. Astronomers also assume that the laws of nature are everywhere the same. In other words, there is nothing unique about our situation in, or our view of, the universe.

The Big Bang – a first look

Most astronomers accept that if the universe really is expanding as observations show, the galaxies must have been closer together in the past, and that at one time all the matter in the universe must have been densely packed together. This belief is behind one of the most widely debated theories in modern science – the theory of the "Big Bang".

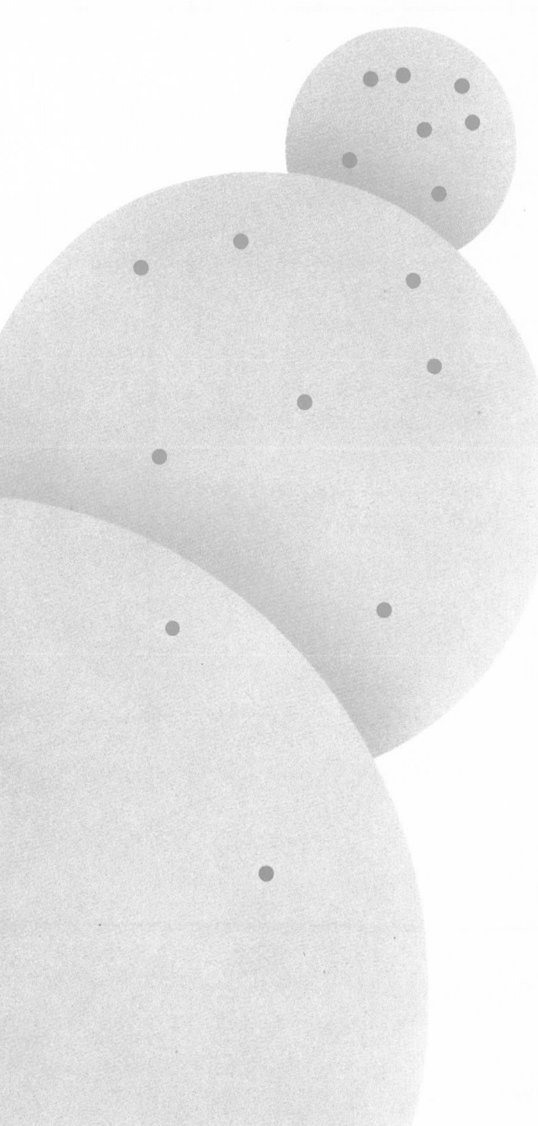

▲ *If the entire universe is represented by the surface of a balloon (ignoring the inside and outside) and galaxies are represented by disks stuck to its surface, it can be seen that as the universe expands each galaxy moves away from every other one.*

The universe is growing larger – but it is not clear whether this growth is speeding up, slowing down or proceeding at a constant rate

According to the Big Bang theory, the universe began by "exploding" from a hot superdense state – possibly from a singularity (a point of infinite density) – and the galaxies are rushing away from each other because of the violence of the initial event.

However, the Big Bang was not like an ordinary explosion which scatters fragments through space. If current ideas are correct, matter did not suddenly erupt forth into a previously empty space; rather, space, time and matter originated with the Big Bang, and space has been expanding ever since. Instead of thinking of galaxies as rushing apart *through* space, it is helpful to think of them as being *at rest in an expanding* space (although there are local random motions). Individual galaxies and clusters of galaxies have sufficient gravitational attraction to hold themselves together, but the space between, where the mean density of matter is lower, continues to expand.

The age of the universe

It is possible to calculate the "age" of the universe – the time since the Big Bang – by dividing the distances of the galaxies by their speeds, assuming that they have been moving apart at their present speeds since the initial event. This method gives an age of about 20 billion years, a result known as the "Hubble Time". If, however, the mutual gravitational attraction between galaxies is slowing down this expansion, galaxies must have been moving apart faster in the past than they are now. Taking this into account, the age of the universe must be less than the Hubble Time, and may be nearer to 13-15 billion years.

Astronomers estimate the distances of galaxies by identifying within them known "standard candles" (◆ page 197). For example, they may compare the apparent brightness of a Cepheid variable or a

Edwin Hubble's Law

The relationship between velocity, V, and distance, D, is known as the Hubble Law, and can be written as V=H×D, where H is a quantity known as the Hubble Constant.

Scientists are unsure of the exact value of H – indeed, reasonable estimates based on the same data may vary widely – but many of them believe that it is about 50km per second per megaparsec (Mpc), which suggests that a galaxy at 1 Mpc will be receding at a speed of 50km per second, and a galaxy at 100Mpc is receding at 50 × 100 = 5,000km per second, and so on. Since 1 Mpc is equal to 1,000,000 parsecs, or 3,260,000 light years, the Hubble Law implies a speed of recession of about 15km per second for every million light years of distance.

The Hubble Law affirms that all the galaxies have taken the same period of time to reach their present distances: if one galaxy is twice as remote as another because its speed is twice as great, it will have taken exactly the same time as the nearer one to reach its present distance. If H is indeed 50km per second per megaparsec, this interval (the Hubble Time) is about 20 billion years.

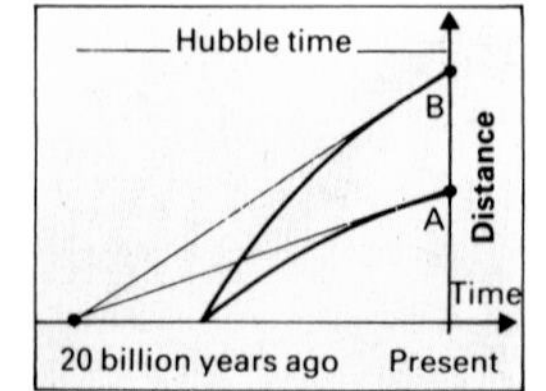

◄ *Galaxy B, receding twice as fast as A, is twice as remote. If galaxies were moving faster in the past (curved lines) they will have reached these distances earlier than if they had been moving at constant speeds (straight lines).*

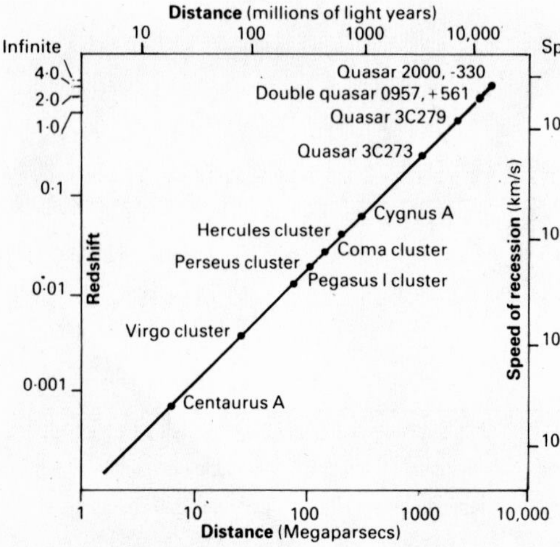

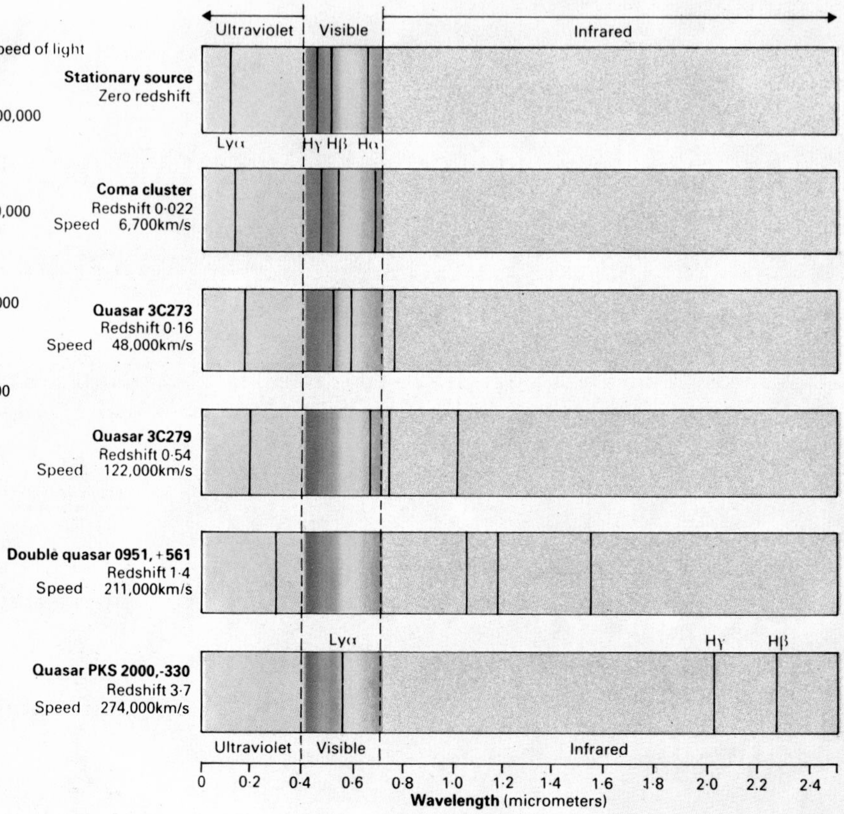

Distance (millions of light years): 10, 100, 1000, 10,000
Redshift: Infinite, 4·0, 2·0, 1·0, 0·1, 0·01, 0·001
Speed of recession (km/s): Speed of light, 100,000, 10,000, 1000, 100
Quasar 2000, -330
Double quasar 0957, +561
Quasar 3C279
Quasar 3C273
Cygnus A
Hercules cluster
Coma cluster
Perseus cluster
Pegasus I cluster
Virgo cluster
Centaurus A
Distance (Megaparsecs): 1, 10, 100, 1000, 10,000

Ultraviolet | Visible | Infrared
Stationary source — Zero redshift (Lyα; Hγ Hβ Hα)
Coma cluster — Redshift 0·022 — Speed 6,700km/s
Quasar 3C273 — Redshift 0·16 — Speed 48,000km/s
Quasar 3C279 — Redshift 0·54 — Speed 122,000km/s
Double quasar 0951, +561 — Redshift 1·4 — Speed 211,000km/s
Quasar PKS 2000,-330 — Redshift 3·7 — Speed 274,000km/s (Lyα; Hγ Hβ)
Wavelength (micrometers): 0, 0·2, 0·4, 0·6, 0·8, 1·0, 1·2, 1·4, 1·6, 1·8, 2·0, 2·2, 2·4

▲ *The redshifts and distances of some remote active galaxies and clusters: redshifts and equivalent recession speeds are on the vertical scales. Distances are shown in millions of light years (top) and megaparsecs (bottom). The line shows the relationship between speed and distance. The effect of redshift on the wavelengths of spectral lines is shown on the right.*

▲ *From left to right: Humason, Hubble, Baade and Minkowski in the 1950s.*

bright supergiant with the assumed luminosity of the galaxy in which it lies. However, as distances grow greater, the number of standard candles that can be used in this way decreases. Supernovae can be seen from distances of several billion light years, but they are rare events; and in order to estimate very large distances, astronomers have to make assumptions about the luminosities or diameters of different classes of galaxies. They assume that the brightest galaxies in rich clusters will be more or less equal in absolute brightness, and that there is a correlation between the rotation rates and the brightnesses of different types of galaxy.

The measurements are difficult to make and hard to interpret, so that astronomers are not sure about the exact relationship between the speeds and distances of remote galaxies. A number – notably the American cosmologist Gérard de Vaucouleurs (b. 1918) – believe that remote galaxies are only half as far away as they are generally thought to be, and, taken at face value, this implies that the universe is no more than 10 billion years old. This poses a problem for the conventional Big Bang theory because the galaxies themselves are estimated to be at least 12-13 billion years old, and they could clearly not be older than the universe which contains them. Therefore, if de Vaucouleurs is correct, then either the ages of the galaxies have been wrongly computed or alternatives to the conventional Big Bang theory will have to be considered – such as those which postulate an accelerating universe in which the rate of expansion increases with time. If the universe was expanding more slowly in the past than it is now, its true age must be greater than the age calculated from the present speeds and distances of the galaxies, and this removes the discrepancy between the ages of the galaxies and that of the universe.

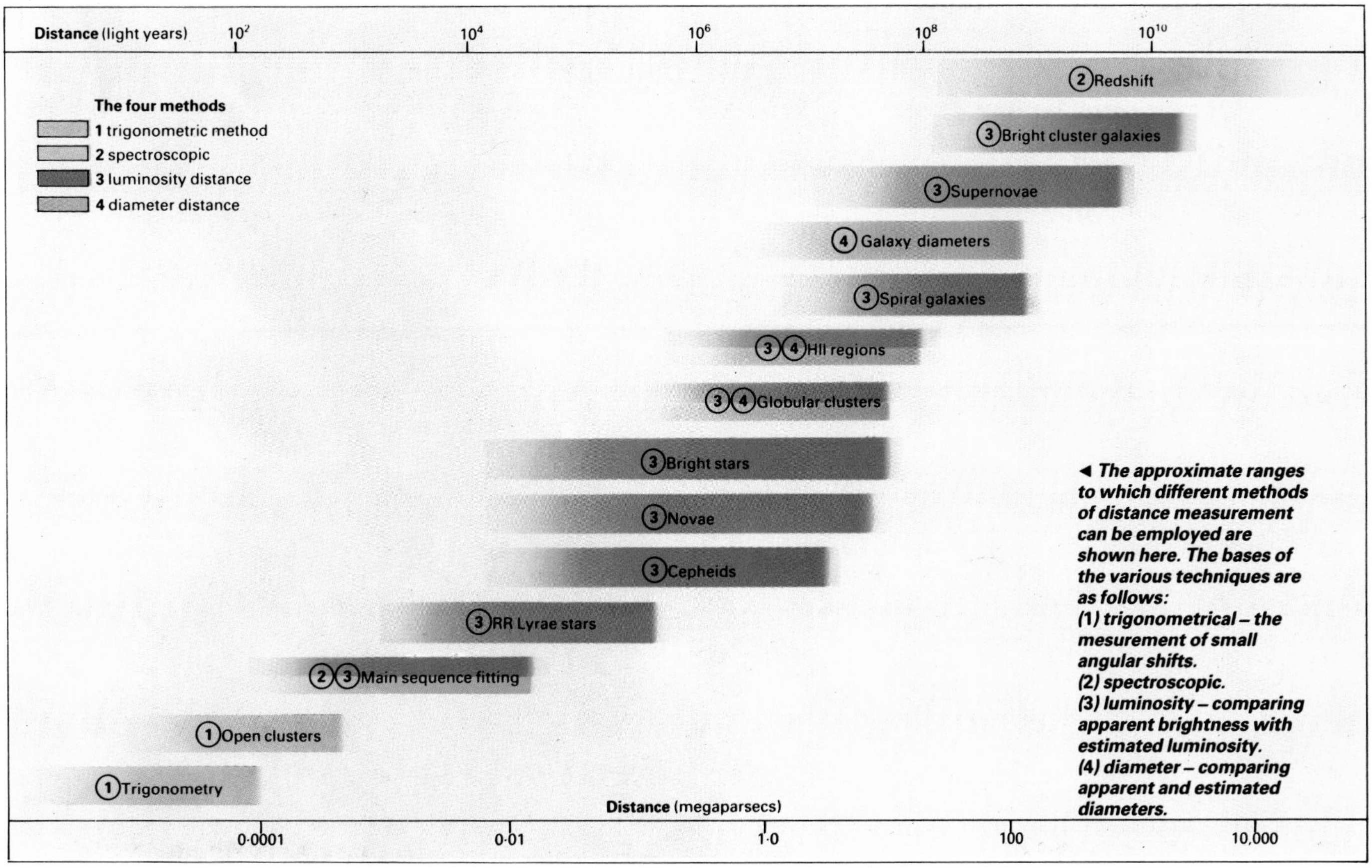

Distance (light years) 10^2 10^4 10^6 10^8 10^{10}

The four methods
1 trigonometric method
2 spectroscopic
3 luminosity distance
4 diameter distance

(2) Redshift
(3) Bright cluster galaxies
(3) Supernovae
(4) Galaxy diameters
(3) Spiral galaxies
(3)(4) HII regions
(3)(4) Globular clusters
(3) Bright stars
(3) Novae
(3) Cepheids
(3) RR Lyrae stars
(2)(3) Main sequence fitting
(1) Open clusters
(1) Trigonometry

Distance (megaparsecs) 0·0001 0·01 1·0 100 10,000

◀ ***The approximate ranges to which different methods of distance measurement can be employed are shown here. The bases of the various techniques are as follows:***
(1) trigonometrical – the mesurement of small angular shifts.
(2) spectroscopic.
(3) luminosity – comparing apparent brightness with estimated luminosity.
(4) diameter – comparing apparent and estimated diameters.

The case for a hot Big Bang, and the case against it – cosmologists weigh up the evidence

Alternative theories

The idea of a force, or property of space, called "cosmic repulsion" which accelerates the galaxies as they move apart, dates back to the early 20th century, before the recession of the galaxies was actually discovered. In order to have a static universe (one which was neither expanding nor contracting), Einstein suggested that at great distances cosmic repulsion precisely balanced gravity and thus prevented the galaxies falling together. However, if this balance were disturbed even slightly, the universe would either collapse or begin an accelerating expansion.

According to another theory – the Lemaître model – the universe originated in a big bang and expanded at a decreasing rate until it had almost stopped, at which point gravity and cosmic repulsion were very nearly balanced. The universe then remained in a nearly static state until the very slow expansion tipped the scales in favor of cosmic repulsion, and there ensued a phase of accelerating expansion. The Eddington-Lemaître model proposed a universe which had no big bang, but which instead accelerated away from a finite size.

The "Steady State" theory, proposed in the late 1940s, was based on the so-called "perfect cosmological principle" that the large-scale appearance of the universe was the same everywhere *at all times*. The Steady State universe was infinite, with no beginning and no end. According to the theory, new material was continuously being created at a very slow rate, and this material formed new galaxies to take the place of those which had moved apart.

Alternative theories, particularly those involving an accelerating universe, cannot be completely discounted. On balance, however, the evidence suggests that the galaxies are indeed younger than the universe – that accelerating universe theories are, therefore, unnecessary, and that the straightforward Big Bang theory is the simplest and best description of nature.

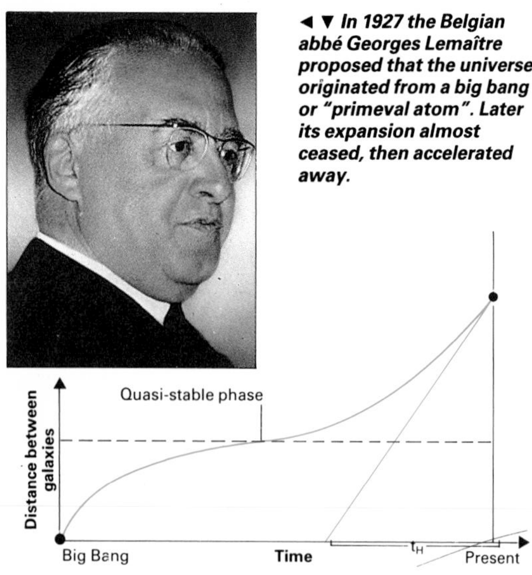

◄ ▼ In 1927 the Belgian abbé Georges Lemaître proposed that the universe originated from a big bang or "primeval atom". Later its expansion almost ceased, then accelerated away.

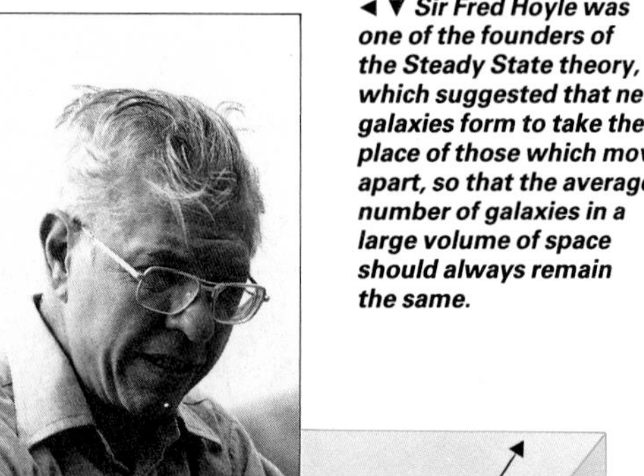

◄ ▼ Sir Fred Hoyle was one of the founders of the Steady State theory, which suggested that new galaxies form to take the place of those which move apart, so that the average number of galaxies in a large volume of space should always remain the same.

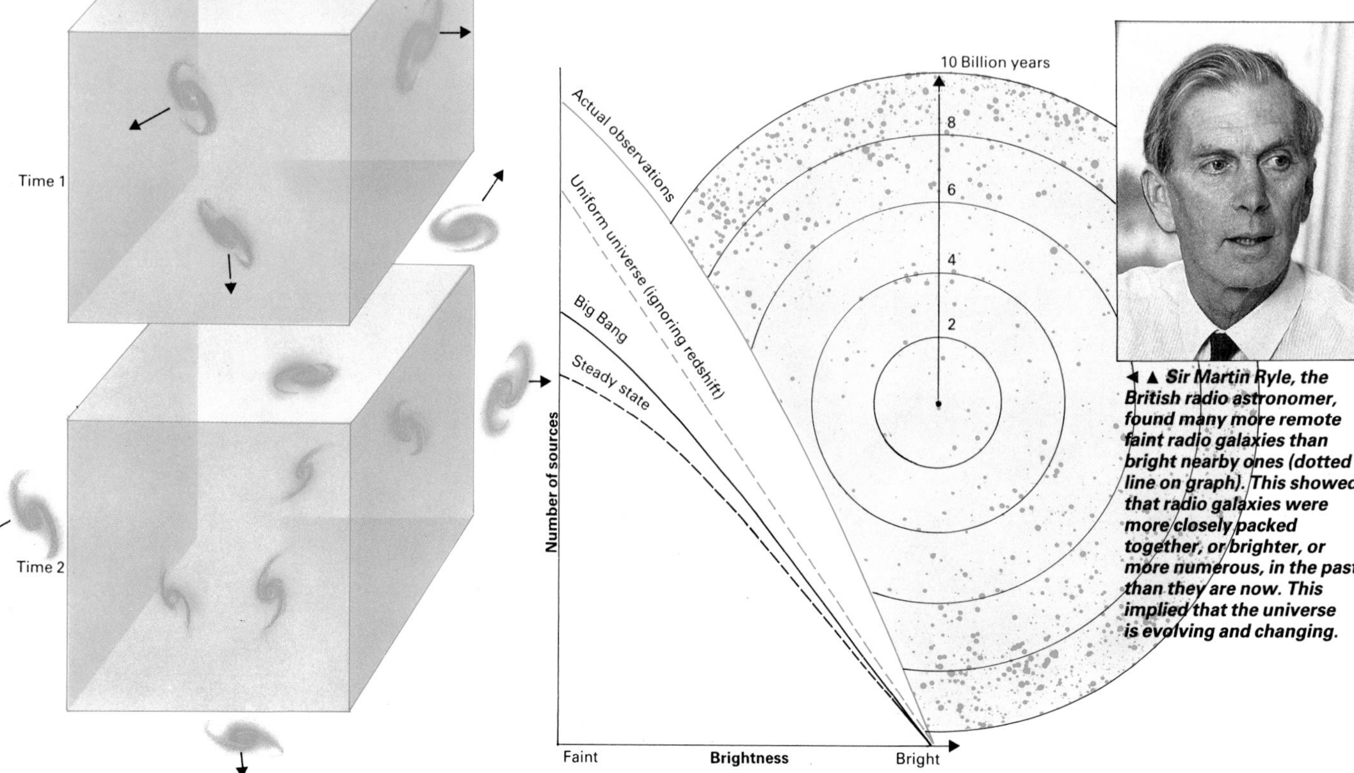

◄ ▲ Sir Martin Ryle, the British radio astronomer, found many more remote faint radio galaxies than bright nearby ones (dotted line on graph). This showed that radio galaxies were more closely packed together, or brighter, or more numerous, in the past than they are now. This implied that the universe is evolving and changing.

Evidence for a hot Big Bang

There are three key pieces of evidence which led astronomers to favor the Big Bang theory over the alternatives such as the Steady State. These are source counts, the microwave background radiation and the so-called "helium problem".

Source counts

The radiation which is now arriving from remote parts of the universe was emitted billions of years ago. By probing to ever greater distances, it is therefore possible to examine progressively earlier states of the universe. Astronomers have counted the number of radio galaxies and quasars to fainter and fainter limits, and (assuming that the fainter the source the greater the distance) have found that these objects were brighter or more numerous – possibly both – in the past than they are now. This "source count" evidence shows that the universe is evolving, and cannot be in a steady state.

The microwave background radiation

In 1965 it was discovered that space is permeated by a weak background of microwave radiation. This is uniform to within 0·01 percent over the whole sky, and is strongest at a wavelength of about 1mm. If the universe had begun in a hot Big Bang, radiation released from the hot dense fireball would have been cooled and diluted by the expansion of the universe, and would by now have reached a temperature of 3K – precisely the temperature of the microwave background radiation. The discovery of this radiation, which could not be accounted for by the Steady State theory, provided one of the strongest pieces of evidence supporting the Big Bang theory.

The helium problem

Spectroscopic evidence shows that by mass about 70 percent of the matter in the universe is hydrogen and about 27 percent is helium, and that the proportion of hydrogen (the lightest element) to helium (the second lightest element) is remarkably constant throughout. If hydrogen is the basic element from which the others were assembled, how was the helium formed, and why is it spread around so evenly? Most of the helium formed in the cores of stars by thermonuclear reactions remains locked up when they die; only a small proportion of them scatter the "ash" from their nuclear furnaces into space, and there cannot have been nearly enough supernovae to account for the observed abundance of helium.

However, if the hot Big Bang theory is correct, the helium would have been created by nuclear reactions which took place during the Big Bang itself. The Steady State theory fails to account for all three sets of observational data.

▲ *Arno Penzias and Robert Wilson beside the radio antenna at Holmdel, New Jersey, with which they discovered the microwave background radiation. The graph (inset) shows that the brightness of this radiation peaks near a wavelength of 1mm and corresponds to a black body temperature of about 3K.*

The Big Bang

► *The main stages in the evolution of the universe, after the first millionth of a second, according to the standard Big Bang theory. The graph shows how the temperature of matter and radiation declined from well over 10^{12}K to about 3,000K in the first 100,000 years or so. At this stage (the "decoupling" of matter from radiation) space became transparent; the radiation released has been diluted and redshifted by the expansion of the universe to become the microwave background.*

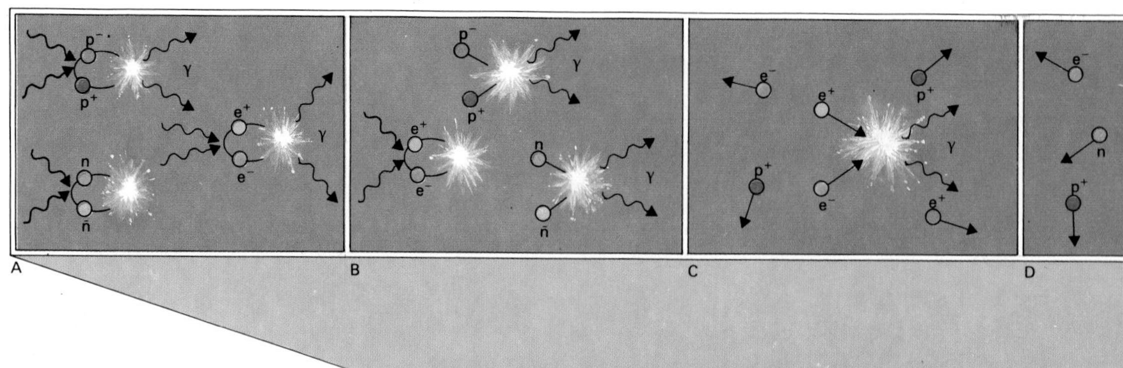

The Big Bang universe – from the first microsecond to the present day

There is no known reason why modern theories of energy, particles and forces should not be applied to processes which occurred a tiny fraction of a second after the Big Bang. According to Einstein's theory of relativity, mass and energy are equivalent and interchangeable. Matter can be transformed into energy, and energy into matter. Two colliding photons of radiation may, if they are of sufficiently high energy, turn into a particle (such as an electron) and its antiparticle (in this case a positron) which has the same mass but opposite charge and spin. Conversely, when a particle collides with its antiparticle, both are annihilated and converted into energy in the form of gamma-ray photons. To produce particles of a given mass (m) the photon energies (E) must be greater than or equal to the mass of the particle multiplied by the speed of light squared (c^2). The higher the energies of the photons, the more massive are the particles which can be created.

In the early stages of the Big Bang, when the temperature was exceedingly high, photon energies were high; but as the universe expanded and cooled, photon energies declined. Less than a millionth of a second after the initial event the temperature of the universe was above 10 million million K, and a great variety of particles and antiparticles was forming and annihilating. At this moment the universe contained particles of matter and photons of radiation in roughly equal numbers. Before the end of the first millisecond the temperature had dropped below 10^{12}K and heavy particles like protons and neutrons could no longer be formed. Most of the ones already in existence were quickly annihilated by collisions with their antiparticles. However, the

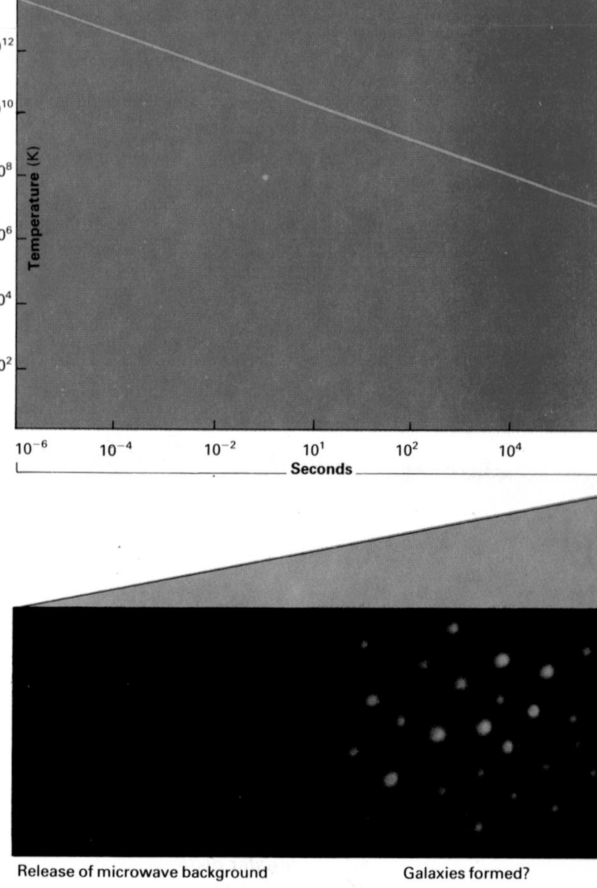

Release of microwave background Galaxies formed?

► *In studying the microwave background, astronomers are looking back to the state of the universe when that radiation was released – about 100,000 years after the initial event. The most remote quasars are seen as they were some 2 billion years after the Big Bang. The most remote galaxies are seen as they were some 5-8 billion years ago. Astronomers have no data relating to the period about 1 billion years after the initial event, when galaxies may have been forming.*

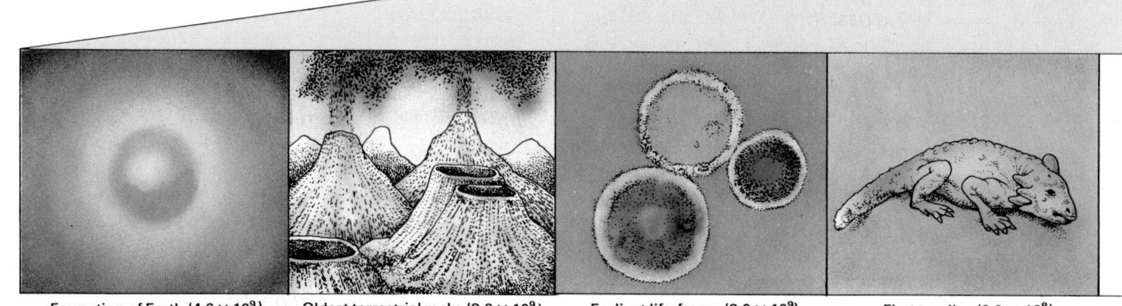

Formation of Earth (4·6 × 10^9) Oldest terrestrial rocks (3·6 × 10^9) Earliest life-forms (3·0 × 10^9) First reptiles (3.0 × 10^8)

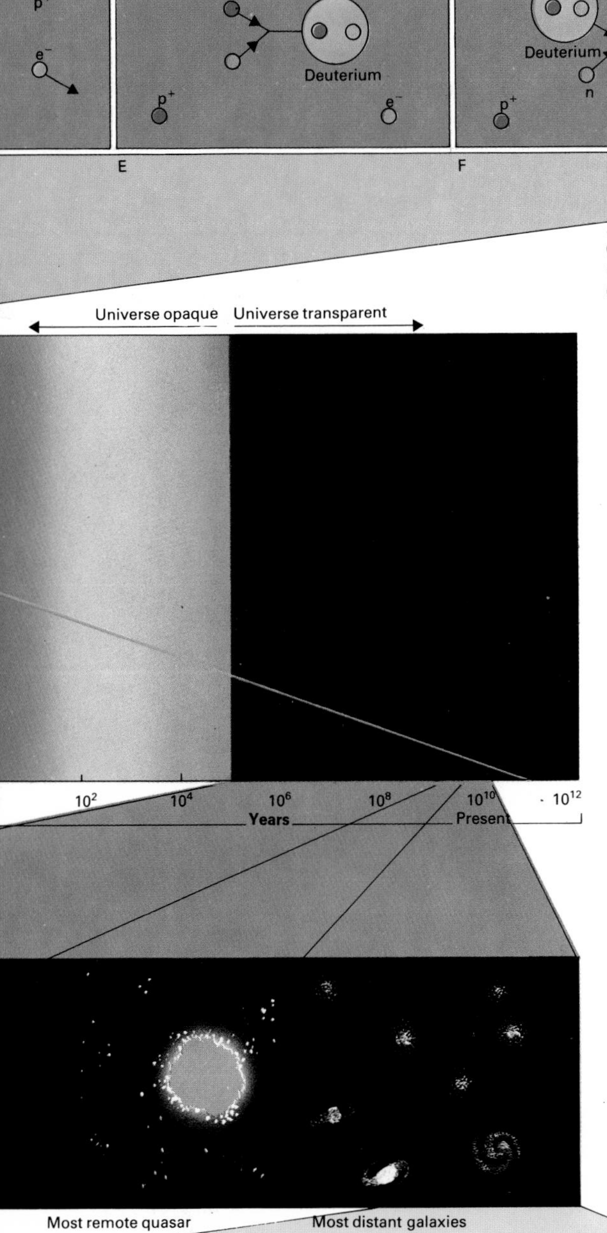

◄ While the temperature was 10^{13}K or more (A) light particles and heavy particles (baryons) like protons, neutrons and their antiparticles were forming and annihilating. After a few millionths of a second (B) baryons could no longer form. Only one in every billion avoided annihilation. Some 10 seconds later, electrons and positrons annihilated (C) leaving only a few electrons (D). After about 3 minutes protons and neutrons began to form deuterium (E), tritium (F) and finally helium (G).

particles outnumbered the antiparticles by a small fraction, so that a residue of protons and neutrons survived. These formed the atoms which comprise the stars and galaxies of the modern universe. Only about one particle in a billion survived, the rest being converted into radiation. A few seconds later, when the temperature had dropped to a few billion K, photon energies became too low to create lightweight particles such as electrons and positrons. Existing electrons and positrons then destroyed each other, leaving just a small residue.

About three minutes after the initial event the universe had cooled to about one billion K, and conditions favored the welding together of protons and neutrons to form helium and small quantities of other light nuclei, such as deuterium and lithium. The present ratio of hydrogen to helium was established at this time. From then on the universe comprised an opaque mixture of matter and radiation, expanding and cooling with the passage of time.

A few hundred thousand years later the temperature had fallen to about 3,000K. Protons became able to capture and hold on to electrons to form atoms of neutral hydrogen, because the radiation was no longer energetic enough to separate (ionize) them. Most of the electrons were quickly mopped up, and it became possible for photons to travel vast distances without being absorbed or suffering collisions. During this stage, known as the "decoupling" of matter and radiation, the universe became transparent.

Since the decoupling the universe has expanded a thousandfold, and the radiation released at that time has been spread ever more thinly over space. It has expanded and cooled, becoming the weak microwave background radiation which is detectable today.

◄ The Solar System was formed 4·6 billion years ago, some 10 billion years after the Big Bang. The oldest terrestrial rocks solidified about a billion years later and the first living organisms appeared soon after that. The earliest species of man dates back about 2 million years. Looking out to the nearby Andromeda galaxy we are looking further back in time than the origin of man, while in looking at the most remote galaxies we are looking back to before the formation of the Earth itself.

The radiation reaching us from the most distant quasars was emitted when the universe was in its infancy – about two billion years old

The formation of the galaxies

The galaxies were formed some time after the decoupling, although astronomers do not yet know exactly how. Some have argued that the galaxies themselves formed first from relatively small concentrations of matter which arose as irregularities in the expanding universe: these then aggregated into clusters, and the superclusters were assembled from encounters between the clusters. Others support the view that very large mass concentrations (superclusters) formed first, and as they collapsed under the action of gravity, they broke up into cluster-sized masses which, in turn, fragmented into individual galaxies. Some support for the first hypothesis comes from the fact that galaxies appear to be nearly as old as the universe itself, and must have formed within one or two billion years of the Big Bang, whereas the irregular and straggly appearance of clusters and superclusters suggests that these huge structures are younger than the galaxies.

In studying the most remote quasars, astronomers are looking back to a time when the universe was about two billion years old, or one-tenth of its present age. The microwave background radiation provides information about the state of the universe when it was less than one million years old. Unfortunately, there are at present no observational data relating to the crucial period between one million and two billion years after the Big Bang during which the galaxies must have been born. One possible clue comes from the fact that recent intensive searches have failed so far to detect many quasars with redshifts much greater than three. This may imply that quasars first "switched on" when the universe was about two billion years old.

New instruments such as the Space Telescope may soon be able to detect even more remote, heavily redshifted galaxies, and so may be able to see far enough back in time to reach the era of galaxy formation; for the moment, however, the birth of galaxies and clusters remains an unsolved problem.

► This image shows how the universe would appear to an observer inside the three-dimensional simulation shown in the cube below right. The distribution of "galaxies" into clusters and filaments broadly resembles the appearance of the real universe.

▼ Adrian L. Melott of the University of Pittsburgh has carried out computer simulations of the growth of large-scale clustering in the universe, assuming that the universe is dominated by "dark matter" such as massive neutrinos, and that its mean density is slightly greater than the critical density. From left to right the diagrams display the simulated appearance of the universe when it was one-eighth and one-third of its present size, and its actual size.

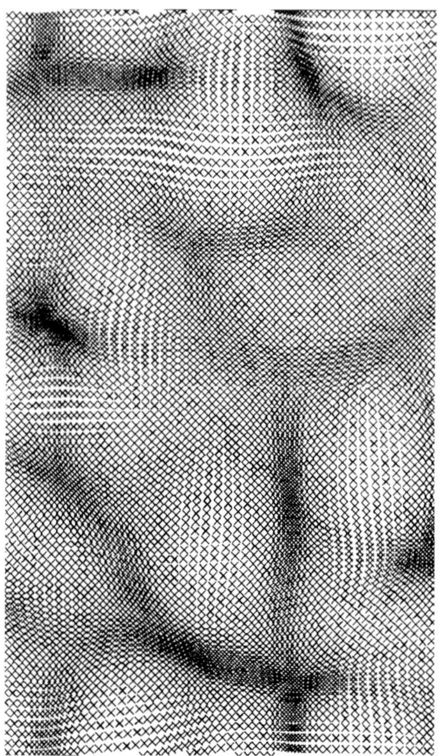

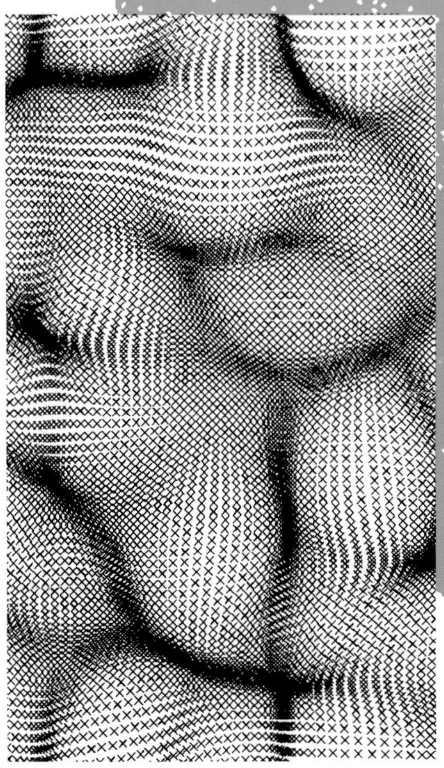

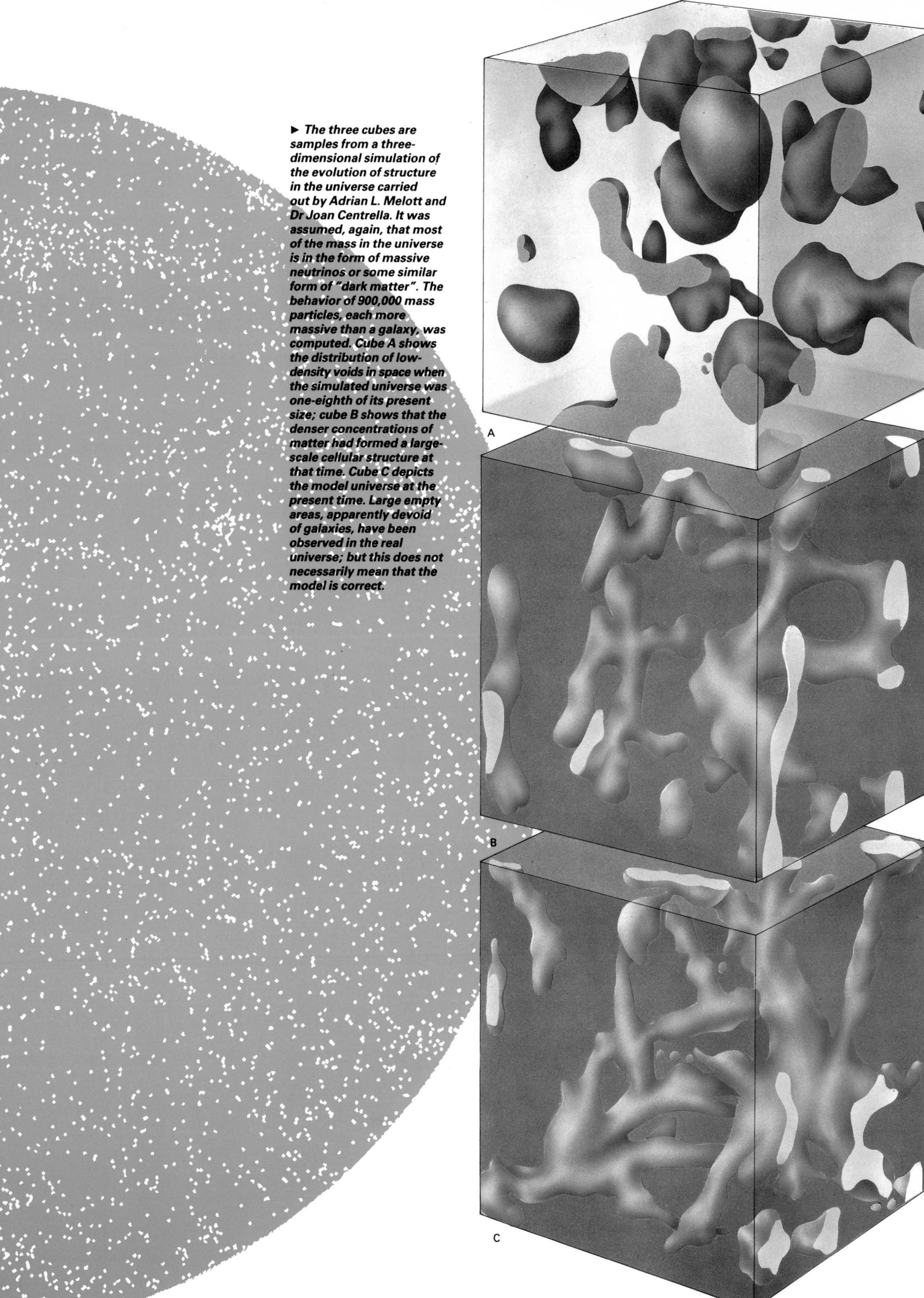

► *The three cubes are samples from a three-dimensional simulation of the evolution of structure in the universe carried out by Adrian L. Melott and Dr Joan Centrella. It was assumed, again, that most of the mass in the universe is in the form of massive neutrinos or some similar form of "dark matter". The behavior of 900,000 mass particles, each more massive than a galaxy, was computed. Cube A shows the distribution of low-density voids in space when the simulated universe was one-eighth of its present size; cube B shows that the denser concentrations of matter had formed a large-scale cellular structure at that time. Cube C depicts the model universe at the present time. Large empty areas, apparently devoid of galaxies, have been observed in the real universe; but this does not necessarily mean that the model is correct.*

The universe which began with a Big Bang may end in a "Big Crunch" or expand into eternity – scientists are trying to find out which

The future of the universe

Will the universe expand forever? The action of gravity is slowing down the rate of expansion, but despite this, if the galaxies are traveling fast enough they will continue to recede forever – with the expansion merely slowing down towards a steady rate. Such an ever-expanding universe is said to be "open". But if the gravitational attraction between galaxies and other forms of matter in the universe is strong enough, the expansion will eventually stop. Gravity will then cause the galaxies to fall together until the universe, which began in a Big Bang, ends in a "Big Crunch". A universe which expands to a maximum size and then collapses is known as a "closed" universe, and in it space is curved into a finite volume which – like the skin of an inflated balloon – has no discernible boundary. Between these two possibilities is the concept of the so-called "flat" universe, in which the galaxies have just enough energy of motion to move apart forever, gradually losing speed but never actually stopping. In such a universe space is "flat" in the sense that on the large scale it is not curved.

In principle, as astronomers probe to fainter and fainter brightness limits or to greater and greater red shifts, the number of sources visible at very great distances should increase by an amount which will depend on whether the universe is open, flat or closed. In practise, measurements of this kind are so difficult both to make and to interpret that results obtained so far have been inconclusive.

However, astronomers should be able to determine whether or not the universe is fated to expand forever by measuring the mean density of matter within it. If the actual density is greater than the critical density which would provide gravitational force strong enough to halt the receding galaxies, the expansion of the universe will eventually cease and its ultimate collapse will ensue; but if the actual density is less than the cricital value, expansion will continue forever.

What the observations show

One way to estimate the mean density of matter in space is to add up the masses of all the observable galaxies and divide the total mass by the volume of space which has been investigated. This method yields a figure that is no more than a few percent (probably 1-2 percent) of the critical density. However, there is growing evidence for the existence of massive dark haloes round the visible galaxies, and for the presence of substantial amounts of intergalactic matter in clusters. When all this is taken into account, it may be necessary to increase the actual mean density by a factor of ten or more – although this still seems to be significantly less than the critical value.

The cosmic abundances of the light elements deuterium and lithium also appear to point to the same conclusion. The theory of the nuclear reactions which took place in the first few minutes of the Big Bang suggests that these elements could only have survived in their observed quantities if the mean density of matter were less than about 20 percent of the critical density. If the density of matter were appreciably greater, only a tiny fraction of the observed amounts of deuterium and lithium would exist in the universe today.

Friedmann models of the universe

If the universe began in a Big Bang, it will either expand for ever (the open universe) or it will expand to a maximum size and then collapse (the closed universe). The dividing line between these possibilities is the so-called "flat" universe.

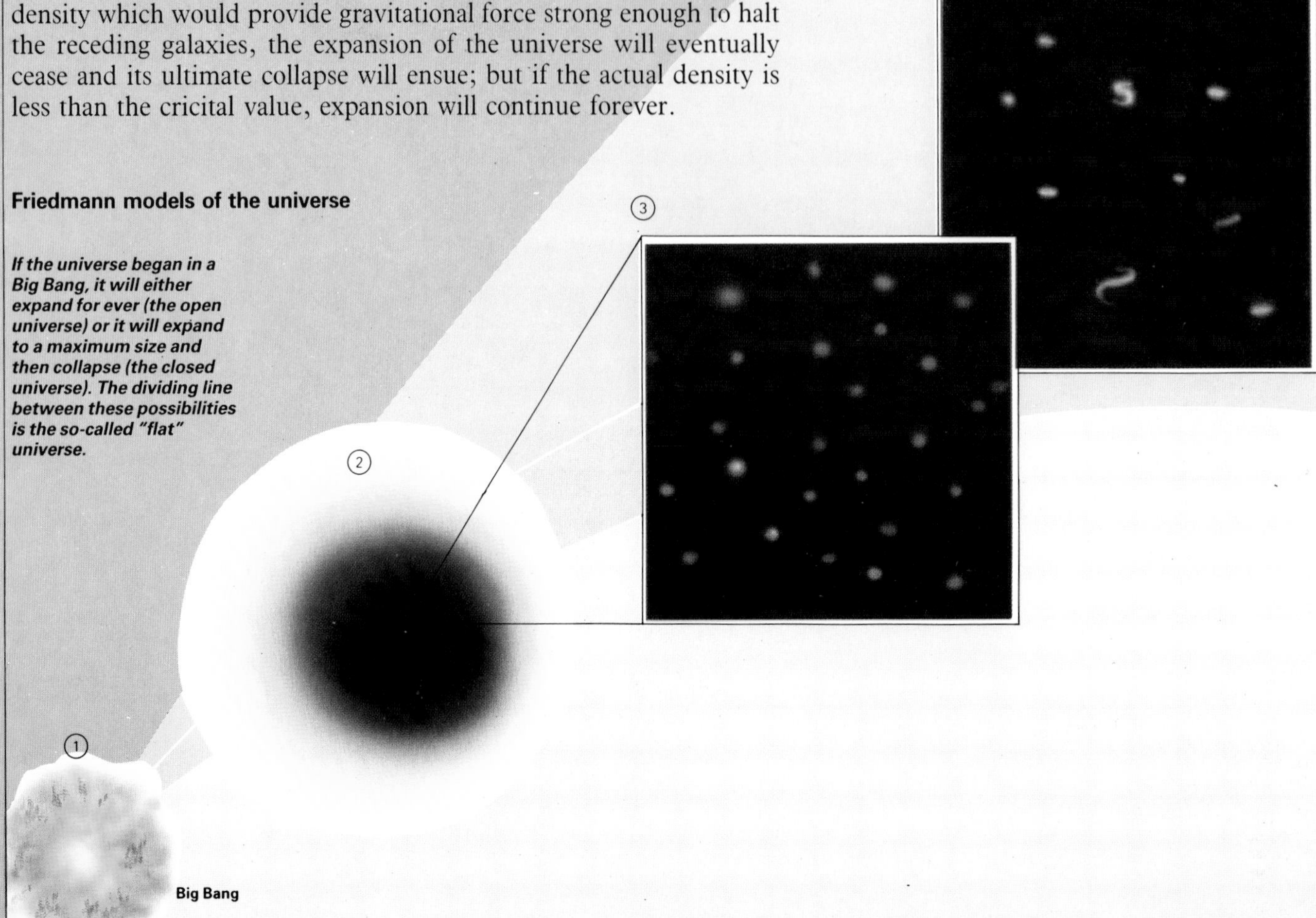

The open universe

1 Big Bang
2 Expanding opaque mass of matter and radiation
3 Galaxies begin to form
4 Galaxies move apart
5a-8a Galaxies move apart from each other for ever, and fade away as their stars run out of fuel

Professor Robert H. Dicke (left) and J.E. Peebles have suggested that if the universe is oscillating, successive cycles will become larger and longer and new material may be created in each successive big bang. If so, there cannot have been an infinite number of cycles.

The closed universe

1 Big Bang
2 Expanding opaque mass of matter and radiation
3 Galaxies begin to form
4 Galaxies move apart
5b Galaxies reach their maximum separation
6b Galaxies begin to fall together
7b Galaxies begin to merge
8b Universe collapses into "Big Crunch"

Eventually the universe will thin out into an unimaginably dilute sea of particles and radiation

Protons
Supergalactic black holes
Galactic black holes
Galaxies
Stellar black holes
Stars

0 Big Bang — Present 10^{10} 10^{12} 10^{15} 10^{20} 10^{24} 10^{27} 10^{30} 10^{33} 10^{40} 10^{50}

Most stars dead, galaxies fading away

Galactic core and dead stars; mutual encounters expel some or move others closer to core

Massive core (black hole) and clustered dead stars; most of rest have been expelled

Last dead stars in core falling into galactic black hole

Encounters between cluster galaxies or galactic black holes form supergalactic black holes

Possible half-life of proton (proton decays into electrons and positrons)

If protons decay, all matter and dead stars (apart from black holes) decay into electrons and positrons

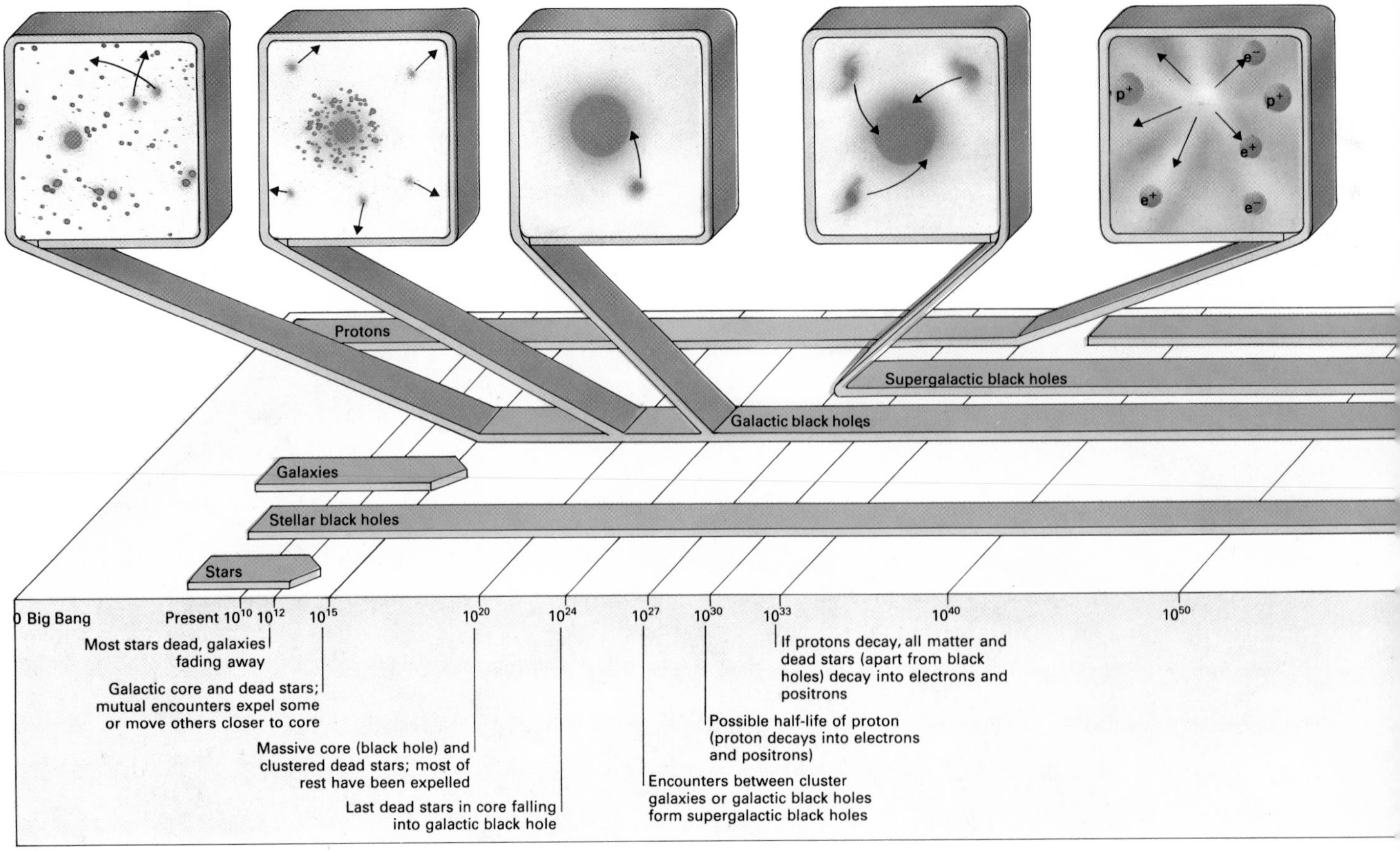

The long-term future of an ever-expanding universe

The balance of evidence at present points towards an open, ever-expanding universe, containing insufficient mass to halt the headlong separation of the galaxies. If this is so, the long-term future looks bleak. Most of the galaxies have already converted most of their gases into stars and the rates of star-formation within them must dwindle. Individual stars will burn out, ending up as black dwarfs, neutron stars or black holes. Even the longest-lived low-mass celestial glow-worms will fade away after a few thousand billion years or so, and the galaxies will fade to dying embers.

Close encounters between dead stars will eventually expel most of them from their parent galaxies, while the rest will fall into the central region to be absorbed into massive galactic black holes. A possible timescale for this is about 10^{27} years (a thousand million million million million years).

Likewise, encounters between dead galaxies will expel many of them from their clusters and cause the rest of them to coagulate into "supergalactic" black holes of, perhaps, hundreds of billions of solar masses.

If, as Stephen Hawking has suggested (◀ page 180), particles can leak very slowly out of a black hole, this will eventually evaporate in a spray of particles and antiparticles. Stellar-mass, galactic and supergalactic black holes, respectively, will require something like 10^{66}, 10^{90} and 10^{100} years to evaporate in this way. Given long enough, even black dwarfs and neutron stars may disintegrate, leaving a universe consisting of an unimaginably rarefied mixture of particles and radiation destined to expand forever towards a state of darkness, zero temperature and infinitely low density.

The missing mass

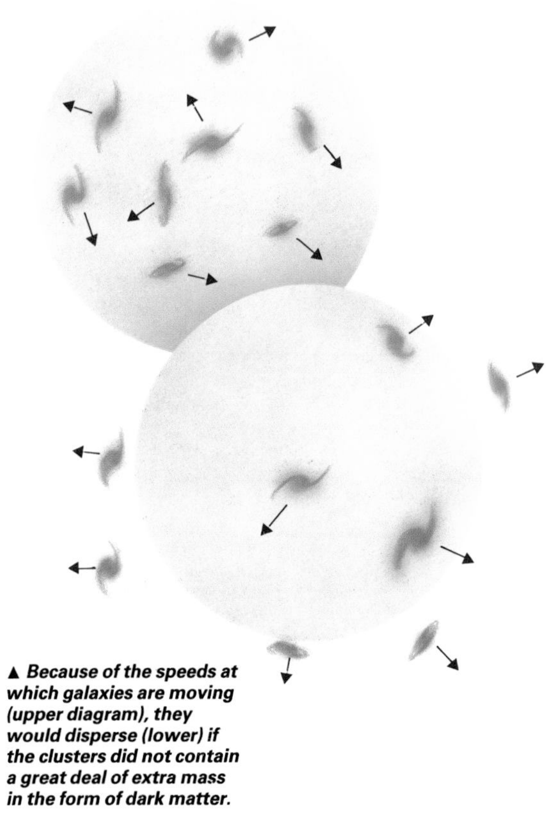

▲ Because of the speeds at which galaxies are moving (upper diagram), they would disperse (lower) if the clusters did not contain a great deal of extra mass in the form of dark matter.

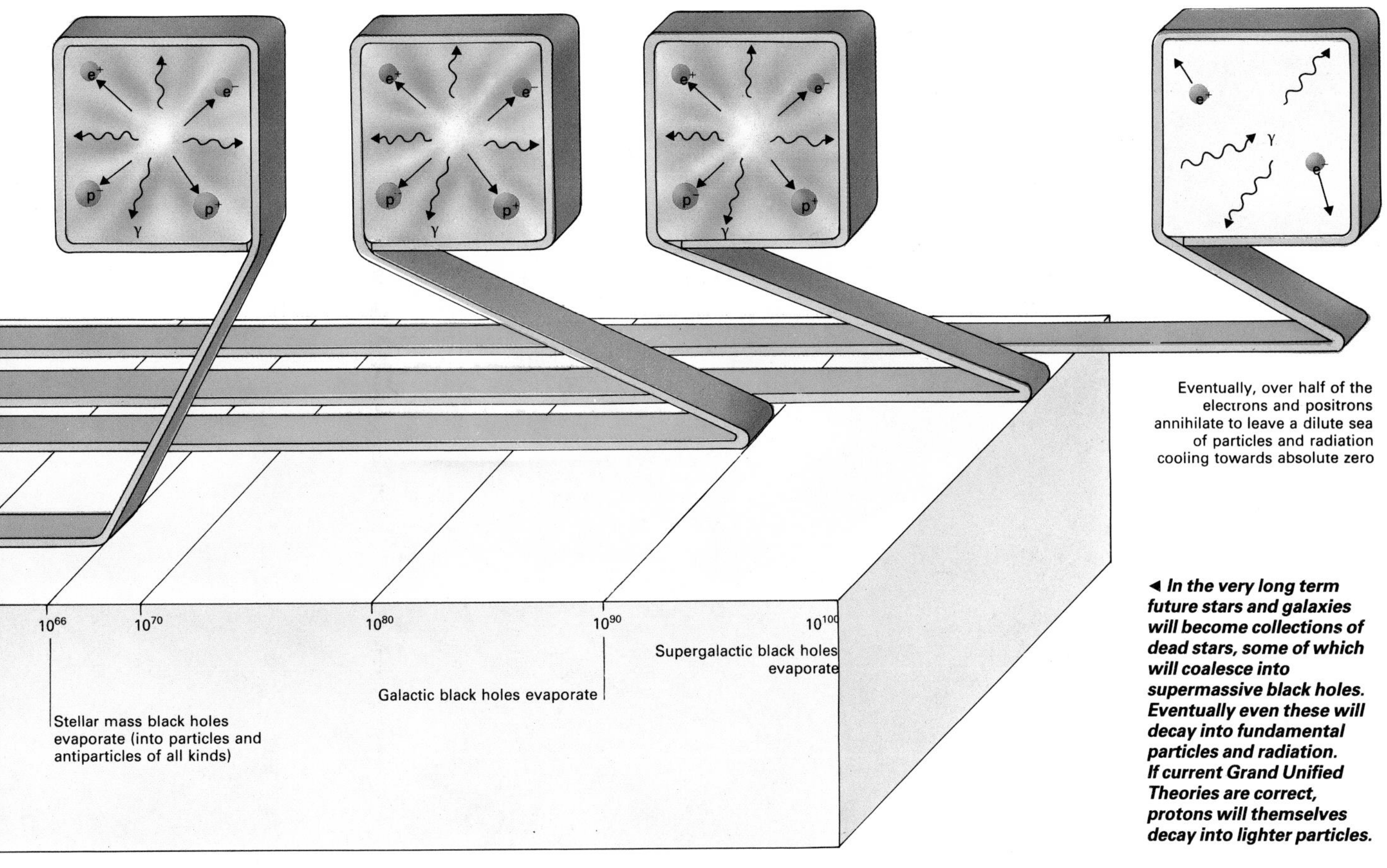

◀ *In the very long term future stars and galaxies will become collections of dead stars, some of which will coalesce into supermassive black holes. Eventually even these will decay into fundamental particles and radiation. If current Grand Unified Theories are correct, protons will themselves decay into lighter particles.*

Weighing the universe

1–2%

10–20%

100%

▲ *If all the mass in the universe is contained in visible galaxies (1), its mean density is only 1-2 percent of the critical density which would halt its expansion. Current estimates of the mean density (2), taking account of dark matter, suggest that the mean density is 10-20 percent of the critical value. The actual value can exceed the critical density (3) only if the bulk of the mass is dark matter of a type different from ordinary matter.*

Dark matter – the universe may yet be closed

In recent years many theoreticians have argued that the universe may contain more than enough "dark matter" – invisible forms of mass – to halt its expansion. So long as this matter does not behave like ordinary "baryonic" matter, which is comprised of particles such as protons and neutrons, its presence would not have affected the amount of deuterium and lithium produced in the Big Bang.

One dark matter candidate is the neutrino (◀ page 102). There are good theoretical reasons for assuming that the neutrino – hitherto considered to have zero mass – in fact has a very small but finite mass. Vast numbers of neutrinos would have been produced in the Big Bang so that, today, there should be between 100 million and one billion neutrinos in every cubic meter of space. If the mass of a neutrino is more than a few hundred thousandths of the mass of an electron, and if they are indeed as numerous as is generally thought, then their total mass will be enough to halt the expansion of the universe. Tentative experimental measurements made in 1980 seemed to show that neutrinos do indeed have mass. However, it remains to be seen whether or not they do have mass, and if so, what that mass may be.

Neutrinos are termed "hot" dark matter because, in the early stages of the universe they would be moving at very high speeds relative to ordinary matter. If most of the mass of the universe were in the form of hot dark matter, the effect of these particles would have been to smooth out any small-scale (galaxy-mass) irregularities and lead to a universe in which superclusters would have been the first units to form. Since observations seem to show that galaxies formed before superclusters, neutrinos are less strongly favored as dark-matter candidates than they were in the early 1980s.

The Universe and the Forces of Nature

Alternative futures

Some cosmologists believe that the presence of "cold" dark matter (invisible matter which moves comparatively slowly relative to ordinary matter) in the early universe would cause ordinary matter to clump into galaxy-masses before superclusters, and would, therefore, lead to the kind of structure which is observed in the universe today. Thus, as yet undetected particles may be so numerous they could halt the expansion of the universe although their masses may be as low as a billionth of the electron mass.

In principle, very large amounts of matter could be locked up in black holes. However, If all the mass needed to close the universe were contained in supermassive black holes of 10 billion solar masses or more, there should be more gravitational lens effects than have actually been observed. But if the universe were filled with primordial "mini" black holes produced in the early stages of the Big Bang, these could indeed provide enough mass to halt the expansion. However, at present there is not a shred of evidence to suggest that objects of this kind actually exist.

On balance, the evidence indicates that the universe will expand forever. But if dark matter does exist in sufficient quantity to close the universe, then the expanion will cease – perhaps tens or hundreds of billions of years hence. Thereafter, slowly at first, but with ever-increasing speed the galaxies (or what is left of them) will converge. Their redshifts will change to blue shifts, the background temperature will rise and eventually matter will dissolve into a "soup" of particles and radiation and the universe will become a collapsing fireball returning towards a state of infinite compression.

The Big Crunch may mark the end of the universe once and for all or, as some cosmologists suggest, it may trigger a new Big Bang and a new cycle of expansion and contraction. The universe may continue to oscillate in this way, with all its matter being reprocessed at the end of each cycle.

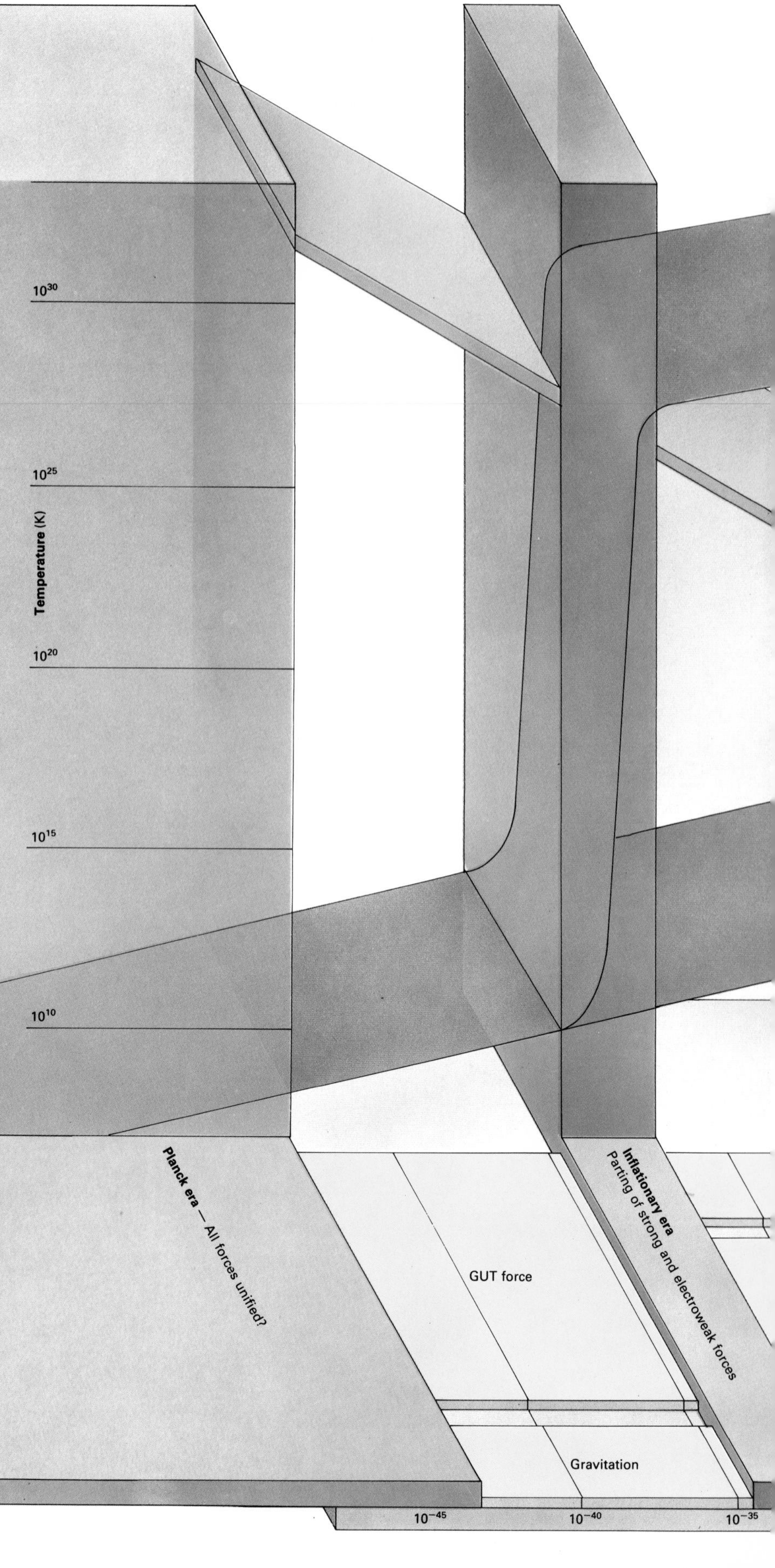

10^{30}
10^{25}
Temperature (K)
10^{20}
10^{15}
10^{10}

Planck era — All forces unified?

Inflationary era
Parting of strong and electroweak forces

GUT force

Gravitation

10^{-45} 10^{-40} 10^{-35}

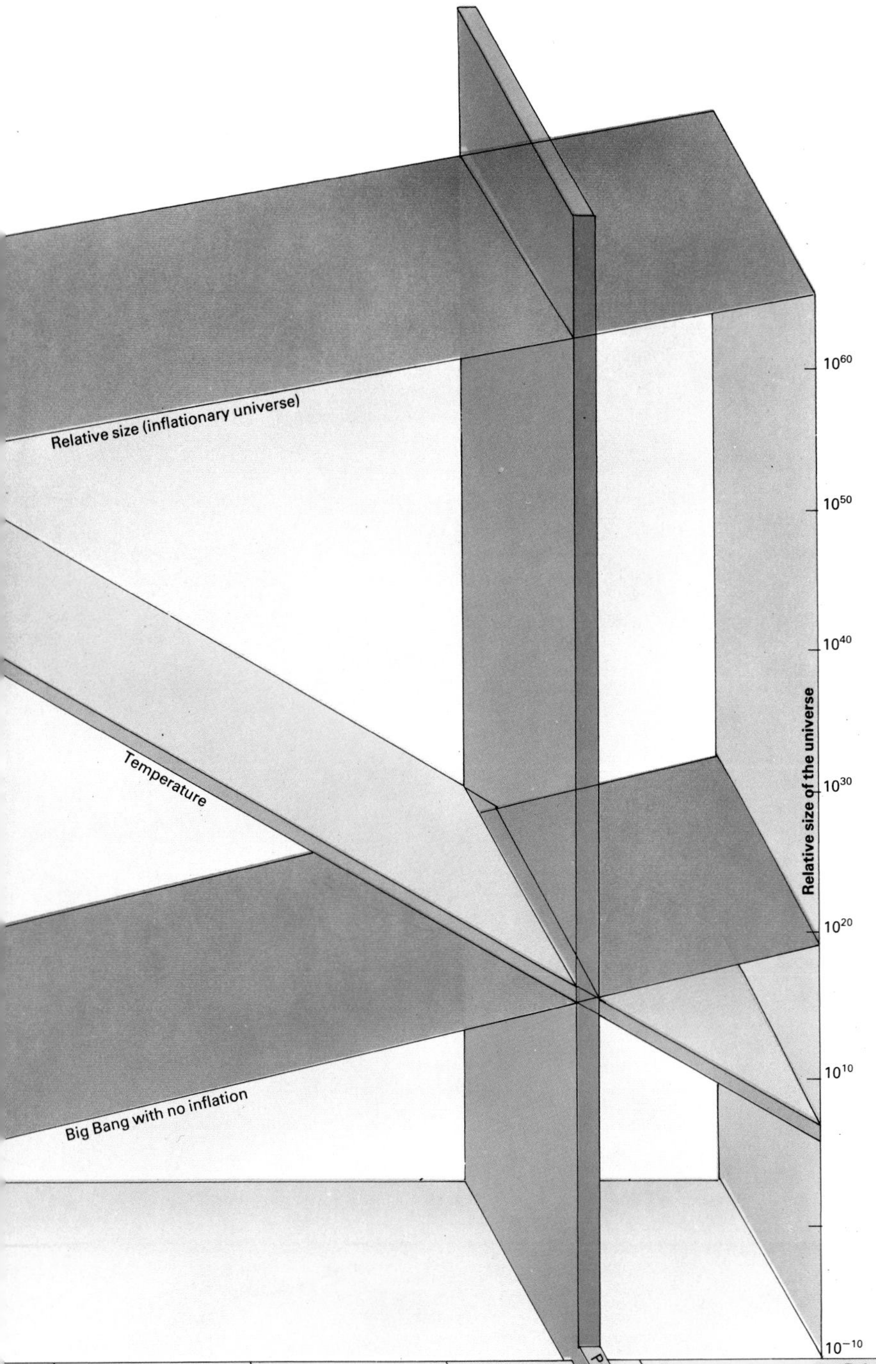

Relative size (inflationary universe)

Temperature

Big Bang with no inflation

Relative size of the universe

10^{60}

10^{50}

10^{40}

10^{30}

10^{20}

10^{10}

10^{-10}

Parting of weak nuclear and electromagnetic forces

Strong force

Electroweak

Gravitation

Force	Role	Relative strength	Range
Strong nuclear	Binds atomic nuclei (prevents like-charged protons flying apart)	1	Short (10^{-15}m)
Electromagnetic	Binds atoms (opposite charges of protons and electrons attract)	10^{-2}	Infinite
Weak nuclear	Controls radioactive decay of some nuclei (decay of neutron to proton and electron and neutrino)	10^{-5}	Short (10^{-17}m)
Gravitation	Mutual attraction between all particles of matter. Controls motion of planets, stars and galaxies	10^{-39}	Infinite

10^{-30} **Time** (seconds) 10^{-25} 10^{-20} 10^{-15} 10^{-10} 10^{-5}

The four forces of nature

Gravity is the force which controls the large-scale behavior of the universe, and yet it is by far the weakest of the four known forces of nature. In order of decreasing strength, these forces are: the strong nuclear interaction (which binds atomic nuclei), the electromagnetic force (which, for example, controls the absorption and emission of light by atoms), the weak nuclear interaction (which governs the radioactive decay of atoms) and gravity. The relative strengths of these forces are 10^{39}:10^{37}: 10^{34}:1. However, both the strong and weak nuclear interactions operate only over tiny distances within atomic nuclei. Only the electromagnetic and gravitational forces operate over infinite distances and, since matter on the large scale is electrically neutral (the total number of positive and negative electrical charges is about the same) gravity dominates matter on the large scale.

Recently, physicists have shown that at high energies the electromagnetic and weak forces act as one "electroweak" force. The Grand Unified Theories, or "GUTs", currently being developed predict that at even higher energies the electroweak and strong nuclear forces will act as one. Many physicists believe that at still higher energies all four forces, including gravity, will merge into one single force.

The significance of the Grand Unified Theories

GUT theories predict that protons, hitherto believed to be the stable building blocks of matter, must eventually disintegrate into lighter particles. The half-life of a proton (the time taken for half of a batch of protons to decay) is believed to be greater than 10^{31} years but, given enough protons, it should be possible at any time to detect the sporadic decay of a few, and experiments are under way to see if this actually happens. If protons do decay, this will lead eventually to the destruction of matter as we know it, and to the disintegration of bodies like black dwarfs after periods of "only" 10^{33} years or so.

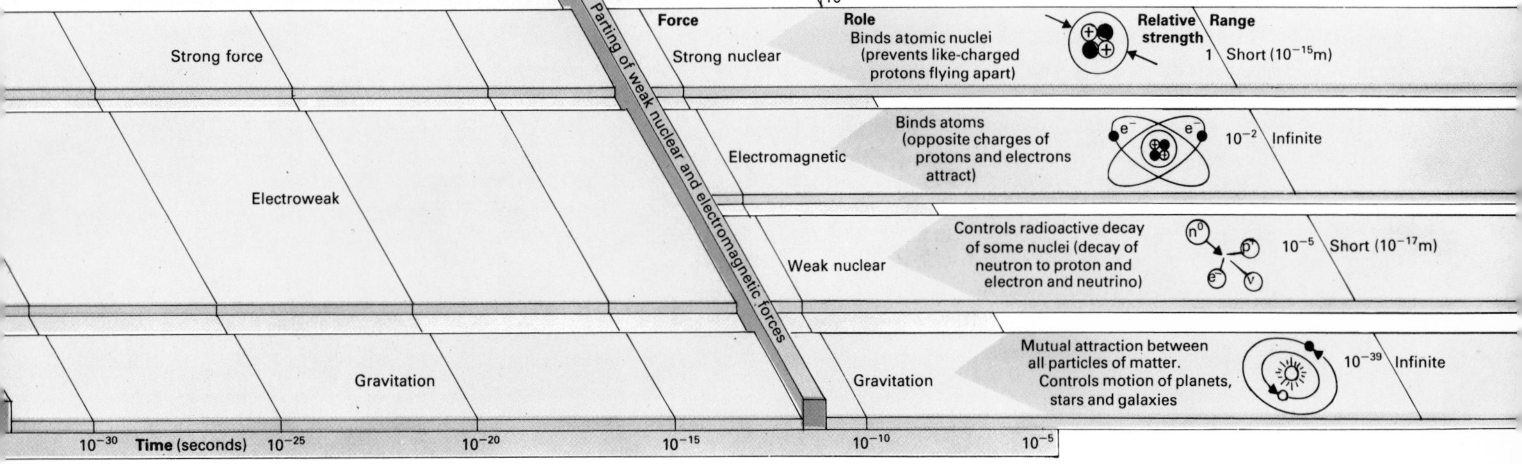

See also
Overview of the Universe 129-32
Observing the Universe 137-44
Black Holes 177-80
Galaxies 197-204
Active Galaxies and Quasars 205-16

The inflationary universe

There is no theory yet which can describe the state of the universe within the first 10^{-43} seconds (the so-called Planck time), but it seems likely that in the first instants the four natural forces behaved as one. As the universe expanded and cooled, and the energy level declined, first gravity, then the strong force and finally the remaining pair split to become separate entities.

Some cosmologists believe that the parting of the strong and electroweak forces which occurred when the universe was about 10^{-35} seconds old and had cooled to a temperature of about 10^{27}K produced a sudden and dramatic change in the state of the universe. Just as water releases heat when it freezes, so the change of state of the universe released vast amounts of energy and produced a sudden "inflation" of the universe, which, for a time, expanded so rapidly that it doubled in size every 10^{-34} seconds. Even if the inflationary period lasted for only just over 10^{-32} seconds, the universe would have grown to about 10^{50} times its previous size before resuming its normal rate of expansion (which slows down with the passage of time, due to the action of gravity).

There are many variants of this theory, but if inflation did occur, the universe must have been blown up to such an extent that it is now vastly larger than the limited volume of space observed so far, and hordes of unseen galaxies may yet lie beyond the horizon.

Another possibility is that the GUT transition – the separation of the strong and electroweak forces – may not have occurred simultaneously at every point in the universe, but may instead have occurred at similar times in a large number of separate centers. If so, spacetime may have divided into a great many separate expanding regions, and the entire observable universe may be no more than a tiny part of just *one* of these expanding bubbles. There may be countless separate and isolated universes.

Such ideas are speculative as yet, but whether or not they prove to be sound, cosmology has entered an exciting phase. Researches into the fundamental nature of particles will be combined with increasingly sophisticated observations of the remote depths of the universe to yield one day, it is hoped, an all-embracing theory of the origin, evolution and ultimate fate of space, time, matter and the universe.

The inflationary universe

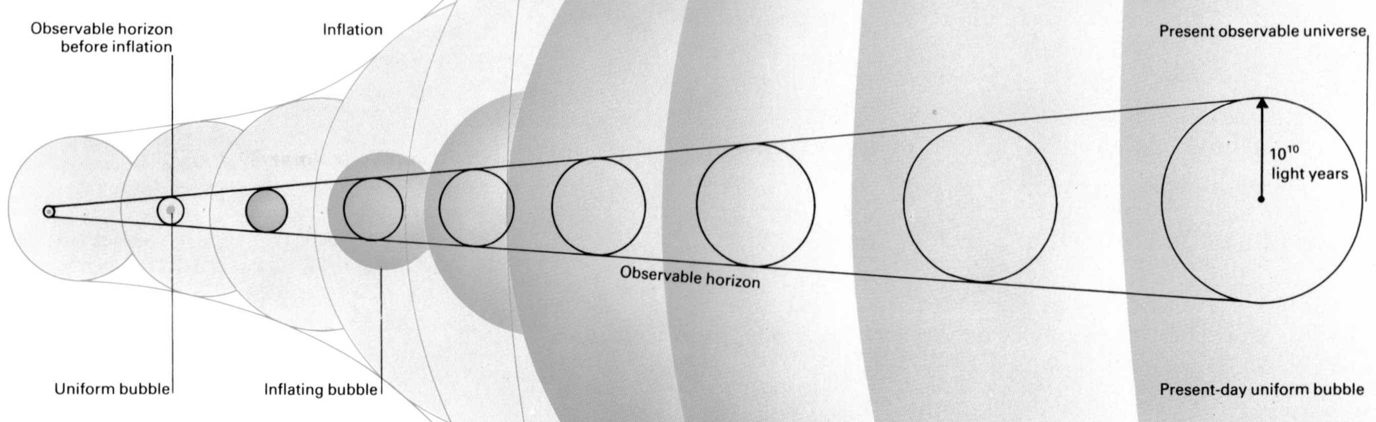

Inflation, uniformity and flatness

The inflationary idea provides natural explanations for two problems which have puzzled Big Bang theorists. These are: why is the universe so smooth and uniform, and why does the universe seem to lie close to the dividing line between the open (ever-expanding) and closed (ultimately collapsing) case?

If the conventional Big Bang theory is correct, the observed universe consists of a large number of regions of space which were isolated from each other when the universe was young because there had not been time for any influence (even traveling at the speed of light) to pass from one to another. This was so even at the time when the microwave background radiation was emitted. Cosmologists are puzzled, therefore, by the uniform nature of the microwave background over the entire sky – any irregularities in the early universe should show up in this radiation.

However, if the inflationary idea is correct, a single tiny region which was small enough to be uniform would have blown up to such an extent that the universe now observed is only a tiny part of a much larger uniform region.

Too much of a coincidence

The second problem, known as the "flatness problem", relates to why the universe is expanding today at a rate close to the "flat" universe case. If, in the early instants of the universe, the rate of expansion had differed by even the slightest amount from the flat case (in other words, if the density were only marginally different from the critical density) then those differences would have been enormously magnified by now.

It perplexes cosmologists that conditions in the early universe could have been so finely balanced. However, if inflation blew up the universe to such an extent that the observable part is only a microscopic fraction of the whole, then space would appear flat to us in the same sort of way as the surface of the Earth appears flat to an observer who can see only a very small fraction of it.

Communication and Travel

The possibility of life elsewhere in the universe...Early searches for extra-terrestrial intelligence...The first broadcasts to the stars...The feasibility of interstellar rocket ships...PERSPECTIVE...The evidence for a planetary system round Beta Pictoris...Would aliens be friendly or hostile?

One of the most fascinating and emotive problems facing modern science is that of finding out whether or not we are the only community of living beings in the cosmos. Is the universe populated with myriad life-forms; is life a rare and isolated phenomenon, or is life unique to Earth?

Although it is possible that life may exist in utterly different forms in wholly alien environments, a reasonable approach to the problem is to limit discussion to life as we know it. On Earth, all living things are constructed from organic molecules which have carbon (C), nitrogen (N), oxygen (O) and hydrogen (H) as their principal constituents; and it is the ability of carbon to form long-chain molecules that enables complex structures such as even the simplest living cell to be assembled. These basic elements are among the most common in the universe, and organic molecules exist in abundance in the dense interstellar clouds in which stars are formed. Scientists have found samples of amino acids (key building blocks of living cells) in a few meteorites, and at least one amino acid – glycine ($C_2H_5O_2N$) – has been identified in an interstellar cloud. Elementary living organisms may have originated on the Earth or, as some astronomers have suggested, in interstellar space; but either way, the materials for life are plentiful in the universe.

Finding the right kind of planet

Life as we know it requires a planet to live on and a suitable star to provide energy. All terrestrial organisms require – for some of the time at least – the presence of liquid water, which comprises about 80 percent of a typical cell. Therefore the temperature on the planet must lie between the freezing and boiling points of water, and this sets limits on the permissible distance between a star and a planet. Hot, highly luminous O, B and A-type stars are probably unsuitable because they do not live long enough to allow life to evolve, and at the other end of the scale, planets would have to be located very close to cool red dwarfs to receive adequate warmth for terrestrial-type life. F, G and K-type stars – broadly similar to the Sun – are probably most suitable.

Most astronomers think that planets form as a natural by-product of the process of star-formation. If this is correct, planetary systems are likely to be common, and there may be as many as ten billion in the Milky Way galaxy alone. Direct evidence is difficult to obtain, but the almost indiscernible "wobbling" motion of several nearby stars (notably Barnard's Star) seems to indicate that they are accompanied by bodies comparable in mass with Jupiter.

New electronic observing techniques, and the superior resolution of the Space Telescope, may lead to the definite detection of extra-solar planets within the next few years.

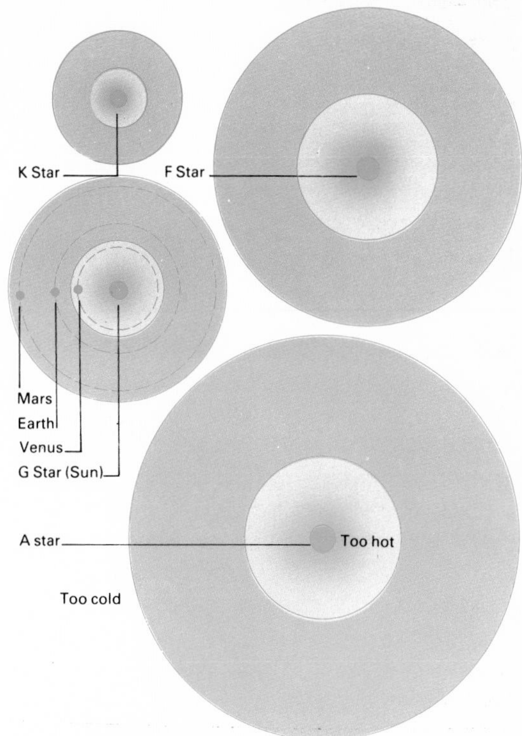

▲ ***The "ecosphere" is the region around a star within which a planet could have a temperature suitable for liquid water and, possibly, life to exist on its surface. Comparative ecospheres are shown for stars of spectral types A, F, G and K.***

Life beyond our Solar System?

Unlike his contemporaries the French author Bernard de Fontenelle (1657-1757), in his book "On the Plurality of Worlds", speculated about life farther out in the universe. "If you were in one of the little Vortexes of the Milky Way, you would find your Sun scarce nearer to you, nor would he sensibly have more force on your eyes, than a hundred thousand other suns of neighboring little Vortexes." Fontenelle wondered whether such peoples might be bathed in perpetual light, and would be astonished to find that there are "unhappy people, who have very dark nights, who fall into profound darkness, and who, when they enjoy light, even then seen only one Sun".

This was certainly a new outlook, and caused a great deal of discussion. Later, William Herschel (1738-1822) was inclined to believe that life might be widespread – he even claimed that there were intelligent beings living in a cool region below the surface of the Sun, and when it became certain that the Sun was a normal star there seemed no reason to doubt that other suns, too, might be attended by inhabited planets.

This outlook has been followed up in modern times by scientists such as Sir Fred Hoyle, who maintains that life was brought to Earth from deep space, probably by a comet. In America, Carl Sagan has also stated his belief in widespread life, and his book on the subject was co-authored by the Soviet astronomer Iosif Shklovsky – though Shklovsky has now changed his mind, and believes life to be unique to Earth.

"If there are so many people out there, where are they?" Nobody can answer that question

Slight wobbling of Barnard's Star may be due to two planets the size of Jupiter. The possible sizes of their orbits are compared with those of Mars and Jupiter.

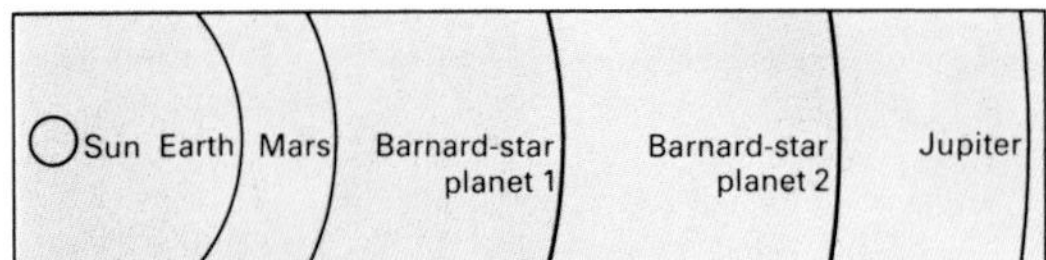

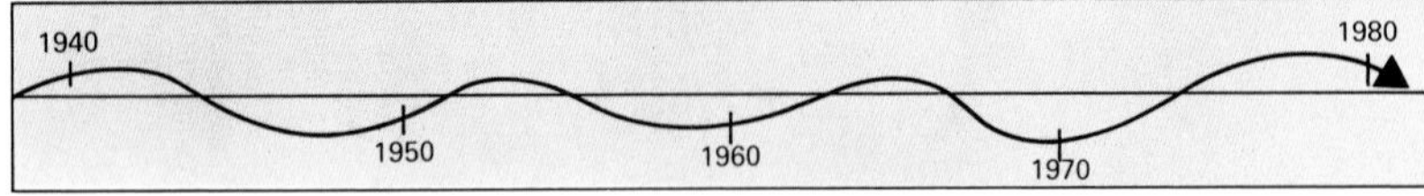

According to one school of thought, if there are billions of planetary systems in the Galaxy, then conditions similar to those on Earth must exist on a great many planets; and *if* life emerges whenever conditions are favorable, then life must be widespread throughout the universe. In that case, intelligent life may also be common, and some intelligent species may have developed the ability to communicate or to travel across interstellar distances. Some astronomers estimate that there should be at least a million advanced technological civilizations capable of interstellar communication in the universe today.

Others contend that the sequence of chance events which led to the evolution of intelligent life on Earth is so unusual that it may not have been repeated anywere else. Yet others argue that if intelligent life is common, there must already exist many species far in advance of our own, in which case it is very odd that we do not see any sign of their activities. If only *one* civilization had developed the means to travel at a few percent of the speed of light, it could have explored the entire galaxy by now. The great physicist Enrico Fermi once remarked, "If there are so many people out there, where are they?". Some scientists argue that we do not see them because they do not exist.

SETI – the search for extra-terrestrial intelligence

If there are numerous advanced species, they may well be sending radio messages into space, and it is possible that some of their broadcasts could be detected by radio telescopes. The problem is to look in the right direction at the right time, and to tune to the correct frequency. The first search for extra-terrestrial intelligence (SETI) was carried out by Frank Drake in 1960. He observed two nearby stars of suitable type, Tau Ceti and Epsilon Eridani, at a frequency of 1420 MHz (21cm wavelength) on the grounds that any alien radio astronomer would choose it because hydrogen emits at that frequency.

There is a great deal of background noise from hydrogen clouds at this frequency, however, and some of the more recent searches have tried different ones.

More than 30 searches have been carried out since 1960, amassing a total of more than 5,000 hours of observing time, but, so far, no signal has been detected. Nevertheless, the search has barely begun, and no conclusions can be drawn from the negative results obtained so far.

The Earth's first deliberate broadcast to the stars was made from the 300m dish at Arecibo in 1974 (◀ page 142). A binary-coded message was beamed towards the globular cluster M13, at a distance of 24,000 light years. In principle, a similar instrument could detect the message when it arrives in 24,000 years' time. Naturally, any reply would take another 24,000 years to get back to Earth, so that this must rank as a very long-term experiment.

The search for other planetary systems

Although we still lack any final proof that planetary systems exist, movements of faint, nearby dwarfs such as Barnard's Star provide some evidence. Recently IRAS, the Infra-Red Astronomical Satellite, detected infrared excesses round at least 40 stars, possibly indicative of planet-forming material.

In 1984 Bradford Smith and Richard Terrile used the large reflector at Las Campanas, in Chile, to study one of these stars – β Pictoris, in the constellation of the Painter – and detected a disk of material extending from the star. The material is thought to be the same as that making up the Earth and the Sun's other planets.

The disk round β Pictoris is seen nearly edge-on, and may not be more than a few hundred million years old (β Pictoris is at least 50 times as luminous as the Sun, and is therefore evolving far more quickly).

Analysis of the density of the material indicates that planets may already have been formed, and that the inner particles, close to the star, have already been swept away – possibly by fully formed orbiting planets.

The dangers of talking to aliens

It has been claimed that we would be unwise to advertise our presence to the inhabitants of other worlds, because of the fear that unfriendly aliens would come to Earth and "take us over". However, this seems absurd on two grounds. First, our radio transmissions have already penetrated far into space: if we assume that major broadcasting began around 1930, any civilization within 45 light years of us would have been able to detect them; thus from Sirius (8·6 light years) the Solar System would already be "radio noisy", though from Rigel (900 light years) it will remain "radio quiet" for another eight centuries.

Secondly, any civilization capable of achieving interstellar travel will have progressed far beyond our own stage of internal warfare; otherwise it would almost certainly have destroyed itself. If any alien civilizations do make contact with us, they are bound to be far more advanced in this respect than we are, so that we would have nothing to fear from them. They would come in a spirit of peace, not war; and this is something which we should welcome, as we could undoubtedly learn a great deal from them.

The Arecibo message

1 Binary numbers 1-40
2 Atomic numbers of hydrogen, carbon, nitrogen, oxygen and phosphorus
3 Chemical formulas for sugars and bases in nucleotides of deoxyribonucleic acid (DNA) molecule
4 Number of nucleotides in human DNA
5 Double helix of DNA
6 Human being
7 Height of human being
8 Size of human population
9 Solar System with Earth displaced towards human being
10 Arecibo telescope
11 Diameter of telescope

▼ Natural sources of radiation in space and in the Earth's atmosphere give a level of background "noise" which could swamp artificial transmissions. The noise is least between frequencies of 1,000 and 10,000MHz. This "window" may be the most logical frequency range for interstellar communication.

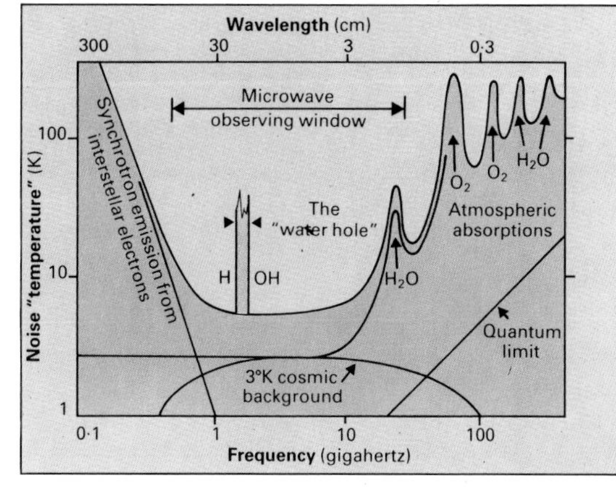

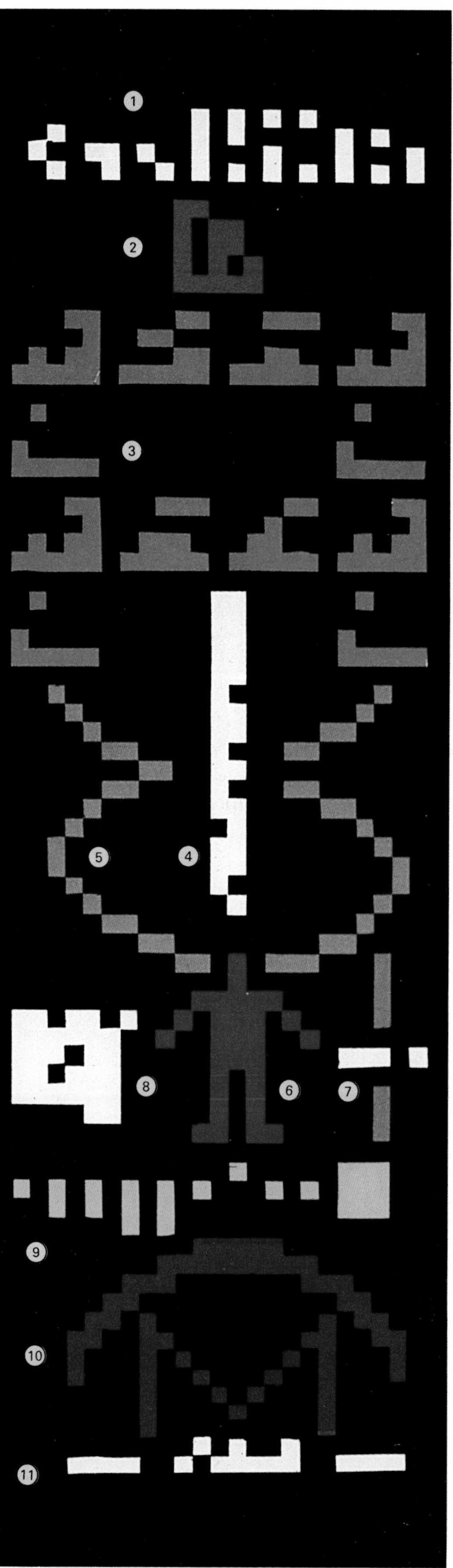

Voyagers 1 and 2 are both carrying an audio-visual disk, containing two hours of sounds (speech and music), digitally encoded pictures and a message from President (at time of launching) Jimmy Carter. Playing instructions accompany the disks.

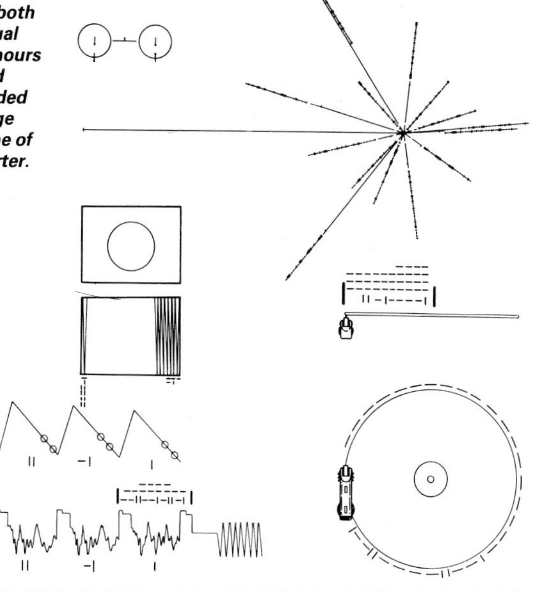

See also
Observing the Planets 21-33
The Origin of the Solar System 121-9
Observing the Universe 137-44
Space, Time and Gravitation 225-8
The Evolving Universe 229-44

Interstellar travel – the ultimate ambition

In one sense, the process of sending rocket probes to the stars has already begun. Pioneers 10 and 11, together with the two Voyagers – each of which explored Jupiter and beyond – have all exceeded the escape velocity of the Solar System, and will eventually enter interstellar space. Pioneer 10 passed beyond the known planets in 1983, but, at its present speed of about 11 kilometers per second, would take about 100,000 years to reach the nearest star, even if it were heading that way – which it is not. Nevertheless, each of these probes carries a plaque to show where it came from.

To accomplish missions in a reasonable period of time, spacecraft will have to attain a significant fraction of the speed of light. Existing rockets achieve only about 1/20,000 of the speed of light, being restricted by very low exhaust velocities of only a few kilometers per second. Fusion rockets, utilizing energy released in nuclear fusion reactions, may be able to attain exhaust velocities a thousand times higher, and it is possible that motors of this kind will be constructed some time in the 21st century. These would open up the possibility of sending probes to the stars at about one-tenth of the speed of light.

The most detailed design study for an interstellar spacecraft so far produced is the Project Daedalus report of the British Interplanetary Society. This envisages an unmanned flight through the Barnard's star system at one-eighth the speed of light, with a flight time of 50 years. The craft would be propelled by a nuclear pulse rocket in which tiny pellets of deuterium and helium-3 would be detonated in miniature thermonuclear explosions by bombardment with laser or electron beams. A detonation rate of 250 per second would ensure a smooth acceleration. The two-stage craft would weigh 54,000 tonnes at launch, 50,000 of which would be propellant, and 30,000 tonnes of helium-3 would have to be "mined" from the atmosphere of Jupiter!

Whether or not such a mission is ever undertaken, the principle of the nuclear pulse rocket appears sound, and the technology is only an extension of existing technology.

If other civilizations exist, we may detect their signals at any time; or if they have undertaken interstellar travel, continued exploration of the Solar System may reveal some of their artefacts. But if intelligent life is unique to Earth, it is surely the duty of the human race to safeguard and propagate the phenomenon of intelligence, and to explore to the full the awesome universe in which we live.

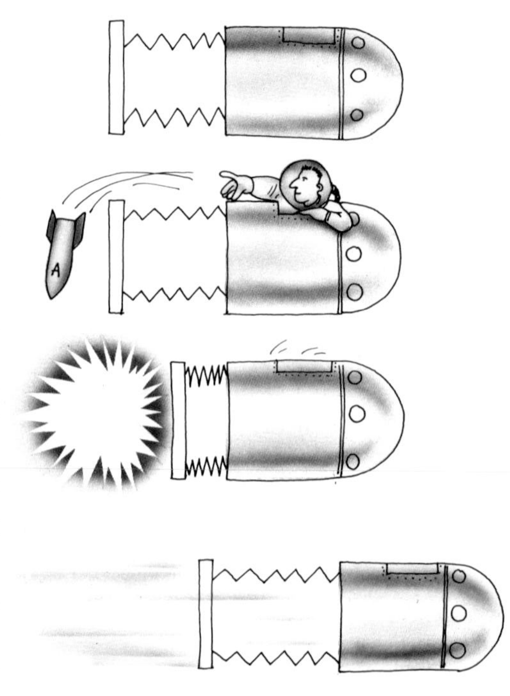

▲ The principle of the nuclear pulse rocket: the spacecraft is accelerated by a series of nuclear detonations, each of which acts on a shock absorber to give an impulse to the craft.

▼ In principle a high-power laser beam could exert enough pressure on a large sail to accelerate a spacecraft to high speeds. Another mechanism would be needed to stop the craft.

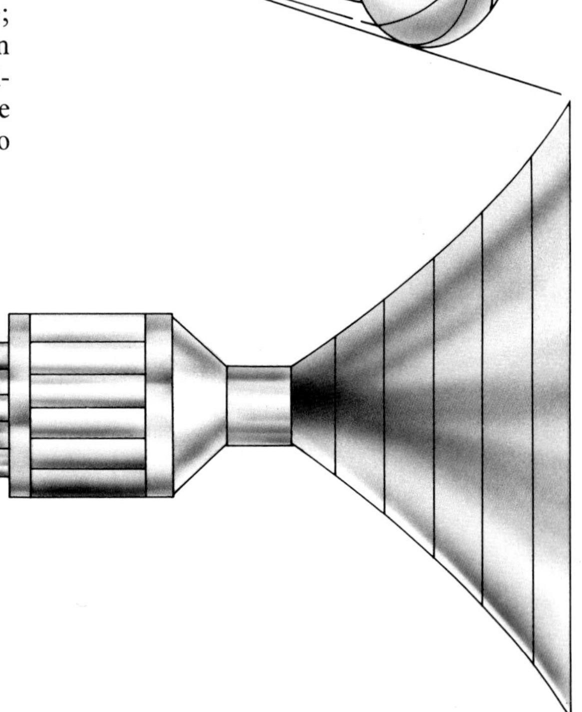

► The interstellar ramjet is a concept first suggested in 1960 by R.W. Bussard. A magnetic scoop would sweep up ionized hydrogen as the craft moved through space, and use this as fuel in a fusion-powered motor. The potential would be almost unlimited, as the faster it traveled, the faster it would sweep up fuel. In principle it could approach ever closer to the speed of light. Doubts have been expressed as to whether the energy gained from the fuel would exceed the energy lost by the drag exerted on the spacecraft as the hydrogen was swept into the scoop. However, it is an elegant concept which deserves to succeed.

Credits

Key to abbreviations: AAT Anglo Australian Telescope Board; BBC HPL Hulton Picture Library, London; NASA National Aero Space Administration, Washington; RAS Royal Astronomical Society, London; SPL Science Photo Library, London. b bottom; bl bottom left; br bottom right; c center; cl center left; cr center right; t top; tl top left; tr top right.

1 Planetarium, Armagh/AAT **2-3**, **4-5** Royal Observatory, Edinburgh **7r** SPL/Kris Davidson **8** High Altitude Observatory National Center for Atmospheric Research/National Science Foundation Colorado **9** Patrick Moore **12tr** Mansell Collection, London **12br** BBC HPL **13t** from Tycho Brahe: *Astronomiae Instaurate Mechanica*, photo Bodleian Library, Oxford **13tr** Mansell Collection, London **13br** from Tycho Brahe: *Astronomiae Instaurate Mechanica*, photo Bodleian Library, Oxford **14tl** National Maritime Museum **14bl,c** Biblioteca Nazionale Centrale, photo Guido Sansoni, Florence **14br** Istituto e Museo di Storia della Scienza **15b** Biblioteca Nazionale Centrale, photo Guido Sansoni, Florence **15r** Istituto e Museo di Storia della Scienza **16t** RAS, photo Eileen Tweedie, London **16bl** Patrick Moore **17b** Royal Society, London **17tr** Trustees of the Portsmouth Estates **20t** Equinox Archives, Oxford **20c** Lowell Observatory **20b** Equinox Archives, Oxford **24** from Hevelius: *Machina Coelestis*, 1673, Science Museum, London **25t** Patrick Moore **25b** Yerkes Observatory/Chicago University **28tl** Sovfoto, NY **28br** NASA, Washington **40** BBC HPL **42t,b** NASA, Washington **43** NASA **44** NASA **45** Private Collection, Leamington Spa, photo Mark Fiennes **46t** Space Frontiers Ltd. **46b** NASA **47** SPL **48t** Patrick Moore **48-49** NASA **49br** Mansell Collection **50-51t** Michael Kobrick/NASA **50c,b** Sovfoto, NY **51l** NASA **51r** D.B. Campbell, Arecibo Observatory, Cornell University (The Arecibo Observatory is part of the National Astronomy and Ionosphere Center which is operated by Cornell University under contract with the National Science Foundation) **52-53** NASA **54tr** SPL **54b** Michael Kobrick/NASA **55t** NASA **55c** US Geological Survey **55bl,r** NASA **56t** Mansell Collection, London **56c** Lowell Observatory **56b** photo: Bodleian Library, Oxford **58t** artwork reference courtesy *Scientific American* **61b** artwork reference courtesy C.F. Capen and *Sky and Telescope* **61c** Charles Capen, Lowell Observatory **62-3t** NASA **62br** NASA **63bc** SPL **63tr** NASA **63c** NASA **63br** SPL/NASA **66** NASA **67** SPL **70-71** NASA **76t** NASA **76cl** Patrick Moore **76cr** NASA **78** NASA **79tl** SPL/NASA **79** NASA **82-83** NASA **84tr** Huygensmuseum "Hofwijck", Voorburg **85** NASA **86-87** NASA **88tr** W.M. Sinton, Institute of Astronomy, University of Hawaii **88bc** CCD Hawaii/Patrick Moore 88br Lowell Observatory **89c** BBC HPL **89b** Archiv Gerstenberg, Wietze **90tl** D. R. Roddy **90br** Paul Brierley **91bl** Paul Brierley, Harlow. **92bl** Patrick Moore **93tr** BBC HPL **93c** Sovfoto, NY **94c,b** SPL/Dr. D. Klinglesmith/NASA **94tr** Royal Society, London, photo: Eileen Tweedie **94tr** Michael Holford **95bl** Scala, Florence **95br** Patrick Moore **96tr** Science Museum, London, photo Butler **96br** CF Capan Planetary Library, Institute for Planetary Research Observatories **97bl** Warden and Fellows of New College Oxford, MS New Coll. 361f.45v., photo Bodleian Library **97r** Patrick Moore **100bl** Deutsches Museum, Munich **100br** Mansell Collection, London **101tl** Bildarchiv Preussischer Kulturbesitz, Berlin **101br** Max–Planck–Institut für Radioastronomie **102t,103** Patrick Moore **105tr** Royal Greenwich Observatory, Herstmonceux Castle **105br** Observatoires du Pic du Midi et de Toulouse **106tr** Kitt Peak National Observatory/Cerro Tololo Inter–American Observatories **107** © 1981 AURA Inc./Kitt Peak National Observatory **108bl** SPL/Hale Observatories **108tr,c,br** Observatoire de Paris–Meudon **109t** SPL/High Altitude Observatory and National Centre for Atmospheric Research **109b** Sacramento Peak Observatory **110tr,c** High Altitude Observatory and National Center for Atmospheric Research, Colorado **110bl** SPL **111t** SPL **111l** IAU/AAS (Coronal hole) **111br** High Altitude Observatory, Colorado **112t** BBC HPL **112bl** from Hevelius: *Machina Coelestis*, 1673, Science Museum, London **113tr** BBC HPL **113cr** by courtesy of Eugene Parker **114** SPL/J. Finch **114** Dr. L. A. Frank, University of Iowa **116t** RAS **117t** Sotheby Parke Bernet **121** RAS **124tr** BBC HPL **124br** Bodleian Library, Oxford 125tr Mansell Collection, London 125b Bodleian Library, Oxford **126t** NASA **127t** Mary Evans Picture Library, London **127br** Popperfoto Ltd., London **128b** University of Toledo, photo William Hartough **129b** Mansell Collection **132b** Patrick Moore **133b** Yerkes Observatory, Chicago **134bl** SPL/Robin Scagell **134tr,br** Hatfield Polytechnic Observatory **136b** Mansell Collection, London **138t** National Maritime Museum, London **138cr** Private Collection, Leamington Spa **138b** Royal Society, photo Eileen Tweedie **139** Mansell Collection, London **138cl** Science Museum, London **139tr** NEI Parsons Ltd./Science and Engineering Research Council **140bl,br** BBC HPL **141t** Kitt Peak National Observatory/Cerro Tololo Inter–American Observatory **141cl** SPL/Hale Observatories **141cr** Planetarium, Armagh/AAT **141br** SPL/John Walsh **142t** Royal Observatory, Edinburgh **142br** Multi–Mirror Telescope **142bl** Centre d'Etudes et de Recherches Geodynamiques et Astronomiques, Grasse **144br** Arecibo Observatory/Cornell University **144cl** Very Large Array/National Radio Astronomy Observatory **144c** European Space Agency/Dr. John Parkinson **147t** BBC HPL **148c** Harvard University Archive **148b** Harvard College Observatory **149t** © 1974 AURA Inc./Kitt Peak National Observatory **149c** Prof. Hanbury Brown/University of Sydney **149b** Lick Observatory, University of California **165tl** Mount Wilson Observatory Photograph/E.P. Hubble Oct. 5/6 1923; reproduction J. Bedlie **165bl** Harvard College Observatory **166** Kitt Peak National Observatory/Cerro Tololo Inter–American Observatoy **167bl** Patrick Moore/Smithsonian Institution **167tr** SPL/Leicester University **167br** SPL/Dr. Kris Davidson **168tr** Patrick Moore **170–1t** The Planetarium, Armagh, © AAT Board **171tr** SPL/US Naval Observatory **171b** The Planetarium, Armagh **174tr** © California Institute of Technology and Carnegie Institution of Washington **174tl** SPL/Dr. J. Dickel **174br** SPL/Dr. Steve Gull and John Fielden **175** Royal Observatory, Edinburgh **176c** Philip Daly, Bath **176br** RAS **180t** Dept. of Applied Mathematics and Theoretical Physics, Cambridge **182b** The Planetarium, Armagh, © AAT Board **182tl** Iain Nicolson **182tr** Smithsonian Institution **183** Royal Observatory, Edinburgh **186tl** The Planetarium, Armagh, © AAT **186bl** SPL/Dr. E.I. Robson **186–7** Royal Observatory, Edinburgh **189tr,cr** RAS/photo Eileen Tweedie **190tl** SPL/Max–Planck Institut **190b** Lund Observatory **191c** European Space Agency **191b** Rutherford–Appleton Laboratory, Oxon **192tr** Harvard College Observatory **192cr** RAS/Hale Observatory 196l D.F. Malin/AAT **197tr** Mount Wilson Observatory © Carnegie Institute of Washington **198tr** Iain Nicolson **198b**, **199** Royal Observatory, Edinburgh **200tr** Huntington Library, California **200cr** Mount Wilson Observatory **201tl,tr,cl,bl** Mount Wilson and Las Campanas Observatory **201** tc, ct, cb, br Clifornia Institute of Technology/Palomar Observatory photograph **201cr** **201b** Kitt Peak National Observatory/CTIAO **202tr** Iain Nicolson **202b** SPL/Dr. Jean Lorre **203** SPL/Mt. Palomar Observatory **204tr,cr** Kitt Peak National Observatory/CTIAO **205tr** Palomar Observatory photograph **205b** Royal Observatory, Edinburgh/UK Schmidt Telescope **206b** The Planetarium, Armagh, © AAT **206tr** © Smithsonian Institution **207tr** SPL/B. Cooper and David Parker **207b** National Radio Astronomy Observatory, operated by Associated Universities, Inc. under contract with the National Science Foundation **208tr** Westerbork Telescope (The Westerbork Synthesis Radio Telescope is operated by the Netherlands Foundation for Radio Astronomy with the financial support of the Netherlands Organisation for the Advancement of Pure Research – ZWO) **208br** SPL **208l** Mullard Radio Astronomy Observatory, University of Cambridge **209r** Westerbork Telescope **210** SPL/Dr. Steve Gull **211b** Susan Wyckoff/Peter A. Wehinger, Arizona State University, using the University of Hawaii 2.2m telescope **211tl** Eric D. Feigelson, Pennsylvania State University **210** SPL/Dr. Steve Gull **212–3** NASA **213tr** RAS **213br** Patrick Moore **216br** RAS/Hale Observatories **217b** Royal Observatory, Edinburgh **219** artwork reference courtesy R. Hess and *Sky and Telescope* **220t** Lick Observatories **220bl,br,221tr** Royal Observatory, Edinburgh **222** Institute for Astronomy, University of Hawaii **223tl** SPL/Hale Observatories **223tr** SPL/Allen **223bl** SPL/Dr. J. W. Sulentic **224tr** Westerbork Telescope **224b** SPL/Kitt Peak National Observatories **225tl,tr** BBC HPL **225cr** T.J. Pearson, from *Nature* vol. 290 pp 365–368 April 1981 **226–7tc** SPL **226bl** SPL/Dr. A. Stockton **227r** Royal Greenwich Observatory, London **228** Dr. J. Weber/University of Maryland, photo Bill Clark **231** Patrick Corvan/© National Geographic Magazine **236-7** S. Matsik, Dept. of Physics and Astronomy, University of Pittsburg **236br** Adrian L. Melott, University of Chicago/Monthly Notices **239** Patrick Corvan, Armagh **246tr** Courtesy of Frank D. Drake and the National Astronomy and Ionosphere Center, Cornell University/National Science Foundation Arecibo Observatory/Cornell University **246–7** NASA.

Artists Robert Burns; Kai Choi; Paul Doherty; Chris Forsey; Mick Gillah; Kevin Maddison; Coral Mulah; Mick Saunders.

Further Reading

The Solar System

Baker, V.R. *The Channels of Mars* (Adam Hilger)
Beatty, J.K., O'Leary B. and Chaikin, A. *The New Solar System* (Cambridge University Press)
Brandt, J.C. and Chapman, R.D. *Introduction to Comets* (Cambridge University Press)
Carr, M.H. *The Surface of Mars* (Yale University Press)
Gehrels, T. *Jupiter* (University of Arizona Press)
Gehrels, T. and Matthews, M. *Saturn* (University of Arizona Press)
Hunten, D (Editor) *Venus* (University of Arizona Press)
Moore, P. *Guide to the Planets* (Lutterworth Press)
Moore, P. and Hunt, G. *The Atlas of the Solar System* (Mitchell Beazley)
Moore, P. and Mason, J. *The Return of Halley's Comet* (Patrick Stephens, London; W.W. Norton, New York)
Noyes, R.W. *The Sun Our Star* (Harvard)
Tombaugh, C. *Out of the Darkness; the Planet Pluto* (Stackpole Books)

Stellar, Galactic and Cosmological

Barrow, J.D. and Silk, J. *The Left Hand of Creation* (Heinemann, London)
Bok, Bart J. *The Milky Way* (Harvard)
Burnham, R. *Burnham's Celestial Handbook* (3 volumes) (Dover Press, New York)
Ferris, T. *Galaxies* (Stewart, Tabori & Chang, New York)
Henbest, N. and Marten, M. *The New Astronomy* (Cambridge University Press)
Islam, J.N. *The Ultimate Fate of the Universe* (Cambridge University Press)
Kippenhahn, R. *100 Billion Suns: The Birth, Life and Death of the Stars* (Weidenfeld & Nicolson, London)
Murdin, P. and Malin, D. *Colours of the Stars* (Cambridge University Press)
Nicolson, I. *Gravity, Black Holes and the Universe* (Halsted Press, New York)
Shapley, H. *Galaxies* (Harvard)
Silk, J. *The Big Bang: Creation and Evolution of the Universe* (W.H. Freeman)
Stephenson, F.R. *The Historical Supernovae* (Pergamon)
Weinberg, S. *The First Three Minutes* (Fontana and Collins, Glasgow; Basic Books, New York)

General

Cohen, D. *In Quest of Telescopes* (Sky Publishing Corporation, Cambridge, Mass.)
Menzel. D.H. and Pasachoff, J.M. *A Field Guide to the Stars and Planets* (Houghton Mifflin)
Moore, P. *The New Atlas of the Universe* (Mitchell Beazley)
Moore, P. *The Unfolding Universe* (Michael Joseph)
Murdin, P. and Allen, D. *Catalogue of the Universe* (Cambridge University Press)
Nicolson, I. *The Road to the Stars* (David & Charles; Wm. Morrow & Co. Inc., New York)
Norton, A.P. *Norton's Star Atlas and Telescopic Handbook* (Gall & Inglis)
Pasachoff, J.M. and Kutner *University Astronomy* (Saunders)
Vehrenberg, H. *Atlas of Deep-sky Splendours* (Cambridge University Press)

Glossary

Abberation of light
The apparent displacement of a star from its true position in the sky, due to the Earth's motion round the Sun and the fact that light has a finite velocity of approximately 200,000km/sec.
Albedo
The reflectivity of a body expressed as the ratio of the amount of light reflected by a body to the amount of light falling on it. Values of albedo range between 1 (for a perfect reflector) and 0 (for a completely black surface).
Aphelion
The orbital position of a planet or other body when at its farthest from the Sun.
Aries, First Point of (vernal equinox)
The point at which the Sun's apparent path in the sky cuts the ECLIPTIC, with the Sun moving from south to north. It used to be in Aries, but due to PRECESSION it now lies in the adjacent CONSTELLATION of Pisces.
Astronomical unit
The distance between the Earth and the Sun: approximately 150,000,000km.
Aurorae
Polar lights, due to electrified particles from the Sun entering the upper atmosphere and causing luminous effects.
Azimuth
The horizontal direction or bearing of a celestial body, reckoned from the north point of the observer's horizon.
Big Bang
The hot, dense explosive event which is widely believed to have been the origin of the universe.
Binary Star See DOUBLE STAR.
Black hole
A region of space surrounding an old, collapsed star, from which not even light can escape.
Brown dwarf
A star of such low mass that it has never become hot enough for nuclear reactions to begin at its core.
Celestial sphere
An imaginary sphere surrounding the Earth, whose center is coincident with the center of the Earth.
Cepheid variables
Short-period variable stars, whose periods are linked with their real LUMINOSITIES; the longer the period, the more luminous the star.
Chromosphere
That part of the Sun's atmosphere lying above the bright surface or photosphere, and below the corona.
Comet
A member of the SOLAR SYSTEM, made up of a nucleus, a coma, and (with large comets) a tail or tails. Most comets have highly eccentric orbits.
Constellation
A group of stars making up a definite pattern. As the individual stars are at very different distances from us, a constellation is a mere line-of-sight effect.
Corona
The outermost part of the Sun's atmosphere. It is visible with the naked eye only during a total solar ECLIPSE.
Declination
The angular distance of a celestial body north or south of the celestial equator on the CELESTIAL SPHERE.
Density
The mass of a given substance per unit volume (eg the mean density of the Earth is 5500kg/m^3). Taking water as unity, the density (specific gravity) of the Earth is 5.5.
Doppler Effect
The apparent change in the wavelength of light due to the motion of the light-source relative to the observer. Redshifts indicate recession; blue shifts, approach.
Double star
A star consisting of two components. If the components are physically associated, the double star is a BINARY SYSTEM.
Eclipse
(A) Solar: the blotting-out of the Sun by the interposition of the Moon.
(B) Lunar: the passage of the Moon through the shadow cast by the Earth.
Ecliptic
The apparent yearly path of the Sun among the stars. It passes through the 12 CONSTELLATIONS of the ZODIAC.
Electromagnetic radiation
Radiation consisting of an electric and a magnetic disturbance which travels in a vacuum at a constant speed known as the speed of light (about 300,000km/s). Visible light and radio waves are examples.
Epicycle
A small circle moving round a "deferent", which in the old cosmologies described a perfect circle around the Earth.
Equinoxes
The points of intersection between the ECLIPTIC and the celestial equator on the CELESTIAL SPHERE: the vernal equinox (or "First point of Aries" - March) and autumnal equinox (or "First point of Libra" - September).
Event horizon
The boundary of a BLACK HOLE.
Flare star
A dwarf star which suddenly and temporarily increases its brightness, presumably because of intense flares on its surface.
Galactic clusters
Clusters of stars in our GALAXY; also known as open or loose clusters. They also occur in external galaxies.
Galaxy
(A) A distinct star system: Galaxies may contain from a few million to a few million million stars together with differing proportions of interstellar matter (gas and dust).
(B) The star-system of which our Sun is a member: it contains about 100,000 million stars.
Hertzsprung-Russell diagram
A diagram in which stars are plotted according to SPECTRAL type and LUMINOSITY.
Hubble Constant
The relationship between the distance of a GALAXY and its velocity of recession: probably about 50km/sec per megaparsec.
Interstellar matter
Thinly-spread material (gas and dust) between the stars. Many interstellar molecules, including organic molecules, have now been identified.
Light-years
The distance traveled by light in one year: 9.46 million million km.
Luminosity, stellar
The total amount of energy emitted by a star in one second. A star's luminosity is often compared to that of the Sun by taking the solar luminosity as a unit.
Magnetosphere
The region of space around a body in which the magnetic field of that body is dominant.
Magnitude, apparent
The apparent brightness of a celestial object as seen from Earth; the lower the magnitude, the brighter the object.
Magnitude, resolute
The APPARENT MAGNITUDE that a celestial body would have if seen from a standard distance of 10 PARSECS.
Main Sequence
The well defined band from upper left to lower right on a HERTZSPRUNG-RUSSELL DIAGRAM.
Meridian
The great circle on the CELESTIAL SPHERE which passes through both poles and the observer's ZENITH.
Meteor
A small particle moving around the Sun. If it enters the Earth's upper atmosphere it becomes heated by friction, and burns away producing the luminous shooting-star appearance.
Meteorite
A small body sufficiently massive to survive entering the atmosphere at speeds of tens of km/sec and which can reach ground level.
Meteoroid
A particle, usually smaller than a grain of sand, but can be as much as a few meters in diameter, in orbit around the Sun.
Milky Way
The luminous band around the sky, made up of the light of millions of stars too faint to be seen individually with the naked eye. It is due to a line-of-sight effect when looking along the main plane of the GALAXY.
Nadir
The point on the CELESTIAL SPHERE directly below the observer, ie opposite to the ZENITH.
Nebula, Galactic
A cloud of dust and gas in space, in which fresh stars are being created.
Neutron star
A star made up chiefly of neutrons. Many neutron stars are emitting radio waves and are rotating rapidly, so that their emissions arrive in pulses and they are, therefore, known as pulsars.
Nova
A star which undergoes a sudden, temporary outburst. All normal novae are BINARY systems.
Occultation
The covering-up of one celestial body by another. Strictly speaking, a solar ECLIPSE is an occultation of the Sun by the Moon.
Orbit
Path.
Parallax
The change in the apparent position of an object when viewed by an observer from two different locations.
Parsec
The distance at which a star would show a parallax of one second of arc; it is equal to 3.26 LIGHT-YEARS.
Perihelion
The point in its orbit in which a member of the SOLAR SYSTEM is at its closest to the Sun.
Perturbation
The disturbance in motion produced by one celestial body upon another.
Planet
A non-luminous body moving around a STAR. There are nine known planets in the SOLAR SYSTEM.
Precession
The apparent slow movement of the celestial poles on the CELESTIAL SPHERE.
Pulsar See NEUTRON STAR.
Quasar
A very remote, superluminous object. Quasars are now believed to be the nuclei inside very active GALAXIES.
Radio galaxy
A GALAXY which emits more strongly at radio than at visible wavelengths.
Red giant star
A star which has evolved off the MAIN SEQUENCE, and has used up the available hydrogen fuel in its core.
Retrograde motion
The opposite of direct motion, ie opposite to the direction in which the Earth spins on its axis, or the direction in which the Earth moves round the Sun. In the sky, a PLANET is said to have retrograde motion when moving from east to west against the starry background.
Right ascension
The time-interval between the transit of the VERNAL EQUINOX and the transit of the celestial object concerned. It is the angle between the VERNAL EQUINOX and a celestial body, measured eastwards from the VERNAL EQUINOX in time units.
Satellite
A secondary body moving around a PLANET.
Sidereal time
The local time reckoned according to the apparent rotation of the CELESTIAL SPHERE. It is 0 hours when the VERNAL EQUINOX crosses the observer's MERIDIAN.
Solar System
The system consisting of the Sun, the PLANETS and their SATELLITES, the ASTEROIDS, COMETS, METEOROIDS and other interplanetary material.
Solar wind
A stream of low-energy atomic particles continuously sent out in all directions by the Sun.
Solstices
The times when the Sun is at its greatest north or south DECLINATION in the sky.

Spectral classification
The classification of stars according to the appearance of their SPECTRUM. The principal classes are denoted by the letters: O, B, A, F, G, K, M.
Spectrum
The distribution of the intensity of ELECTROMAGNETIC RADIATION with wavelength. The visible spectrum encompasses those wavelengths to which the human eye responds, the different wavelengths corresponding to the different colors.
Star
A self-luminous celestial object: our Sun is an ordinary MAIN SEQUENCE star.
Star cluster
A genuine grouping of stars. Clusters may be open (loose) or globular (symmetrical).
Sunspots
Darker patches on the Sun's bright surface. They are associated with powerful magnetic fields and are cooler than the rest of the surface.
Supergiant stars Stars of exceptionally high LUMINOSITY.
Supernova
A cataclysmic stellar outburst. Some supernovae are due to the collapse of a massive star; others are due to the complete disruption of the WHITE DWARF component of a BINARY system.
Transit
The passage of a celestial body across the observer's MERIDIAN. (The term is also used for Mercury and Venus when they pass across the Sun's disk as seen from the Earth.)
Variable stars
Stars which fluctuate in brilliancy, either regularly or irregularly, over comparatively short periods.
Vernal equinox
The point of intersection between the ECLIPTIC and the celestial equator at which the Sun passes from south to north of the celestial equator on the CELESTIAL SPHERE.
White dwarf star
A star which has evolved off the MAIN SEQUENCE and has passed through its giant stage. White dwarfs have no nuclear reserves left, and are very dense and feeble.
Zenith
The observer's overhead point.
Zodiac
A belt stretching round the sky to 8 degrees on either side of the ECLIPTIC, in which the Sun, Moon and bright PLANETS are always to be found.
Zodiacal light
A cone of light rising from the horizon and stretching along the ECLIPTIC. It is due to small particles scattered throughout the main plane of the SOLAR SYSTEM.

Units

Powers of Ten
Very large and very small numbers are often encountered in astronomy. To avoid writing them out in full a convenient shorthand is "index notation" or "powers of ten" whereby 10^n (ten to the power "n") denotes the number "1" followed by n zeros. For example, one thousand (1,000) is 1 followed by three zeros, and would be written as 10^3 (ie ten to the power three). When very large numbers are encountered it is obviously easier, for example, to write 10^{40} than laboriously to write out "1" followed by forty zeros. Very small numbers can be expressed in a similar way by using a negative index: 10^{-n} (ten to the power $^{-n}$). Thus one thousandth (1/1000, or 0·001 is "1 divided by 1000", or $1/10^3$, and would be written as 10^{-3}. Again it is easier to write, for example, 10^{-12}, than to write out the number as 0·000000000001.

Common examples are:

one	1	10^0
ten	10	10^1
hundred	100	10^2
thousand	1,000	10^3
million	1,000,000	10^6
billion	1,000,000,000	10^9
tenth	0·1	10^{-1}
hundredth	0·01	10^{-2}
thousandth	0·001	10^{-3}
millionth	0·000001	10^{-6}
billionth	0·000000001	10^{-9}

A number such as 250 is 2·5 multiplied by 100, and would be written as $2·5 \times 10^2$; 250,000,000,000 (two hundred and fifty billion) would be written as $2·5 \times 10^{11}$ and so on. The number 0·25 would be written as $2·5 \times 10^{-1}$; 0·0000025 (2·5 millionths) would be $2·5 \times 10^{-6}$, and so on.

Units of measurement
In general the International System of Units – SI units – is used throughout this book. However, certain "astronomical" units, such as "light year" are employed where their usage is commonplace. The Basic SI units which will be encountered in this book are:

mass: kilogram (kg)*
length: meter (m)
time: second (s)
temperature: kelvin(K)**

* *Mass* is the quantity of matter contained in a body. *Weight* is the force of gravity exerted on a mass. The mass of a body is a property of that body, but its weight depends on the strength of gravity at the surface of the planet on which it is sitting. For example, a 100 kilogram mass taken to the surface of the Moon would have only about one-sixth of the weight which it has on the Earth's surface because the force of gravity at the Moon's surface is about one-sixth of the strength of gravity on the Earth's surface. Its mass (the amount of matter which it contains) would, however, be completely unchanged.

** The *kelvin* relates to the absolute temperature scale which has its zero level at *absolute zero*, the lowest possible temperature (the temperature at which all motion of atoms and molecules ceases). Absolute zero occurs at a temperature of −273°C (degrees Celsius). Thus 0K = −273°C, 273K = 0°C (the freezing point of water), 373K = 100°C (the boiling point of water), and so on.

Certain multiples and submultiples of units are given prefixes, as follows:

Prefix	*Symbol*	*Multiple/submultiple*
tera	T	10^{12}
giga	G	10^9
mega	M	10^6
kilo	k	10^3
hecto	h	10^2
deca	da	10^1
deci	d	10^{-1}
centi	c	10^{-2}
milli	m	10^{-3}
micro	μ	10^{-6}
nano	n	10^{-9}
pico	p	10^{-12}

For example, 1 kilometer (km) = 10^3 meters (m)
1 millimeter (mm) = 10^{-3} meters (m)
1 micrometer (μm) = 10^{-6} meters (m)
1 nanometer (nm) = 10^{-9} meters (m)

Other SI units frequently encountered in this book include:

energy: joule (J)
(1 joule is the kinetic energy – energy of motion – possessed by a 2kilogram mass moving at a speed of 1 meter per second)
power: watt (w)
(power = energy per second. 1 watt = 1 joule per second)
(1 kilowatt (kw) = 1,000 watts)
force: newton (N)
(1 newton = the force which will accelerate 1 kilogram at a rate of 1 meter per second per second)

Units and conversion factors
Although SI units are becoming widely adopted for scientific purposes, some other units are more familiar in everyday experience. The conversion factors given below will allow a reader to convert SI units to units which may be more familiar, with the aid of a pocket calculator.

mass 1 kilogram (kg) = 2.205 pounds (lb)
1 tonne (t) = 1,000kg = 0·984 tons

length 1 meter (m) = 39·37 inches = 3·281 feet (ft)
1 kilometer (km) = 0·621 miles

area 1 square meter (m^2) = 10·765 square feet (ft^2)

volume 1 cubic meter (m^3) = 35·32 cubic feet (ft^3)
1 liter (l) = 0·220 UK gallons
= 0·264 US gallons

density (the quantity of mass contained in a unit volume)
1 kilogram per cubic meter (kg m^{-3}, or kg/m^3)
= 0·062 pounds per cubic foot (lb ft^{-3} or lb/ft^3)
(water has a density of 1,000kg/m^3)

time 1 year = $3·16 \times 10^7$ seconds

speed 1 meter per second (m s^{-1} or m/s)
= 3·281 feet per second
= 3·600 kilometers per hour
= 2·236 miles per hour

force 1 newton (N) = 0·225 pounds-force

energy 1 joule (J) = 10^7 ergs = 0·239 calories

power 1 watt (w) = 0·00134 horsepower (hp)

magnetism 1 Tesla (T) = 10,000 gauss

temperature absolute temperature (K) = °C + 273
Fahrenheit temperature (°F) = 9/5 × (celsius temperature) + 32

Angles
Angles are measured in the following units:
1 degree (°) = 1/360th part of a circle
1 minute (′) = 1/60th of 1 degree; often written as "arcmin" ("minute of arc") to avoid confusion with "minute" of time.
1 second (″) = 1/60th part of 1 arcmin; often written as "arcsec".

Astronomical distance units
astronomical unit (AU) = $1·496 \times 10^8$km
light year (ly) = $9·461 \times 10^{12}$km
parsec (pc) = $3·086 \times 10^{13}$km

Selected physical and astronomical constants
speed of light (c) = $2·998 \times 10^8$m s^{-1} = $2·998 \times 10^5$ kilometers per second
gravitational constant (G) = $6·672 \times 10^{-11}$ N m^2kg^{-2}
mass of hydrogen atom = $1·6735 \times 10^{-27}$kg
mass of electron = $9·1096 \times 10^{-31}$kg
mass of proton = $1·6726 \times 10^{-27}$kg
mass of neutron = $1·6749 \times 10^{-27}$kg
mass of the Earth = $5·974 \times 10^{24}$kg
mass of the Sun = $1·989 \times 10^{30}$kg

Index

Page numbers in *italics* refer to illustrations, captions or side text

L

M

N

P

Q

R

S

T

U

V

W

Y

Z